W9-BZF-698

NATIONAL GEOGRAPHIC

ENCYCLOPEDIA *of* SPACE

In 1997 the Hale-Bopp comet put on a show in the clear skies above Mount Whitney in California. The comet's blue tail was created as solar wind particles struck ions emitted from its nucleus.

NATIONAL GEOGRAPHIC

ENCYCLOPEDIA *of* SPACE

Linda K. Glover

with Andrew Chaikin, Patricia S. Daniels, Andrea Gianopoulos, *and* Jonathan T. Malay

Foreword by Buzz Aldrin

WASHINGTON D.C.

CONTRIBUTING ESSAYISTS

GHASSEM R. ASRAR
National Aeronautics and Space Administration

J. KELLY BEATTY
Sky & Telescope

CARISSA BRYCE CHRISTENSEN
The Tauri Group

LEONARD DAVID
SPACE.com

DAVID DEVORKIN
National Air and Space Museum, Smithsonian Institution

SYLVIA A. EARLE
National Geographic Explorer-in-Residence

DIANE L. EVANS
Jet Propulsion Laboratory, California Institute of Technology

GARY A. FEDERICI
Center for Naval Analyses

REAR ADM. RAND FISHER
National Reconnaissance Office

SENATOR JAKE GARN
U.S. Senate 1974-1993

WILLIAM HARWOOD
Independent consultant

SEAN O'KEEFE
National Aeronautics and Space Administration

SARA SCHECHNER
Harvard University

ROBERT W. SMITH
University of Alberta

KATHRYN D. SULLIVAN
COSI Columbus

JAMES TREFIL
George Mason University

J. ANTHONY TYSON
Lucent Technologies, Bell Labs

CHRISTOPHER WANJEK
Independent consultant

DEBORAH JEAN WARNER
National Museum of American History, Smithsonian Institution

DAVID WILKINSON
Princeton University

ROBERT W. WILSON
Harvard-Smithsonian Center for Astrophysics

Contents

3 | REACHING & MANEUVERING IN SPACE

Patricia S. Daniels & Linda K. Glover

4 | HUMAN SPACEFLIGHT

ANDREW CHAIKIN

5 | EARTH SCIENCE & COMMERCE FROM SPACE

JONATHAN T. MALAY

6 | MILITARY & INTELLIGENCE USES OF SPACE

LINDA K. GLOVER

APPENDIX OF MAPS

The immense nebula NGC 604, 2.7 million light-years away, explodes with activity as more than 200 newly formed, massive stars at its core heat the surrounding gas, producing a characteristic nebular glow. This image was taken by the Hubble Space Telescope in 1995.

FOREWORD

Buzz Aldrin

IT IS AN ENDURING HONOR TO BE ONE OF THE VERY FEW HUMANS WHO HAVE STOOD on the moon and looked back at our Earth from there. My career as an astrophysical scientist and astronaut seems almost predestined in retrospect, since my father was an early aviation pioneer and my mother's maiden name was Moon. My journey to the moon began with my first airplane flight, in a single-engine Lockheed Vega my father was piloting, when I was just two years old. But what did it take to get us to the moon?

Just 12 years after the launch of the Soviet Union's first Earth-orbiting artificial satellite, Sputnik 1, Neil Armstrong and I set foot on the moon on July 20, 1969. This was an extraordinary decade of technological development and space exploration. An unwavering focus on long-term goals was needed to develop new materials, rockets, orbits, space capsules, and space suits, and to make the thousands of scientific breakthroughs necessary to get humans to the moon and back alive. In the 1960s this impetus was provided by the space race for military advantage and national prestige between the world's two emerging superpowers—the United States and the Soviet Union. But it was also supported by how much the newness and boldness of spaceflight captured the imagination of scientists in all fields and the general public around the world.

What did we accomplish in this endeavor of reaching the moon? We overcame immense technological obstacles, and built new industries that developed many new materials, processes, and products having applications to our everyday life on Earth. We fired the imagination of a generation in the United States and across the world, which led to an increased attention in schools—and increased interest of students—in the basics of mathematics and science. And we gathered information vital to further human exploration of space.

Why has it been so long since we pressed human exploration beyond the moon? The greatest threat to the progress of our civilization today is our increasing focus on short-term objectives and rewards. In the 35 years since the first moon walk, and aside from five follow-on Apollo moon landings that ended in 1972, more than 250 astronauts have lived and worked in space, all near the Earth in low Earth orbits. Human spaceflight, while still complex, dangerous, and expensive, has become common enough that serious interest in and plans for space tourism are growing. Many brilliantly designed unmanned systems—from planetary probes and rovers to deep-space probes and telescopes—have made spectacular discoveries. These studies have found evidence pointing to the current and/or past presence of water on our moon, the moons of Jupiter, and on Mars, and have found more than a hundred planets orbiting other stars like our sun. It is time for a long-term, ambitious

Astronaut Buzz Aldrin walks through the Sea of Tranquility region of the moon on July 20, 1969, as his Apollo 11 crewmate Neil Armstrong, reflected in Aldrin's visor, snaps this image with a 70-mm lunar-surface camera.

plan to acquire the intelligence, resourcefulness, and ability to adapt to new situations that will allow humans back into the space-exploration business.

I am very encouraged by the U.S. space-exploration vision announced by President George W. Bush on January 14, 2004, and the subsequent plan published by NASA in February 2004. Finally, 35 years after we first reached the moon, we have a viable, focused plan for expanding human exploration of space.

The broad strokes of the plan are both sobering and immensely exciting. We'll use the International Space Station to experiment with the effects of long-distance spaceflight on humans. Meanwhile non-manned missions to the moon will prepare us for a permanent manned station on the moon. Tests there can ready us for a Mars landing, but also look into the exciting possibilities of finding and using resources on the moon. For example, perhaps frozen water at the moon's poles can be broken down into hydrogen and oxygen to make rocket propellant in space, reducing the need to carry so much propellant from Earth.

Envisioned by many of us are fixed facilities at a place in space called L1—a location near the moon where the gravitational pull of the Earth and moon are equal, so an object can stay in place indefinitely using virtually no power; since we would not need to overcome Earth's gravity, launching probes to Mars from these stations would require far less energy and power. We will also send more probes to Mars, focused on the opportunities and threats associated with a human landing on the planet. A manned landing may be attempted first on one of the moons of Mars. And finally we will send humans to explore the surface of Mars itself—humans who can react quickly to what they see and change their daily scientific plans without the hours of computer reprogramming now required with robotic rovers.

There are many scientific and technological breakthroughs needed and many decisions that must be made along the way to manned exploration of Mars. My take on a few of them has been long documented. We should use liquid, not solid, propellants. The liquid propellant boosters should be reusable, with the tanks designed to separate when they are expended and "fly" back to Earth for reuse. We should investigate multiple-crew modules on the same launch vehicle. When we are ready for Mars landings, we should plan for sustained presence there; astronauts should sign up for multiyear missions to encompass the time on Mars and the round-trip. And for the trip itself we could use what I call the "Mars Cycler"—a transportation system comprised of an orbital cycling spacecraft in an extended and stable orbit between Earth and Mars—and a smaller ferrying vehicle to transfer people and supplies to and from the Cycler as it approaches each planet every 26 months.

We will need new fuels, propulsion systems, spacecraft, space suits, life-support systems, launch vehicles, and communication and scientific tools. Despite all the unknowns, the current crop of American astronauts is already in long-duration spaceflight training for the journey!

Success in this ambitious plan will require strong and continued political will that stays on course through economic upturns and downturns and transcends political parties and presidential administrations. It will also require increased cooperation between government and industry. And we must work more effectively with other nations. China has entered the human spaceflight arena—successfully completing its first manned orbital flight (21.5 hours) on October 16, 2003. France continues to expand and perfect its space-launch capabilities. The Soviet—now the Russian—space program has long been an innovator in space systems, beating the Americans to most space "firsts" until our moon landing. It will take a concerted effort by all of us working together to make it to Mars. Most of all, it will require incredible human ingenuity and tenacity to overcome the very fundamentals of time, space, mass, and energy that rule our universe. But I know we can do it.

There's also the growing interest in space tourism, with companies already selling tickets for space rides. I believe it will take government and industry investments in the moon-Mars mission to develop the cost-effective technologies that will get our children into space. I anticipate a first phase with quite a few people on a "space tour bus" orbiting the Earth for a day. Perhaps a second stage would have space tourists orbiting Earth for five days or so in a space-capsule-like "space hotel." And in time we should see tourists traveling at 40,200 kilometers an hour in a path that allows them to swing by the moon's surface for a close look and then return to Earth.

These myriad space activities will open up broad new areas of research science, new industry jobs, new areas of law and social studies, and new career opportunities for the coming generations. While in graduate school in astronautics at MIT, I developed new techniques for rendezvous of manned space vehicles with space stations that were later used in the U.S. and Russian space programs. And as a young astronaut, I pioneered the use of the new underwater field of scuba diving to train for the weightlessness of spaceflight. I also designed the "orbital cycler" approach to travel in the solar system—using a stable orbit around two nearby bodies that brings you close to both on a routine basis—that I hope will be used in the missions to Mars. If the students among you study hard, work to understand new technologies, think about making new connections between things you learn, and embrace new challenges, you can be part of the incredible adventures in space discovery that lie in the coming decades.

All of these topics and many more are presented in this book. It is designed as a broad introduction to a full range of space-related topics. I encourage you to learn more about the vast universe, your own planet, and the tools used to study both. And I encourage you to dream—about life on other planets, about new space vehicles and propulsion techniques, and about traveling into space yourself someday. As you navigate through these pages, enjoy the "flight"! ■

Preface

Linda K. Glover

PEOPLE HAVE ALWAYS BEEN FASCINATED BY SPACE—ITS VASTNESS AND MYSTERY, ITS order, its brilliant clean beauty. For some, this fascination has been life altering. Early astronomers risked imprisonment to pursue scientific truths about the heavens. Astronauts and cosmonauts have shown the incalculable courage required to sit atop a controlled bomb and be launched into space. These men and women have passed on to us a new appreciation for the smallness and fragility of our Earth in this environment of space that is almost entirely unsupportive of human existence.

Many individuals work outside the public spotlight, overcoming great mathematical and technical challenges in a field that *really is* rocket science. Others pursue commercial opportunities in space and ways to use the electromagnetic spectrum to measure, "sense," and predict conditions in the universe and on Earth. And still more people work in the shadows making unheralded breakthroughs in classified military and intelligence projects.

All of us use space in our daily activities—in long distance phone calls, ATM transactions, using maps, checking the weather—without having any idea of our dependence on satellites. Most of us have unclear memories or a vague understanding of the great moments in space science and conquest during our lifetimes.

I had both vague and poignant references to space in my childhood. I've never seen what my maternal grandparents looked liked—not even an old photograph. They and their home were completely destroyed by the first German V-2 rocket that hit London in the Blitz of World War II. "Muv" married a U.S. Army Air Corps officer heading the first Mobile Weather Unit supporting the D-day invasion. Dad was later deeply involved in satellites—on Vice President Humphrey's Space Council, improving weather satellites, and in many classified programs I didn't learn about until decades later when my own job opened the books. But I still remember vividly the whole family in the backyard in October 1957 watching Sputnik, the first satellite, cross the sky. My parents named the first family cat "Sputnik."

I didn't embrace space in those days, but went to the oceans instead—a more accessible medium I could feel, touch, smell, be immersed in. But, in one of those great arcs in life that bring you back to your roots, I suddenly had a job at the National Reconnaissance Office with a very "steep learning curve" about space. It was a great adventure, but frustrating because I couldn't find a "SPACE 101" book anywhere that covered all the subjects I needed at an entry level.

Most books touch on a narrow range of topics, without really introducing them all—deep space, life cycle of stars, our solar system, rockets, satellites, orbits, human spaceflight, international space treaties, and research, commercial, military, and intelligence uses of space. It is this broad range of space knowledge that we offer you here.

About This Book

The wonders and uses of space—an area of great interest to scientists, explorers, entrepreneurs, military tacticians, and dreamers to name but a few—are explored within this book in six easy-to-use thematic chapters, organized in an encyclopedic format.

The first chapter, "Deep Space," explains the basics of astronomy, revealing the intricacies of the stars, galaxies, and universe. Chapter two, "Our Solar System," brings space closer to home, focusing on the planets and other celestial bodies that belong to our solar system. Chapter three, "Reaching and Maneuvering in Space," details the technology and physics of getting into and staying in space. The fourth chapter, "Human Spaceflight," covers humankind's exploration of space. Chapter five, "Earth Sciences and Commerce from Space," reveals the means by which we are using space to our advantage and to better understand our planet. And finally, chapter six, "Military and Intelligence Uses of Space," describes the United States' strategic uses of space.

To follow an encyclopedic format, each chapter is broken into subchapters that further break down into entries and subentries. You can read each subchapter (such as Nebulae in the example shown) as a single entity, or use the extensive index to find the specific entry (Reflection Nebulae) or subentry (Eagle Nebula) within the subchapter that discusses your particular interest. Each entry can also stand alone. Within each entry, boldfaced words cross-reference to other entries or subentries in the book that explain the term or subject that is bolded.

Between the entries are essays written by leaders in their fields. These provide personal insight and an additional level of expertise on a variety of subject matters.

Throughout, tables, charts, and time lines accompany the text to provide additional information and quick references. All measurements in the book use Standard International units (the metric system). Tables of useful references and measurements appear at the end of the maps appendix. Photographs, illustrations, and artwork provide a beautiful and educational view of a variety of subjects; diagrams illustrate complex processes and interactions.

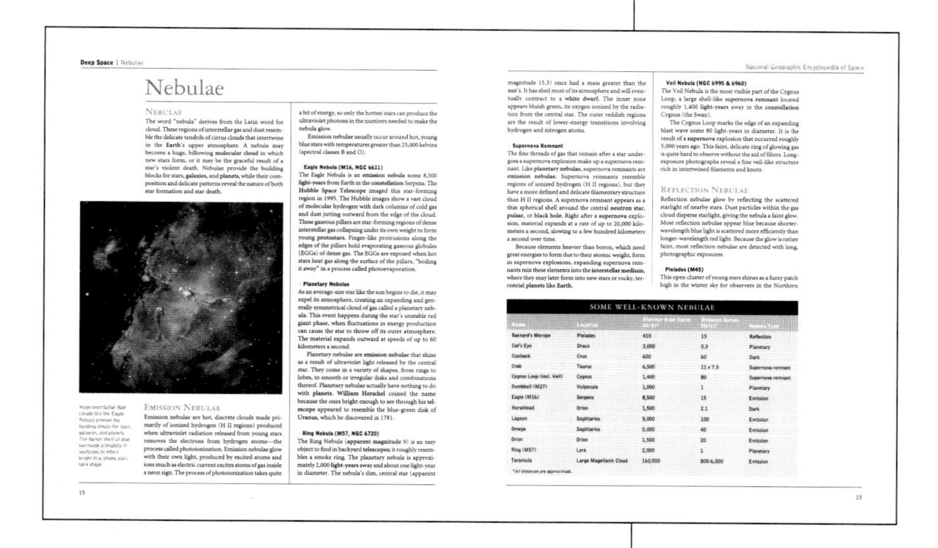
Deep Space | Nebulae

National Geographic Encyclopedia of Space

Nebulae

Nebulae

The word "nebula" derives from the Latin word for cloud. These regions of interstellar gas and dust resemble the delicate tendrils of cirrus clouds that intertwine in the **Earth**'s upper atmosphere. A nebula may become a huge, billowing **molecular cloud** in which new stars form, or it may be the graceful result of a star's violent death. Nebulae provide the building blocks for stars, **galaxies**, and **planets**, while their composition and delicate patterns reveal the nature of both star formation and star death.

Huge interstellar dust clouds like the Eagle Nebula provide the building blocks for stars, galaxies, and planets. The darker shell of dust surrounds a brightly lit workspace in which bright blue young stars take shape.

Emission Nebulae

Emission nebulae are hot, discrete clouds made primarily of ionized hydrogen (H II regions) produced when ultraviolet radiation released from young stars removes the electrons from hydrogen atoms—the process called photoionization. Emission nebulae glow with their own light, produced by excited atoms and ions much as electric current excites atoms of gas inside a neon sign. The process of photoionization takes quite a bit of energy, so only the hottest stars can produce the ultraviolet photons in the numbers needed to make the nebula glow.

Emission nebulae usually occur around hot, young blue stars with temperatures greater than 25,000 kelvins (spectral classes B and O).

Eagle Nebula (M16, NGC 6611)

The Eagle Nebula is an **emission nebula** some 8,500 **light-years** from Earth in the **constellation** Serpens. The **Hubble Space Telescope** imaged this star-forming region in 1995. The Hubble images show a vast cloud of molecular hydrogen with dark columns of cold gas and dust jutting outward from the edge of the cloud. These gaseous pillars are star-forming regions of dense interstellar gas collapsing under its own weight to form young **protostars**. Finger-like protrusions along the edges of the pillars hold evaporating gaseous globules (EGGs) of dense gas. The EGGs are exposed when hot stars heat gas along the surface of the pillars, "boiling it away" in a process called photoevaporation.

Planetary Nebulae

As an average-size star like the sun begins to die, it may expel its atmosphere, creating an expanding and generally symmetrical cloud of gas called a planetary nebula. This event happens during the star's unstable red giant phase, when fluctuations in energy production can cause the star to throw off its outer atmosphere. The material expands outward at speeds of up to 60 kilometers a second.

Planetary nebulae are **emission nebulae** that shine as a result of ultraviolet light released by the central star. They come in a variety of shapes, from rings to lobes, to smooth or irregular disks and combinations thereof. Planetary nebulae actually have nothing to do with **planets**. **William Herschel** coined the name because the ones bright enough to see through his **telescope** appeared to resemble the blue-green disk of **Uranus**, which he discovered in 1781.

Ring Nebula (M57, NGC 6720)

The Ring Nebula (**apparent magnitude** 9) is an easy object to find in backyard **telescopes**; it roughly resembles a smoke ring. The planetary nebula is approximately 2,000 **light-years** away and about one light-year in diameter. The nebula's dim, central star (apparent magnitude 15.3) once had a mass greater than the sun's. It has shed most of its atmosphere and will eventually contract to a **white dwarf**. The inner zone appears bluish green, its oxygen ionized by the radiation from the central star. The outer reddish regions are the result of lower-energy transitions involving hydrogen and nitrogen atoms.

Supernova Remnant

The fine threads of gas that remain after a star undergoes a supernova explosion make up a supernova remnant. Like **planetary nebulae**, supernova remnants are **emission nebulae**. Supernova remnants resemble regions of ionized hydrogen (H II regions), but they have a more defined and delicate filamentary structure than H II regions. A supernova remnant appears as a thin spherical shell around the central **neutron star**, **pulsar**, or **black hole**. Right after a **supernova** explosion, material expands at a rate of up to 20,000 kilometers a second, slowing to a few hundred kilometers a second over time.

Because elements heavier than boron, which need great energies to form due to their atomic weight, form in supernova explosions, expanding supernova remnants mix these elements into the **interstellar medium**, where they may later form into new stars or rocky, terrestrial **planets like Earth**.

Veil Nebula (NGC 6995 & 6960)

The Veil Nebula is the most visible part of the Cygnus Loop, a large shell-like **supernova remnant** located roughly 1,400 **light-years** away in the **constellation** Cygnus (the Swan).

The Cygnus Loop marks the edge of an expanding blast wave some 80 light-years in diameter. It is the result of a **supernova** explosion that occurred roughly 5,000 years ago. This faint, delicate ring of glowing gas is quite hard to observe without the aid of filters. Long-exposure photographs reveal a fine veil-like structure rich in intertwined filaments and knots.

Reflection Nebulae

Reflection nebulae glow by reflecting the scattered starlight of nearby stars. Dust particles within the gas cloud disperse starlight, giving the nebula a faint glow. Most reflection nebulae appear blue because shorter-wavelength blue light is scattered more efficiently than longer-wavelength red light. Because the glow is rather faint, most reflection nebulae are detected with long, photographic exposures.

Pleiades (M45)

This open cluster of young stars shines as a fuzzy patch high in the winter sky for observers in the Northern

SOME WELL-KNOWN NEBULAE

[illegible]	[illegible]	[illegible]	[illegible]	[illegible]
Barnard's Merope	Pleiades	410	13	Reflection
Cat's Eye	Draco	3,000	0.3	Planetary
Coalsack	Crux	600	60	Dark
Crab	Taurus	6,500	11 x 7.5	Supernova remnant
Cygnus Loop (incl. Veil)	Cygnus	1,400	80	Supernova remnant
Dumbbell (M27)	Vulpecula	1,000	1	Planetary
Eagle (M16)	Serpens	8,500	15	Emission
Horsehead	Orion	1,500	2.1	Dark
Lagoon	Sagittarius	5,000	100	Emission
Omega	Sagittarius	5,000	40	Emission
Orion	Orion	1,500	20	Emission
Ring (M57)	Lyra	2,000	1	Planetary
Tarantula	Large Magellanic Cloud	160,000	800-6,000	Emission

*All distances are approximate.

15

15

The appendix of maps contains star charts of the northern and southern skies, Earth's moon, Mars, the solar system, the Milky Way, and the universe.

A list of additional recommended reading and a comprehensive easy-to-use index round out the *National Geographic Encyclopedia of Space.*

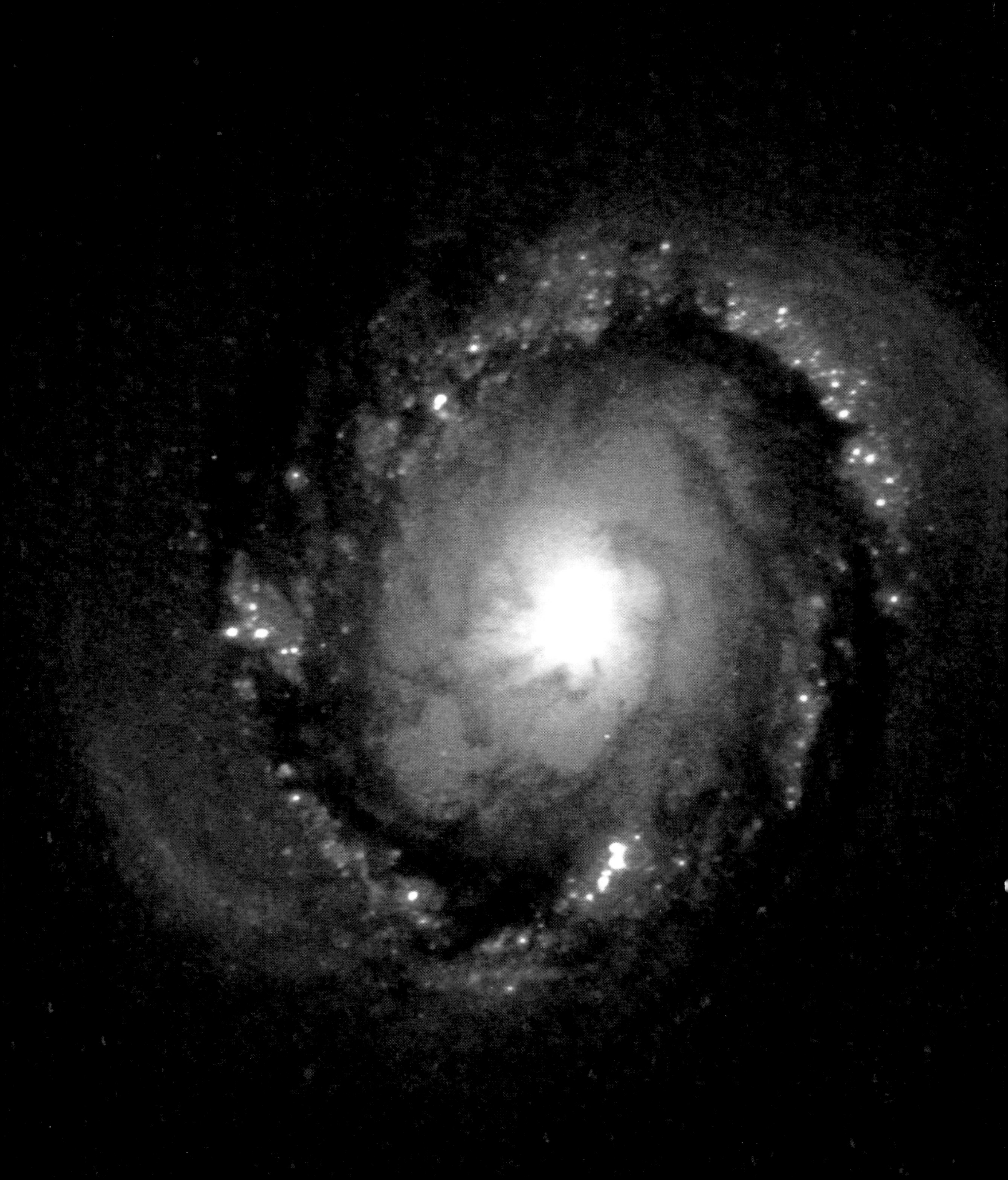

1 | Deep Space

Clusters of bright infant stars, only about five million years old, have formed a ring around the core of the barred spiral galaxy NGC 4314. The set of spiral arms separated by dust lanes circumscribing the stellar ring contributes to the cloud of dust and gas within. This image, shot from the Hubble Space Telescope in 1995, shows only the core of a much larger and much older galaxy.

THE SKY AND ITS MOVEMENTS HAVE FASCINATED HUMANS FOR MILlennia. This fascination and a desire to understand the heavens and our place in them have made astronomy the oldest science in the world. Some of the earliest stargazers were shepherds who watched their flocks by night. They kept an eye on the sky for signs of the changing seasons that indicated which pasture their flocks should be grazing. As they viewed the heavens, these early astronomers began to recognize patterns and movements of the brightest objects. The earliest astronomical records, ancient clay tablets created by the Sumerians some 5,000 years ago, hold the names of these familiar star patterns. They include the constellations Taurus (the Bull), Leo (the Lion), and Scorpius (the Scorpion).

Stargazing eventually gave way to frequent observation. The Babylonians were one of the first cultures to regularly chart the movements of the sun, moon, and planets. Some of the oldest astronomical calculations, found on clay tablets dating from the fourth century B.C., were based on data collected through generations of observations. Early sky-watchers collected data and made calculations to predict the motions of the moon, planets, and sun because they believed they could understand the past and the future if they could only decode the message and movements of the heavens. Astrology is the belief that the movement and position of the celestial bodies can predict or influence our fortunes—either good or bad. Today, we know astrology is superstition, but that early desire to understand the future kept the science of observational astronomy alive. The early records of lunar, solar, and planetary positions offered the first astronomers a foundation on which to build.

From humble beginnings, modern astronomy has developed into a broad science that not only offers an understanding of the sun, moon, planets, and stars, but also the nature of the universe. Like our ancestors, we are eager to understand our origins and our place. But unlike them, we have an arsenal of ground- and space-based telescopes that peer through the vast expanse of space and time to detect the faint glow of the earliest galaxies. We launch robotic probes that orbit and penetrate the atmospheres of our planetary neighbors. We develop instruments that allow us to view the sky at wavelengths that far exceed what the human eye can detect.

This chapter outlines astronomy's early history as well as its more recent discoveries, offering information about the great astronomers, telescopes, observatories, specific astronomical objects such as galaxies, nebulae, stars, black holes, and more.

What did the first galaxies look like? To answer this question, the Hubble Space Telescope cut across billions of light-years to capture a collage of the oldest galaxies in what is known as the Hubble Ultra Deep Field. The brightest galaxies in this close-up thrived one billion years ago.

The Celestial Sphere

Celestial Sphere

Gaze skyward on any clear night from a dark-sky site and you'll see a myriad of **stars, planets**, and fuzzy patches of light. As the night progresses, you'll spy new stars popping into view in the east while others disappear below the western horizon. From your earthly vantage point, it's easy to think of the sky as a large, hollow, spherical shell slowly turning with a stationary **Earth** at its center. The stars appear to be fastened to the slowly rotating sphere, never moving with respect to one another. Ancient Greek astronomers believed this celestial sphere was a crystalline orb embedded with jewels that were the stars.

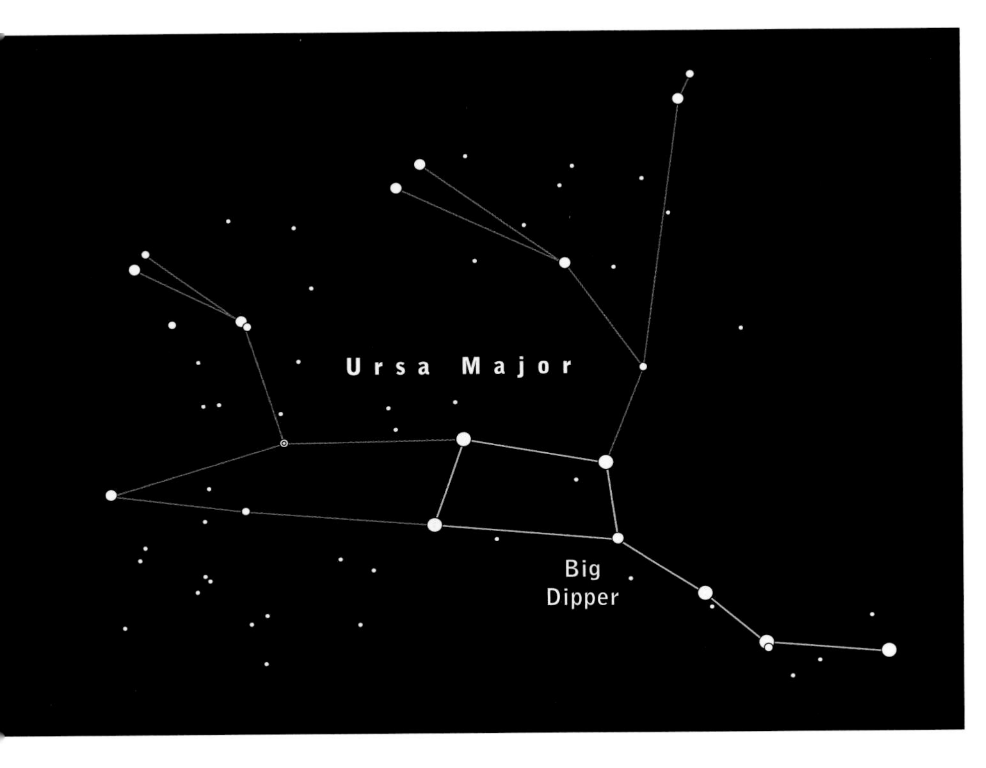

Seven bright stars outline the bowl and handle known as the Big Dipper. The familiar sight is an asterism, a pattern of stars, within the constellation Ursa Major, or Great Bear.

Today, we know that Earth is not at the center of a great celestial sphere, and the apparent rising and setting of the stars is due to Earth's motion. As Earth spins on its **axis** (west-to-east or counterclockwise as viewed above the North Pole), objects in our sky appear to rise in the east and set in the west. For those of us living in the Northern Hemisphere, some stars never set. These stars are called circumpolar stars because they appear to circle the north polar point (a projection of Earth's North Pole up into the sky) near the star **Polaris.** There is a similar point in the Southern Hemisphere near the star Sigma Octantis.

Although the stars appear to be fixed to an imaginary sphere, the **sun, moon**, and planets all move against the backdrop of stars. The moon's motion is the easiest to detect, as it orbits Earth once every 27.3 days and moves roughly 13 degrees across the sky every 24 hours. The sun's motion with respect to the stars is due to Earth's yearly orbit around the sun. As we travel through the **solar system**, the sun appears to move through the **constellations** of the **zodiac.** The planets also travel around the sun. Their orbital motion, along with Earth's, generates a more complex path against the backdrop of stars.

Constellations

If you take time to enjoy the bright twinkles of an evening sky, your brain will begin connecting the dots to form patterns. Ancient sky-watchers, who didn't have television or the Internet to distract them, celebrated these stellar patterns by using them to pass on their history and culture. Most civilizations placed their heroes, gods, and legends in the sky, recounting the stories over and over to their children.

Constellations and their related mythology are a testament to human imagination. The sky holds bears, kings, queens, a dragon, a centaur, a lion, and more. Unfortunately, the majority of these star patterns possess little resemblance to their mythological figures. Native American sky legends not only hold spirit, power, and ritual, but also values. Their constellation stories often center on a moral lesson that provides a celestial reminder of how to behave.

Today, the Western world is most familiar with constellations that originated in Mesopotamia more than 5,000 years ago. Babylonian, Egyptian, and Greek astronomers also made contributions during the classical age. In 1928 the International Astronomical Union (IAU) established the official list of 88 constellations we recognize today. Of those, 48 are from ancient times. The remaining 40 were added in more recent centuries. They include relatively modern instruments such as a microscope, **telescope,** and compasses. The IAU also defined each constellation's border so that constellations not only represent star patterns, but also specific regions of the sky.

Asterism

An asterism is a group of stars that form a distinctive pattern, but these patterns are not **constellations.** Instead, the stars that form an asterism are part of one or more constellations. The most famous asterism is the Big Dipper or Plow. The seven stars that define this shape are actually some of the brighter stars in Ursa Major. Its smaller neighbor, the Little Dipper, is part of Ursa Minor. Other asterisms include the Sickle of Leo, the Teapot in Sagittarius, the Northern Cross in Cygnus, and the "W" or "M" of Cassiopeia. Sometimes asterisms include stars from several constellations, like the Summer and Winter Triangles or the Winter Hexagon. The Summer Triangle holds the star Deneb in Cygnus, Altair in Aquila, and Vega in Lyra. Betelgeuse in Orion, Procyon in Canis Minor, and Sirius in Canis Major form the Winter Triangle. The Winter Hexagon is roughly centered on the star Betelgeuse and also includes the other two stars of the Winter Triangle as well as Rigel in Orion, Aldebaran in Taurus, Capella in Auriga, and Pollux in Gemini. The Great Square of Pegasus is another multiconstellation pattern. It holds three stars from Pegasus and one from Andromeda.

Zodiac

This band of **constellations** extends roughly 9 degrees on each side of the **ecliptic**, the sun's apparent yearly path through our sky. The apparent paths of the **moon** and **planets**—with the exception of **Pluto**—can also be found here. Ancient Greek astronomers divided the zodiac into 12 parts.

Known as the signs of the zodiac, these 12 sections of sky originally referred to the constellations Aries, Taurus, Gemini, Cancer, Leo, Virgo, Libra, Scorpius, Sagittarius, Capricornus, Aquarius, and Pisces. Over time, Earth's slight rotational wobble has shifted the ecliptic by more than 30 degrees against the backdrop of stars. Ophiuchus is now one of the constellations along the sun's yearly path, and the 18-degree-wide zodiac includes parts of Cetus, Orion, and Sextans.

Determining Brightness

A casual glance at the night sky reveals that stars do not have the same brightness. Closer inspection unveils the dim, fuzzy patches of star clusters and **nebulae** while bright **planets** outshine them all. Astronomers determine the brightness of these objects by measuring their intensity or luminous flux—the amount of light energy passing through one square meter in one second, as measured from Earth.

An entire branch of observational astronomy, called photometry, deals solely with the measurement of light energy from stars. One type of photometry is called photoelectric photometry. It uses a metal plate with a very small hole placed in the focal plane of a **telescope.** As light from a star enters the telescope and comes to a focus, it passes through the small hole in the metal plate and lands on an electronic device (photomultiplier) that measures the light's intensity. The electric current generated by the photomultiplier gives an accurate measure of the amount of light passing through the small hole in the metal plate.

Distance plays a key role in star brightness. Imagine a row of streetlights, each one emitting the same amount of light. If you stand directly under one of the lights, it appears far brighter than another lamp a block away. The brightness decreases by the "inverse square law"—an equation that divides brightness by the distance squared (or distance multiplied by itself).

Magnitude Scale

The Greek astronomer Hipparchus (160-127 B.C.) developed a method to measure visual brightness of stars more than two millennia ago. Hipparchus divided the stars into six categories. He dubbed the brightest

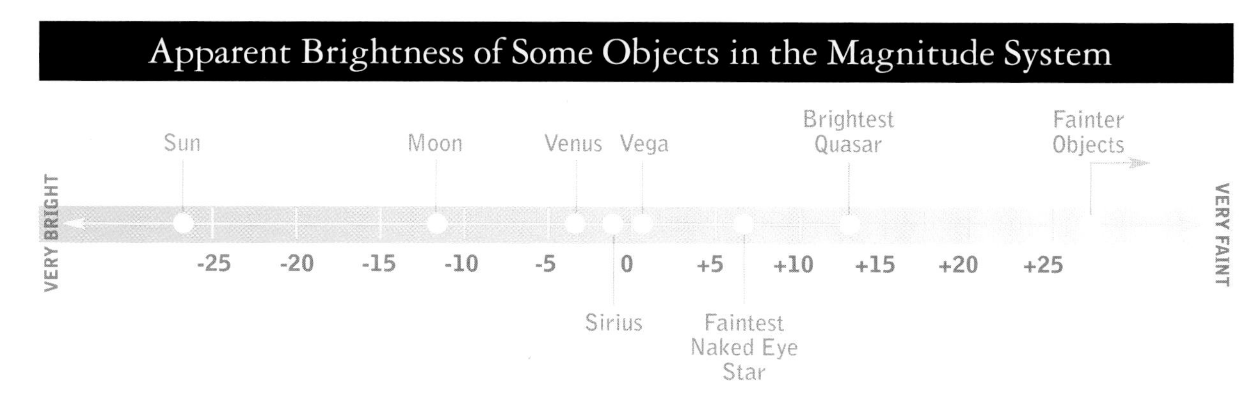

CONSTELLATIONS

Constellation	Pronunciation	Name	Genitive	Abbreviation	First-magnitude Stars
Andromeda	an-DROM-eh-duh	Chained Lady	Andromedae	And	
Antlia	ANT-lih-uh	Air Pump	Antliae	Ant	
Apus	AY-pus	Bird of Paradise	Apodis	Aps	
Aquarius	ack-KWAIR-ee-us	Water Carrier/Bearer	Aquarii	Aqu	
Aquila	ACK-will-ah	Eagle	Aquilae	Aql	Altair
Ara	AY-ruh	Altar	Arae	Ara	
Aries	AIR-eez	Ram	Arietis	Ari	
Auriga	or-EYE-gah	Charioteer	Aurigae	Aur	Capella
Boötes	boe-OH-teez	Herdsman	Boötis	Boo	Arcturus
Caelum	SEE-lum	Chisel	Caeli	Cae	
Camelopardalis	ka-MEL-oh-PAR-da-lis	Giraffe	Camelopardalis	Cam	
Cancer	KAN-surr	Crab	Cancri	Cnc	
Canes Venatici	KAY-neex ven-AT-iss-see	Hunting Dogs	Canum Venaticorum	CVn	
Canis Major	KAY-nis MAY-jer	Great Dog	Canis Majoris	CMa	Sirius, Adhara
Canis Minor	KAY-nis MY-ner	Little Dog	Canis Minoris	CMi	Procyon
Capricornus	kap-rih-KORN-nus	Goat	Capricorni	Cap	
Carina	ka-REEN-uh	Ship's Keel	Carinae	Car	Canopus
Cassiopeia	kass-ee-oh-PEE-yah	Queen	Cassiopeiae	Cas	
Centaurus	sen-TAW-rus	Centaur	Centauri	Cen	Alpha (Rigil Kentaurus), Beta Centauri (Hadar)
Cepheus	SEE-fee-us	King	Cephei	Cep	
Cetus	SEE-tuss	Whale	Ceti	Cet	
Chamaeleon	ka-MEEL-eon	Chameleon	Chamaeleontis	Cha	
Circinus	SUR-sin-us	Compasses	Circini	Cir	
Columba	ko-LUM-bah	Dove	Columbae	Col	
Coma Berenices	KO-mah bear-en-EYE-sees	Berenice's Hair	Comae Berenices	Com	
Corona Australis	kor-OH-nah oss-TRAY-liss	Southern Crown	Coronae Australis	CrA	
Corona Borealis	kor-OH-nah bo-ree-ALICE	Northern Crown	Coronae Borealis	CrB	
Corvus	CORE-vuss	Crow	Corvi	Crv	
Crater	KRAY-turr	Cup	Crateris	Crt	
Crux	KRUX	Southern Cross	Crucis	Cru	Apha Crucis , Beta Crucis
Cygnus	SIG-nuss	Swan	Cygni	Cyg	Deneb
Delphinus	del-FINE-uss	Dolphin	Delphini	Del	
Dorado	dough-RAH-dough	Swordfish	Doradus	Dor	
Draco	DRAY-ko	Dragon	Draconis	Dra	
Equuleus	ek-KWOO-lee-us	Foal	Equulei	Equ	
Eridanus	eh-RID-ah-nuss	River	Eridani	Eri	Achernar
Fornax	for-NAX	Furnace	Fornacis	For	
Gemini	JEM-in-eye	Twins	Geminorum	Gem	Pollux
Grus	GRUSS	Crane	Gruis	Gru	
Hercules	HER-kyou-leez	Hercules	Herculis	Her	
Horologium	hor-oh-LO-jee-um	Pendulum Clock	Horologii	Hor	
Hydra	HIGH-druh	Water Snake	Hydrae	Hya	
Hydrus	HIGH-drus	Little Water Snake	Hydri	Hyi	

CONSTELLATIONS

Constellation	Pronunciation	Name	Genitive	Abbreviation	First-magnitude Stars
Indus	IN-duss	Indian	Indi	Ind	
Lacerta	la-SIR-tah	Lizard	Lacertae	Lac	
Leo	LEE-oh	Lion	Leonis	Leo	Regulus
Leo Minor	LEE-oh MY-ner	Little Lion	Leonis Minoris	LMi	
Lepus	LEE-pus	Hare	Leporis	Lep	
Libra	LEE-brah	Balance	Librae	Lib	
Lupus	LEW-puss	Wolf	Lupi	Lub	
Lynx	LINKS	Lynx	Lyncis	Lyn	
Lyra	LYE-ruh	Lyre	Lyrae	Lyr	Vega
Mensa	MEN-sah	Table	Mensae	Men	
Microscopium	my-kro-SKO-pee-um	Microscope	Microscopii	Mic	
Monoceros	mon-OSS-er-us	Unicorn	Monocerotis	Mon	
Musca	MUS-kah	Fly	Muscae	Mus	
Norma	NOR-mah	Set Square	Normae	Nor	
Octans	AHKK-tanz	Octant	Octantis	Oct	
Ophiuchus	off-ih-YOU-kuss	Serpent Bearer	Ophiuchi	Oph	
Orion	oh-RYE-en	Hunter	Orionis	Ori	Rigel, Betelgeuse
Pavo	PAY-voh	Peacock	Pavonis	Pav	
Pegasus	PEG-uh-uss	Winged Horse	Pegasi	Peg	
Perseus	PUR-see-us	Hero	Persei	Per	
Phoenix	FEE-nix	Phoenix	Phoenicis	Phe	
Pictor	PICK-torr	Painter	Pictoris	Pic	
Pisces	PIE-sees	Fishes	Piscium	Psc	
Piscis Austrinus	PIE-siss OSS-TRY-nus	Southern Fish	Piscis Austrini	PsA	Fomalhaut
Puppis	PUPP-iss	Ship's Stern	Puppis	Pup	
Pyxis	PICK-sis	Compass	Pyxidis	Pyx	
Reticulum	reh-TIC-you-lum	Net	Reticuli	Ret	
Sagitta	sah-JIT-tah	Arrow	Sagittae	Sge	
Sagittarius	saj-ih-TAY-rih-us	Archer	Sagittarii	Sgr	
Scorpius	SKOR-pih-uss	Scorpion	Scorpii	Sco	Antares
Sculptor	SKULPT-tor	Sculptor	Sculptoris	Scl	
Scutum	SKYOU-tum	Shield	Scuti	Sct	
Serpens	SIR-pens	Serpent	Serpentis	Ser	
Sextans	SEX-tans	Sextant	Sextantis	Sex	
Taurus	TAW-rus	Bull	Tauri	Tau	Aldebaran
Telescopium	tell-ih-SKO-pee-um	Telescope	Telescopii	Tel	
Triangulum	try-ANGH-gu-lum	Triangle	Trianguli	Tri	
Triangulum Australe	try-ANGH-gu-lum OSS-TRAY-lee	Southern Triangle	Trianguli Australis	TrA	
Tucana	too-KAH-nah	Toucan	Tucanae	Tuc	
Ursa Major	URR-sah MAY-jer	Great Bear	Ursae Majoris	UMa	
Ursa Minor	URR-sah MY-ner	Little Bear	Ursae Minoris	UMi	
Vela	VEE-lah	Sails	Velorum	Vel	
Virgo	VER-go	Virgin	Virginis	Vir	Spica
Volans	VO-lanz	Flying Fish	Volantis	Vol	
Vulpecula	vul-PECK-you-lah	Fox	Vulpeculae	Vul	

stars visible to the human eye as first-class or magnitude 1 stars, while the faintest were sixth-class or magnitude 6 stars, which are barely visible to the human eye. Hipparchus used his brightness scale to catalog some 850 stars. The magnitude scale Hipparchus developed was extended by **Ptolemy** (second century A.D.) and is still used by astronomers. Using Hipparchus' original scale, magnitude 1 stars are roughly 100 times brighter than magnitude 6 stars. However, modern astronomers know that some stars are far brighter than magnitude 1. The modern scale still defines a jump of 5 magnitudes as a brightness difference of 100, but the scale has been extended in both directions. Vega is so bright that it measures nearly zero magnitude at 0.04. The brightest star in the night sky, Sirius sparkles at magnitude -1.42, while the brilliance of the sun ranks at magnitude -26.5. Instruments such as the **Hubble Space Telescope** have extended the scale in the other direction—detecting objects fainter than magnitude +28.

An important concept to note about the magnitude scale is that larger (more positive) magnitude numbers mean fainter objects, while brighter objects have smaller (more negative) magnitude numbers.

Apparent Magnitude

The apparent magnitude of an object refers to its brightness as seen with the human eye. Apparent magnitude is a subjective measurement that depends on eye physiology as well as perception. It tells us little about a **star**'s intrinsic brightness. A very luminous star will appear faint if it is farther away, while a much less luminous star can appear bright because it is closer.

Absolute Magnitude

Absolute magnitude is a way of determining a star's visual magnitude by calculating how bright the star would be if it were a standard distance of 10 **parsecs** (32.6 **light-years**) away. Absolute magnitude takes into account a star's intrinsic brightness or luminosity.

Light Pollution

Drive away from bright lights of a city toward dark skies of the country and you'll see the effects of light pollution. Street and domestic lighting wash the sky of dimmer objects so that only the brightest stars and **planets** are visible. Although man-made lighting is the worst culprit, natural sources of light pollution can also be a problem. These sources include the **moon**, the background light of our **galaxy**, the scattering of sunlight by **solar system** dust, and airglow—a faint, pervasive glow in **Earth's atmosphere**.

A cluster of more than 500 stars in a span of two degrees, the Pleiades, or Seven Sisters, light up the sky from some 400 light-years away. Surrounded by halo-like blue reflection nebulae, the jewel-like stars of the Pleiades appear as a faint fuzzy patch in the night sky.

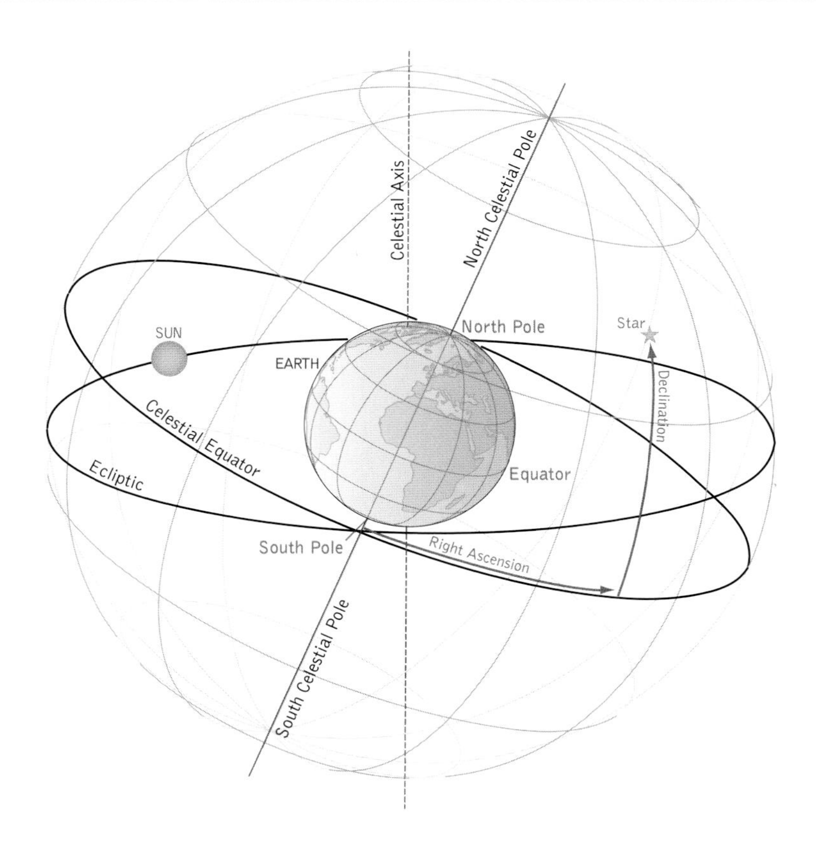

Rising and setting according to Earth's motion, the stars follow a system of coordinates that mimics the one used to define our planet's surface. The celestial sphere features the equivalents of Earth's Equator, latitude, and longitude lines. An extra feature is the ecliptic—the path of the sun against the backdrop of the stars.

Measuring Positions

Like maps of Earth, maps of the sky have an imposed grid system. The grid on a map of Earth holds lines of longitude that mark a point's east-west position, and latitude that mark its north-south position. Celestial cartography's imposed grid system is similar, but uses circles of **right ascension** and **declination.** Every celestial object has specific coordinates, measured from the vantage point of **Earth.**

Right Ascension

An object's right ascension is its angular position along the **celestial equator**—equivalent to **Earth** longitude. The grid begins at the vernal, or spring, equinox and moves eastward along the celestial equator. Because Earth spins on its **axis** once every 24 hours, right ascension is usually measured in hours, minutes, and seconds. One hour of right ascension is equal to 15 degrees of arc.

Declination

This is an object's angular position north or south of the **celestial equator**—equivalent to **Earth** latitude. Objects between the celestial equator and the north **celestial pole** have declinations between 0 and 90 degrees; objects between the celestial equator and the southern celestial pole have declinations between 0 and minus 90 degrees.

Azimuth

The azimuth is measured relative to the observer's place on **Earth.** It is the angular distance from 0 to 360 degrees along the horizon eastward from due north. North is 0 degrees azimuth, east is 90 degrees, south is 180 degrees, and west is 270 degrees.

Altitude

Altitude is measured relative to the observer's position. It is the angular distance to an object above or below the observer's horizon. A point directly overhead is 90 degrees altitude. Altitude is usually expressed as a measure of angular "height" of an object above the horizon.

Zenith

Zenith is the point on the **celestial sphere** that lies directly overhead—90 degrees above the horizon.

Nadir

Nadir is the point on the **celestial sphere** directly under a given point, 180 degrees from the **zenith.** It is 90 degrees below the horizon—actually on the other side of the world— and cannot be seen.

Celestial Poles

The celestial poles are projections of **Earth**'s geographical North and South Poles onto the **celestial sphere.** From the North Pole, the north celestial pole is directly overhead; the south celestial pole is directly overhead at the South Pole. The celestial poles represent the **axis** about which the sky appears to rotate, east to west, a motion actually caused by Earth's west-to-east rotation.

Polaris

Better known as the North Star, Polaris currently lies less than one degree away from the north **celestial pole.** Although often thought to be the brightest star in our sky, this creamy yellow supergiant has a magnitude of only 1.97—far dimmer than magnitude -1.42 Sirius. Polaris is a **Cepheid variable star** that subtly pulsates with a period—the interval of time in which it completes one cycle—of roughly four days. Close inspection with a small **telescope** reveals that Polaris has an 8.2 magnitude (fairly dim) companion star.

Celestial Equator

The celestial equator, like its polar counterparts, is a projection of **Earth**'s Equator onto the **celestial sphere.**

Ecliptic

The plane of **Earth**'s orbit around the **sun** is called the ecliptic. This is offset from the plane of Earth's Equator because Earth does not "stand up straight" as it orbits the sun, but tilts on its **axis** at 23.5 degrees. Because the **celestial sphere** is referenced to the Earth's Equator, the ecliptic plane is inclined to the celestial sphere by 23.5 degrees. The apparent path of the sun against the Earth-referenced sphere is also defined by the ecliptic. The ecliptic intersects the **celestial equator** in two places. The first, in the **constellation** Pisces, marks the sun's position on the first day of spring (vernal equinox). The second, in Virgo, marks the sun's position on the first day of fall (autumnal equinox). The point at which the sun is at its highest above the celestial equator's plane—over the Tropic of Cancer—denotes the first day of summer (summer solstice) in the Northern Hemisphere. Its highest point below the celestial equator—over the Tropic of Capricorn—marks the first day of winter (winter solstice).

Angular Distance

On **Earth** directions are usually given in terms of miles or kilometers; astronomers give directions to celestial objects by using angular distances or separations. The angular separation between two objects is the angle between the objects. Angles are measured in degrees, and there are 360 degrees (360°) in a circle. Each degree in that circle holds 60 minutes of arc (60'), and each minute holds 60 seconds of arc (60"). An astronomer might say that **Venus** is three degrees north of the crescent moon, or the two stars he sees through his **telescope** are separated by eight arc seconds (8").

One way of measuring celestial angles is by using your own hands. Because human bodies are relatively proportional, the length of an adult's arm is proportional to the size of the hand. An index finger held at arm's length is roughly one degree wide, while a tightly closed fist is about ten degrees across. An outstretched hand with fingers spread measures roughly 18 degrees from the tip of the thumb to the tip of the index finger.

Clockwork Movements

Early astronomers were dedicated observers. By watching the cyclic movements of the **sun, moon, planets,** and stars, they realized that these recurring movements could be used to develop a clock and calendar. During daylight hours, the east-west position of the sun provides the time of day, while the stars can be used to tell time during nighttime hours. By watching the movements of stars, people in the Northern Hemisphere noticed that the handle of the Little Dipper moves once around the star **Polaris** every day. The stars of the Little Dipper are circumpolar because they circle the north **celestial pole.** As a result, the Little Dipper's handle is somewhat like the hour hand of a clock, and provides us with an estimate of the time. It circles around Polaris once every 24 hours, so the time it takes to complete one-quarter of its revolution around the North Star is six hours. If the Little Dipper appears overhead at midnight, it would be about half way between overhead and the horizon at 6 a.m. and just above the horizon at noon.

During winter months in the Northern Hemisphere, the sun rises in the southeast and sets in the southwest, making for a short day. Summer finds the sun rising in the northeast, moving to a position high in the south at midday, then setting in the northwest. The long arc the sun travels across our sky yields long summer days. By watching the sun along the horizon, ancient peoples knew when to sow and harvest crops.

Conjunction

When two orbiting bodies are aligned so that they have relatively close right ascensions, they are in conjunction. There are two types of conjunctions. Superior conjunction occurs when a planet is directly behind the **sun** as seen from **Earth.** Inferior conjunction occurs when a planet aligns between Earth and the sun. Only **Mercury** and **Venus,** because their orbits are inside Earth's, can be at inferior conjunction. **Earth's moon** is at inferior conjunction during a **solar eclipse** and at new moon.

Opposition

Opposition occurs when an object is on the opposite side of **Earth** from the **sun. Earth's moon** is at opposition when it is full, and as the sun sets, the moon rises. The inner planets, **Mercury** and **Venus,** can never be in opposition in our sky because their orbits around the sun are inside Earth's. Near opposition is the most favorable time to observe the outer planets—**Mars, Jupiter, Saturn, Uranus, Neptune, and Pluto**—because, like our full moon, the planets at opposition have the full light of the sun on them as seen from Earth.

Occultation

An occultation occurs when one object completely or partially obscures the light of another object. As it orbits **Earth, Earth's moon** occults countless stars and occasionally a **planet** or the **sun** as in a **solar eclipse.** In an average year, the moon may occult more than 4,000 stars in the range of a backyard *(continued on page 31)*

MAPPING THE HEAVENS

Deborah Jean Warner

THE WESTERN TRADITION OF CELESTIAL CARTOGRAPHY HAS TWO ROOTS. ONE lies in the stories that the ancients told about the constellations. The best known constellation stories are those presented in the *Phaenomena,* a Greek poem written by Aratus of Soli in the third century B.C., and in the *Poeticon Astronomicon,* a Latin poem written by Caius Julius Hyginus about 200 years earlier. These stories were also presented in graphic form, as can be seen on the so-called "Farnese Atlas," a Roman copy of a Greek statue of Atlas holding a celestial globe. This statue, unearthed in Italy in the early 16th century and acquired by the Roman art patron, scholar, and Cardinal, Alessandro Farnese, is the only surviving celestial globe from classical antiquity.

The other root of Western celestial cartography lies in the early star catalogs, particularly the one Claudius Ptolemy of Alexandria compiled in the second century and included in his *Almagest,* written around A.D. 150. Ptolemy cataloged 1,025 stars that were visible from the Mediterranean and bright enough to be distinguished by the naked eye. He grouped these stars into 48 constellations and indicated the position of each star in these constellations (by its celestial latitude and longitude), as well as its magnitude (apparent brightness), and its place within its constellation figure. This Ptolemaic catalog formed the basis for stellar astronomy for the next 1,400 years. It was not until the late 16th century that any astronomer recorded both coordinates of the Ptolemaic stars, or systematically charted any non-Ptolemaic ones.

Celestial Charts

In the final quarter of the 16th century, Danish astronomer Tycho Brahe redetermined the positions of many Ptolemaic stars and charted some stars that Ptolemy had overlooked. Dutch cartographer Willem Janszoon Blaeu visited Tycho at his observatory on the island of Hven in the winter of 1595-96 and obtained a copy of his new but as yet unpublished star catalog. Returning to Amsterdam, Blaeu published a celestial globe based on Tycho's work. On it, Blaeu also depicted the nova in Cassiopeia, a bright star that Tycho and others had observed in 1572. The discovery of the nova led to a heated discussion of whether celestial objects were as immutable as previously believed.

In the 16th century, as northern sailors began exploring the southern seas, they began identifying stars in the southern skies that Ptolemy had not been able to see and therefore did not chart. Blaeu knew about these southern stars, but decided not to include them on his globe, as the accuracy of their positions was substantially inferior to Tycho's northern star positions. But he soon changed his mind. In the 1603 revised edition of his globe, Blaeu included the two new southern constellations formed of Ptolemaic stars, Columba and the Southern Cross (which he called El Cruzero Hispanis), as well as the 196 southern stars and 12 southern constellations that Dutch navigators had recently charted. These Dutch constellations included Apus, Chamaeleon, Dorado, Grus, Hydrus, Indus, Musca, Pavo, Phoenix, Piscis Volucris, Triangulum Australe, and Toucan. Around that same time, Johann Bayer, a lawyer in Augsburg, Germany, produced the first important celestial atlas, the *Uranometria.* Published in 1603 and based on Tycho Brahe's catalog, the *Uranometria* contains a copper-engraved chart of each Ptolemaic constellation, as well as a chart of the new southern stars and two planispheres. On the back of each chart Bayer printed a discussion of the various names for the pictured constellation and a catalog of its stars. This work was immensely influential. The text was reissued five times, and the charts—without the accompanying text—were reissued eight times.

Bayer's most enduring innovation is his method of identifying the stars by letters, Greek for the brighter and Roman for the fainter, with the alphabetical order

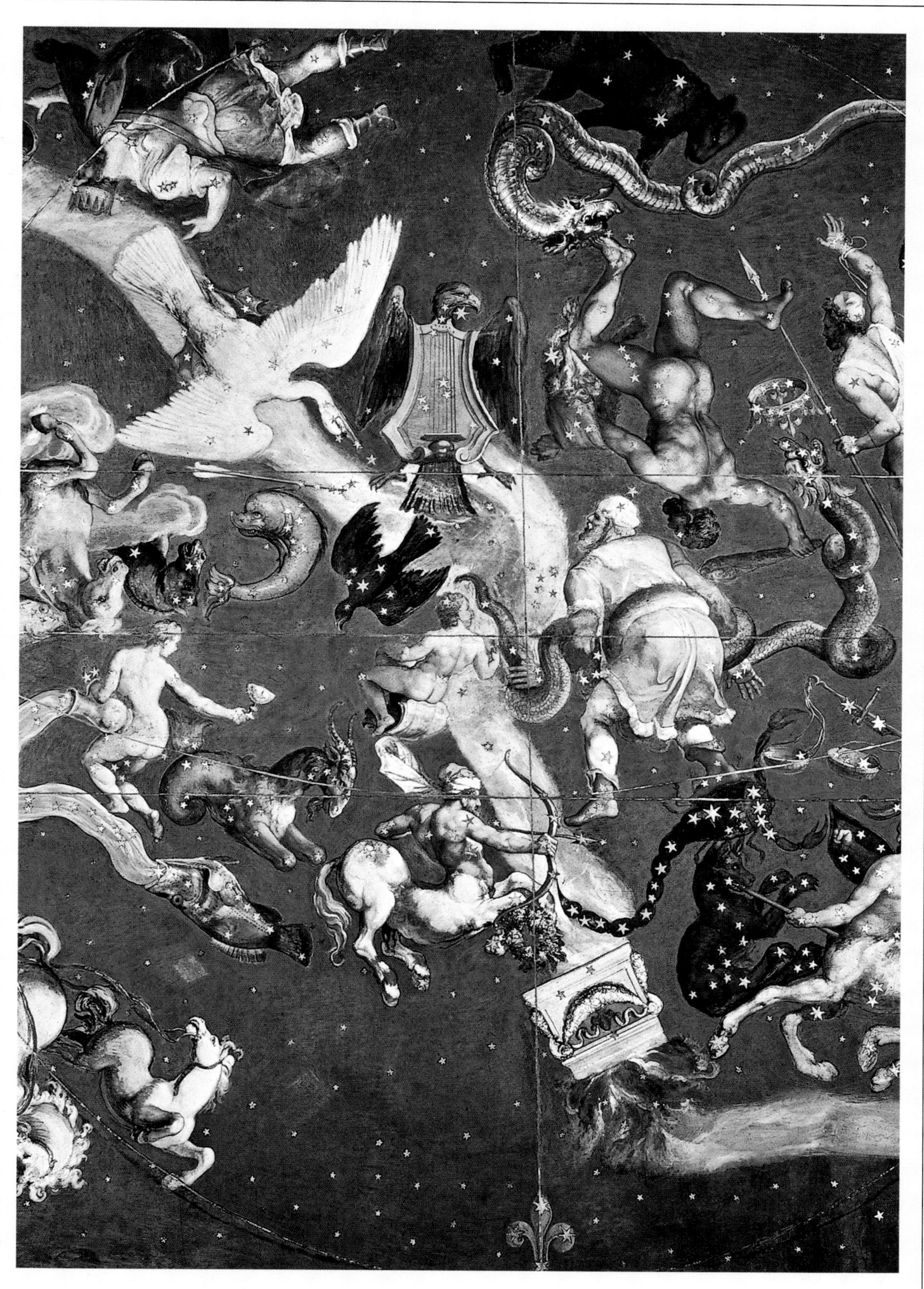

This ceiling fresco in the Palazzo Farnese of Caprarola, Italy, from the late 16th century groups the night sky into interacting constellations based on mythological and allegorical themes. Similar representations—widely varying in style—were used on detailed star charts until the mid-19th century.

corresponding, for the most part, with decreasing brightness. Perhaps because of his new method of star identification, Bayer felt free to disregard the Ptolemaic convention of how constellations were to be shown. While Bayer's charts are like Ptolemy's in that they are geocentric, showing the skies as they would be seen from the Earth, they differ in that some figures face the Earth and some face away. As a result, astronomers could no longer easily understand what Ptolemy meant when he described stars as being in the right arm of one constellation, or the left knee of another.

This same problem arose in the *Firmamentum sobiescianum sive uranographia,* the beautiful star atlas produced by Johannes Hevelius, of Poland, and published in 1687. Unlike Bayer, Hevelius was a real astronomer. Working in his private observatory in Danzig (now Gdansk), he charted the features of the moon, the paths of comets, and the positions of the stars. His *Prodromus astronomiae,* the star catalog on which his atlas was based, appeared in 1690. In addition to the traditional Ptolemaic constellations, Hevelius included several new constellations surrounding stars discovered during the course of the 17th century. Hevelius himself introduced nine star groups, most of which are still in use.

Johannes Hevelius produced the first fairly detailed maps of the moon and its phases in his 1647 *Selenographia sive lunae descriptio.* This map of the moon's near side shows defined craters as well as other geographic features, mistakenly assumed at the time to be seas and continents similar to those found on Earth.

John Flamsteed was another prominent astronomer who contributed to the evolution of celestial maps. He was appointed England's first Astronomer Royal in 1675 and spent much of the rest of his life producing a star catalog and set of charts based on telescopic observations.

As an observer, Flamsteed clearly understood that an astronomer working in a modern and well-funded observatory—such as those at Paris and Greenwich—could probably locate a celestial object from knowing its coordinates, but that most rank-and-file observers at smaller installations and outposts could not. Most observers, he knew, described the positions of planets, comets, or new stars by reference to other well-known and easily located objects. Accordingly, he described his charts as "the glory of the work and, next [to] the catalogue, the usefullest part of it." For this same reason, Flamsteed criticized Bayer and other cartographers who neglected the Ptolemaic conventions concerning star positions within the constellations, and he made sure that his own charts were correct by Ptolemaic standards. Therefore, while Flamsteed's charts are geocentric, his constellation figures all face in toward Earth.

Flamsteed's reluctance to publish his star catalog without the accompanying atlas led to a well-publicized and acrimonious dispute with Isaac Newton, a mathematician who needed the star positions, but who had little appreciation for the pictorial charts. Flamsteed won in the end. His catalog was published posthumously in 1725 as the *Historia coelestis britannicae,* while the *Atlas coelestis* appeared in 1729.

The *Historia coelestis britannicae* was the last great celestial atlas. In time, as telescopes proliferated and became more powerful, the number of known stars increased dramatically, and the number of constellations became unmanageable.

So, at the same time that technology freed astronomers from their dependence on the visual spatial relationships of traditional star charts, the information to be charted overwhelmed the traditional chart format. By the mid-19th century, sensible astronomers called for reform. They reduced the number of constellations in use, rationalized their boundaries, and eventually omitted the constellation figures altogether. The result was maps devoid of charm or beauty, serving simply as functional tools of modern science.

Today, although professional astronomers have moved away from the colorful and fanciful star charts of the past, these magnificent documents of art and science are still preserved and venerated as relics of the visual era in astronomy. ■

telescope. On occasion planets occult stars. As a star passes behind a planet, its light shines through the planet's atmosphere. The illumination of a planet's atmosphere offers astronomers the opportunity to gather information about its different layers. When the orbital planes of **Jupiter** and **Saturn** align with Earth, backyard astronomers may be able to spot the moons of those planets occulting each other. The four largest moons of Jupiter often undergo mutual occultation.

Sidereal Period

The sidereal period is the time it takes a **planet** or **moon** to move from a position back to the same position relative to the **stars,** as viewed from the **sun.** A sidereal month is the time it takes our moon to complete one revolution around **Earth** relative to the stars, roughly 27 days. Earth's sidereal day is 23 hours, 56 minutes, and 4 seconds, while a sidereal day on **Venus** is more than 243 Earth days.

Synodic Period

The synodic period is measured with respect to the **sun** as viewed from **Earth** and measures the time required for a **planet** to return to a certain alignment with the sun, such as from **opposition** to opposition.

Precession

More than 2,000 years ago, Hipparchus compared the position of stars during his time to measurements made nearly two centuries before. Hipparchus found that the **celestial pole** and **equator** had moved ever so slightly. This movement, called precession, is the result of a slight wobble in **Earth**'s **axis** of rotation.

Earth spins around an imaginary line through the North and South Poles called its axis. This axis is tilted 23.5 degrees from the **ecliptic**—or the plane of the Earth's orbit around the **sun.** This tilt, along with the ever changing directions of gravitational pulls from the sun, **moon,** and **planets,** causes Earth to wobble as it spins. This wobble creates precession.

Today, Earth's North Pole points almost directly to the star **Polaris.** Egyptian records show that the star Thuban in Draco was near the north celestial pole in 3,000 B.C. In 12,000 years the North Pole will have precessed toward the star Vega (within 5 degrees).

Proper Motion

Proper motion is the apparent annual movement of a star on the **celestial sphere.** The stars may seem to be fixed points of light gleaming on the celestial sphere. In reality, all stars are moving in orbits around the center of the **galaxy.** This motion is not noticeable on a daily basis, but over centuries it can completely alter the pattern formed by a **constellation.**

Proper motion is one way astronomers can determine stellar distances. A star with large proper motion is closer to our **sun** than one with a very small proper motion. As an example, the star Altair has a proper motion of 0.662 arc seconds per year and lies 5 parsecs (16.8 light-years) away. Albireo has a proper motion of 0.002 arc seconds per year and lies 18 **parsecs** (386 light-years) away.

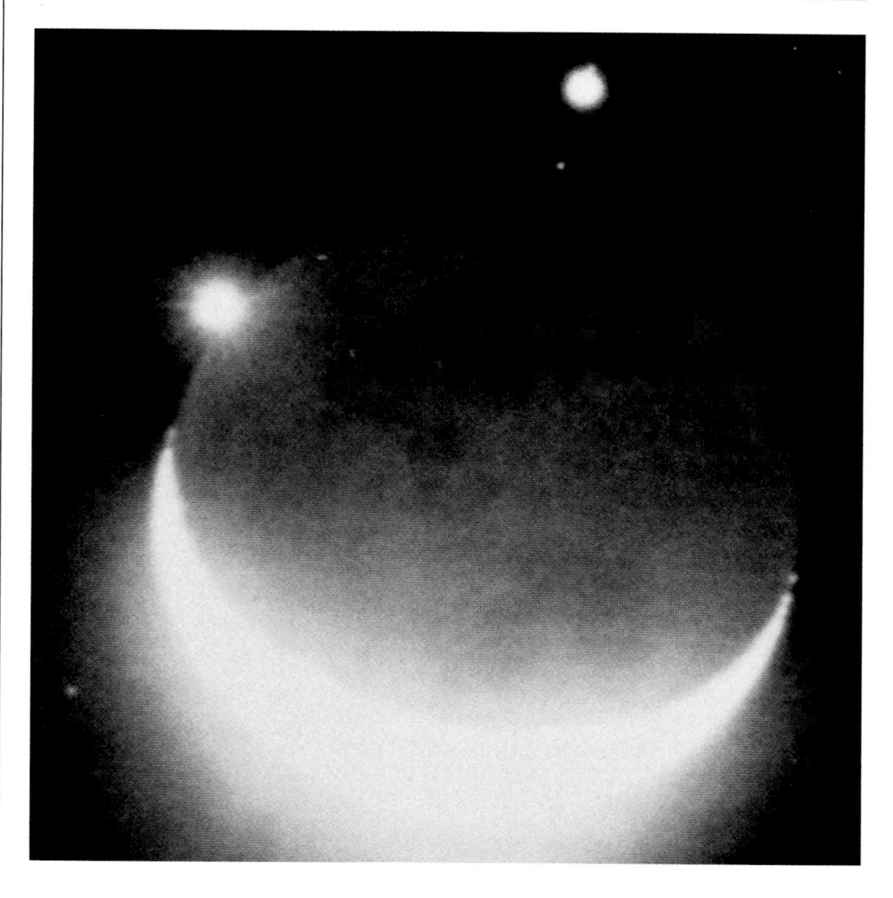

In this photograph taken when Jupiter and Venus were both occulted by our moon, Venus (left) seems to be clinging to the side of our Earth-shined moon, while Jupiter (top) hovers above. Ganymede, one of Jupiter's moons, stands between our moon and Jupiter, and Io, another of the planet's moons, appears as a tiny speck on top of Jupiter.

Universal Time

Universal time (UT) is the standard time scale used worldwide based upon a precise measurement of the **Earth**'s rotation rate and the daily motion of stars in our sky. Greenwich mean time (GMT) is UT.

Because Earth's rotation is gradually slowing down, universal time is changing. Coordinated universal time (UTC) is a more accurate measure of time based on cesium-beam and hydrogen-maser "atomic" clocks. To keep the accurate UTC and the slowing UT aligned within 0.09 seconds, a "leap second" is occasionally deleted from UTC.

The Origins of Modern Astronomy

The Origins of Modern Astronomy

The foundations of observational astronomy go back thousands of years. Monuments such as England's Stonehenge and the Maya pyramids in South America are testaments to the importance ancient cultures gave to the heavens. Every ancient culture developed its own legends of how the universe began. In most cultures, priests were keepers of astronomical knowledge, telling their people when to plant and harvest. The importance of heavenly cycles to agriculture is most evident in ancient Egypt, where the position of certain stars was used to predict the annual flooding of the Nile River. Other astronomical events—such as planetary **conjunctions,** eclipses, or the appearance of a **comet**—were thought to be portents of earthly events to come.

The Greeks were the first to study the sky scientifically. **Aristotle** (384-322 B.C.) observed the cycles of the **planets** and **stars** to develop an Earth-centered (geocentric) model of the universe. Aristarchus (320-250 B.C.) estimated the distances of the **sun** and **moon.** He is also credited with developing the first sun-centered (heliocentric) model of the universe. Eratosthenes (276-194 B.C.) calculated Earth's circumference to within 15 percent of its true value. He also estimated Earth's tilt on its **axis** and cataloged 675 stars. Hipparchus (146-127 B.C.) not only cataloged stars by their brightness, but also discovered the slow wobble of Earth's axis of rotation and developed mathematical models to describe the motion of the sun and moon. The last of the great ancient astronomers, **Ptolemy** (second century A.D.), used the work of his predecessors to develop his own geocentric model, which became the standard for almost 1,500 years.

It was not until the religious, political, and intellectual upheavals of the Renaissance that our modern view of the universe began to take shape. **Nicolaus Copernicus** (1473-1543) was the instigator of this astronomical revolution. Copernicus developed ausable heliocentric model of the universe. Decades later, **Galileo Galilei**'s (1564-1642) observations supported this model. Meanwhile, in northern Europe, two other astronomers were diligently working: **Tycho Brahe** (1546-1601) and **Johannes Kepler** (1571-1630). The work of Brahe and Kepler further developed the Copernican model into something that accurately predicted planetary positions. Brahe, however, remained a confirmed geocentrist.

Copernicus, Galileo, Brahe, and Kepler lived during a time of great intellectual change. Their work helped shift our view of the universe from the long-held geocentric model of perfect spheres and uniform motion to a dynamic heliocentric model. They paved the way for one of the most influential figures in Western science, **Isaac Newton** (1642-1727).

Notable Dates in the Origins of Modern Astronomy

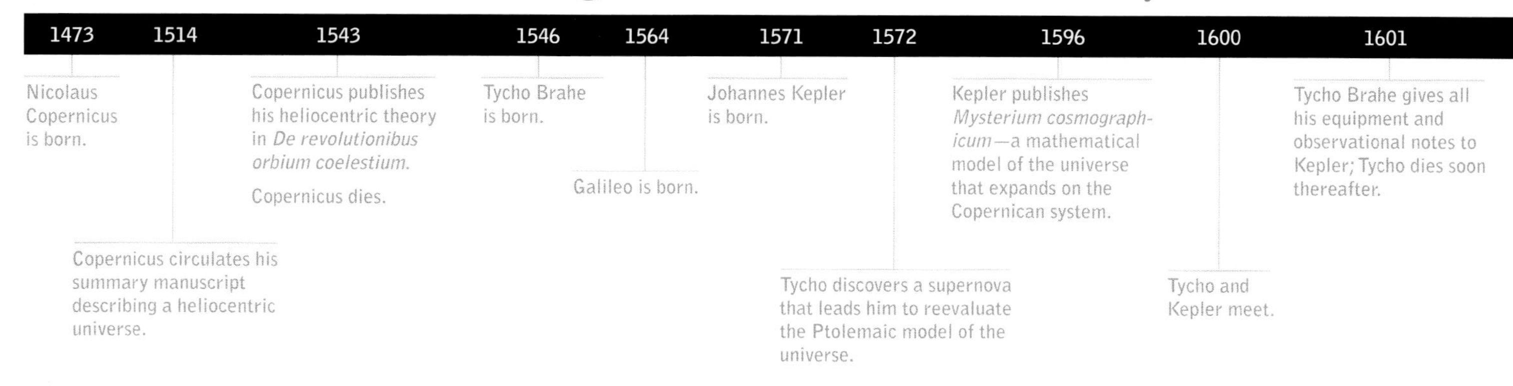

1473
Nicolaus Copernicus is born.
1514
Copernicus circulates his summary manuscript describing a heliocentric universe.
1543
Copernicus publishes his heliocentric theory in *De revolutionibus orbium coelestium.*
Copernicus dies.
1546
Tycho Brahe is born.
1564
Galileo is born.
1571
Johannes Kepler is born.
1572
Tycho discovers a supernova that leads him to reevaluate the Ptolemaic model of the universe.
1596
Kepler publishes *Mysterium cosmographicum*—a mathematical model of the universe that expands on the Copernican system.
1600
Tycho and Kepler meet.
1601
Tycho Brahe gives all his equipment and observational notes to Kepler; Tycho dies soon thereafter.

Like Galileo, Newton studied optics and the motions of falling bodies. He also invented calculus, built the first **reflecting telescope,** discovered gravity, and developed three very important laws of motion. Newton outlined his revolutionary new physics of motion and the concept of gravity in 1687.

William Herschel (1738-1822) was born eleven years after Newton's death. Herschel, along with his sister, Caroline, and later his son, John, took astronomy beyond the **solar system.** Their work cataloging stars, star clusters, and **nebulae** greatly expanded our understanding of the known **universe.**

The next great revolutionary was **Albert Einstein** (1879-1955). Just as Newtonian physics had transformed the Aristotelian world, Einstein's special and general theories of relativity transformed the Newtonian view, catapulting us into the age of space-time and quantum physics. A contemporary of Einstein, **Arthur Eddington** (1882-1944) not only proved Einstein's general theory but also developed a theory of energy production in a star's interior as well as a theory of how stars evolve.

During the same time period, **Edwin Hubble** (1889-1953) was busy setting the stage for a new cosmology. Hubble's discovery that the universe is expanding was as revolutionary as the Copernican model during its time. Before Hubble, our understanding of the universe was limited to the **Milky Way.** His work spurred others to enlarge their cosmological view, developing theories that endeavored to explain the history of the entire universe.

Aristotle (384-322 B.C.)

Aristotle was a Greek philosopher and the most famous and productive of Plato's students. A consummate observer, Aristotle insisted that observation was the guiding principle in the study of nature. In his books, *On the Heavens* and *Meterologica,* Aristotle tried to explain the apparent motions of **planets,** the **moon,** and **stars.** Unfortunately, he didn't have a true understanding of the nature of motion. His model of the universe was based on the idea that everything moved in perfectly uniform, circular motions; he knew nothing of momentum or gravity.

Aristotle's observations led him to develop a complicated cosmological model of 49 spheres centered on **Earth.** He believed that Earth was stationary and at the center of the **universe.** If Earth were spinning, he hypothesized, an object thrown upward would not drop back to the point where it left Earth, which it does. Aristotle also reasoned that if Earth rotated around the **sun,** the stars would display an annual shift and not appear to be fixed. No one had ever observed such a change because the stars are so far away. Today, we can measure this slight shift, or stellar **parallax.**

Although Aristotle's detailed observations led him to an inaccurate cosmological model, they did produce some correct conclusions. Aristotle deduced that Earth must be a sphere because it casts a curved shadow on the moon during a **lunar eclipse.** This assumption led him to calculate the size of Earth's diameter to be 5,100 kilometers. Although Aristotle was way off, his later counterpart, Eratosthenes, came much closer in 200 B.C. by calculating a diameter of 13,400 kilometers—Earth's diameter is actually 12,756 kilometers.

Claudius Ptolemy (Second Century A.D.)

Claudius Ptolemaeus (Ptolemy), of Alexandria, Egypt, was the last great astronomer of antiquity. He is best

1609 Galileo builds a telescope.

Kepler publishes *Astronomia nova,* in which he outlines the first two laws of planetary motion.

1610 Galileo publishes his observations that support Copernicus' heliocentric view of the universe. Written in Italian, *Siderius nuncius* reaches a wide audience.

1613 Galileo declares himself a Copernican.

1616 The Catholic Church bans Copernicus' *De revolutionibus orbium coelestium.*

1619 Kepler publishes *Harmonice mundi,* in which he outlines the third law of planetary motion.

1630 Johannes Kepler dies.

1633 Galileo is tried before the Roman Inquisition.

1642 Galileo dies.

Isaac Newton is born.

1687 Newton outlines his three laws of motion and the concept of gravity in his book *Philosophiae naturalis principia mathematica.*

1727 Isaac Newton dies.

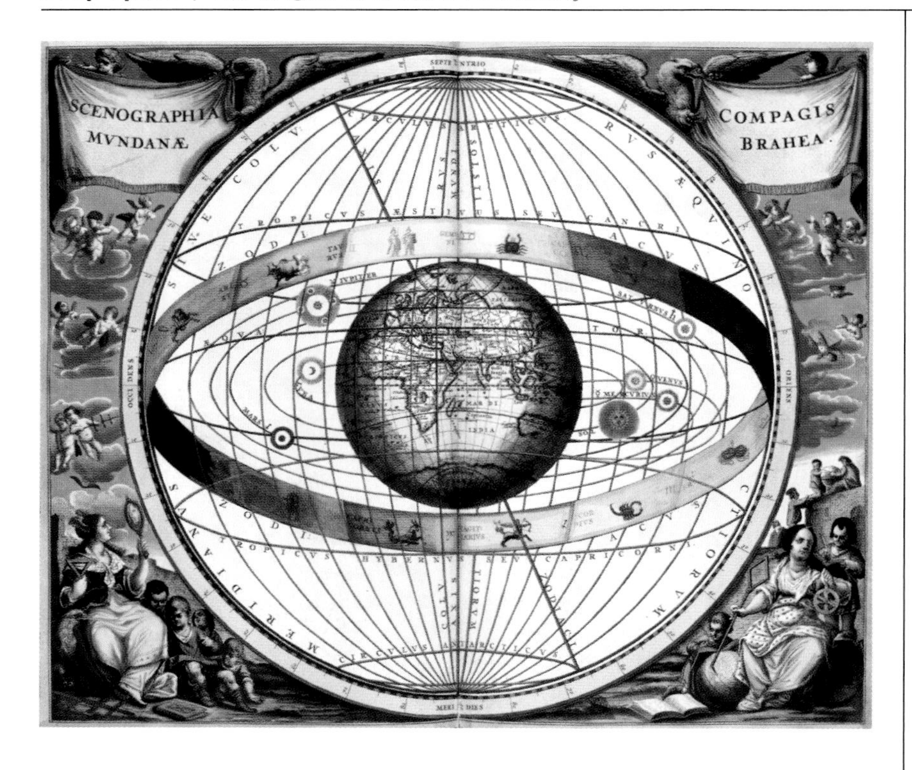

A skilled observer of the skies, 16th-century astronomer Tycho Brahe mapped the orbits of the stars and planets. Brahe's model, like Ptolemy's, was geocentric; however, the appearance of a brilliant supernova in the constellation Cassiopeia convinced him that stars lay beyond the moon—not, as earlier astronomers thought, between Earth and the moon.

known for the Ptolemaic model of the solar system. Like **Aristotle**'s model, this geometrical representation of the **planets, sun,** and **moon** put **Earth** at the center of the universe. But unlike Aristotle's model, the Ptolemaic model predicted the motion of these bodies with considerable accuracy—to within one degree.

Apollonius of Perga originally proposed what became the Ptolemaic model in the third century B.C.; Hipparchus expanded on it; and Ptolemy completed it. In this model each planet moved in a small circular **orbit,** called an epicycle, along the larger Earth orbit. The center of each epicycle moved around Earth in circular orbits (called the deferent) of increasing radius. But these simple circles did not quite predict the motions seen in the sky. To compensate, Ptolemy put Earth slightly off center. He also specified the position from which the center of an epicycle would appear to move at a constant speed. This point was called the equant. The Ptolemaic model was an intricate system of several dozen circles with varying sizes rotating at different rates. As the centuries progressed, errors began to accumulate and the Ptolemaic model became a relatively inaccurate predictor of planetary motion.

Ptolemy's greatest labor was the 13-volume *Almagest* (140 A.D.), which includes much of his own work as well as thoughts and accomplishments of ancient Greek astronomers, principally Hipparchus. Ptolemy's ideas dominated astronomical thought for nearly 1,500 years. By the 16th century, however, Ptolemy's system began to fade as the Copernican model began to rise.

Nicolaus Copernicus (1473-1543)

Born to a prosperous family on February 19, 1473, in Toru, Poland, Nicolaus Copernicus created a revolution when he proposed a sun-centered (heliocentric) model of the universe. The commonly accepted Ptolemaic model of the day was based on the work of **Aristotle** and then Hipparchus. The heavens, Aristotle believed, were the most perfect of regions; conditions worsened downward from the sky, with **Earth**'s center being the most imperfect. This idea fit perfectly with Christian beliefs of heaven and hell. Challenging Aristotle was equivalent to challenging the church. Yet Copernicus was well connected to the church. His uncle was a bishop in Poland, and Copernicus became a canon of the church at the age of 24. Nonetheless, he vehemently, though quietly, opposed the argument for a stationary Earth.

Although his training was in law and medicine, Copernicus's main interests were astronomy and mathematics. While in school, he read the works of the ancient Greeks. By 1514 he had formulated his own ideas and circulated a summary manuscript that outlined them to his friends.

Over the next 30 years, Copernicus refined his heliocentric model but held off publishing his ideas until near the end of his life. His greatest work, *De revolutionibus orbium coelestium (On the Revolution of the Heavenly Bodies)* was published in Nuremberg, Germany, in 1543.

The Copernican system required that the **planets** move in circular paths around the **sun.** It also assumed that the closer a planet lies to the sun, the greater the speed of its revolution. What was radical about Copernicus's cosmic view was his placement of the sun at the center of the universe. In a sun-centered model, Earth orbits the sun faster than the planets that lie farther from the sun. As a consequence, Earth periodically overtakes and passes these planets, making them appear to move backward (retrograde) in our sky. Copernicus accounted for this motion by explaining that the planets farther out took longer to orbit the sun and did not follow the epicycles described in the Ptolemaic model.

Although the Copernican model was more elegant than the Ptolemaic one, it still had problems. The uniform circular motion of the Copernican system made it an inaccurate predictor of the motions of the planets. Both models would generate position errors as large as two degrees—four times the diameter of a full **moon.**

Galileo Galilei (1564-1642)

Galileo Galilei, whose father was a musician and tradesman, was born in Pisa, Italy, on February 15, 1564. Contrary to popular belief, Galileo did not invent the **telescope.** In 1609 he learned of a Dutchman who had developed a spyglass that made things appear closer.

Galileo built his own version, which made objects seem roughly 30 times closer than viewed with the unaided eye. One of the first objects he studied was **Earth's moon,** which appeared pockmarked and rugged. Through the telescope Galileo saw peaks, valleys, and what he thought were seas, or maria. He used the length of the shadows cast by lunar mountains to calculate their height. Galileo also trained his telescope on the faint band of light called the **Milky Way** that stretches across the sky. Innumerable stars, too faint to see with the unaided eye, popped into view. Finally, Galileo pointed his telescope toward the **planets.** When he viewed bright **Jupiter,** he found what he thought were four more planets circling it. We know these objects as the Galilean moons, the four largest satellites of Jupiter.

Galileo's observations were inconsistent with the Aristotelian idea that celestial objects were perfect. The moon was not a smooth sphere. Its surface held features similar to those found on "imperfect" **Earth.** Galileo's observations supported the sun-centered Copernican model. Critics of the Copernican model said Earth could not move because it would leave behind the moon if it did. When Galileo saw Jupiter's moons orbiting the planet, which in turn orbited the **sun,** he realized Earth could travel around the sun with its moon. The motion of the Galilean moons around Jupiter shattered the Aristotelian belief that the motion of all celestial objects was centered on Earth.

On March 12, 1610, Galileo published his observations in *Siderius nuncius* (*The Starry Messenger*). With the exception of the title, the book was written in Italian, not Latin, which made it accessible to a wide audience. The book became a huge success—within five years there was even a Chinese version.

Galileo continued making observations. He discovered **sunspots**—dark, cooler regions of sun's **photosphere**—on the supposedly perfect sun. He also found that **Venus,** like the moon, has phases. Because Venus is closer to the sun than Earth, we see it at different angles relative to the sun, just as we see the moon. From our earthly viewpoint, Venus appears to phase from crescent to full. However, when Venus is "full" it lies on the opposite side of the sun and cannot be seen from Earth.

By 1613 Galileo publicly declared himself a Copernican. In 1616 an official of the Inquisition warned Galileo to stop teaching the Copernican model as fact. At about the same time, the Catholic Church placed **Copernicus**' book, *De revolutionibus orbium coelestium,* in an *Index of Forbidden Books.* Galileo became the center of a firestorm that lasted nearly 20 years.

On June 22, 1633, at the age of 70, Galileo knelt before the Roman Inquisition and recanted. He was sentenced to life in prison, but was confined to his villa until his death on January 8, 1642—99 years after that of Copernicus.

By placing the sun at the center of the universe, Polish astronomer Nicolaus Copernicus broke with tradition and the teachings of the Catholic Church. "In the center of all rests the sun," wrote Copernicus. "For who would place this lamp of a most beautiful temple in a better place?"

EARLY ASTRONOMICAL INSTRUMENTS

Sara Schechner

THE DEBATE OVER THE GEOMETRIC AND COMPUTATIONAL MODELS OF THE COSmos was carried out not only in lively discussion and written tracts, but also with scientific instruments. This part of the story has always held particular fascination for me. While working in the collections at Harvard, the Adler Planetarium and Astronomy Museum, and elsewhere, I began to examine preserved records and surviving instruments, hoping to discover who designed and produced the instruments, how they were used and by whom, what functions they served, and how their forms not only followed their functions, but also how they embodied the scientific theories that they were used to explain. I also began to create replicas of historical instruments in order to make observations in the old ways. These hands-on activities helped me understand the sophistication of early astronomy and the challenges its practitioners faced.

Deceptively simple, the gnomon (a shadow-casting stick) was a sophisticated astronomical instrument for its time. Anaximander of Miletus set up a gnomon in Sparta, Greece, in the sixth century B.C. By tracing the lengths and angles of the gnomon's shadow, Greek philosophers could mark the times of the solstices and equinoxes, measure the number of days between them, establish calendars, and project the celestial sphere onto a flat or curved surface. The combination of the gnomon and the mathematical projection created the sundial. By the third century B.C. hour-finding sundials were common in Greece and Rome. They shed light on the place of mathematics and astronomy in people's daily lives.

The celestial globe was another projection of the heavenly sphere. Early globes documented the positions of major star groups, along with the principal celestial circles—the Equator, the Tropics of Cancer and Capricorn, the Arctic and Antarctic Circles, and the ecliptic (the path of the sun against the fixed stars). These globes were used to solve basic problems of positional astronomy. Eudoxus of Cnidus possessed and used a globe, as did Ptolemy five centuries later.

Globes were teaching tools as early as the third century B.C. in Greece. Globes embodied the essential astronomical and cultural cosmology of the day. They represented the universe as spherical and closed, as if it were being viewed by a spectator beyond the universe. Moreover, globes delineated the constellations, which were used both as scientific maps and allusions to the traditional gods of Greek mythology. This rich combination of information made globes prized scientific instruments as well as cultural symbols of power and prestige. The earliest surviving globe is the one borne on the shoulders of the "Farnese Atlas," a Roman statue dating from about 200 B.C. and preserved in the Museo Archeologico Nazionale in Naples.

Another model of the universe and a close relative to the celestial globe was the armillary sphere, composed of rings representing the great circles of the celestial sphere. Two forms of the armillary sphere descended from antiquity; one was an instrument for observation, and the other was a tool for teaching. The observational instrument was devised by Ptolemy to measure the positions of stars and planets in the coordinate system of his choice. Observational armillaries were a standard fixture in medieval Islamic observatories. Information on the construction and use of this instrument was transmitted to the West in the late 12th century. Nicolaus Copernicus was among those who used an observing armillary sphere in the early 16th century. Tycho Brahe refined its design and use at the end of that century.

The teaching armillary sphere was a scaled-down version of the larger observational instrument. Between the outer rings representing the sphere of

The astrolabe was developed near Alexandria, Egypt, in the third century A.D. as a means of sighting the positions of celestial objects and telling time. It was used extensively in the Islamic world before entering Western Europe through Islamic Spain. A quibla indicator (bottom) is an instrument used for calculating the direction to Mecca when time for Muslim prayer.

fixed stars and a central globe representing the Earth, this instrument had a series of nested, movable rings or spheres for the moon, the sun, and the planets. Thus, the armillary embodied the Earth-centered cosmology of Aristotle and represented our place in the universe. It was a standard piece of equipment for astronomers and scholars from the medieval period through the 18th century.

The quadrant was an important angle-measuring instrument for the astronomer. The plinth, a forerunner described by Ptolemy, used a peg to cast a shadow on a vertically standing 90-degree arc to measure the altitude of the sun. Al-Battānī, a Muslim astronomer, had a large quadrant fixed to a north-south wall in his observatory. He used an alidade, or sighting rule, to make his observations. Nasīr al-Dīn al-Tusīn Maragha had movable quadrants that could measure angles of altitude or azimuth (the bearing of an object). These quadrants were adopted by Western astronomers during the Renaissance. Small portable quadrants were used in medieval times for surveying and determining time.

The queen of ancient scientific instruments, however, was the astrolabe, which simulated the apparent rotation of the stars around the celestial North Pole. An astrolabe consisted of a "see-through" star map that rotated upon a stack of engraved plates, each representing the sky as seen from a different latitude. Once the user selected the plate for his particular latitude, he could determine where and when celestial bodies would move in the sky from his vantage point. The astrolabe was used for locating stars in the sky and finding the times of their rising and setting; for determining the hour, day or night; for making astrological calculations; for finding latitude or longitude; for surveying; and, in Islamic regions, for trigonometric calculations and determining the hours of prayer as well as the direction of Mecca. In short, the astrolabe was an analog computer and portable model of the heavens. Developed near Alexandria, Egypt, before the fourth century A.D., it was widely used and improved in the Islamic world before arriving in the Latin West through Spain in the tenth century.

To the inventors of the astrolabe and the quadrant, to the brilliant minds who saw how a simple stick could teach them about the movement of the sun, to the philosophers and mathematicians, and the peoples of the Earth who found purpose and order in their observations of the night sky, the scientists of today owe a great debt. Our appreciation and understanding of the universe has been built on these culturally rich foundations. ■

Tycho Brahe (1546-1601)

Tygre (Latinized to Tycho) Brahe was born to nobility on December 14, 1546, in what was then Skåne, Denmark, but is now part of Sweden. Tycho was one of ten children, but was raised by his uncle Jörgen Brahe and his wife. At age 13 Tycho began studying at the Lutheran University of Copenhagen. His uncle wanted him to become a lawyer, but Tycho became enamored with astronomy after watching the **solar eclipse** of August 12, 1560. His family was not pleased, so Tycho followed his passion secretly—spending his allowance on astronomy books and sneaking out late at night to observe.

Master of the sky—and master of two observatories funded by King Frederick II of Denmark—Tycho Brahe is shown directing his assistants in observations of the heavens. In these observatories, he mapped the positions of 777 stars.

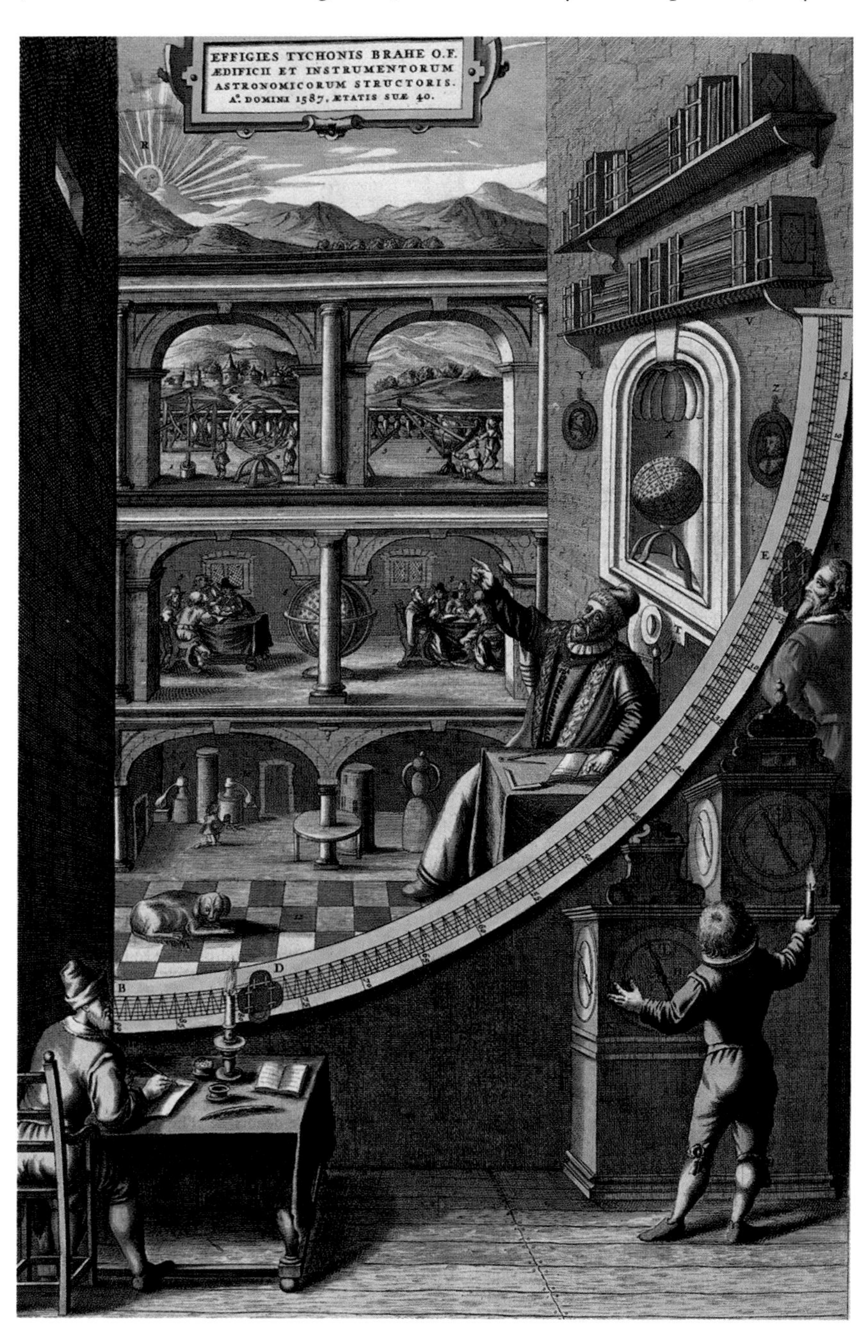

In August 1563 Brahe watched the **conjunction** of the **planets Jupiter** and **Saturn.** Night after night he charted the paths of the two planets as they moved closer to each other, nearly merging into a single point on August 24. As he made his observations, Tycho became increasingly aware of errors in the astronomical tables based on the Ptolemaic model. He wanted to correct these errors and, as a result, became a skillful observer.

Brahe's uncle died in 1565. Tycho, only 19, began openly studying astronomy at the University of Wittenberg. One year later his personality, which was said to be abrasive and arrogant, got the best of him. The brash young man engaged in a duel with his cousin that cost him part of his nose. For the rest of his life he wore a prosthesis of gold, silver, and copper stuck on with wax.

On November 11, 1572, Brahe noticed that a "new and unusual **star,** surpassing the other stars in brilliance," was shining directly over his head. This new star was a brilliant **supernova** (now called Tycho's star) in the **constellation** Cassiopeia. It shone for 18 months, reached a magnitude of -4, and puzzled the classically trained astronomer. He subscribed to the Aristotelian view that the stars exist in a perfect, unchanging sphere, so that such new stars had to be lying closer to **Earth** than the **moon.**

A dedicated observer, Brahe carefully watched the brilliant supernova until it faded from view and collected data from other observers all over Europe. He determined that the star's position did not change. It was in the same place in Cassiopeia, regardless of where observations were made. Because the star did not shift relative to other stars, it had to be part of that outer, perfect sphere of stars and not between Earth and the moon. These results forced Brahe to reevaluate the Ptolemaic model. His results were published in a small book, *De stella nova (The New Star),* in 1573.

Fame followed Tycho and his star. In 1576 King Frederick II of Denmark offered Brahe the island of Hven, in the Danish sound, and funds to build an observatory there. He built Uraniborg (Castle of the Heavens) and later Stjerneborg (Castle of the Stars), where he lived and observed the night sky with a myriad of grand instruments for more than 20 years.

While on Hven, Brahe tried to measure stellar **parallax** (a slight shift in a star's position), and when he failed to detect it, he concluded that the Copernican model was wrong. Today, we know that stellar parallaxes caused by Earth's **orbit** of the **sun** are a hundred times smaller than Brahe could have detected.

Brahe measured the positions of 777 stars without the aid of a **telescope.** His meticulous observing habits and the large instruments he designed provided him with a high degree of accuracy, better than four arc minutes.

When Frederick II died in 1588, Brahe's tempestuous personality lost him favor with the new king. He packed up his instruments and observations and took the position of imperial mathematician to the Holy Roman Emperor, Rudolph II, in Prague. Based on his years of data, Brahe intended to develop his own "Tychonic" model of the universe. He hired several astronomers and mathematicians to help with the calculations. One of them was **Johannes Kepler.**

In November 1601 Tycho Brahe collapsed. While on his deathbed he convinced Rudolph II to hire Kepler as his replacement.

Johannes Kepler (1571-1630)

See page 116.

Sir Isaac Newton (1642-1727)

An English physicist and mathematician, Isaac Newton was born at Woolsthorpe Manor, in Lincolnshire, England, on December 25, 1642. He could also claim January 4, 1643, as his birth date. In 1582 the Gregorian calendar was introduced in Europe by papal decree. However, Protestant England continued to follow the Julian calendar. So December 25, 1642, in England was January 4, 1643, in Europe.

As a child Newton was inquisitive and mechanically inclined. He entered Trinity College in Cambridge in 1661. The core curriculum at Trinity was weighted heavily toward an Aristotelian point of view. Newton wrote in his notebook: "Plato and Aristotle are my friends, but my best friend is truth."

The black death overwhelmed Britain in 1665, and Trinity College closed. Newton returned to Woolsthorpe, where he developed his ideas on mechanics and optics. During this time Newton invented calculus, investigated the nature of light, and began to develop his three laws of motion. He also had his legendary flash of insight that related a falling apple to the **orbit** of the **moon.**

Newton wondered if the force that causes an apple to fall to the ground could be the same force that holds the moon in orbit around **Earth.** He deduced that the strength of this force (gravity) would decrease as the square of the distance increased.

Upon returning to Trinity, Newton shared his findings with his mentor, Isaac Barrow, who recognized his genius. In 1669 Newton was appointed the Lucasian Professor of Mathematics when Barrow retired. By 1671 Newton had built the first **reflecting telescope,** which used mirrors instead of lenses to bring an image into focus. He published a few papers on light and color. They generated such controversy that Newton vowed never to publish again.

Newton continued to hone his theories in isolation until his friend Edmund Halley (1656-1742) asked for advice with a problem related to **Kepler**'s elliptical orbits. Halley, interested in the force between the sun and the planets, was surprised to find that Newton had already identified the force of gravity.

Halley convinced Newton to publish his work, launching the beginning of modern celestial mechanics. *Philosophiae naturalis principia mathematica (The Mathematical Principles of Natural Philosophy)* was published in 1687. It outlined Newton's new physics of motion and the concept of gravity.

The book opens with his three laws of motion, which include the inertial law, the force law, and the reaction law. The inertial law is more commonly known as the conservation of momentum. Essentially, it states that a motionless object stays put while a moving object keeps moving. The force law says that the momentum of an object can change only if an outside action influences it. The amount and direction of that change are directly proportional to the outside force and inversely proportional to the object's mass: Force equals mass times acceleration ($F=ma$). Newton's reaction law states that for every action there is an equal and opposite reaction. This third rule implies that all forces occur in pairs that are mutually equal to and opposite each other.

Newton used these laws to attack the problem of planetary orbits and subsequently devised his universal law of gravity. It states that all objects universally attract each other, and the amount of gravitational force is proportional to an object's mass and inversely proportional to the square of the distance of the two objects from one another.

Queen Anne knighted Newton in 1705. The latter half of his life was darkened by controversy, but his reputation survived. Newton died on March 20, 1727, in London. His influence on mathematics and astronomy has been called the "Newtonian revolution."

Herschel Family: Sir William (1738-1822), Caroline (1750-1848), Sir John (1792-1871)

Known as the founder of stellar astronomy, Sir William Herschel was a **telescope** maker, planet finder, and theoretician. He was born to a musical family on November 15, 1738, in Hanover, Germany. As a child, Friedrich Wilhelm (anglicized to William) spent hours gazing at the night sky with his father. He learned to play the violin and oboe and in 1753 joined his father as a member of the regimental band of the Hanoverian Guards. Four years later he moved to England, where he became a music teacher, then organist of the Octagon Chapel in the city of Bath. In 1772 Herschel's sister, Caroline, joined him in England.

Through his years of teaching and playing music Herschel remained fascinated by the night sky. In 1773 he began to build telescopes, turning nearly every room in the house into a workshop. On March 4, 1774, Herschel turned his newly completed 1.6-meter **reflecting telescope** toward the Orion Nebula. It was the beginning of his astronomical career.

Herschel built several more telescopes, the largest measuring 12 meters and holding a 1.2-meter mirror. On March 13, 1781, Herschel spotted an object that he first thought was a **comet.** After watching it for several nights, he realized it was a new **planet** moving beyond the orbit of **Saturn.** Herschel had discovered **Uranus**—the first new planet found since ancient times. He originally named his planet Georgium Sidus to honor King George III, but adopted the name Uranus in keeping with the tradition of naming planets after gods from Greek mythology. The following year the king gave Herschel a modest pension and appointed him Royal Astronomer. Herschel devoted himself to astronomical research and began to methodically study and catalog **stars.** In 1783 he published an index of double and multiple stars and began a systematic search for nebulae—eventually finding nearly 2,500 of them over a period of 20 years. In 1783 Caroline discovered three new **nebulae** (at the time, distant galaxies were also thought of as nebulae). William and Caroline were the perfect pair of observers—William sitting at the eyepiece describing his views to his sister, who took copious notes.

During times that William was away, Caroline searched for comets with a small telescope William had given her. In 1786 she discovered the first of eight comets she would find during her lifetime. Caroline Herschel was not only a devoted sister, but also a magnificent observer. Finding comets requires tremendous depth knowledge of the sky. Most star clusters, nebulae, and **galaxies** would have been faint, fuzzy, comet-like patches in Caroline's telescope.

During this prolific period, William Herschel discovered two moons of Uranus (1787) and two of Saturn (1789). He and Caroline also mapped the stars in three dimensions. By systematically sweeping across the sky and counting the number of stars in any given field of view, Herschel realized that our **solar system** was inside a disk-shaped cloud of stars, the **Milky Way.**

On May 8, 1788, William married Mary Pitt. For 16 years Caroline had cared for William and his household, and she remained devoted to him as well as to their work. William and Mary gave birth to a son, John Fredrick William Herschel, in 1792. John succeeded William and Caroline in their research, compiling the *General Catalogue of Nebulae and Clusters* from their work. While in England, he discovered several hundred new star clusters and nebulae, and he took one of his father's telescopes to Africa's Cape of Good Hope. There he surveyed the southern skies, cataloging 1,200 new stars and 1,700 new nebulae and star clusters. In 1849 John wrote *Outlines of Astronomy,* which became the standard astronomy textbook for decades.

When William Herschel died at age 83 on August 25, 1822, Caroline was devastated. William and their work together had been the center of her life. After more than 50 years in England, she returned to Hanover, Germany, in October 1823. There she arranged all the star clusters and nebulae discovered by William into zones. For this she was awarded the Gold Medal of the Royal Astronomical Society in 1828. In 1846 the King of Prussia awarded her the Gold Medal for Science. Caroline kept in close contact with her nephew and often wished to be out observing with him. She died on January 9, 1848. Her nephew John died 23 years later—bringing to a close an astronomical dynasty that lasted almost a century.

Star Catalogs

Many astronomers preceded the Herschels in cataloging stars. **Tycho Brahe** produced a star catalog in the final quarter of the 16th century. In 1781 Charles Messier cataloged his 87th nebulous object, **M87.** Published in 1888, the New General Catalog (NGC) gives each object an NGC prefix and number. Index Catalogs (IC), two supplements published at the turn of the century, added more than 5,000 objects to the general NGC. Some star catalogs, such as Edward E. Barnard's catalog of dark **nebulae,** are very specialized.

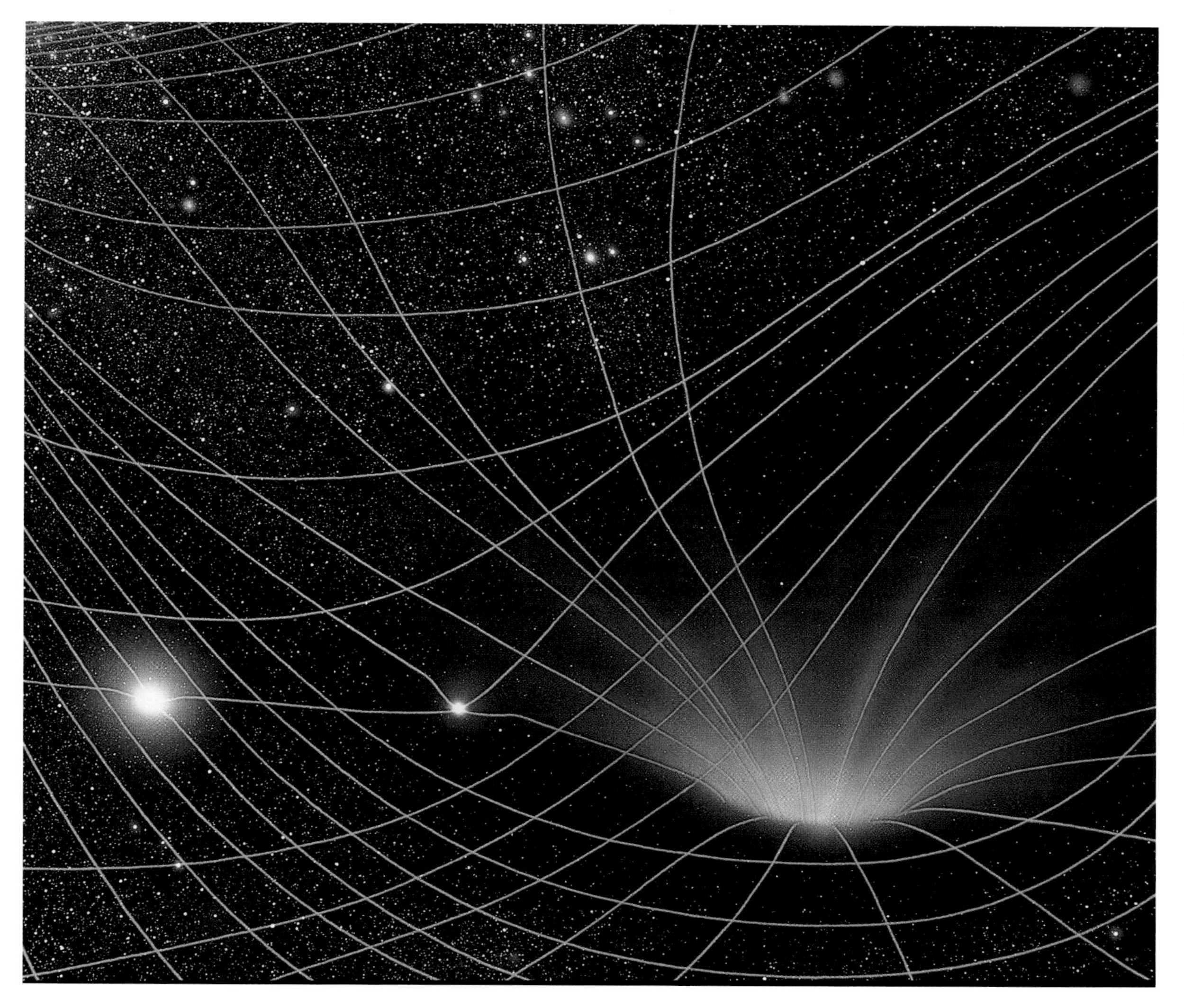

The superheavy gravitational pull of a black hole warps the thin fabric of space, illustrating Albert Einstein's theory that space is distorted by gravity. Our sun, far left, barely makes a dent, while a denser more massive neutron star to its right creates a slight distortion. Space is like a rubber sheet on which objects of varying weight, such as planets and stars, produce smaller or larger dents.

Albert Einstein (1879-1955)

Born on March 14, 1879, in Ulm, Germany, Albert Einstein is the most widely known scientist of the 20th century. Although he taught himself geometry at age 12, Einstein showed no particular aptitude for learning in the highly structured German school system. Nonetheless, Einstein graduated with a degree in physics from the Swiss Polytechnic Institute in 1900.

In 1902 he began working as a junior patent examiner in the Swiss Patent Office. While scrutinizing applications, Einstein began to examine the properties of space and time and how motion and gravity interact. In 1905 he introduced his special theory of relativity, which held that motion, time, and distance are not absolute, but are relative to moving frames of reference. Imagine sitting on a train that is parked in the station. Looking out the window, you see the train next to you slowly moving. For a moment, you cannot tell if it's your train or the train next to you that is moving. At low speeds, Einstein's theory yields the same predictions as do **Newton**'s laws of motion. It is only at very high velocities (near the speed of light) that the predictions of the two theories diverge.

In 1916 Einstein put forward a more general theory of relativity, which provided a new description of gravity. Basically Einstein's theory states that gravity curves **space-time.**

This curvature controls the natural motions of bodies in space. Imagine a large sheet of rubber stretched across a frame. If a bowling ball is placed in the center of the sheet, it will bend to the mass of the ball. If you tried to roll a golf ball in a straight line across the sheet, you could not. It rolls toward the bowling ball. Just as that rubber sheet flexes and the bowling ball influences the movement of the introduced golf ball, matter curves space-time while

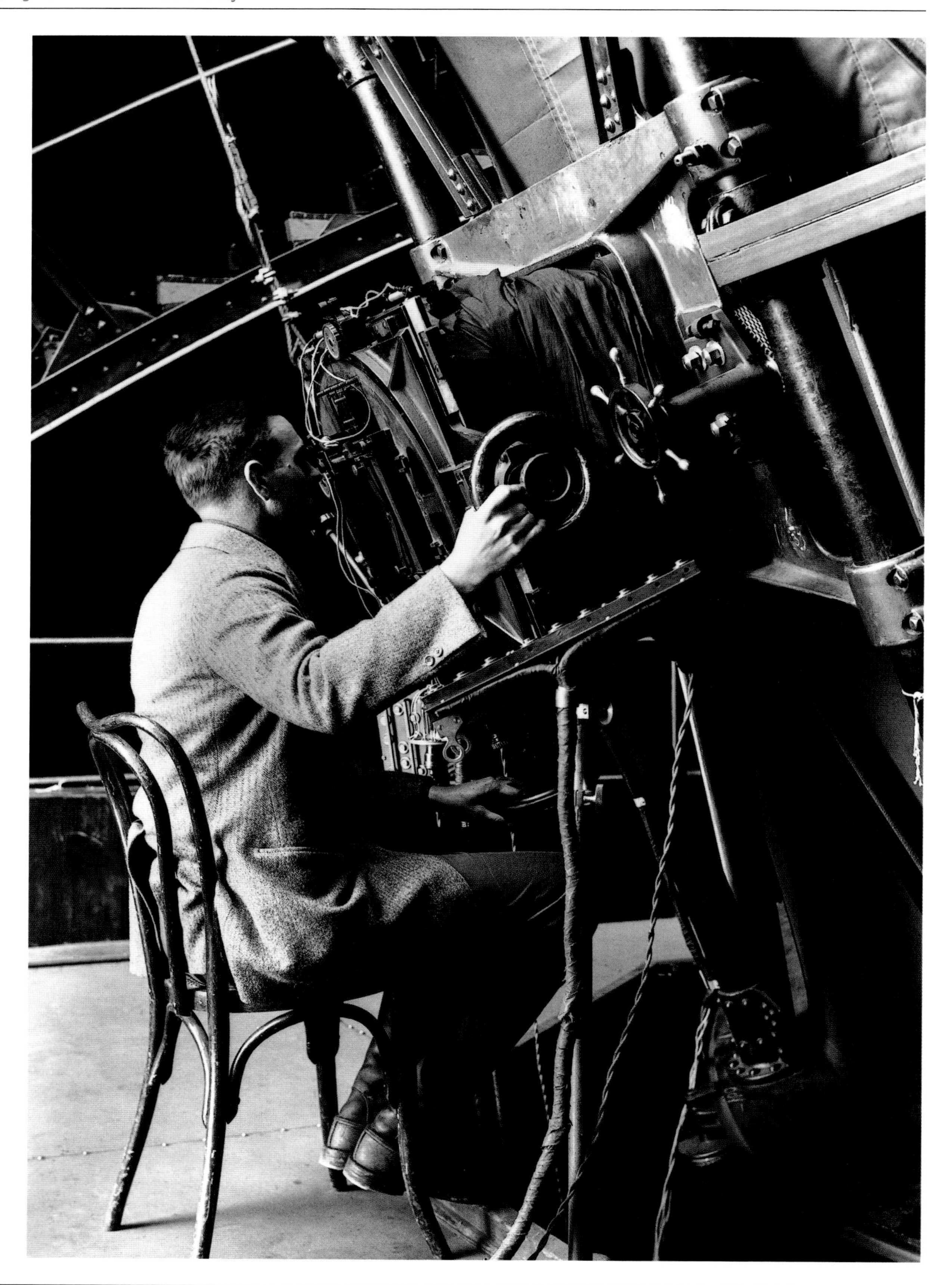

Using the Hooker Telescope at California's Mount Wilson Observatory, Edwin Hubble proved that most nebulae are actually galaxies rushing away from one another in deep space beyond our Milky Way. His discoveries showed that the universe is expanding, rather than staying the same size.

space-time determines how matter moves. Einstein predicted that light from a distant star passing by the **sun** would be bent by the sun's large mass. **Arthur Eddington** confirmed this effect during the total **solar eclipse** of 1919.

Sir Arthur Stanley Eddington (1882-1944)

Arthur Stanley Eddington was born in Westmoreland, England, on December 28, 1882. Considered the father of modern astrophysics, Eddington is best known for his contributions to stellar astronomy. Graduating with a strong background in physics and mathematics, Eddington was appointed chief assistant at the Royal Observatory at Greenwich in 1906. Here he learned observational astronomy.

In 1910 Eddington published a catalog of some 6,000 **stars.** Four years later, in *Stellar Movements and the Structure of the Universe,* he correctly suggested that distant "spiral **nebulae**" were actually **galaxies** far outside the **Milky Way.**

By 1917 Eddington was trying to develop a theory of a star's energy production and evolution. Armed with his knowledge of atomic physics and **Einstein**'s special theory of relativity, Eddington demonstrated that heat is transported by radiation in stars. He also reasoned that at the high temperatures in stellar interiors, electrons would be stripped from their nuclei, forming what physicists today call a plasma. Finally, Eddington developed a relationship between a star's mass and its luminosity. In 1926 he outlined these findings in *The Internal Constitution of the Stars.*

Eddington wrote a report on the theory of general relativity in 1918. One year later he led a **solar eclipse** expedition to Principe Island, near the coast of West Africa. During the darkest part of the eclipse, Eddington obtained several photographic plates of stars near the rim of the darkened **sun.** Upon returning to England, he carefully examined the images, measuring the stars' positions. The slight change in position that Eddington detected confirmed Einstein's general theory.

Edwin Powell Hubble (1889-1953)

The son of a lawyer, Edwin Hubble was born in Marshfield, Missouri, on November 20, 1889. Hubble began attending the University of Chicago in 1906 and received a Rhodes scholarship to Queen's College in Oxford, England, in 1910. In 1914 he returned to the University of Chicago to earn his Ph.D. Hubble pursued his graduate work as a research assistant studying faint **nebulae** at Yerkes Observatory. Following a two-year stint in the United States Infantry during World War I, Hubble joined the staff of **Mount Wilson Observatory** in 1919. He remained on the staff of the observatory for the rest of his career.

Hubble used Mount Wilson's 2.5-meter Hooker Telescope to develop a new classification of nebulae. At the time there were two theories as to the nature of nebulae. One camp believed they were clouds of interstellar gas within the **Milky Way,** while the other camp thought they were **galaxies** beyond the Milky Way. It turns out that both camps were right. Hubble, in 1922, was the first to classify diffuse nebulae of the Milky Way into **reflection** and **emission nebulae.** The following year, on October 4, 1923, Hubble distinguished stars in the Andromeda Nebula—known today as the **Andromeda galaxy.** By studying **Cepheid variable stars** in Andromeda and other so-called nebulae, Hubble was able to prove that the nebulae were actually galaxies that lay far beyond any object in the Milky Way.

Hubble's discovery of galaxies had even greater implications because of **Einstein**'s general theory of relativity. At the time, astronomers questioned whether the universe was static, expanding, or contracting. Hubble measured the spectra of 46 galaxies. He found that the light from these galaxies had shifted toward the red end of the **spectrum**—redshifted. The light was at longer, red wavelengths, proving that the galaxies were receding, or moving away, from **Earth,** and that the **universe** was expanding. Hubble found that the farther away a galaxy was, the faster it was receding. He concluded that there was a rate of expansion that could be calculated using what is now known as the Hubble constant—the rate at which the expansion velocity changes with distance. Its precise value is still not known.

His observations of galaxies led Hubble to classify them by their shape. Known as the Hubble classification, this system splits galaxies into four main classes: **elliptical** (E), **spiral** (S), barred spiral (SB), and **irregular** (Irr).

Toward the end of his career, Edwin Hubble was involved with the construction of the five-meter Hale Telescope on Mount Palomar, in California. In 1948 he was the first to operate the instrument. The Hale Telescope became his primary instrument until his death as the result of a stroke on September 28, 1953.

Eyes on the Sky

Spectrum

Only a small fraction of the energy available for study is visible to the human eye. The visible spectrum is the continuum of color formed when the wavelengths in a beam of white light are dispersed and arranged in order. A wide range of instruments has been developed to view the sky in wavelengths that go far beyond the visible spectrum. Although astronomy is historically centered on visible-light observations, today astronomers look at the sky through radio, infrared, ultraviolet, gamma ray, and x-ray "eyes."

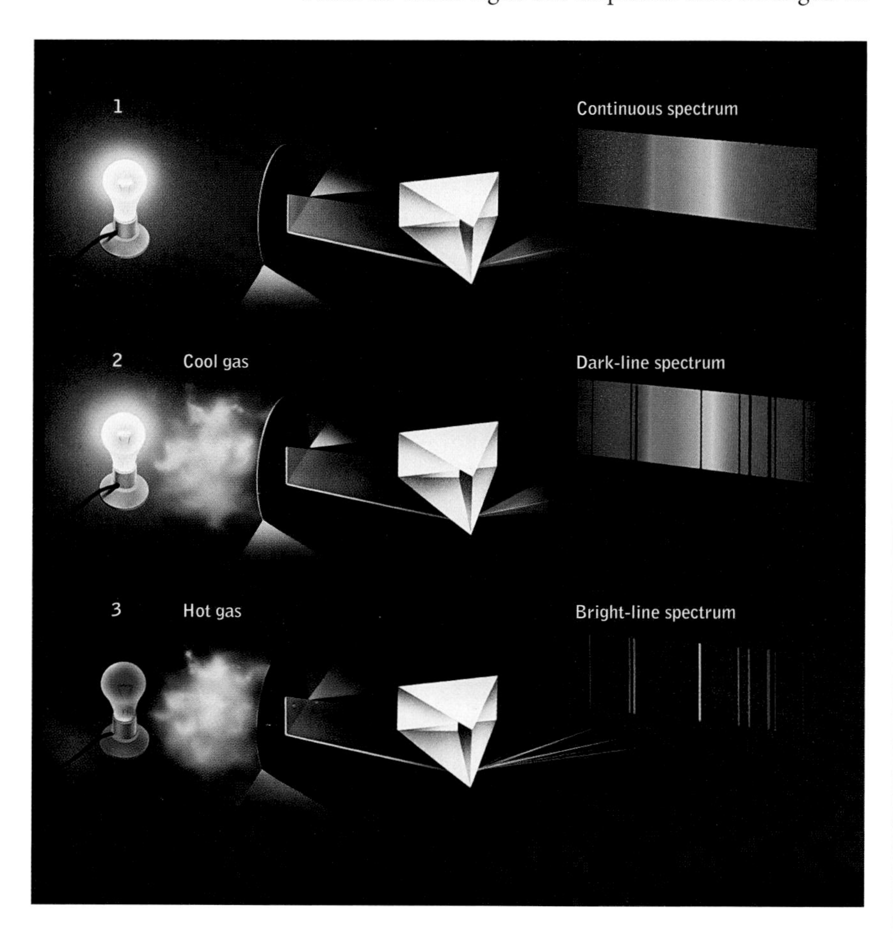

Spectra help scientists decode starlight. Light from a bulb's hot filament passing through a prism produces a continuous band of colors (1). If the light goes through a cool gas, certain wavelengths are absorbed, making dark lines (2). In contrast, a hot gas produces bright lines (3).

Capturing this wide array of wavelengths requires an assortment of different types of **telescopes.** Because our atmosphere filters most **electromagnetic radiation, ground-based telescopes** operate at visual and radio wavelengths. High-flying aircraft, balloons, rockets, and Earth-orbiting **satellites** allow us to extend our sight beyond the shield of our atmosphere, while **infrared, ultraviolet, x-ray,** and **gamma ray astronomy** are done primarily from space.

A rainbow is sunlight dispersed through raindrops into a spectrum visible to the human eye. When astronomers examine the light from **galaxies,** stars, and other celestial objects, they often record the distribution and intensity of an object's radiation by creating a spectrum. Similar to the way a rainbow separates light, dispersing a celestial object's radiation into its constituent wavelengths creates the spectrum of that object. A spectrum can reveal how hot an object is, what elements are present, and how far away it might be. The method astronomers use to obtain an object's spectrum depends on the wavelength in which they are observing.

Continuous Spectrum

A continuous spectrum is an unbroken distribution of light over a broad range of wavelengths, say from red to violet. Such a spectrum is created when the particles that make up a hot and opaque solid, liquid, or dense gas are excited to a point at which they emit light. The object radiates or produces photons at all wavelengths, giving a continuous spectrum.

Emission Spectrum

An emission spectrum appears as a pattern or series of bright lines formed by discrete wavelengths of energy. They occur when a hot, low-density gas is excited to emit photons at specific wavelengths.

Absorption Spectrum

An absorption spectrum is a series or pattern of dark lines denoting missing wavelengths of light superimposed on a continuous spectrum. These dark lines form when light from an object radiating a continuous spectrum passes through a cool, low-density gas. The gas absorbs photons at certain wavelengths, creating the dark lines. The dark lines in a solar spectrum are called Fraunhofer lines, and they tell us what elements are present in the **sun**'s atmosphere.

Spectroscopy

Spectroscopy is the method by which astronomers acquire a **spectrum.** A spectrum is obtained with an instrument called a spectrograph. There are several varieties of spectrographs. One type, an Echelle spectrograph, obtains high-resolution spectra over a very narrow range of wavelengths, and a spectroheliograph allows solar astronomers to probe the **sun.**

Diffraction Grating

A diffraction grating is a device inside a spectrograph that disperses white light into its constituent colors. The surface of the diffraction grating holds a series of closely spaced, equidistant, parallel grooves. When white light hits the little grooves of the grating, the different wavelengths that make up white light are separated and spread out, creating the visible **spectrum.** Each centimeter of a typical diffraction grating has six thousand of these little grooves.

Blackbody

A blackbody is a theoretical object that absorbs all of the energy—at all wavelengths—that falls upon it. A blackbody is also the perfect emitter of radiation. Although such an ideal absorber of radiation is hypothetical, the inner workings of stars and their related colors and temperatures can be described using this hypothetical ideal.

Blackbody radiation is the heat energy emitted by a blackbody at a particular temperature. The **spectrum** of such a radiator is continuous, and the peak wavelength emitted depends only on the blackbody's temperature. Because a blackbody is a perfect absorber of energy, it is also a perfect emitter of heat energy, which is where stars come in. We know that atoms at higher temperatures move faster than atoms at lower temperatures. That means that the hotter an object is, the faster its constituent atoms move. When these agitated particles collide with electrons, they may accelerate the electrons releasing energy in the form of photons.

The incandescent lightbulb you may be reading this by also radiates as a blackbody. The bulb's filament is heated by electricity flowing through it. As this heat energy is released, the filament glows.

Electromagnetic Radiation

The word "light" evokes images of visible light or the range of wavelengths to which human eyes are sensitive. Each color we see represents a different wavelength, from red to violet. The combination of all wavelengths of the visible **spectrum** results in white light. The electromagnetic spectrum—the complete range of electromagnetic radiation from the longest to shortest wavelength—extends far beyond these colors in both directions. Electromagnetic radiation is an oscillating disturbance that travels through a vacuum or matter. Although electromagnetic radiation behaves as both a particle (photon) and a wave, most of its properties can be described in terms of a propagating wave—a wave that feeds on its own energy and continues into infinity. A propagating wave has an electric field and a magnetic field. The electric and magnetic fields are bound together perpendicularly and travel as waves at the speed of light. They are directly related: A changing electric field will produce a changing magnetic field and vice versa.

The propagation of electromagnetic energy follows the inverse-square law—that is, as energy is released, it expands outward, covering an area that is proportional to the square of the distance it has traveled, and the signal grows weaker. This means that the strength of a signal reaching **Earth** decreases at a rate of one over the square of the distance. If an object is ten **astronomical units** from Earth, the signal received is one one-hundredth its original strength.

Types of electromagnetic energy differ from each other based on their wavelengths. Radio waves are the longest. They range from 30 kilometers down to roughly one millimeter in length. Continuing through successively shorter wavelengths, infrared radiation is next (roughly one millimeter to roughly one micrometer in length), then visible light (red, at 700 nanometers, or nm, to violet, at 400 nm), ultraviolet (400 nm to 10 nm), x-rays (10 nm to 0.01 nm), and finally gamma rays (less than 0.01 nm). Several of these categories are further subdivided. These divisions include hard x-rays; soft x-rays; extreme ultraviolet; near, middle, and far infrared; submillimeter and millimeter wave.

Interferometry

Interferometry is a process that combines two or more electromagnetic signals from an astronomical object to gather information about it. An interferometer combines observations of the same object made by separate instruments to obtain a higher resolution image than either instrument could obtain on its own. The signals, when combined correctly, either reinforce or cancel each other out. This process is easy to understand when you think of the wave nature of light. If two signals are in phase so that each crest and trough is aligned, the combined signal is stronger. If the two are out of phase, they will cancel each other out. Interferometers allow astronomers to better study weak signals.

THE BLACKBODY SPECTRUM

David Wilkinson

THE COBE SATELLITE MEASURED THE INTENSITY OF THE COSMIC MICROWAVE background radiation (CMBR)—the big bang fireball—over a range of infrared and microwave wavelengths, indicating an extremely cold radiation. The resulting graphed curve looked like a blackbody spectrum, which relates the variation of the intensity of radiation (vertical axis) to wavelength (horizontal axis). Any object that absorbs all incident radiation, regardless of the wavelength, is called a blackbody. It also emits radiation wavelengths according to a formula that graphically takes a universal shape, regardless of its composition. A heated iron bar has the properties of a blackbody because all the energy it radiates is thermal energy. The line in the figure below shows the universal shape of a blackbody spectrum. The position of the peak and the height of the curve depend on the temperature of the emitter, but the shape is the same for all blackbodies. The cosmic background radiation spectrum peaks at a wavelength of about 2 millimeters. Your body radiates heat with a spectrum that peaks at about 0.02 millimeters—a shorter wavelength, hence a higher frequency and a higher energy density.

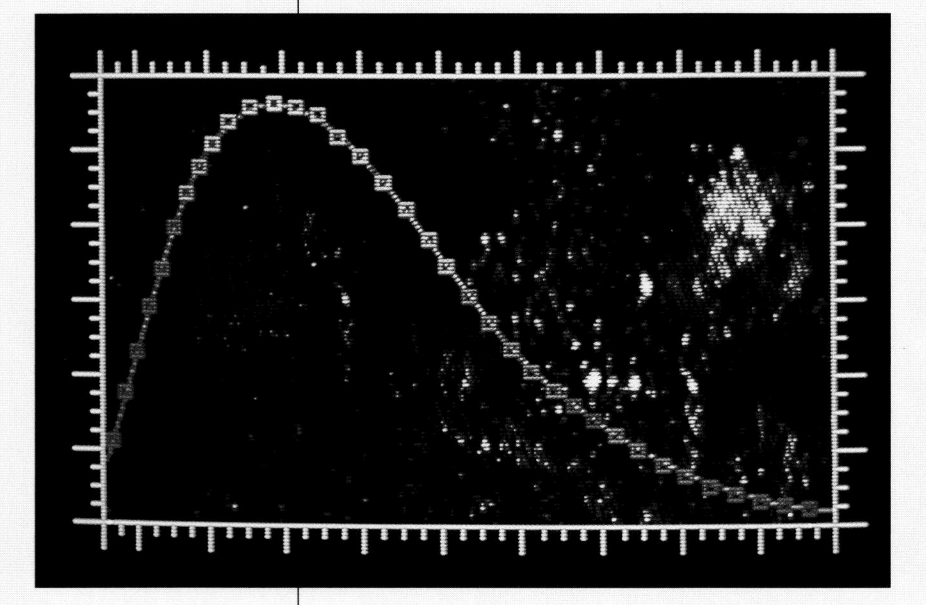

A graph of a blackbody spectrum, generated from information gathered by the Cosmic Background Explorer (COBE) satellite, provides support for the big bang theory of the formation of the universe.

The blackbody spectrum figured in two important 20th-century discoveries: quantum mechanics and big bang cosmology. In 1900 Max Planck developed the quantum to explain the shape of blackbody spectra. Classical mechanics explained the shape of the curve at long and short wavelengths, but failed to explain its shape around the peak of the spectrum. Planck found that by assuming that radiation energy comes in packets (quanta), he could calculate the measured shape of the blackbody spectrum. Quantum mechanics was thus introduced to explain the blackbody radiation spectrum. Fifty years later, George Gamow found that blackbody radiation was a consequence of the big bang model, and that the radiation would still be filling the universe, though it would be much colder now. Measurements of the spectrum of the CMBR, especially those from the COBE satellite, fit the spectrum of a blackbody at a temperature of 2.728 kelvins above absolute zero.

Verifying the blackbody character of the remnants of the big bang helped determine what the big bang was. Bodies that radiate energy with purely thermal properties look exactly like the radiation properties of the big bang. Comparison between the observations of Arno Penzias and Robert Wilson and the Robert Dicke group hinted at this, but the difference in the wavelength ranges was small. Other physicists studied parts of the radiation distribution at different wavelengths. Balloons, aircraft, and eventually the COBE filled in the picture. They confirmed that the big bang was a thermal process—the radiation came from an extremely hot body, not from some other source of radiation. ■

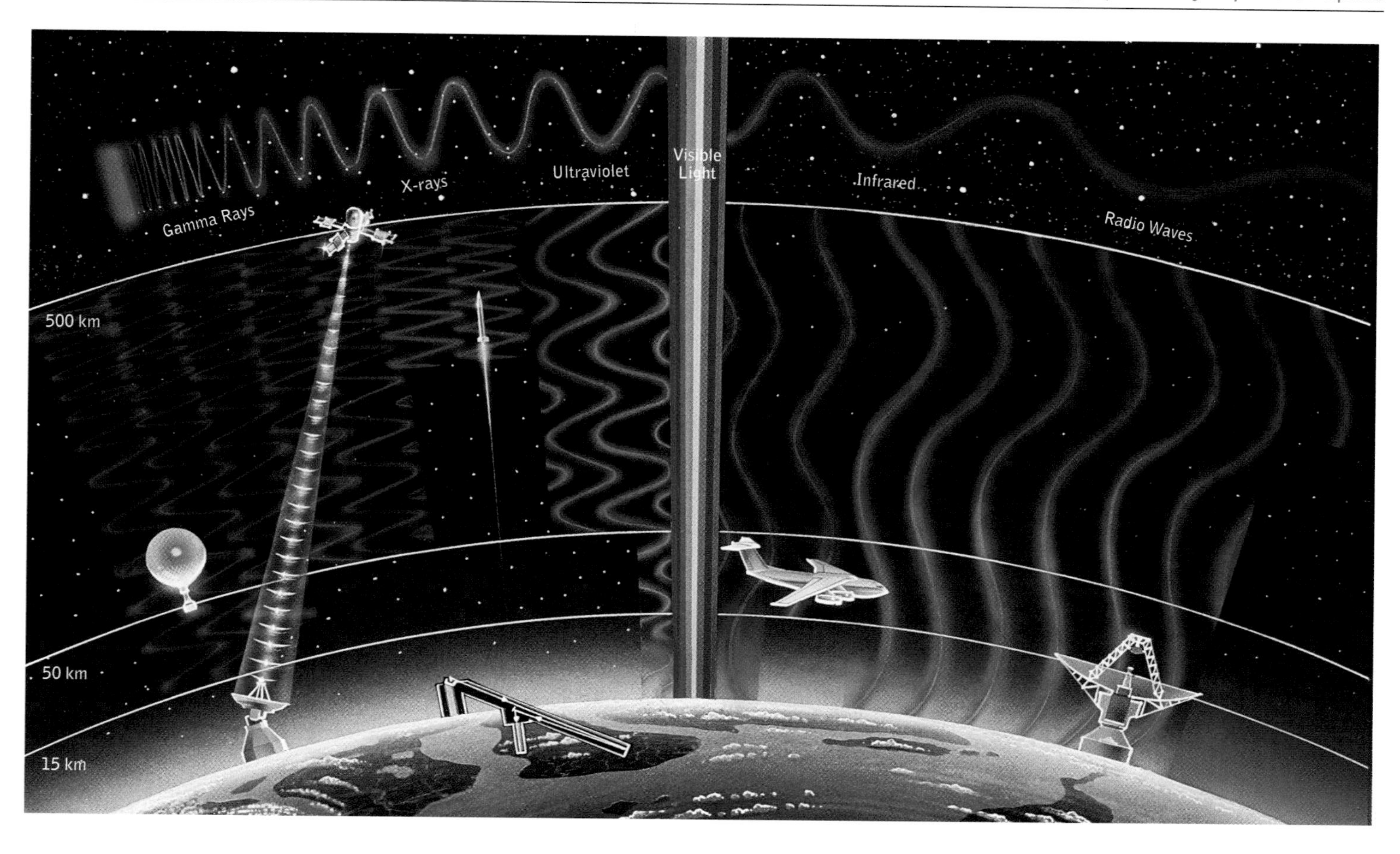

Optical interferometers such as the Keck Telescopes on Mauna Kea in Hawaii or the Very Large Telescopes (VLT) on Cerro Paranal in the Atacama Desert of South America use **laser** beams to synchronize the signals from each individual **telescope.** The combined signals effectively yield a telescope that has a diameter equal to their separation. In the case of the VLT, that is a light-collecting area of 200 square meters and a resolution of 0.001 arc seconds. The VLT could resolve a quarter some 500,000 kilometers away! Interferometry is very useful for observing long wavelengths, which have very weak signals. Because the energy of a photon depends on its wavelength, long wavelengths such as radio waves have very low energy signals.

Radio Astronomy

Radio astronomy, which uses radio waves to study celestial phenomena, began in the 1930s when Karl Jansky (1905-1950), a scientist for Bell Laboratories, started looking for sources of static that were disrupting long distance calls. One of the first noisy signals he located originated in the **constellation** Sagittarius. Jansky had found the radio signature of the center of the **Milky Way.** He published his findings in 1932, but most astronomers took little interest. Fortunately, a radio engineer named Grote Reber focused on Jansky's work. Reber began searching the sky for radio signals and by the 1940s had mapped a large portion of the radio sky. Unfortunately, most astronomers still took little interest in radio astronomy. It did not come into its own until after World War II, when advances in electronics and solid-state physics spurred improvements in their instrumentation.

In 1964 Arno Penzias and Robert Wilson discovered the radio signal left by the formation of the **universe.** Like Jansky, both were working for Bell Labs and looking for radio static that interfered with telecommunications. The pair was surveying the Milky Way's galactic halo with a six-meter horn antenna when they detected a persistent 7.35-centimeter wavelength signal. No matter where they pointed their **telescope,** this signal was there—it had origins beyond the Milky Way. Their discovery of this **cosmic microwave background radiation** brought Penzias and Wilson the 1978 Nobel Prize for physics.

Only certain wavelengths of the electromagnetic spectrum—such as visible light—can reach the Earth's surface; the planet's atmosphere blocks most other waves from fully entering. To observe stars, galaxies, and other objects in the universe in the full electromagnetic spectrum, scientists must send up planes, balloons, rockets, and satellites to take the pulse of space.

Today, every type of astronomical object has been studied at radio wavelengths. The **sun, planets, galaxies, nebulae,** and **quasars** are just a few of the objects radio astronomers have examined. Radio telescopes are particularly useful for looking at large clouds of cool hydrogen. These regions of star formation are invisible to conventional optical telescopes because they produce no visible light and reflect too little to be detected. Radio signals also give us a view of distant objects. The universe is a dusty place, and visible light gets scattered. Radio waves penetrate these dusty regions, giving radio astronomers a view beyond the dust clouds.

The radio range of the **spectrum** includes wavelengths from roughly one millimeter to thousands of kilometers in length; radio waves longer than 30 meters are blocked by **Earth's atmosphere.** Because a telescope's resolution depends on its light-gathering surface and the wavelength being gathered, radio telescopes need to be very large. A typical radio wave is 100,000 times longer than a visible light wave, so radio telescopes need to be 100,000 times larger than their optical counterparts to achieve the same resolution. To circumvent this, astronomers use **interferometry** to increase the signal-gathering power of their telescopes.

Arecibo Observatory

Located in the Guarionex Mountains of northwestern Puerto Rico, the Arecibo Observatory has the largest single-dish radio **telescope** in the world. Cornell University began construction of the telescope in 1960, and it was dedicated three years later. The telescope is located in a natural depression. Its 305-meter-wide dish is 51 meters deep and covers eight hectares. Although the telescope is fixed and pointing upward, the **antennas** that feed signals to the receiver can be moved north or south. This capability, along with its location (18° N), allows Arecibo to view objects located 70 degrees north or south of the Equator.

In 1974 the dish was used to send a signal rather than collect one. A coded message holding information about **Earth** was sent toward the globular cluster M13, roughly 25,000 **light-years** away.

Very Large Array (VLA)

Completed in 1980, the Very Large Array is on the plains of San Augustin, west of Socorro, New Mexico. The VLA consists of 27 radio antennas arranged in a Y-shaped pattern; each rests on railroad tracks that allow astronomers to change the distance between the instruments. At their highest resolution, the **telescopes**

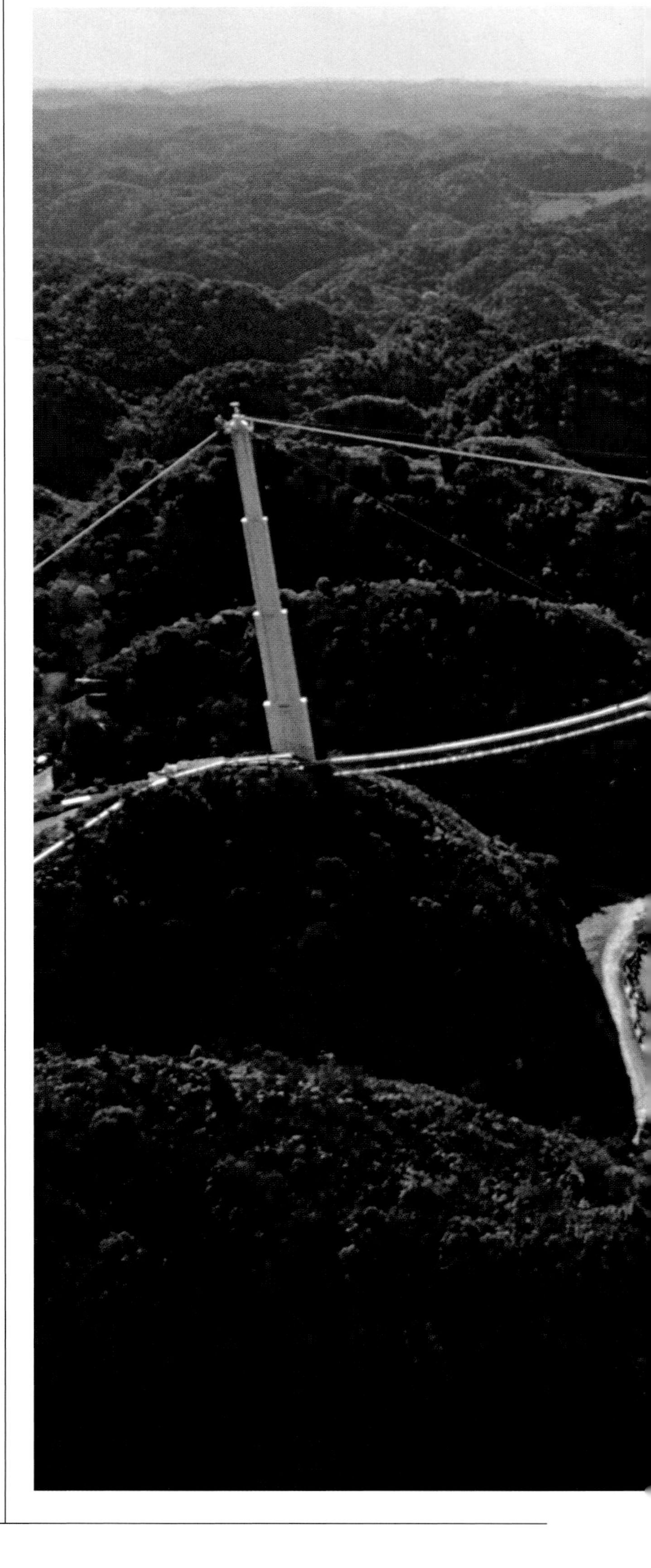

A deep dish, the Arecibo 305-meter radio telescope covers about eight hectares and plunges 51 meters into a natural sinkhole in northwestern Puerto Rico. The reflecting surface is made up of almost 40,000 panels of perforated aluminum, each measuring one meter by two meters.

MAJOR RADIO ASTRONOMY TELESCOPES

Name	Wavelength	Size	Location	Date Operational
SINGLE DISH				
Arecibo	Centimeter and meter	305 m (fixed)	Puerto Rico	1963
Green Bank	Centimeter and meter	110 x 100 m	U.S.	2000
Effelsberg	Millimeter and centimeter	100 m	Germany	1972
Jodrell Bank	Centimeter and meter	76 m	U.K.	1957
Parkes	Centimeter and meter	64 m	Australia	1961
Nobeyama	Millimeter	45 m	Japan	1978
IRAM	Millimeter	30 m	Spain	1985
James Clerk Maxwell	Submillimeter and millimeter	15 m	U.S.	1987
Swedish-ESO	Submillimeter	15 m	Chile	1987
Kitt Peak 12-Meter Telescope	Millimeter	12 m	U.S.	1984
ARRAYS				
Very Long Baseline Array		8,000 km / 10 dishes	U.S.	1993
Australia Telescope		320 km / 8 dishes	Australia	1988
MERLIN		230 km / 7 dishes	U.K.	1980
Very Large Array		36 km / 27 dishes	U.S.	1980
BIMA		2 km / 10 dishes	U.S.	1996
IRAM Interferometer		15 m / 6 dishes	France	1996
Submillimeter Array		6 m / 8 dishes	Hawaii	2003

are at their maximum spacing and operate like a telescope some 36 kilometers across. Each radio dish is 25 meters wide. With the planned addition of new dishes, the VLA's baseline will increase to 300 kilometers.

Very Long Baseline Array (VLBA)

This radio **telescope** holds ten identical radio dishes each 25 meters across. They are placed no more than 8,000 kilometers apart and are located in Saint Croix, U.S. Virgin Islands; Hancock, NH; Liberty, IA; McDonald Observatory, outside Fort Davis, TX; Los Alamos, NM; Pie Town, NM; **Kitt Peak,** AZ; Owens Valley, CA; Brewster, WA; and **Mauna Kea Observatory** in Hawaii.

When the VLBA makes an observation, technicians at each dish record and send their data to the operations center in Socorro, New Mexico. There the signals are combined, giving the VLBA a resolution of less than one one-thousandth of an arc second—the equivalent of reading a newspaper in New York from Los Angeles.

Very Long Baseline Interferometer (VLBI)

Expanding on the concept behind the **VLBA,** the VLBI consists of radio **telescopes** located around the world and separated by thousands of kilometers. With baselines of more than 10,000 kilometers, the VLBI effectively creates a radio telescope that has **Earth**'s diameter as its aperture. The VLBI allows astronomers to study the fine structure of distant, faint radio sources such as **quasars.** This work has also been used to study Earth's continental drift. By closely monitoring an extremely distant quasar, astronomers have been able to track the position of the VLBI telescopes—detecting slight motions in the movement of Earth's continental plates and the wandering of its North and South Poles.

Infrared Astronomy

Infrared astronomy is the study of the universe using infrared radiation. Infrared radiation is sometimes called heat radiation. Warmth from the **sun** and heat from the embers of a campfire are part of the infrared **spectrum.** The longer wavelengths of infrared radiation penetrate dust clouds more easily than visible wavelengths. This property allows astronomers to study dense dust clouds where stars form as well as the center of our own **Milky Way** and other **galaxies.** Stellar nurseries produce sizeable amounts of infrared radiation. Objects such as starburst galaxies can be observed at very large distances. Learning how **stars** form in these galaxies provides us with clues to the formation of objects in our own corner of the **universe.** Galaxies with an active nucleus also produce large quantities of infrared radiation. Studying a galaxy's infrared signature lets astronomers probe deep into the galaxy's core, where a suspected **black hole** may be generating energy.

Our planetary neighbors also reveal information at infrared wavelengths. By selecting the specific infrared wavelength in which to observe, scientists can probe different layers of a **planet**'s atmosphere. The resulting **absorption spectrum** gives an indication of the temperature and composition of the planet's atmosphere. Water and carbon dioxide in Earth's atmosphere absorb most infrared radiation. Although some **telescopes** placed on high mountaintops are able to perform infrared astronomy, most infrared observations are made from space.

Spitzer Space Telescope

Launched on August 25, 2003, the Spitzer Space Telescope is the last in a family of four Great Observatories launched by **NASA** to study the universe at four different wavelengths. The other three missions are the **Hubble Space Telescope, Compton Gamma Ray Observatory,** and the **Chandra X-ray Observatory.**

The Spitzer holds an 85-centimeter-wide telescope with three cryogenically cooled cameras and spectrographs. The energy Spitzer detects is faint heat energy with wavelengths of 3 to 180 microns. To avoid the heat generated by **Earth,** which would fry the instruments, Spitzer had to be launched into a unique orbit that trails Earth as it goes around the **sun.** The telescope's primary mission is to expand our knowledge of the early **universe** by providing us with information about when objects first formed and what they are made of. Because the universe is expanding, light from the most distant objects in space is shifted toward the red end of the **spectrum.** Astronomers using Spitzer plan to observe the most distant **galaxies** through the dust of the early universe.

Spitzer will also spend time looking at objects closer to home. **Planets, nebulae,** and **stars** are among its assignments. One type of star Spitzer will study is the **brown dwarf.** These cool, "failed stars" may hold the key to much of the missing matter believed to be prevalent somewhere in the universe. The orbiting observatory will also look at giant **molecular clouds** scattered between the stars. These stellar nurseries hide the birth of new stars behind a wall of dense gas and dust.

Ultraviolet Astronomy

Ultraviolet radiation may be a primary contributor to skin cancer, but it provides us with a view of hot, excited regions of space. Ultraviolet astronomy tells us something about the physics and chemistry of hot gases in and around stars, **nebulae,** and the **interstellar medium.** Most of the atoms and ions that make up the interstellar medium have their strongest absorption lines at ultraviolet wavelengths. By studying these spectra, we get a better idea of the chemistry and processes that make up the material between stars.

Ultraviolet astronomy also helps astronomers discern the components that make up **solar system** objects. The International Ultraviolet Explorer (IUE) **satellite** observed the release of gas from **comets,** while the Far Ultraviolet Spectroscopic Explorer (FUSE) discovered molecular hydrogen in the atmosphere of **Mars.** The satellite FUSE was launched in 1999 with a mission to help us better understand the first few minutes after the **big bang.** Deuterium is an isotope that is a by-product of the big bang. By studying the distribution of deuterium relative to hydrogen and other heavier elements, astronomers can better understand how **galaxies** evolve and what the universe might have looked like just after it was born.

The evolution of stars figures prominently in ultraviolet exploration. The Extreme Ultraviolet Explorer (EUVE), launched in 1992, studied hot **white dwarf** stars, cataclysmic **variable stars,** and the hot outer atmospheres of cool red dwarf stars. EUVE cataloged more than 400 white dwarfs. Several of these were part of a binary system, with material from the secondary star falling into the white dwarf. Astronomers study the ultraviolet emission lines generated by this transfer of stellar material to determine how the material flows and its effects on **stellar evolution.**

Most ultraviolet radiation is absorbed high in the **stratosphere** by **Earth**'s protective ozone layer. Wavelengths shorter than roughly 290 nanometers, known as the far ultraviolet region, are completely absorbed by the ozone layer, so these studies must be done from space.

X-ray Astronomy

X-ray astronomy is used to study high-energy objects in the **universe. Supernovas, black holes,** and binary stars, for example, generate extreme energy processes, such as nuclear reactions, and have extreme physical conditions such as exceedingly high temperatures or very strong magnetic fields that radiate energy in x-ray

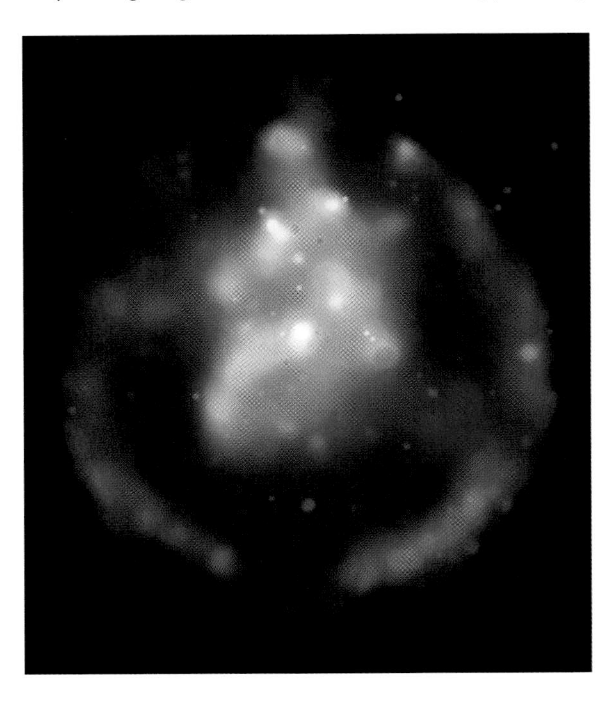

A ring around a rosy cloud of enriched gas marks the outer limit of a shock wave created by ejected material from the explosion of a supernova about 190,000 light-years from Earth. The powerful x-ray telescope known as Chandra captured this image some 200,000 years after it occurred.

wavelengths. One such high-energy system is the death of a star. The violent remnant of a supernova explosion is suited for high-energy x-ray observations. Shock waves from the blast plow through the surrounding **interstellar medium** at speeds of hundreds of kilometers a second, compressing and heating the surrounding gas and releasing vast amounts of x-rays. By studying the x-ray signature, astronomers can determine the rate at which the shock wave is expanding and the composition of the surrounding gas.

Another violent process may be occurring at the heart of our own **galaxy.** The **Milky Way**'s core emits large amounts of x-rays from a quickly rotating source roughly 2 degrees in size. Astronomers believe the source may be a supermassive black hole lurking behind the dust of our galaxy's core. As material succumbs to the black hole's enormous gravitational pull, it swirls into the black hole—much as water spirals down a drain—and releases x-rays along the way.

X-ray astronomy began in 1948, when small rockets and balloons equipped with x-ray detectors were launched. By 1970 nearly 30 x-ray sources had been identified. In December of that year, **NASA** launched the **satellite** Uhuru (Swahili for "freedom") into orbit. Uhuru found more than 300 sources of celestial x-rays. Advancements in equipment increased that number by the late 1970s. Today, more than 60,000 x-ray sources are known. X-rays have been detected in a wide range of objects, which has greatly extended our understanding of this branch of the **electromagnetic spectrum.**

Chandra X-ray Observatory

Launched from the **space shuttle *Columbia*** on July 23, 1999, the Chandra X-ray Observatory is greatly improving our understanding of the most turbulent regions of space. It is the world's most powerful x-ray **telescope,** producing high-resolution x-ray images as detailed as 0.5 arc seconds—the equivalent of reading the letters of a stop sign some 30 kilometers away. Chandra can detect sources more than 20 times fainter than earlier x-ray telescopes.

Gamma Ray Astronomy

With short wavelengths and extremely high energies, gamma rays are not well understood. Their energies are so high (10,000 times greater than visible light photons) that not many gamma rays are produced. Those that are around are incredibly hard to capture; they pass through most materials without interacting. Silicon, germanium, mercuric iodide, and cadmium telluride semiconductors are materials with which gamma rays will interact. The resulting flash created by the interaction is used to record the passage, arrival time, trajectory, and energy of a gamma ray.

Gamma rays appear to be produced by the same violent objects that generate x-rays. Objects such as **supernovas, black holes, neutron stars, solar flares,** and erupting **galaxies** all have gamma ray signatures, but many gamma ray sources have completely unknown identities. Most are steady emitters that appear to be associated with the **Milky Way.** Some of these sources may turn out to be neutron stars or **supernova remnants.** Gamma rays are almost totally absorbed by **Earth's atmosphere.** We didn't detect them until the launch of Explorer 11 in 1961, which found only 22 cosmic gamma ray events. The Orbiting Solar Observatory III satellite detected some 621 gamma-ray events some six years later. Designed to monitor violations of the Nuclear Test Ban Treaty, the **satellite**'s **detectors** discovered the first **gamma ray burst.** Several other satellites were launched during the 1970s, but gamma ray astronomy did not reach its prime until the launch of the **Compton Gamma Ray Observatory** in 1991.

Compton Gamma Ray Observatory

The 17-ton Compton Gamma Ray Observatory was launched from the **space shuttle** *Atlantis* on April 5, 1991. It was the second in **NASA**'s Great Observatory program. Compton's mission included the study of **solar flares, gamma ray bursts, pulsars, nova** and **supernova** explosions, **black holes, quasars,** and **cosmic ray** interactions with the **interstellar medium.**

To achieve this broad range of observations, the Compton observatory carried a collection of four high-technology instrument packages: the Burst and Transient Source Experiment (BATSE), the Oriented Scintillation Spectrometer Experiment (OSSE), the Imaging Compton Telescope (COMPTEL), and the Energetic Gamma Ray Experiment Telescope (EGRET). BATSE measured the brightness of gamma-ray bursts and solar flares. It also generated an all-sky map of gamma ray burst positions that confirmed their existence outside the Milky Way. OSSE found the gamma ray signature generated when positively and negatively charged particles—positrons and electrons—annihilate each other in the interstellar medium. COMPTEL mapped the entire sky at medium gamma ray energies. Last but not least, EGRET discovered blazars—a type of quasar.

Cosmic Rays

Cosmic rays are subatomic particles traveling at incredible speeds. As cosmic rays hit **Earth's atmosphere,** the particles crash into gas atoms that fragment, creating secondary cosmic rays that rain down on Earth. Because these subatomic particles are charged, they are bounced around by magnetic fields throughout the **galaxy** so that it is impossible to determine exactly where they are coming from. The **interstellar medium** is aglow with cosmic rays. Like gamma rays, cosmic rays are high-energy signals that are best viewed from above Earth's atmosphere. They appear to originate in some of the same objects as gamma rays, such as a galaxy's center. The elliptical galaxy **M87**, in the Virgo cluster, produces some of the highest-energy cosmic rays. The **Hubble telescope** detected a rapidly rotating disk of hot gas at M87's core—an indication of a supermassive **black hole.**

Cosmic rays were first detected in 1912, when Victor Hess, of the Viennese Academy of Sciences, flew balloons carrying electrometers to roughly 4,900 meters. The higher the balloons rose, day or night, the higher their readings, indicating a cosmic, rather than a solar, radiation source.

Exploring space above the snow-covered surface of Antarctica, the balloon-borne BOOMERANG Telescope maps subtle variations in cosmic background radiation across a small area of sky. In 1998 BOOMERANG obtained detailed images of the structure of the infant universe some 14 billion years ago.

Stars

Stellar Evolution

Stars are gaseous balls, primarily fueled by hydrogen and helium, that emit radiation. Hydrogen and helium are the basic building blocks of matter and the most common elements in the universe. As the American astronomer Carl Sagan said: "We are all star stuff." The path a star takes through life depends on its size and therefore its gravitational pull. As material streams into the core of a forming star, the temperature and pressure increase. Once temperatures are high enough to ignite nuclear reactions, a star is born. Newborn stars survive as long as they can fend off the relentless force of gravity trying to collapse them.

Hunter and nurturer, the constellation Orion holds one of the nearest and most active stellar nurseries in the Milky Way. Tens of thousands of new stars were formed in the Orion Nebula during the past ten million years or so.

Stars like our **sun** generate their energy by a proton-proton chain reaction (PP reaction) that fuses hydrogen into helium, the process called nuclear fission. Because a helium nucleus has roughly 0.7 percent less mass than the four hydrogen nuclei that created it, the reaction directly converts the mass difference into energy. The sun transforms roughly 700 million tons of hydrogen into 695 million tons of helium every second. The remaining five million tons of matter is converted directly into energy. This energy radiates outward from the core, counteracting the star's gravitational collapse. As long as a star shines, its radiation pressure will keep it alive. This gravity-radiation pressure balance is called hydrostatic equilibrium.

Stars greater than 1.5 solar masses produce most of their energy in another reaction called the carbon-nitrogen-oxygen cycle (CNO cycle). Like the PP reaction, the CNO cycle essentially fuses four hydrogen nuclei into one helium nucleus, but the CNO cycle uses carbon as a catalyst. Oxygen and nitrogen are also involved in the reaction. The CNO cycle is highly dependent on temperature and is the dominant energy-producing mechanism at temperatures higher than 18 million kelvins.

Once a star converts all the hydrogen in its core into helium, the fire goes out. Without nuclear reactions there is no outward radiation pressure, and gravity begins to collapse the star. As gravitational potential energy is converted to kinetic energy, the core contracts and begins to heat up. Fresh hydrogen outside the star's core ignites to form a hydrogen-burning shell, but the star is still contracting. This process heats the hydrogen-burning shell and excites the reaction to produce more energy. At this stage of its life, the star is overproducing energy, and radiation pressure exceeds its gravitational collapse. The star expands into a red giant. More massive stars expand further to become red supergiants. The bloated supergiant spends the next few million years burning its hydrogen shell while its core gradually collapses and slowly heats up. If core temperatures reach 100 million kelvins, helium burning begins in an explosive process called helium flash. This chain of nuclear reactions that fuses three helium nuclei (alpha particles) to form one carbon nucleus is called the triple-alpha process. Although the star shrinks and dims , the helium flash is unseen because it occurs deep within the star's core.

Eventually the helium is depleted, and the core shuts down again; a helium-burning shell surrounds the carbon core. At this stage the star is a bit unstable. It pulses and outgases matter that forms a **planetary nebula** around its hot core. For stars as large as our sun, this is the end. The core slowly contracts to become a carbon **white dwarf.** Over the next thousands of millions of years, the white dwarf cools to a black dwarf.

Once a star of at least 20 solar masses (20 times the mass of the sun) converts all its core helium to carbon, it begins to contract. When the core reaches 600 million kelvins, carbon burning begins in a flash that can rip the star apart. This cataclysmic explosion thrusts the star's outer layers into space while its core contracts to a **neutron star.** If the star is even more massive, the core may contract into a **black hole.**

Interstellar Medium

The interstellar medium (ISM) is the matter between the stars of a galaxy. It comprises roughly 10 percent of the **Milky Way**'s mass and is concentrated in the spiral arms and disk of the **galaxy.** Various forms of hydrogen and some helium make up this matter, which includes hot, discrete clouds of ionized hydrogen (H II regions), as well as cooler clouds of neutral hydrogen (H I regions). The ISM also holds **cosmic ray** particles and a fair amount of cosmic dust that cause dimming (interstellar extinction) and reddening (interstellar reddening) of distant starlight.

For every 1,000 **parsecs** (3,260 light-years) of distance, a star appears two magnitudes dimmer than it would if space were completely empty. The ISM also scatters a distant star's blue photons, making it appear slightly redder than it would in empty space. The nanometer- to micrometer-size particles of cosmic dust are made of carbon (diamond dust and graphite),

BRIGHT STARS *(star classification page 59)*

Star	Constellation	Right Ascension (RA) hours	minutes	Declination (Dec.) degrees	minutes	Apparent Magnitude	Absolute Magnitude	Spectral Type	Distance (in ly)
Achernar	Eridanus	01	38	-57	14	0.45	-2.77	B3V	144
Alpha Crucis	Crux	12	27	-63	06	0.76	-4.19	B1V	321
Adhara	Canis Major	6	59	-28	58	1.50	-4.10	B2II	431
Aldebaran	Taurus	04	36	16	31	0.87	-0.6	K5III	65
Altair	Aquila	19	51	08	52	0.76	2.20	A7V	17
Antares	Scorpius	16	29	-26	26	1.06	-5.29	M1.5Iab	604
Arcturus	Boötes	14	16	19	11	-0.05	0.31	K2III	37
Betelgeuse	Orion	05	55	07	24	0.45	-5.14	M1Ia–M2Iab	427
Canopus	Carina	06	24	-52	41	-0.62	-5.53	F0Ib	314
Capella	Auriga	05	17	46	0	0.08	-0.48	G6III & G2III	42
Castor	Gemini	07	35	31	53	1.58	1.9	A1V & A2V	52
Deneb	Cygnus	20	41	45	17	1.25	-8.73	A2Ia	3,230
Fomalhaut	Piscis Austrinus	22	58	-29	37	1.16	1.73	A3V	25
Pollux	Gemini	07	45	28	02	1.16	1.09	K0III	34
Procyon	Canis Minor	07	39	05	14	0.40	2.68	F5IV–V	11
Regulus	Leo	10	08	11	58	1.36	-0.52	B7V	77.5
Rigel	Orion	05	15	-08	12	0.18	-6.69	B8Ia	773
Rigil Kentaurus	Centaurus	14	40	-60	50	-0.28	4.07	G2V & K1V	4
Sirius	Canis Major	06	45	-16	43	-1.44	1.45	A1V	9
Spica	Virgo	13	25	-11	10	0.98	-3.50	B1V	262
Vega	Lyra	18	37	38	47	0.03	0.58	A0V	25

Life Cycle of a Star

James Trefil

The Milky Way galaxy contains a few hundred billion stars, only a few thousand of them visible to the naked eye. Like our sun, they arose from collapsing clouds of interstellar gas and dust and managed to counter the pull of gravity and stave off their collapse by causing hydrogen to fuse into helium in their cores. Every star, from the largest to the smallest, begins life as a glowing ball containing primordial hydrogen, which is consumed to produce the energy that the star needs. All stars are not the same, however. Depending upon the density, temperature, and structure of matter in the cloud from which it sprang, a star can be much smaller than our sun or much larger. In fact, if you imagine the sun as being the size of a basketball, we know of stars in the sky as small as BBs and as big as 50-story buildings. In addition, some stars are bright and some are dim, some are faintly bluish while others are white, and so on.

Apart from huge variations in their distances from the Earth, stars differ from each other in two basic ways: They have different masses (that is, different amounts of material in them), and they are at different stages in their life cycles.

Main-Sequence Stars

You might expect that the bigger a star is, the longer it will last—after all, a larger star has a lot more hydrogen, a lot more fuel to burn. As it turns out, however, the situation is more involved. Although a big star has more hydrogen, it also exerts a larger gravitational force on its components. Therefore it has to burn its fuel more rapidly to keep from collapsing. Because of this fact, we have the seemingly paradoxical result that the more fuel a star has, the shorter its lifetime. Our sun is a medium-size star that started out with enough hydrogen to burn for about 11 billion years. A star 40 times larger than our sun may live only a little over a million years, while a star with half the sun's mass can be expected to putter along for 200 billion years as it frugally expends its supply of hydrogen. The watchword for stars seems to be: Live fast, die young—and make a spectacular corpse.

Four-and-a-half billion years ago, when the sun condensed from a cloud of interstellar dust and gas and first began to shine, that dust cloud would have absorbed much of its visible light. But as the fusion reactions in its core increased in strength and the dust cloud dissipated, the sun shone brighter and brighter, quickly becoming almost as bright as we see it today.

Stars like the sun, which are still burning hydrogen in their cores, are called main-sequence stars. One of the most interesting questions about them concerns what happens when they run out of hydrogen fuel. Our sun, for example, will run out of hydrogen in its core in about six billion years. At that point, its outpouring of energy will begin to diminish and will no longer counterbalance the force of gravity, which will have been waiting in the wings for 11 billion years. Material will be pulled in, and the sun will start to contract.

This contraction, in turn, will raise inner temperatures, so that any hydrogen left in layers outside the core will begin to burn. More importantly, the helium at the sun's very center—the "ashes" left from the fusion of hydrogen—will itself begin to fuse, forming carbon. Thus the ash of one nuclear fire will becomes the fuel for the next.

During this period the sun's outer layer will again expand, throwing some of its mass into space in the form of a strong solar wind. At its largest, the sun's surface will extend past the present orbit of Venus. Its surface will be cooler, however, because the same amount of energy will be passing across a much larger surface. Stars that emit their energy through this sort of large, cool surface are called red giants. When our sun enters this stage, it is possible that its considerable loss of mass and gravitational pull by then will have caused the planets to move outward. Consequently, only Mercury would be swallowed up, while Earth and Venus would survive in orbits farther out than their current ones. This is small consolation for living things, however, because Earth's oceans would have boiled away long before the sun attained its maximum expansion.

Once all the helium and hydrogen atoms in the sun's core have been used up, the sun will again start to collapse, this time shrinking down to an object about the size of the Earth—less than one percent of its current diameter. At this point, the electrons in the sun will be so crowded together that they cannot be compressed any more. The contraction will stop, and the sun will stabilize. Although no longer generating energy, it will be very hot. A small, hot star like this, one that is no longer undergoing fusion, is called a white dwarf. Think of it as a cooling cinder floating in space for billions of years.

Giant Stars

More massive stars—those ten times greater than our sun—have a very different end from the relatively sedate red-giant/white-dwarf pattern. Like the sun, these massive stars burn hydrogen fuel early in their lives. But they consume their hydrogen much more quickly. Once it is gone, the inevitable collapse occurs, and the helium ash is itself burned to create carbon, as it will be in our sun. But in much heavier stars, the greater force of gravity behind their collapse drives the temperature up to the point where carbon itself is burned. Elements such as oxygen, neon, magnesium, and silicon are created by fusion as the star tries to stave off the effects of its own gravity. Eventually, these nuclear reactions begin to produce iron. Iron has the most tightly bound nucleus of any element. It requires energy to break iron into smaller nuclei, and it requires energy to make iron undergo fusion reactions to make heavier nuclei. Iron, in other words, makes a lousy fuel for a star—there is just no way to get energy from it.

Once a star reaches this stage, the unburnable iron ash begins to accumulate in its core. Just as in a white dwarf, electrons become so crowded together that at first they keep the core from collapsing further. Under the inexorable influence of gravity, however, more of the star's material is burned to make iron, which accumulates until the pressure in the core is high enough to force the electrons to merge with the protons in the iron nuclei. This creates neutrons—particles as heavy as protons, but with no electrical charge. Once this process begins, it rapidly snowballs. Soon the entire mass of the iron core becomes converted into neutrons.

With the electrons gone there is nothing to counteract the effect of gravity, so the neutron core collapses. In the space of a few minutes, the entire center of the star implodes, creating an incredibly dense object about ten miles across—from something that started out much larger than our sun. Titanic shock waves rip through the star's outer envelope, tearing it apart. For a few days the nuclear reactions in the dying star may emit more light than an entire galaxy! This event, the death of a giant star, is called a supernova.

In most galaxies a supernova appears about every 30 years. In our own Milky Way, supernovas are not easy to observe, because most of the stars that produce them are located in the plane of the galaxy, and thus are shrouded by clouds of dust and gas. We can, however, see supernovas in other galaxies, and they are often truly spectacular. ■

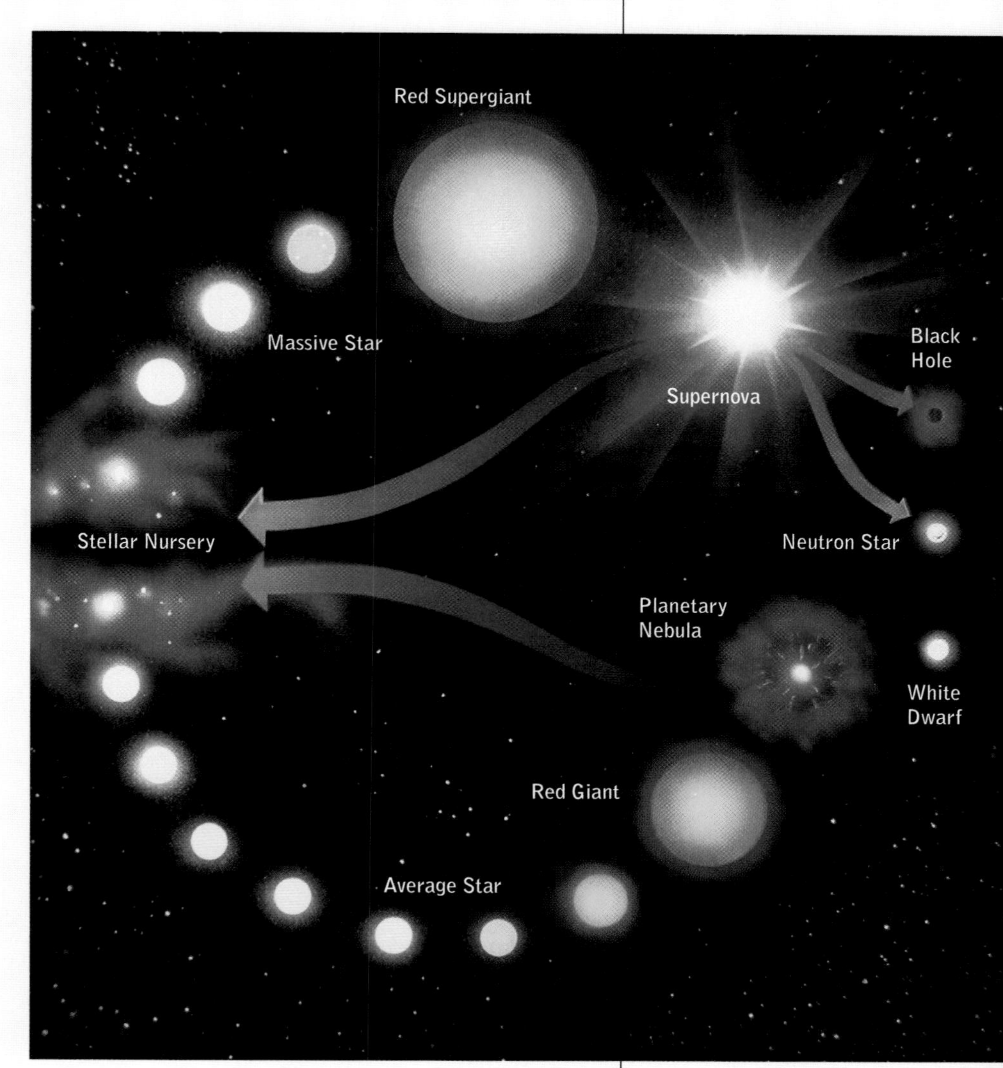

Out of hydrogen, an average-size star expands into a red giant, then shrinks to a white dwarf. A massive star (more than three times the mass of the sun) swells into a red supergiant before it collapses and implodes—a supernova—resulting in a black hole or a neutron star.

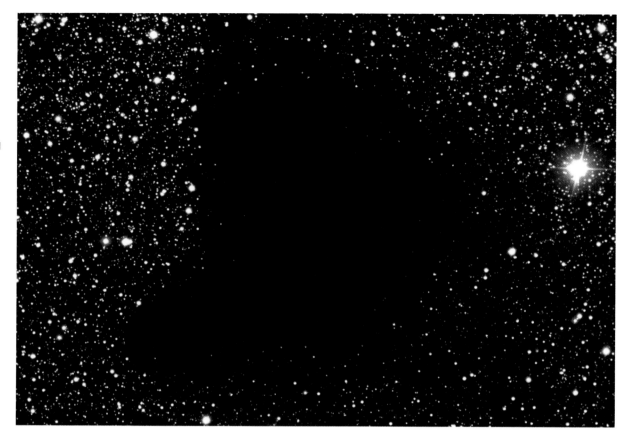

Blotting out the stars, a dark molecular cloud known as Barnard 68 absorbs almost all the visible starlight in a section of the constellation Ophiuchus, creating an eerie visual vacuum. Molecular clouds, or dark absorption nebulae, consist of high concentrations of dust and molecular gas that absorb surrounding light.

silicates, and iron. Most of these particles are released into the galaxy when a star goes **supernova.**

Molecular Cloud

Molecular clouds are cool, dense regions of interstellar matter. They have a million times the density of the **interstellar medium** and are made primarily of hydrogen molecules (H2) and cosmic dust (about one percent of the cloud's mass). Molecular clouds range in size from less than one solar mass to several hundred solar masses across. The largest molecular clouds, dubbed giant molecular clouds (GMCs), have masses up to ten million times the **sun**'s mass. GMCs are stellar nurseries that lie within the plane of a **galaxy.** There are four to five thousand GMCs distributed within the disk of the **Milky Way.** The most active GMCs are found within the galaxy's spiral arms, while cooler clouds are randomly distributed throughout the disk.

Orion Molecular Cloud

The Orion Molecular Cloud (OMC) is a vast complex of giant molecular clouds in the **constellation** Orion (the Hunter). They may have masses 100,000 times that of the **sun** and be more than a hundred **light-years** across. These huge stellar nurseries churn out young stars, especially in their dense cores. One such core, OMC-1, is associated with the Orion Nebula (M42, or NGC 1976).

Gleaming brightly in the winter sky just below the three stars of Orion's belt, the Orion Nebula marks the great hunter's sword. At only 1,500 light-years away, this dazzling expanse of ionized hydrogen (H II) is one of the brightest **emission nebulae** in the sky. At the heart of the Orion Nebula lies the Trapezium, a young, open cluster of stars, **protostars,** and dust. No more than one million years old, the four hot, massive stars of the Trapezium release a torrent of ultraviolet light that illuminates the cloud. The Orion Nebula holds roughly 700 other young stars at various stages of formation. Some of these stellar infants shoot off high-speed jets of hot gas that tear through the nebula at supersonic speeds of 160,900 kilometers an hour, creating thin, curved loops of hot gas.

Maser

The word "maser" is an acronym for "microwave amplification by the stimulated emission of radiation."

Masers are celestial objects whose molecules amplify surrounding radiation. As molecules of a celestial object are excited by radiation, they are stimulated to release photons at the same energy level. This enhances or amplifies the amount of radiation coming from the source. Masers have been detected in old variable stars and in the star-forming regions of **molecular clouds.** The first maser source was discovered in the Orion Nebula in 1965.

Protostar

Protostars represent the earliest stage of **stellar evolution.** Shock waves generated by a **supernova** explosion or the nuclear ignition of a newly formed star trigger star-formation events. As a passing shock wave moves through a giant **molecular cloud,** it compresses the cloud's molecules and dust. Once these dense cores begin to contract, gravity draws the atoms toward the center. As the atoms begin this free-fall contraction, they gather speed and randomly collide with each other—causing them to heat up. As the cloud continues to contract, it develops a hot, high-density core surrounded by a cooler, low-density cocoon. The denser the core becomes, the faster it rotates, eventually flattening the outer cocoon to a protoplanetary disk. The disk funnels material to the center, where a protostar is forming. Protostars radiate energy, but are not yet producing it by nuclear fusion. Once a protostar is optically visible, it is called a pre-main-sequence star.

Brown Dwarfs

Brown dwarfs are failed stars. They are similar to **Jupiter** in that they were born with too little mass to ignite nuclear fusion reactions in their cores. The energy released by brown dwarfs is the result of gravitational contraction. Their sizes are no larger than 75 times the mass of Jupiter, and their spectra show an abundance of methane. The first confirmed brown dwarf, Gliese 229B, was discovered in 1995.

Star Classification

Some of the brightest stars are visible on a winter evening in the Northern Hemisphere. These bright stars have very different colors. Two of the brightest stars in the **constellation** Orion are at opposite ends of the visible spectrum. Rigel is a hot, blue star; Betelgeuse is a cool, red star. Capella, in the constellation Auriga, is yellow-white. Although these stars have distinct colors, they are not emitting only one color of light. Stars emit light over the full range of the visible spectrum; they appear more red or blue due to their temperature. Stars are classified by their color, temperature, and size. Color directly relates to temperature, which in turn plays a key role in the spectral signature a star has. The atoms and molecules present in a star determine the absorption lines in its spectrum and hence its spectral type. Hot, blue stars have weak hydrogen lines but very strong ionized helium lines, while cooler, red stars show strong absorption bands of titanium oxide. Most stars can be divided into one of seven spectral types, in order of decreasing surface temperature—O, B, A, F, G, K, and M. Each type is further divided into subclasses from 0 (hottest) to 9 (coolest). Stars of a given spectral type are further subdivided into luminosity classes that relate directly to their size: Ia (bright supergiants), Ib (supergiants), II (bright giants), III (giants), IV (subgiants), V (main sequence dwarfs).

Hertzsprung-Russell Diagram

Named for its originators, Ejnar Hertzsprung and Henry Norris Russell, the Hertzsprung-Russell (H-R) diagram demonstrates the relationship between a star's spectral type (temperature) and its absolute magnitude (luminosity). Luminosity is charted along the vertical axis, and temperature along the horizontal. Each star's position on the diagram reveals a great deal about the star.

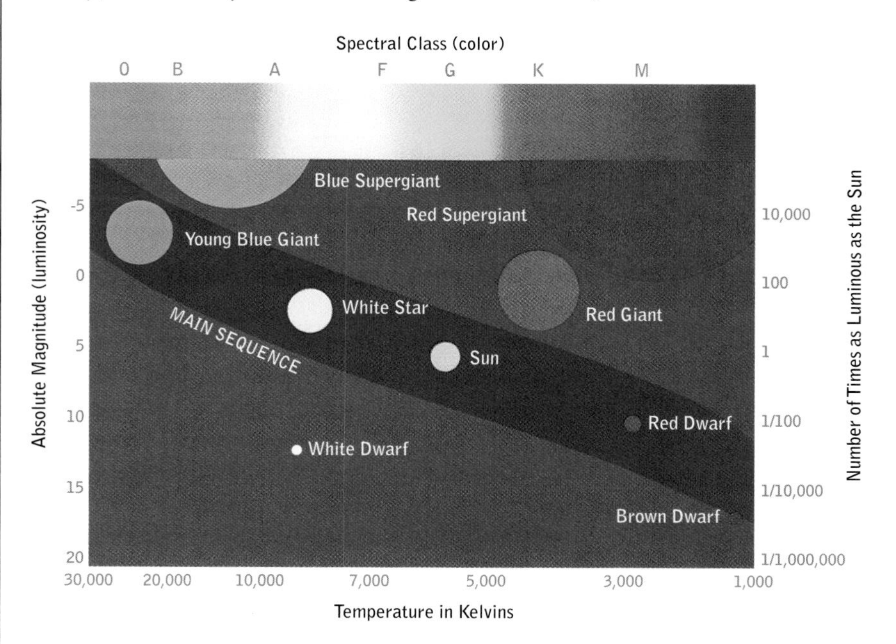

The Hertzsprung-Russell diagram shows the relationship between a star's temperature and luminosity. Most stars fall along the main sequence.

Stars in the upper left of the chart are young, hot, giant blue stars, while those in the upper right are cooler red giants and supergiants near the end of their lives. Stars located on the lower left are **white dwarfs.** These hot, small stars are very faint and at the end of their life cycles. The middle of the diagram, from the upper left to the lower right, marks the main sequence. Roughly 90 percent of all stars are on the main sequence.

A star spends most of its life on the main sequence. Once a newly formed star's core temperature is high enough to begin nuclear reactions, it enters this phase. The new star is called a zero-age main-sequence star. As hydrogen is converted to helium in the star's core, its composition and structure change, shifting it from its zero-age main-sequence position. Later in its evolution, the star leaves the main sequence, moving to the red giant, supergiant, or white dwarf zones. Although a star spends most of its life on the main sequence, once it stops burning hydrogen, it leaves the main sequence, evolves, and dies.

The time a star spends on the main sequence depends on its size. Low-mass stars burn fuel slowly and shine for billions of years. High-mass stars of 25 solar masses live hard, fast lives, burning their fuel rapidly, and die in 700 million years.

Double stars, or binaries, orbit each other, taking turns blocking the other's light. A double star called Algol (top), "winks" every 69 hours as its fainter star obscures its brighter sister. Other stars, such as Mira (bottom), expand and contract. When Mira shrinks, its light shines brightest.

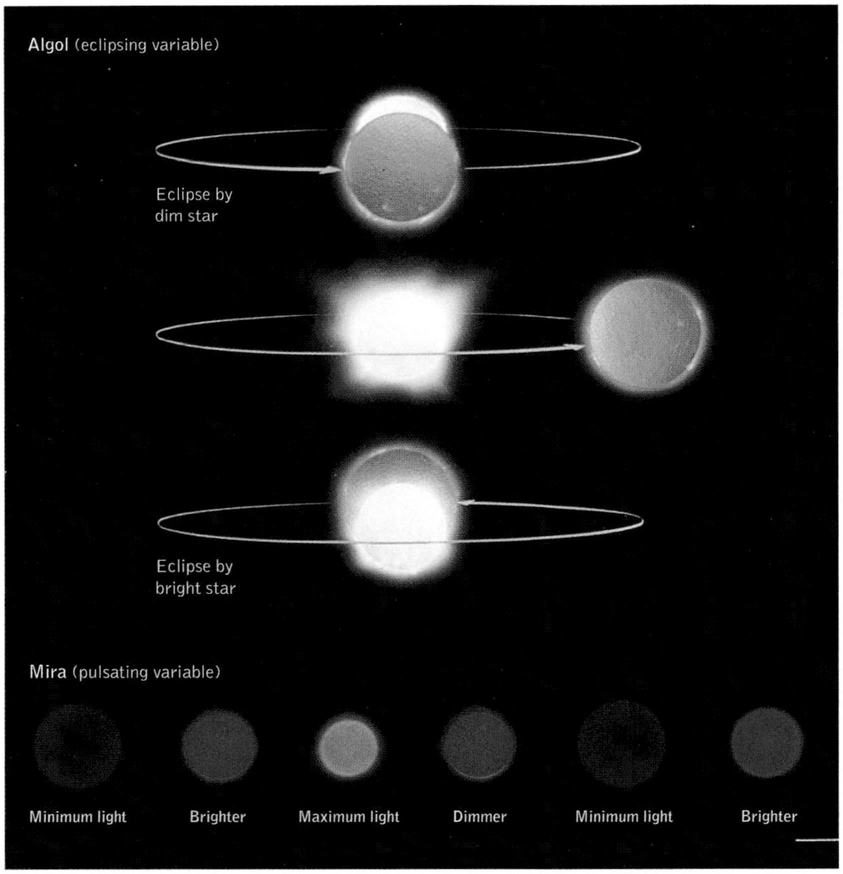

Variable Stars

Stars whose brightness appears to change over time are called variable stars. There are two broad categories of variables: intrinsic and extrinsic. Intrinsic variables have some fundamental property that causes their luminosity to change. Extrinsic variables change their brightness as a result of an outside cause, such as a passing dust cloud or an eclipsing companion star, or even as a result of a mechanical cause, such as rotation.

Pulsating stars are a type of intrinsic variable. Once a star lands on the main sequence, its luminosity changes very little and very slowly. Pulsating variables change their brightness relatively quickly. Most pulsating variables are post-main-sequence stars that lie above the main sequence on the **Hertzsprung-Russell diagram**. They tend to be in the helium-core-burning stage at the end of their lives. Types of pulsating variables include **Cepheid variables** and **RR Lyrae stars.** Cataclysmic variables are stars that undergo sudden outbursts that cause their brightness to change abruptly. These stars include **novas** and **supernovas** and are often part of a close binary-star system in which one star is losing material to a dense white dwarf. Eruptive variables change brightness suddenly through surface eruptions or flares.

Eclipsing Binary Stars

Because stars form in large clouds, they are likely to form in groups. Most stars are double, or binary, stars that orbit each other. Eclipsing binaries have orbits that are inclined so that the stars appear to cross in front of one another. Because these systems usually hold large stars with small orbits, we cannot resolve the individual stars of the system: They're just too close to one another. What we do see is a brightening or dimming as one star passes in front or behind the other. As one star passes in front of the other, both appear dimmer. When each is visible, the system is brighter.

Algol (Beta Persei)

Algol is one of the best-known **eclipsing binary stars** visible to the naked eye. Italian mathematician Geminiano Montanari first studied it in 1669, but not until 1783 did British astronomer John Goodricke explain Algol's periodic brightness variations. Goodricke found that Algol brightened, dimmed, and brightened again every 68.82 hours. He reasoned two possible explanations: It might be a star with dark spots that rotated in and out of view, or the star might have a dark companion orbiting around it. Today, we know that

Algol's "dark companion" is a fainter star. Algol normally shines at magnitude +2.1, but during eclipses dims roughly 68 percent to magnitude +3.4.

Algol is the second brightest star in the **constellation** Perseus. Its variability was puzzling to the ancients, who associated it with demons. According to Greek mythology, Perseus killed Medusa—the Gorgon who had a head full of serpents instead of hair. In the constellation's mythological outline, Algol marks Medusa's severed head. The name Algol comes from the Arabic *al ghul,* which means "the demon."

Cepheid Variable Stars

British astronomer John Goodricke discovered Cepheid variable stars in 1784, two years before his death from pneumonia at 21. They are large yellow stars of spectral type F or G. Their brightness can change 0.1 to 2 magnitudes while their periods—bright to faint to bright—can be from 2 to as many as 60 days. Over 700 Cepheids have been identified in the **Milky Way** and several thousand in our **Local Group** of **galaxies.**

There are two types of Cepheid variables. Classic Cepheids (Type I Cepheids) are hot, young **giant stars** found in the galaxy's spiral arms. They have periods that range from five to ten days and a brightness change that averages 0.5 magnitude. The star Delta Cepheid typifies Type I Cepheids. This yellow supergiant pulsates between magnitude 3.5 and 3.3 with a period of 5.4 days. Type II Cepheids are also giant stars, but they're much older than Type I's. Type II's are found toward the center and in the halo of a galaxy as well as in globular clusters. They have magnitude fluctuations similar to Type I Cepheids and periods of 1 to 35 days. The star W Virginis, in the constellation Virgo, typifies Type II Cepheids.

Cepheid variable stars are relatively rare but very important to astronomy. Their periods of pulsation directly relate to their average luminosity. If we know a Cepheid's period, we can determine its absolute magnitude. Astronomers use Cepheid variables in distant galaxies to determine that galaxy's distance.

RR Lyrae Stars

RR Lyrae stars belong to a large group of pulsating variables similar to, but fainter than, **Cepheid variables.** Like Cepheids, RR Lyrae stars change their luminosity in a regular cycle. They have typical periods of less than one day, and their luminosity changes by 0.2 to 2 magnitudes. RR Lyrae stars are old giants of spectral class A and F. They are primarily found in globular clusters but also occur at the center and in the halo of **galaxies.**

Named for their prototype in the **constellation** Lyra, RR Lyrae have roughly the same mean absolute magnitude (0.6) and are often used to determine the distances of galaxies. Solon I. Bailey, an astronomer with the Harvard College Observatory, discovered RR Lyrae stars in 1895.

T Tauri Stars

T Tauri stars are very young, irregular, **variable stars.** They are usually embedded in dense, star-forming clouds and are typically found in groups called "T associations." These groups provide important clues to deciphering the **sun**'s early childhood. T Tauri stars fall into the same spectral class as the sun (G), with a mass similar or somewhat less than the sun's and a diameter several times the sun's. Studying the individual members of T associations affords a glimpse of what the sun was like at various stages of its youth.

Unlike most pulsating variable stars, T Tauri stars are pre-main-sequence stars as opposed to post-main-sequence. They lie above the main sequence on the **Hertzsprung-Russell diagram.** These young **protostars** are still contracting and shedding their natal envelope of gas and dust at speeds of 300 kilometers a second. T Tauri stars lose nearly ten million solar masses each year. They have an abundance of lithium, an element destroyed early in a star's life.

Henrietta Swan Leavitt (1868-1921)

The daughter of a minister, Henrietta Leavitt, born in Lancaster, Massachusetts, revolutionized our understanding of the brightness and variability of stars. After graduating from Radcliffe College (then the Society for the Collegiate Instruction of Women), Leavitt began volunteering at Harvard College Observatory in 1895. In 1902 the observatory's director, Edward Charles Pickering (1816-1919), offered her a permanent staff position. In 1907, as head of Harvard's Department of Photographic Photometry—a job that paid 30 cents an hour—Leavitt began the task of cataloging **variable stars.** In 1912 Leavitt was determining the number of **Cepheid variable stars** in the Small Magellanic Cloud by examining photographic plates taken in Peru. She noticed that the periods of Cepheid variables were related to their average brightness. She found that the longer the period, the brighter the star. Since all these stars were at relatively the same distance from **Earth,** their intrinsic brightness had to be directly related to their periods. This period-to-luminosity relationship provides astronomers with a star's **absolute magnitude** while observations yield its **apparent magnitude.** If you know how bright a star is and how bright it should be, you can calculate its distance.

Leavitt was one of several women doing calculations and data reduction at the observatory, essentially performing the job of computers. She worked at the observatory until her death in 1921.

Degenerate Matter

Degenerate matter is matter compressed so tightly that electrons are stripped from their nuclei. The free electrons and nuclei exist in a closely packed mass. Compressing this matter further means pushing against the electrons, which cannot be packed any tighter, forming a degenerate gas as strong as hardened steel. Unlike normal gas pressure, degenerate pressure is dependent on density and not temperature. Degenerate matter provides the main support against gravity in **white dwarfs** and **neutron stars.**

White Dwarf

White dwarfs are exceedingly dense, compact, faint stars at the end of their lives. They represent the final stage in the life cycle of all but the most massive stars. As the star's nuclear fuel is exhausted, the core collapses, and the star expels its outer layers to form a **planetary nebula.**

A white dwarf's mass shrinks down to less than 1.44 solar masses. Because the matter in its core is degenerate, a white dwarf shines dimly, but has no nuclear fusion reactions. Surrounding the core is a thin gaseous atmosphere that slowly leaks any residual heat into space. As the star continues to cool, it eventually loses all its energy and becomes a black dwarf. The **sun** will eventually become a white dwarf that is roughly a hundred times fainter than the sun is today.

Neutron Star

Neutron stars are the smallest, most dense stars known. Their mass is 1.4 to 3 times the sun's mass—not great enough to collapse into a **black hole.** Neutron stars begin their lives as main-sequence stars of roughly ten solar masses. As a neutron star approaches the end of its life, the gravitational collapse it undergoes creates pressures great enough to compress electrons and protons into neutrons.

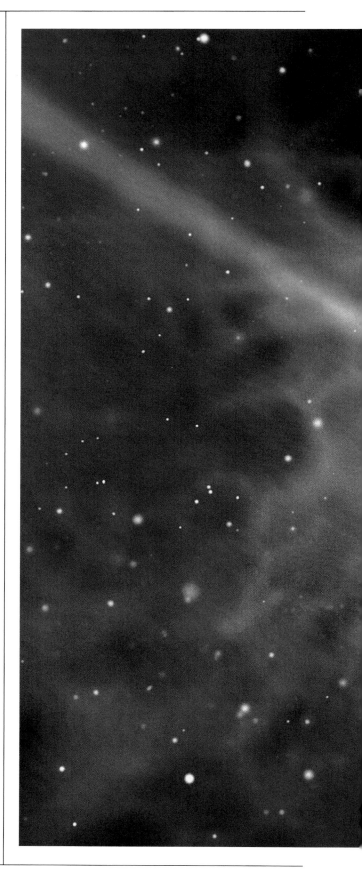

A pulsar—a rapidly spinning neutron star—shoots out a narrow band of energy from its north and south magnetic poles. When visible from Earth, these beams look like the regularly spaced signals from a flashing lighthouse.

Neutron stars are so densely packed that a sugar-cube-size lump of neutron star material would weigh a hundred million tons! Typical neutron stars have a very thin atmosphere and a surface made of elements such as iron. But the iron differs from **Earth**'s. Its atomic particles are packed so tightly that the iron encasing a neutron star is 10,000 times stronger than iron on Earth. Neutron stars are probably the remnants of **supernova** explosions.

Pulsar

Pulsars are rapidly spinning **neutron stars,** and they transmit energy much as a lighthouse sends out beams of light. A pulsar spins so fast that its strong magnetic field generates an electric field around itself. As the star rotates, a beam of radiation is shot out at its north and south magnetic poles. If **Earth** happens to be in the line of sight, we see a pulse as the beam sweeps past us. Pulsars have periods that range from a few milliseconds to five seconds. The average pulsar spins once every second, while the average pulse lasts only a few tenths of a millisecond. Our **galaxy** alone may hold nearly 100,000 pulsars.

Young pulsars spin very rapidly, nearly a hundred times per second. As it ages, a pulsar's magnetic field weakens, and its rotation slows. Eventually it rotates

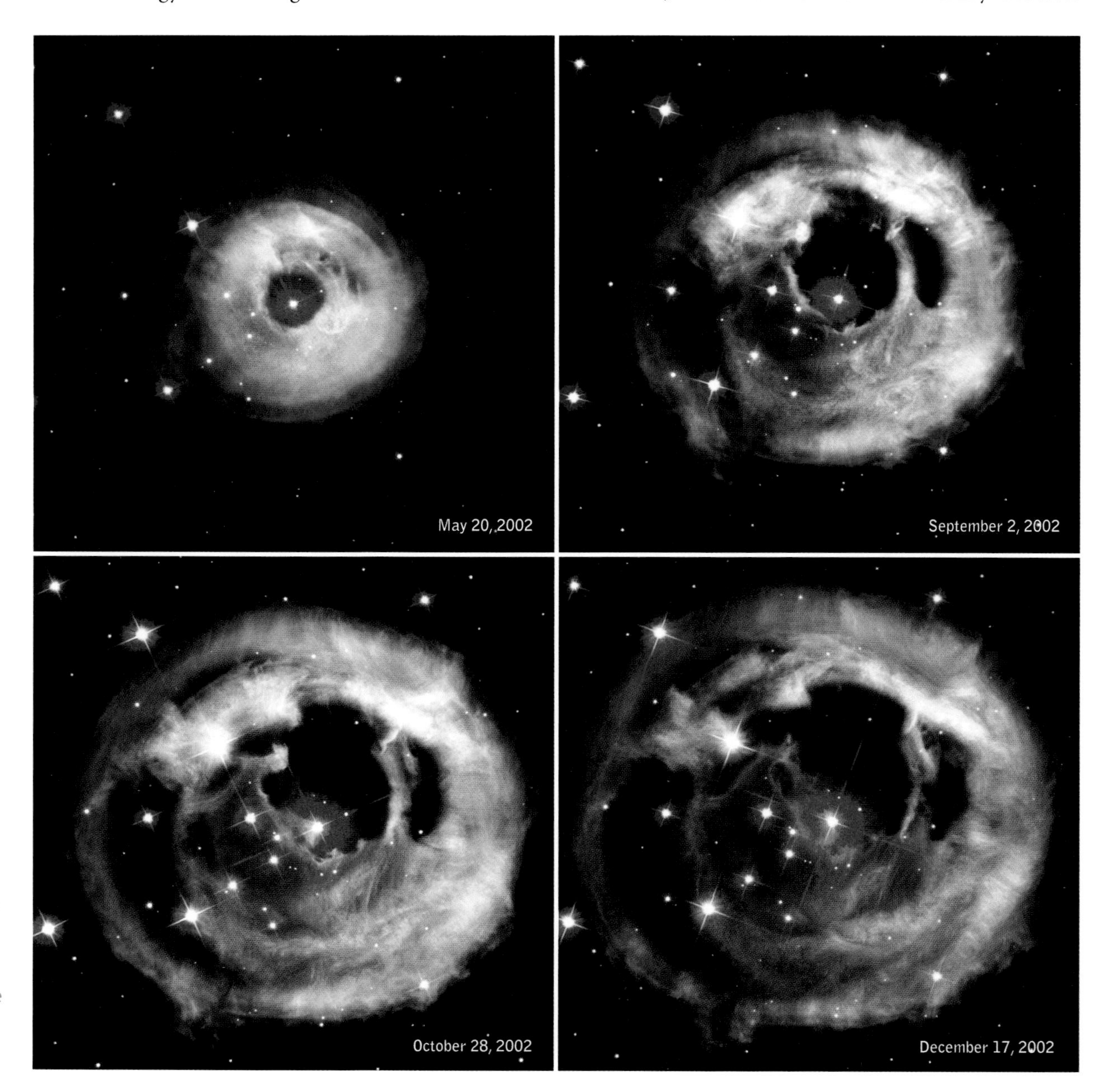

Once an unremarkable tiny twinkle in the sky, V838 Monocerotis suddenly brightened and swelled to many times its former size in early 2002. Later that year, the Hubble Space Telescope took several photographs of the dusty nebula surrounding V838 Mon made visible by "light echoes" from the star's dramatic outburst.

too slowly to generate a beam of radiation. Millisecond pulsars spin even more rapidly. PSR 1937+21, in the **constellation** Vulpecula, has a period of 1.56 milliseconds. That means it spins 642 times per second or at roughly 40,000 kilometers per second (one-tenth the speed of light). Such breakneck speeds are nearly at the point of breaking PSR 1937+21 apart. Millisecond pulsars have a very stable rotational period so they keep extremely accurate time. In fact, they are more accurate than an atomic clock. Roughly 50 of these quickly spinning balls of neutrons have been identified. Many are found in **globular clusters,** which means they are very old.

Gamma Ray Burst

Gamma ray bursts are intense flashes of gamma rays. They occur without warning and fade in a matter of seconds to a few minutes. Most occur in very distant galaxies. They were discovered in the late 1960s by U.S. Air Force **satellites** monitoring the 1963 nuclear test ban treaty. Detonated nuclear bombs emit gamma rays, so the satellites were looking for bursts of these high-energy particles. Astronomers were surprised to detect roughly one burst each day. Fortunately these bursts were coming from space and not **Earth.** If one occurred in a nearby star system, Earth could receive an amount of gamma radiation equal to a 10,000-megaton nuclear explosion.

The Air Force declassified its gamma ray burst observations in 1973. In 1991 **NASA** launched the **Compton Gamma Ray Observatory,** which discovered a few bursts each day. The origin of gamma rays remains a mystery, but there is evidence that points to a relationship with **neutron stars.**

Magnetar

Young **neutron stars** with magnetic fields 100 to 1,000 times stronger than the average neutron star are called magnetars. These objects produce repeated bursts of low-energy (soft) gamma rays (SGRs). Such bursts are generated when shifts in a magnetar's magnetic field break the iron crust of a neutron star.

Nova

The word "nova," contracted from the phrase "stella nova," is Latin for new star. A nova is a type of cataclysmic **variable star** that unexpectedly and rapidly increases in brightness. The star may shine 50,000 to one million times brighter in just a few hours or days. Then it slowly begins to decrease in brightness over a period of several hundred days. A typical nova releases the same amount of energy the **sun** produces in 100,000 years! Most novas occur in binary or double star systems in which one star has evolved to a **white dwarf.** Typically, the companion star to the white dwarf has evolved to a red giant star. As the red giant expands, it reaches the outer limits of its own gravitational field—a region of space called the Roche lobe. Matter within the Roche lobe cannot escape the star's gravity, but in a binary system the Roche lobes of each star touch one another. Hydrogen from the outer atmosphere of the red giant begins streaming toward the white dwarf. As hydrogen spirals into the white dwarf, it forms a new hydrogen envelope around the degenerate star. Temperatures and pressures continue to rise as more material funnels in. When the temperature in the bottom layer reaches a few million kelvins, the hydrogen ignites, fueling a massive explosion that sends material into space at speeds of hundreds of kilometers per second.

A recurrent nova is a type of cataclysmic variable star that undergoes a series of violent outbursts. In this system, the red giant loses material to the white dwarf 1,000 times faster than a typical nova. Hydrogen builds up on the surface of the white dwarf much faster, and the outburst can recur after a few decades.

Supernova

A supernova is a dying star even more explosive than a **nova.** At its peak luminosity, a supernova shines more than 100,000 times brighter than a nova. There are two general types of supernovas. Type I have a sharp luminosity curve. They quickly reach a maximum brightness of about four billion times the **sun**'s luminosity, fade quickly, then gradually die. Typical Type II supernovas reach a maximum luminosity quickly, about 0.6 billion times that of the sun, but their decline is not as smooth as that of Type I supernovas. Type II's initially stay brighter longer than Type I's, giving them a broader luminosity curve. Once Type II's do start fading, however, they do so more rapidly than Type I supernovas. Type I supernovas have no hydrogen lines in their spectra, while Type II supernovas do. This means that Type I supernovas involve highly evolved stars that lack hydrogen. They also seem to occur in stars that have a mass similar to the sun's.

Type II supernovas are thought to originate in stars that spend most of their lives as massive (more than ten times the sun's mass) O and B stars. As these stars die, they become red supergiants with iron cores that

A dramatic before and after of the Tarantula Nebula reveals the appearance of the first supernova visible to the naked eye since 1604—SN1987A. The top picture was taken in 1984, the bottom one in February 1987. SN1987A glows brightly to the right of the Tarantula's red cloud in the bottom picture.

exceed 1.4 solar masses (the **white dwarf** limit). Because iron is so stable, the core can't fuse it into heavier elements; the star's hydrostatic balancing act falls apart. The star no longer generates energy, so its radiation pressure goes to zero and gravity takes over. In minutes the core collapses; material from the outer shell rushes toward the dense core, where it bounces. The resulting shock wave pushes the star's outer envelope outward in a massive supernova explosion that rapidly increases its brightness. The dead star leaves behind a supernova remnant of expelled gas, and in some cases a **neutron star, pulsar,** or even a **black hole.**

SN1987A

On the night of February 24, 1987, light from a dying, blue supergiant star named Sanduleak reached Earth. High atop a mountain in Chile, Ian Shelton was preparing a photographic plate at the Las Campanas Observatory. The resident observer at the University of Toronto's Southern Observatory, he was trying to image one of the **Milky Way**'s companion galaxies, the Large Magellanic Cloud. Instead, he captured the first naked-eye **supernova** since **Kepler**'s star in 1604.

Sanduleak was a blue supergiant and not a red supergiant; this made SN1987A an uncommon Type II supernova. Because it was much hotter than the typical supernova progenitor, it evolved much more rapidly. In addition, SN1987A's peak luminosity was lower than a typical Type II. It reached a visual magnitude of 2.8 by mid-May 1987.

Sanduleak was some 160,000 **light-years** away from **Earth.** It took radiation created by the explosive blast 160,000 years to reach our **planet.** Nearly 20 years after that initial brightening, **Hubble Space Telescope** views of SN1987A show a complex system of three bright rings that appear to be intertwined.

Crab Nebula (M1, NGC 1952)

Imagine yourself as a Chinese astronomer in 1054. You've been observing all night, and the **sun** is about to rise. Before the first rays of dawn, you take one last look at the stars rising in the east. There you spot a bright, though unfamiliar, object. This "guest star" was the **supernova** blast of a star some 6,500 light-years away. The supernova quickly became bright enough to see during daylight hours—about magnitude -6. After a month it slowly began to fade, but took another two years to vanish completely. When we point our modern **telescopes** toward the region of sky once occupied by the "guest star," we find the Crab Nebula. This intricate array of filaments shines with the light of 100,000 suns. Today, it is roughly 13 light-years across and continues to expand at a rate of 1,000 kilometers a second.

At the heart of the expanding cloud lies the Crab Pulsar. Spinning 30 times each second, this ten-kilometer-wide **neutron star** is one of the very few pulsars to be identified at all wavelengths, including the visible. In May 1996 the **Hubble Space Telescope** captured an image of the pulsar surrounded by wisps of gas.

Black Hole

If the core of a dying star goes beyond three solar masses (the **neutron star** limit), gravity prevails; nothing will stop the complete collapse of the star. The force of gravity crushes any outward forces, including the repulsive force between particles. As the object continues to collapse, its density and gravity will increase to infinity while the star's size shrinks to zero. Although it's difficult to picture, it becomes an object whose constituent particles are packed so tightly that it takes up no volume of space yet it still exists. Such a point in space is called a singularity.

Many physicists have long questioned the existence of an object whose density is infinite and volume is zero. In the spring of 2004, one group of physicists published a model of black holes as large balls of tangled cosmic strings. String theory portrays the universe as a symphony of vibrating strings in 10 dimensions. It describes all energy and matter as filaments of energy. When described in terms of string theory, black holes no longer become objects without volume. The singularity disappears and black holes become extremely compressed strings. Like neutron stars, black holes described with string theory behave like any other highly compressed ball of matter.

The size of a black hole is determined by its mass—the more matter it has, the larger its gravity. In 1916, **Albert Einstein** (1879-1955) published his general theory of relativity in which he postulated that space and time were a single entity. Einstein's equations showed that gravity distorts **space-time.** German astrophysicist Karl Schwarzchild (1873-1916) used Einstein's general theory to show that if matter is packed tightly enough space-time curves back in on itself. He essentially described the conditions around a black hole. Beyond a certain point, nothing can escape the gravity of a black hole. Today we call the boundary between a black hole and the rest of the universe the event horizon. The size of the event horizon is called the Schwarzchild radius.

Because a black hole's gravity dominates its region of space, any matter near the black hole will be drawn toward it. Like a cosmic drain, the black hole sucks matter toward itself. A vast whirling accretion disk forms just outside the event horizon. As matter draws closer, the gas and dust heat up and begin to ionize. The ever-increasing gravity accelerates the ionized gas and it emits x-rays. One of the best black hole candidates is the x-ray source Cygnus X-1 in the **constellation** Cygnus (the Swan). This massive and very energetic object has a hot, blue supergiant (spectral type O) companion that appears to be transferring some of its matter to the suspected black hole. If the corpse of a star is larger than the neutron star limit, it must become a black hole at the end of its life. No force can stop the unrestrained collapse caused by the star's gravity. However, an object of any mass can form a black hole if the force to compress it is large enough. One such force is found in supernova explosions, which can smash matter down to a size below its Schwarzchild radius. If the sun were compressed into a black hole its Schwarzchild radius would be three kilometers across—smaller than a typical sunspot!

Wormhole

A wormhole is a theoretical portal—lasting only a brief moment—consisting of two **black holes** in different locations that are joined. Wormholes could connect two points in our own universe or perhaps in another time. Matter falling into a black hole at one point should emerge through a proposed "white hole" (the reverse of a black hole) at the other end. The concept is highly speculative.

Stephen W. Hawking (1942-)

British theoretical physicist and cosmologist Stephen Hawking has developed many revolutionary theories that involve **black holes** and the origins of the universe. His doctoral dissertation in *(continued on page 71)*

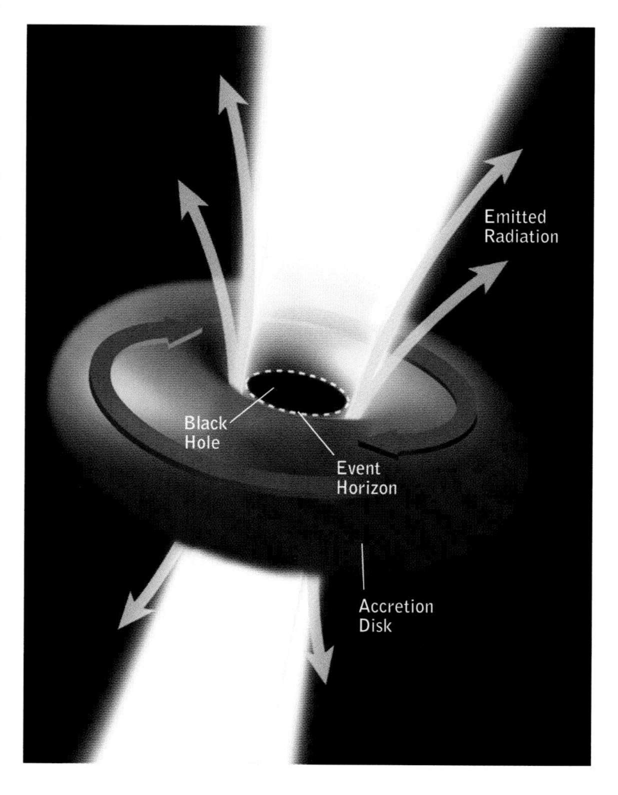

If a black hole left by a dying star emits no light, how can we see it? The hole's super-strong gravitational pull sucks gas and dust toward itself, forming a whirling accretion disk around the hole. The disk heats any matter that crosses it, emitting x-rays.

SEARCHING FOR OTHER WORLDS: EXTRASOLAR PLANETS

William Harwood

WHEN I WAS A BOY, MY BROTHER BOB AND I OCCASIONALLY WHILED away a summer evening under the stars. We speculated about how many of the stars we saw might harbor solar systems or even alien civilizations. It never occurred to either of us that, in a galaxy of 200 billion suns, our solar system might be the only one.

We were not alone in that assumption, of course. In 1968, about the time scientists were looking for meteors and debating the reality of aliens, Swarthmore College astronomer Peter Van de Kamp announced the results of a painstaking reanalysis of data that, he said, indicated the presence of a Jupiter-class planet in orbit around a star just six light-years from Earth. Van de Kamp had spent decades carefully monitoring the target star against the background of interstellar space, measuring a tiny back-and-forth wobble he believed was caused by the gravity of some unseen companion tugging on the star. Others using similar techniques had made similar claims for other stars during the early 1940s, but their observations could not be confirmed. Van de Kamp cited evidence for an extrasolar companion of some sort in 1943. Twenty years later, he felt confident enough to claim it was a planet, and five years after that, he bolstered his earlier conclusion with additional analysis.

But other astronomers later challenged his results and, in 1973, proved to the satisfaction of the astronomical community that no such planets existed. There matters stood until 1991, when Alex Wolszczan, using the great radio telescope at Arecibo, Puerto Rico, noticed unexpected changes in the timing of signals from a spinning pulsar. He ultimately concluded that the changes were caused by the gravitational effects of three planets orbiting the pulsar, a discovery that made headlines around the world. But because pulsars—collapsed stars—are strange beasts, it was not immediately clear what his discovery meant about the likelihood of planets circling more sun-like stars.

In 1995 astronomers Michel Mayor and Didier Queloz used spectroscopic techniques to discover a Jupiter-class planet orbiting within a few million kilometers of the star 51 Pegasi, far closer than current theories of planetary evolution predicted. Shortly thereafter, a team lead by Geoffrey Marcy and R. Paul Butler found massive planets around two more stars. Another major surprise was that one of these planets followed a highly elliptical path, in stark contrast to the more circular orbits of planets in our own solar system.

Since then, finding extrasolar planets has become one of modern astronomy's hottest fields. By March 2004, a database maintained by the Paris Observatory listed 120 confirmed extrasolar planets orbiting main sequence stars, 2 pulsar-type planetary systems, and another 20 "unconfirmed, controversial or retracted planets."

Search Methods

Of the distant solar systems that have been confirmed, astronomers have yet to find one very similar to ours, in which Jupiter-like planets orbit parent stars in nearly circular paths and at relatively large distances. Instead, astronomers have been astonished to find huge Jupiter-size planets orbiting fairly near their suns. This is most likely a result of the observation methods used, not an indication of some universal law governing planetary formation. With current techniques, the nearer a planet is to its primary star, the easier it is to detect. Many astronomers believe the unusual number of "hot Jupiters"—large planets that travel so close to their sun that they are heated to several thousand degrees—discovered to date is an artifact; as instrumentation improves, more sun-like solar systems should turn up. But making predictions in a field as dynamic as this is a risky business, and no one really knows what will be discovered in the years ahead.

A group of scientists working with the Keck 1 Telescope at the Mauna Kea Observatory, in Hawaii, during the late 1990s discovered two planets orbiting within the solar system of Gliese 876, a red giant star 15 light-years from Earth. Here, an artist has imagined the view from a hypothetical moon of one of the planets.

The basic method for indirectly detecting extrasolar planets was developed more than three centuries ago by Isaac Newton, who put gravitational interactions on a mathematical footing, and the German mathematician Johannes Kepler, who formulated a set of laws governing orbital motion. The velocity of a planet, moon, or any other orbiting body depends only on the radius of the orbit. The closer a body is to its "sun," the faster it travels. The relationship between an object's velocity and its orbital radius is easily demonstrated within our own solar system. Mercury needs only 88 days to circle the sun at an average distance of 58 million kilometers, while Pluto—5.87 billion kilometers out—takes 248 years to orbit the sun once. Interestingly, the orbiting body's mass does not significantly affect either its orbital period or its velocity.

Gravity also varies with distance. Isaac Newton discovered that the gravitational force between two bodies is inversely proportional to the square of their distance from each other. Thus a planet orbiting four units from its sun will experience one-sixteenth of the gravitational attraction that a planet only one unit away experiences. Consider the sun and Jupiter, ignoring for the moment all other components of our solar system. The sun's gravity keeps Jupiter in an 11.9-year orbit, exactly as predicted by Newton. But Jupiter's gravity also tugs on the sun, pulling it ever so slightly toward the planet. In fact, the sun and Jupiter actually orbit a common center of mass, and the amount of their mutual tugging is directly proportional to the masses involved. Because the sun outweighs Jupiter about a thousand to one, the common center of mass is one thousand times closer to the sun, a point about 49,890 kilometers beyond the sun's upper atmosphere. Viewed directly above the plane of the solar system, the sun moves around this point as Jupiter sails along in its orbit. The picture grows more complicated, of course, when one adds the effects of the other planets, but the basic result is the same. The sun wobbles ever so slightly as the planets wheel about. We can detect the motion because we are so close. Viewed from 33 light-years away, however, the sun's motion due to Jupiter's gravitational influence would be comparable to the size of a dime seen from 1,610 kilometers away. How to detect so subtle a motion from afar?

One way is to measure the actual movement of a star in space, the same technique Van de Kamp used. Another is to measure the intensity of light from a broad population of stars. When an extrasolar planet moves in front of its star as viewed from Earth, the star's light will dim slightly and then brighten. By detecting and measuring such periodic fluctuations in intensity, astronomers can determine the nature of the objects that cause the dimming.

So far, the most productive technique for finding extrasolar planets uses ultra-sensitive spectroscopes to measure a star's radial velocity around the star-planet system's center of mass. Just as the pitch of a siren changes as it approaches and then recedes, so does the wavelength of a star's light change as it approaches or moves away from the observer. Such changes in stellar wavelength are on the order of one part in a hundred million, but they are detectable. With enough observations, radial velocity data can provide the planet's minimum mass, its distance from the parent star, and its orbital time. Other data are needed to determine characteristics such as the tilt of the planet's orbit relative to the star's equatorial plane.

Hot Jupiters

Current technology clearly favors the discovery of massive planets that orbit relatively near their stars. Whether such relationships ultimately prove to be common or uncommon is not yet known, but just realizing that hot Jupiters exist has been a revelation for planetary theorists trying to figure out just how solar systems evolve. It is not surprising that a planet's orbit might change over time; theorists have considered that possibility since the 1980s, and data from the Galileo probe indicate that Jupiter actually may have taken shape much farther from the sun than it is now, and only later moved to its present location. What is surprising about the hot Jupiters discovered so far is how close many of them are to their suns. It is possible that gravitational interactions among other massive, nearby planets—planets yet to be directly detected—play a role. A more likely scenario involves gravitational interactions between a growing Jupiter-class planet and the solar nebula that spawns it. Such interactions can kick a planet into a higher orbit or even eject it from its solar system. More often, the end result is a loss of angular momentum, causing the planet to migrate closer to its sun.

All this begs the question of why Jupiter and Saturn are in the outer regions of our solar system, not closer to the sun. Given Earth's history, it's not merely an academic question. If hot Jupiters are the rule rather than the exception, life in this universe could be rare indeed. Only time will tell whether solar systems like ours are common or rare. Based on past experience, however, it is a safe bet that the universe will turn out to be even more surprising than we currently imagine. ■

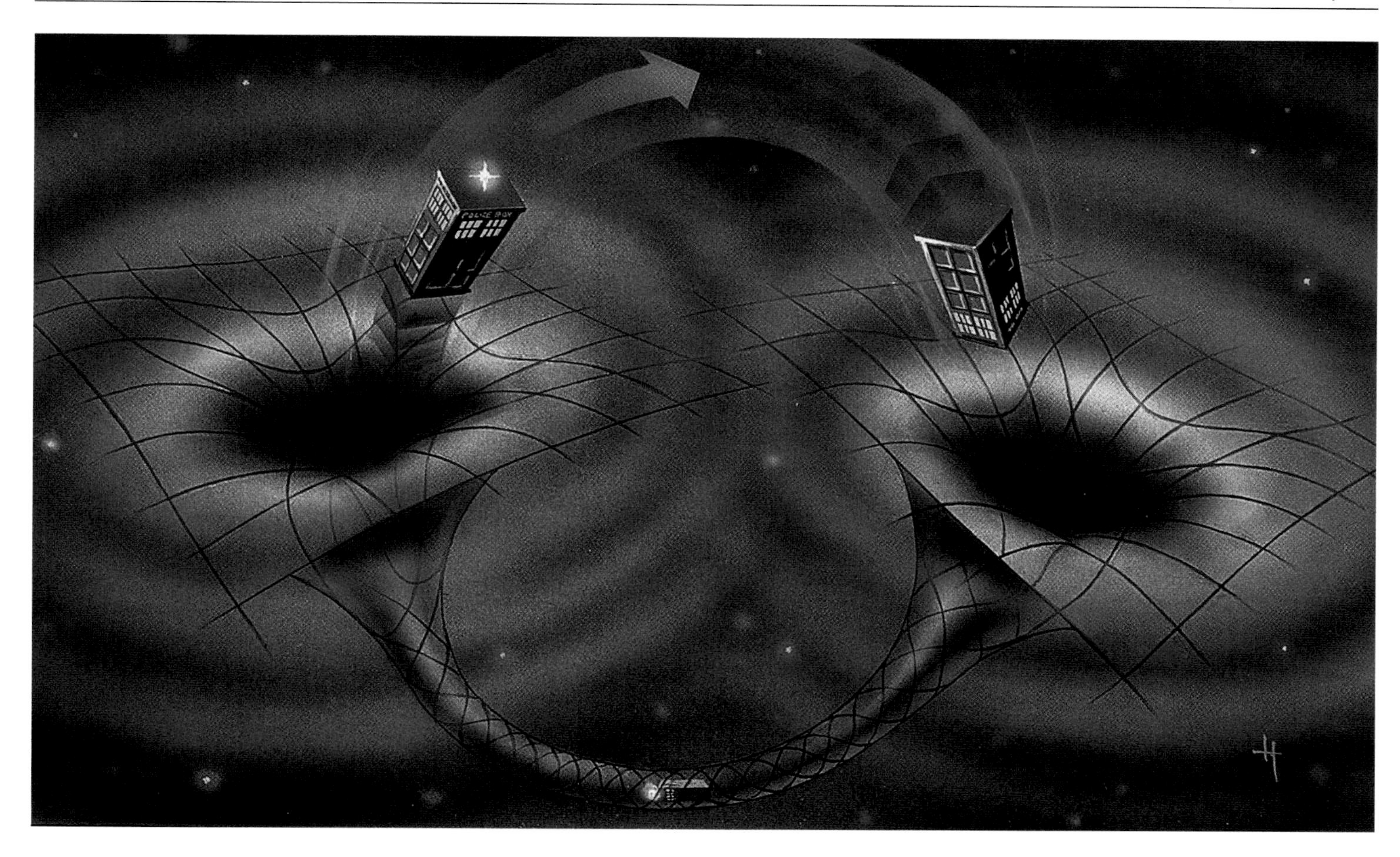

the late 1960s demonstrated that if **Einstein**'s general theory of relativity holds true, a singularity (an object of infinite density and no radius) represented conditions at the beginning of the universe in the **big bang.** Hawking then showed that the big bang could have produced many "mini black holes" with masses of a billion tons that were no larger than a proton. Much of Hawking's work has centered on black holes. In his 1988 book, *A Brief History of Time,* Hawking theorized that a black hole could grow smaller and eventually evaporate. Black holes might emit a type of thermal radiation, called Hawking radiation, by simultaneously creating pairs of matter and antimatter particles. Along the event horizon, the distortion of **space-time** is so great that these particles could change their positive or negative energy for the opposite charge. The positive energy leaving the black hole would be balanced by the flow of negative energy into it. Using Einstein's equation $E=mc^2$, Hawking showed that the loss of energy (E) through Hawking radiation would decrease the black hole's mass (m). Eventually the black hole would evaporate to nothing. In 1963 Hawking was diagnosed with amylotrophic lateral sclerosis. In 1979 he was appointed Lucasian Professor of Mathematics at Cambridge University, a chair once held by Isaac Newton.

Extrasolar Planet

Planets outside our **solar system** are called extrasolar planets. **Ground-based telescopes** are not powerful enough to directly observe them. Astronomers detect planets around stars by their Doppler shift. The gravity of an orbiting planet tugs ever so slightly on the star. This causes a periodic shift in the star's **spectrum.** Roughly 5 percent of main-sequence stars appear to have planets around them. The majority of extrasolar planets found to date are just under the size of two **Jupiters** to about the size of **Saturn.** Most are fairly close to their parent stars, so they orbit rather quickly. There are many other planetary systems in the **Milky Way,** and most of them seem to be **gas giants** like Jupiter. The larger gas planets could be absorbing or flinging out any smaller terrestrial planets similar to Earth from these extrasolar systems. So far it seems our solar system, with its nearly equal division of terrestrial and Jovian planets, may be unique in the **universe.**

In theory, travelers could move faster than the speed of light through both space and time within wormholes—theoretical tunnels linking one black hole to another, maybe in another time. But wormholes, if they exist, would do so for only a fraction of a second, making such trips impractical.

Nebulae

Nebulae

The word "nebula" derives from the Latin word for cloud. These regions of interstellar gas and dust resemble the delicate tendrils of cirrus clouds that intertwine in the **Earth**'s upper atmosphere. A nebula may become a huge, billowing **molecular cloud** in which new stars form, or it may be the graceful result of a star's violent death. Nebulae provide the building blocks for stars, **galaxies,** and **planets,** while their composition and delicate patterns reveal the nature of both star formation and star death.

Huge interstellar dust clouds like the Eagle Nebula provide the building blocks for stars, galaxies, and planets. The darker shell of dust surrounds a brightly lit workshop in which bright blue young stars take shape.

Emission Nebulae

Emission nebulae are hot, discrete clouds made primarily of ionized hydrogen (H II regions) produced when ultraviolet radiation released from young stars removes the electrons from hydrogen atoms—the process called photoionization. Emission nebulae glow with their own light, produced by excited atoms and ions much as electric current excites atoms of gas inside a neon sign. The process of photoionization takes quite a bit of energy, so only the hottest stars can produce the ultraviolet photons in the numbers needed to make the nebula glow.

Emission nebulae usually occur around hot, young blue stars with temperatures greater than 25,000 kelvins (spectral classes B and O).

Eagle Nebula (M16, NGC 6611)

The Eagle Nebula is an **emission nebula** some 8,500 **light-years** from Earth in the **constellation** Serpens. The **Hubble Space Telescope** imaged this star-forming region in 1995. The Hubble images show a vast cloud of molecular hydrogen with dark columns of cold gas and dust jutting outward from the edge of the cloud. These gaseous pillars are star-forming regions of dense interstellar gas collapsing under its own weight to form young **protostars.** Finger-like protrusions along the edges of the pillars hold evaporating gaseous globules (EGGs) of dense gas. The EGGs are exposed when hot stars heat gas along the surface of the pillars, "boiling it away" in a process called photoevaporation.

Planetary Nebulae

As an average-size star like the sun begins to die, it may expel its atmosphere, creating an expanding and generally symmetrical cloud of gas called a planetary nebula. This event happens during the star's unstable red giant phase, when fluctuations in energy production can cause the star to throw off its outer atmosphere. The material expands outward at speeds of up to 60 kilometers a second.

Planetary nebulae are **emission nebulae** that shine as a result of ultraviolet light released by the central star. They come in a variety of shapes, from rings to lobes, to smooth or irregular disks and combinations thereof. Planetary nebulae actually have nothing to do with **planets. William Herschel** coined the name because the ones bright enough to see through his **telescope** appeared to resemble the blue-green disk of **Uranus,** which he discovered in 1781.

Ring Nebula (M57, NGC 6720)

The Ring Nebula (**apparent magnitude** 9) is an easy object to find in backyard **telescopes;** it roughly resembles a smoke ring. The planetary nebula is approximately 2,000 **light-years** away and about one light-year in diameter. The nebula's dim, central star (apparent magnitude 15.3) once had a mass greater than the **sun**'s. It has shed most of its atmosphere and will eventually contract to a **white dwarf.** The inner zone appears bluish green, its oxygen ionized by the radiation from the central star. The outer reddish regions are the result of lower-energy transitions involving hydrogen and nitrogen atoms.

Supernova Remnant

The fine threads of gas that remain after a star undergoes a supernova explosion make up a supernova remnant. Like **planetary nebulae,** supernova remnants are **emission nebulae.** Supernova remnants resemble regions of ionized hydrogen (H II regions), but they have a more defined and delicate filamentary structure than H II regions. A supernova remnant appears as a thin spherical shell around the central **neutron star, pulsar,** or **black hole.** Right after a **supernova** explosion, material expands at a rate of up to 20,000 kilometers a second, slowing to a few hundred kilometers a second over time.

Because elements heavier than boron, which need great energies to form due to their atomic weight, form in supernova explosions, expanding supernova remnants mix these elements into the **interstellar medium,** where they may later form into new stars or rocky, terrestrial **planets** like **Earth.**

Veil Nebula (NGC 6995 & 6960)

The Veil Nebula is the most visible part of the Cygnus Loop, a large shell-like **supernova remnant** located roughly 1,400 **light-years** away in the **constellation** Cygnus (the Swan).

The Cygnus Loop marks the edge of an expanding blast wave some 80 light-years in diameter. It is the result of a **supernova** explosion that occurred roughly 5,000 years ago. This faint, delicate ring of glowing gas is quite hard to observe without the aid of filters. Long-exposure photographs reveal a fine veil-like structure rich in intertwined filaments and knots.

Reflection Nebulae

Reflection nebulae glow by reflecting the scattered starlight of nearby stars. Dust particles within the gas cloud disperse starlight, giving the nebula a faint glow. Most reflection nebulae appear blue because shorter-wavelength blue light is scattered more efficiently than longer-wavelength red light. Because the glow is rather faint, most reflection nebulae are detected with long, photographic exposures.

Pleiades (M45)

This open cluster of young stars shines as a fuzzy patch high in the winter sky for observers in the Northern

SOME WELL-KNOWN NEBULAE

Name	Location	Distance from Earth (in ly)*	Distance Across (in ly)*	Nebula Type
Barnard's Merope	Pleiades	410	13	Reflection
Cat's Eye	Draco	3,000	0.3	Planetary
Coalsack	Crux	600	60	Dark
Crab	Taurus	6,500	11 x 7.5	Supernova remnant
Cygnus Loop (incl. Veil)	Cygnus	1,400	80	Supernova remnant
Dumbbell (M27)	Vulpecula	1,000	1	Planetary
Eagle (M16)	Serpens	8,500	15	Emission
Horsehead	Orion	1,500	2.1	Dark
Lagoon	Sagittarius	5,000	100	Emission
Omega	Sagittarius	5,000	40	Emission
Orion	Orion	1,500	20	Emission
Ring (M57)	Lyra	2,000	1	Planetary
Tarantula	Large Magellanic Cloud	160,000	800-6,000	Emission

**All distances are approximate.*

Hemisphere. It marks the shoulder of the **constellation** Taurus (the Bull) and is easily spotted with the unaided eye. The cluster is roughly 400 **light-years** away and about 76 million years old.

The brightest Pleiades are hot, young, blue stars of spectral class B, while dimmer stars are of spectral classes A and F.

Photographs reveal detail in the blue, wispy **reflection nebula** surrounding the brightest stars. The star Merope illuminates one of the most extensive reflection nebulae. In 2000 the **Hubble Space Telescope** imaged part of the cloud around Merope. This intricate patch of rippled gas and dust, called IC 349 or Barnard's Merope Nebula, has been shaped by Merope's radiation pressure. At only 0.06 light-years from Merope, IC 349 has been slowly moving closer to Merope at about 11 kilometers a second.

The Pleiades are also known as the Seven Sisters. The cluster's component stars were named after the Ancient Greek mythological figure Atlas and his daughters. Although only the six bright stars are visible to the unaided eye, the cluster probably holds as many as 100 stars.

Absorption or Dark Nebulae

Visible as dark regions along the **Milky Way,** dark nebulae are dense clouds of gas and dust that have no hot, young stars within to ionize the gas or nearby stars whose starlight they could reflect. Instead, dark nebulae appear as silhouettes against the backdrop of the Milky Way or bright **nebulae.**

The smallest dark nebulae are called Bok Globules after the Dutch-American astronomer Bartholomeus (Bart) Jan Bok, who first studied these objects in the 1930s. They are usually less than one **light-year** in diameter with a mass less than one solar mass.

Horsehead Nebula (Barnard 33)

Like an ethereal version of the knight chess piece, the dark profile of the Horsehead Nebula is one segment of Orion's giant **molecular cloud.** We can see the Horsehead because of its proximity to the **emission nebula** IC 434, near the star Alnitak (Zeta Orionis)—one of the three stars that mark the belt of Orion (the Hunter).

The Horsehead Nebula is roughly 1,500 **light-years** away, and its horizontal diameter, from the tip of its nose to the back of its mane, is just over two light-years across.

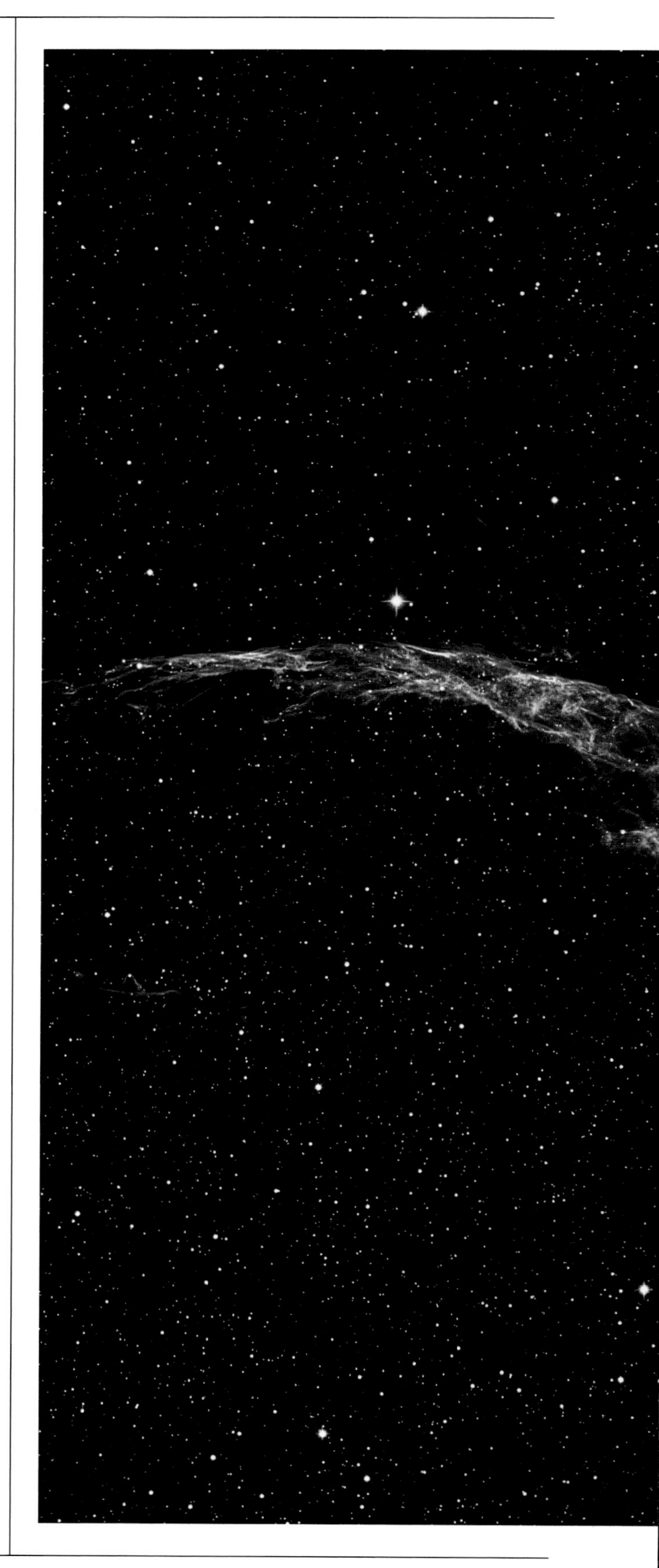

A glowing star known as 52 Cygnus seems to rest on a gaseous cloud called the Witch's Broom Nebula. More than 15,000 years ago, a supernova exploded, leaving a macramé-like nebula threaded with knots and filaments. The Witch's Broom Nebula trails into space at the western end of the larger Veil Nebula.

Milky Way

Milky Way

On a clear summer night you may spy a faint band of light stretching across the sky. The Milky Way is our home **spiral galaxy,** and it's a bit of a giant as **galaxies** go. Its mass is thought to be between 400 billion and one trillion solar masses, and its immense disk spans more than 100,000 **light-years.** Though broad, on average the Milky Way is quite thin—roughly 3,000 light-years thick. The disk is thickest (about 13,000 light-years) near the nucleus and thinnest at the outer edge (a few hundred light years). Our part of the galaxy is estimated to be roughly 1,000 light-years thick. The galaxy's rather thin disk has graceful spiral arms that curve outward from a small bar piercing the galaxy's center. The arms hold relatively young stars and giant **molecular clouds** that churn out these new suns.

A luminous pathway, the Milky Way—Earth's home galaxy—spills stars, dark dust, and glowing red gases into the evening sky. Yellow clouds at lower left reflect sunlight clinging to Earth at day's end.

Our small, yellow star is only one of at least 200 billion stars that make up the Milky Way. The **sun** is roughly 26,000 light-years from the center of the Milky Way and 14 light-years above the central plane of the disk. This position places us in a smaller spiral arm called the Orion Arm. The Orion Arm is located between two larger arms: the Perseus (Outer) and Sagittarius (Inner) Arms. There is at least one arm farther out than the Perseus Arm and at least two inside the Sagittarius Arm—the Scutum-Crux Arm and the Norma Arm. As the galaxy slowly spins, the sun and our **solar system** travel along with it at a speed of about 250 kilometers per second. The sun completes one revolution around the galactic center every 200 million years. The last time our solar system occupied its current spot in the Milky Way, **Earth** was in the Triassic Period, an age of reptiles and continental breakup.

Surrounding the galaxy's center is the so-called galactic bulge, a core of older, yellowish stars. The disk and bulge are encapsulated within a huge halo of old stars and globular star clusters that stretches some 300,000 light-years from the center. All of this structure may be floating within a vast halo of unseen dark matter, many times larger and more massive than all the visible parts of the Milky Way.

Because Earth is inside the Milky Way, our view of this structure is skewed by our position. Visually, we see

MILKY WAY	
Type of galaxy	Spiral
Total mass (including dark matter in billion solar masses)	~400
Disk diameter	~100,000 ly
Disk thickness (near nucleus)	13,000 ly
Disk thickness (near outer edges)	~1,000 ly
Number of stars	~200 billion
Age of oldest star clusters	~14 billion years
Distance of sun from nucleus	~26,000 ly

Solar mass: 1.99 x 10^{30} kg

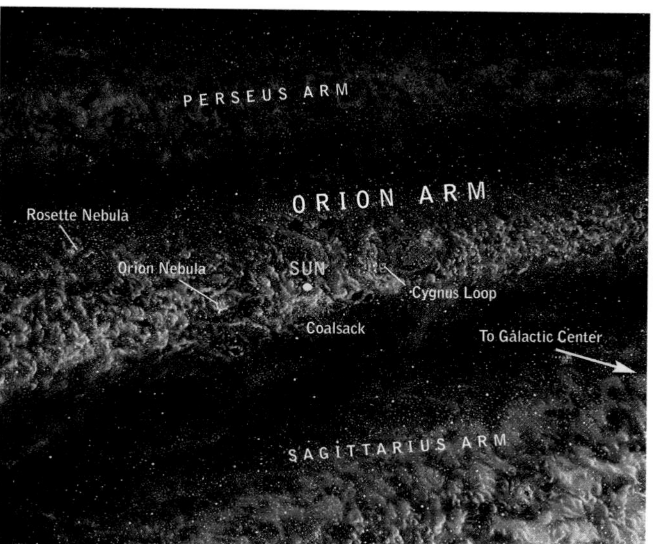

The white box on the first map outlines our little corner of the Milky Way—an area called the Orion Arm, shown in larger detail on the second map. The spiral arm, sandwiched between two larger arms, centers around our sun, one of at least 200 billion stars in the Milky Way.

the disk edge-on as the faint band of light that stretches across our sky. But if we look closer, we find clues to the Milky Way's spiral structure. By studying other spiral galaxies, we know that stars form along the inner portion of the spiral arms. Therefore bright, young O- and B-type stars are used as spiral tracers. The cool, hydrogen clouds in which stars form are another indicator. Radio **telescopes** can pick up the 21-centimeter wavelength signal these regions emit.

Roughly five degrees west-northwest of the star Gamma Sagittarii (in the **constellation** Sagittarius) is the most mysterious region of our galaxy—its nucleus. Gas and dust obscure our home spiral; its center is visible only by looking at it with radio and infrared telescopes. These reveal a region of densely packed stars moving at high velocity around a large center of mass. A massive **black hole,** some three million times the mass of the sun, may be at the Milky Way's heart.

The Milky Way's fluid, glowing arc across the night sky has inspired many different myths. To Native American Seminoles it is the path that leads good souls to heaven. In Norse legend it is the road to Valhalla, while Chinese and Arab myths represent the Milky Way as a river. The name "Milky Way" comes from the ancient Greeks, who saw this faint band of white light as a spray of milk or "gala," from which the word "galaxy" originated.

Interstellar Extinction

As starlight travels through the **galaxy,** dust between the stars absorbs and scatters the stars's radiation, making it appear dimmer. This phenomenon, interstellar extinction, is most significant toward the center of the galaxy. Stars in this direction lose about one magnitude of brightness for every 3,200 **light-years** the light must travel. Starlight appears reddened when observed through a dust cloud because scattering varies with wavelength—shorter (blue) wavelengths are scattered more than longer (red) wavelengths.

Open Clusters

Open clusters are loose or open collections of stars in the disk of the **galaxy.** They may hold ten to a few thousand stars. A typical open cluster has a radius of about ten **light-years.** Stars in open clusters have a common origin. Because stars form in large clouds or **nebulae,** they tend to form in groups or irregularly shaped clusters and are likely to be young to middle-aged. The oldest open clusters are only a few billion years old. The **Milky Way** holds roughly 1,200 known open clusters within its spiral arms, but this is a small fraction of the total—some estimates put the number as high as 100,000. One of the best-known open clusters is the collection known as the **Pleiades.**

Beehive Cluster (Praesepe, M44, NGC 2632)

This large **open cluster** in the **constellation** Cancer (the Crab) looks somewhat like a hive; its "bees" are more than 200 stars. Hipparchus cataloged the cluster as the "little cloud" in the second century B.C. The Beehive has an overall magnitude of 3.1 and can easily be seen with the unaided eye. Pliny, the Roman, wrote in the first century B.C. that the Beehive was used to predict the weather. High clouds, which precede bad weather, obscure the faint, fuzzy patch, while brighter stars in that part of the sky remain visible. The Beehive is roughly 500 **light-years** away and about 400 million years old.

Globular Clusters

Through a **telescope,** a globular cluster looks like a globe of densely packed stars. Each can hold tens of thousands to a million stars in an area only 10 to 300 light-years in diameter. Such a region is 1,000 times more densely packed than our part of the **Milky Way.** At a cluster's center, there may be a hundred stars jammed in one cubic **light-year.**

Stars in older globular clusters do not hold an abundance of heavy elements. Such heavy elements are formed when stars die. This tells us that those globular cluster stars are very old. They formed before there was a wealth of starbirth and **supernovas** to produce the heavy elements that have since enriched star-forming clouds. The oldest globular clusters probably formed 13 to 14 billion years ago.

All globular clusters were once thought to be remnants of the earliest generations of star formation in galaxies. In recent years the **Hubble Space Telescope** has provided us with views of globular cluster star formations. Globular clusters form in giant **molecular clouds** of cold hydrogen and dust. When two galaxies collide, the gas outside the cloud is ionized and begins squeezing the cold gas inside the cloud. Eventually high-density pockets inside the cloud begin to collapse under their own gravity to form stars. Like swarms of insects, globular clusters hover around a galaxy's center. The Milky Way has roughly 150 globular clusters in its halo. The majority of them can be seen in the constellations Sagittarius, Scorpius, and Ophiuchus.

Omega Centauri (NGC 5139)

Omega Centauri is the queen of **Milky Way globular clusters.** This majestic globe holds more than ten million stars packed into a span of 150 **light-years.** At five million solar masses, it is ten times more massive than other large globulars and nearly the same mass as the smallest galaxies. It lies some 1,500 light-years away and at magnitude 3.7 can be spotted with the unaided eye in the **constellation** Centaurus.

Edmond Halley first cataloged Omega Centauri during his 1677 journey to Saint Helena. He listed it as a "luminous spot or patch in Centaurus."

Magellanic Clouds

The Large and Small Magellanic Clouds (LMC, SMC) are companion **galaxies** to our own **Milky Way.** They are **irregular galaxies** that orbit our home spiral.

The LMC is 160,000 light-years away in the **constellations** Dorado (the Swordfish) and Mensa. The LMC has a diameter of roughly 20,000 **light-years** and a mass of 47 billion stars. It holds many noteworthy objects, including the Tarantula Nebula.

NASA's COBE satellite created this view of the spiral-shaped Milky Way galaxy with its central bulge—a core of older, yellowish stars. Normally, gas and dust obscure visible wavelengths of light from the galaxy's center, but COBE was equipped to detect infrared light from surrounding stars to produce this image.

This diffuse nebula is so bright that it was first cataloged as the star, 30 Doradus. Today, we know that the Tarantula is one of the most active starburst regions in the local **universe.**

Sanduleak, a blue supergiant star in the LMC, died in a spectacular **supernova** explosion whose light reached Earth in February 1987. Sanduleak was the first naked-eye supernova observed since **Johannes Kepler** spotted one in 1604.

The SMC is some 200,000 light-years away from **Earth** in the constellation Tucana (the Toucan). It has a diameter of about 9,000 light-years and may have a distorted, barred, disk-like structure deformed by gravitational forces of the Milky Way and LMC. The SMC lies just north of the **globular cluster** 47 Tucanae (NGC 104); that globular cluster lies only 13,000 light-years away and is 120 light-years across.

Intergalactic Medium

The matter between **galaxies,** the intergalactic medium, unlike the matter between stars, holds little dust but may hold gas. Between galaxies that are part of a rich cluster, observations reveal substantial amounts of highly ionized hydrogen and helium. Low concentrations of heavier elements, such as oxygen, formed during **supernova** explosions, are also present. These same elements are also visible between galactic clusters, but in much smaller concentrations.

Dark matter

Dark matter is unseen matter whose gravitational presence is indicated in systems such as **galaxies.** The movement of galaxies indicates that there must be more matter than we can see to keep them together. By watching a galaxy's rotation speed, astronomers can determine a galaxy's mass. They can also estimate what the rotation speed should be by calculating the amount of visible mass a galaxy has.

Astronomers have found galaxies moving twice as fast as they should. The material we see just is not enough to keep the fast-moving galaxies together. There must be some other mass contributing to overall gravity of galaxies or they would fly apart. This invisible or dark matter probably comprises at least 75 percent or more of the mass of the universe. A galaxy may have 10 to 100 times more dark matter than visible matter. Some of this missing mass may come from **brown dwarf stars** and **planets,** or it may be in the form of baryons (subatomic particles such as protons and neutrons). However, baryonic matter cannot account for all the missing mass, or we would find more helium in the **universe.**

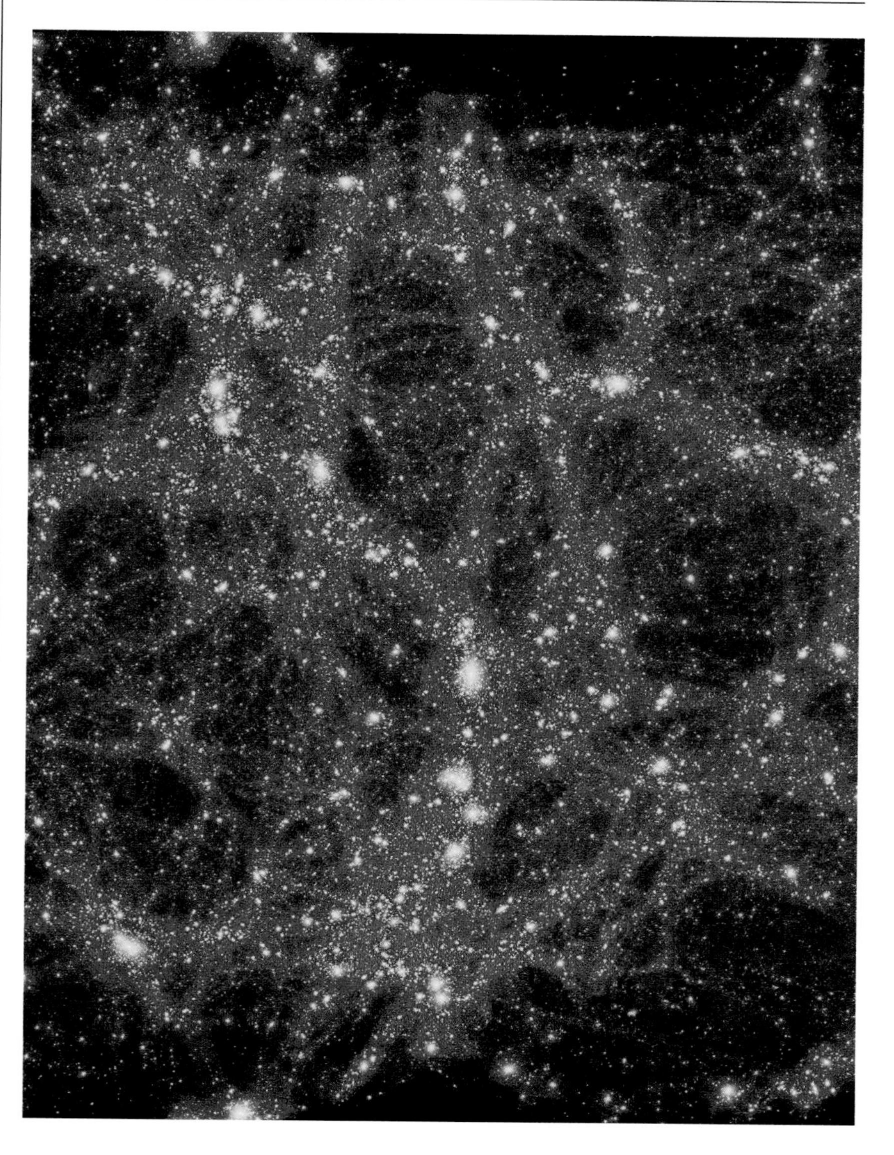

Cold dark matter—matter we cannot see because it emits little or no radiation—helps hold galaxies together. This hypothetical mass adds to the universe's gravity and keeps it from flying apart. Dark matter probably comprises almost 75 percent of the universe. The brighter colors in this simulation represent denser matter—stars, galaxies, and galaxy clusters.

One alternative is slow moving, exotic elementary particles called cold dark matter (CDM). WIMPs (Weakly Interacting Massive Particles) are a form of dark matter. They are heavy, hypothetical particles that rarely interact with other matter. WIMPs include such exotic particles as axions, hot dark matter, massive neutrinos, and photinos. Theoretically, WIMPs could have formed in the early universe. However, we have yet to discover any. MACHOs (Massive Compact Halo Objects) are another form of dark matter. They include objects like brown dwarfs, **white dwarfs, neutron stars, planets,** and **black holes.** These objects are presumed to be in the halos of galaxies, adding the missing mass or dark matter needed to explain a galaxy's rapid rotation.

Galaxies

Galaxy

A galaxy is an enormous aggregation of stars, gas, dust, and **dark matter** bound together by its own gravity. Galaxies vary in size, luminosity, and mass. The largest galaxies are a million times brighter than dwarf galaxies, the faintest known. Galaxies exist in three primary shapes, elliptical, spiral, and irregular. Each variety provides us with clues to how galaxies form and evolve.

In 1924 Edwin Hubble began taking photographic plates of bright "spiral nebulae." **Telescopes** of the time were unable to resolve individual stars within galaxies, and photographic plates were not sensitive enough to pick up individual stars. The **spiral galaxies** appeared fuzzy or cloudlike and were dubbed spiral nebulae. Hubble, making his observations with the recently completed 2.5-meter Hooker Telescope on Mount Wilson in California, not only detected individual stars, but **Cepheid variables** as well. He estimated the brightness of these Cepheids to be roughly magnitude 18. Using the period-to-luminosity relationship **Henrietta Leavitt** had developed 12 years earlier, Hubble determined that the "spiral nebulae" were far too distant to be inside the **Milky Way.** Instead, they were distant galaxies outside our home spiral. Hubble continued to study the fuzzy patches in the sky and discovered two other types of galaxies, elliptical and irregular. He developed a method of classifying galaxies by their shape and structure. The system identified three main shapes—elliptical, spiral, and barred spiral. Hubble arranged them in a tuning-fork diagram (opposite) that holds **elliptical galaxies** on the fork's handle. As ellipticals gradually develop spiral arms, they diverge to create the two forks of the diagram. The upper fork holds **spiral galaxies;** the lower holds barred spirals. **Irregular galaxies** were not included in Hubble's original diagram.Within each category, galaxies are classified further by their structure, concentration of dust, gas, and star formation.

Elliptical Galaxies

Elliptical galaxies appear round or elliptical. Edwin Hubble's classification system represents them by the letter E, with the numbers 0 through 7 denoting the degree of their ellipsoidal shape: E0 is spherical, E4 is moderately elliptical, E7 is the most elliptical. Elliptical galaxies have almost no visible internal structure

SELECTED GALAXIES

Galaxy	Constellation	Distance to Galaxy (in ly)*	Type	Apparent Magnitude	Year Discovered
Milky Way	n/a	n/a	Spiral	n/a	Prehistory
Large Magellanic Cloud	Dorado/Mensa	160,000	Irregular	+0.1	Prehistory
Small Magellanic Cloud	Tucana	200,000	Irregular	+2.3	Prehistory
Andromeda (M31, NGC 224)	Andromeda	2,500,000	Spiral	+3.4	905
And VIII	Andromeda	2,700,000	Spherical	+9.1	2003
M110	Andromeda	2,900,000	Spherical	+8.0	1773
Pinwheel (M33)	Triangulum	3,000,000	Spiral	+5.7	1654
NGC 3109	Hydra	4,500,000	Irregular	+10.4	1835
Whirlpool (M51, NGC 5194)	Canes Venatici	25,000,000	Spiral	+8.4	1773
Great Barred Spiral (NGC 1365)	Fornax	60,000,000	Spiral	+9.0	ca 1835
Antennae (NGC 4038 & 4039)	Corvus	60,000,000	Interacting	+10.5	ca 1790
Sombrero	Virgo	65,000,000	Spiral	+8.0	1781

**All distances are approximate.*

and lack **nebulae** and hot, bright stars. Because they hold almost no gas and dust, they lack the basic building blocks for star formation. As a result, elliptical galaxies hold older red giant stars, which give them a reddish hue.

The **sun** and its neighbors all revolve around the **Milky Way**'s nucleus. Unlike the Milky Way or **spiral galaxies,** elliptical galaxies have no defined axis of rotation. Stars in elliptical galaxies all follow individual orbits around the galaxy's center of mass.

M87 (NGC 4486, Virgo A)

Some 55 million **light-years** away, M87 is an enormous, **elliptical galaxy** at the heart of the Virgo cluster of galaxies. It is symmetrical and holds no distinct structure, so it is classified as type E1. French astronomer Charles Joseph Messier (1730-1817) first cataloged M87 on March 18, 1781. It was the 87th nebulous object he wrote about, hence its name: "M," for Messier, "87." Messier called M87 a "nebula without a star." Through his **telescope** the massive **galaxy** appeared as a bright, fuzzy ball with no external detail. Messier catalogued seven other "bright nebulae" in that observation; all are galaxies of the Coma-Virgo cluster.

M87 appears spherical and densely packed with stars. It has a diameter of roughly 120,000 light-years and a mass estimated at nearly three trillion times the mass of the **sun.** The galaxy is very active and bright. It has an **absolute magnitude** of about -22. Long-exposure photographic plates show that M87 has a very wide reach. Its outer regions extend more than half a million light-years. Long-exposure images also show the galaxy's outer regions to be highly elongated and no longer spherical. The outer layers are distorted by gravitational interactions with other galaxies in the cluster. This material is most likely from smaller galaxies that have been absorbed by massive M87.

M87 has two famous attributes. The first is the number of **globular clusters** that swarm around the galaxy. There are thousands of these round star clusters around M87. Some estimates place the number as high as 15,000. In contrast, the **Milky Way** has roughly 200 globular clusters in its galactic halo. M87 is also famous for its 4,000-light-year-long jet that spews energetic particles away from the core. **Hubble Space Telescope** observations of M87's core reveal a small disk of gas surrounding a starlike nucleus. Material in the disk is just 60 light-years from the core and orbits at 450 kilometers a second. The core of M87 probably holds a supermassive **black hole** roughly two to three billion times the mass of the sun. As material is drawn toward the black hole, it forms a rapidly spinning, flat disk (similar to the shape of water flowing down a drain). The rapidly moving charged particles in the disk create strong magnetic fields that are perpendicular to the axis of the rotating disk. The jet forms when high-energy particles get caught in these fields and are shot away from the disk at nearly the speed of light. M87's jet is ten million times brighter than the sun!

All the rapidly moving, high-energy particles make M87 a strong radio source called Virgo A. Radio observations of the jet show blobs and filaments that change slowly over time.

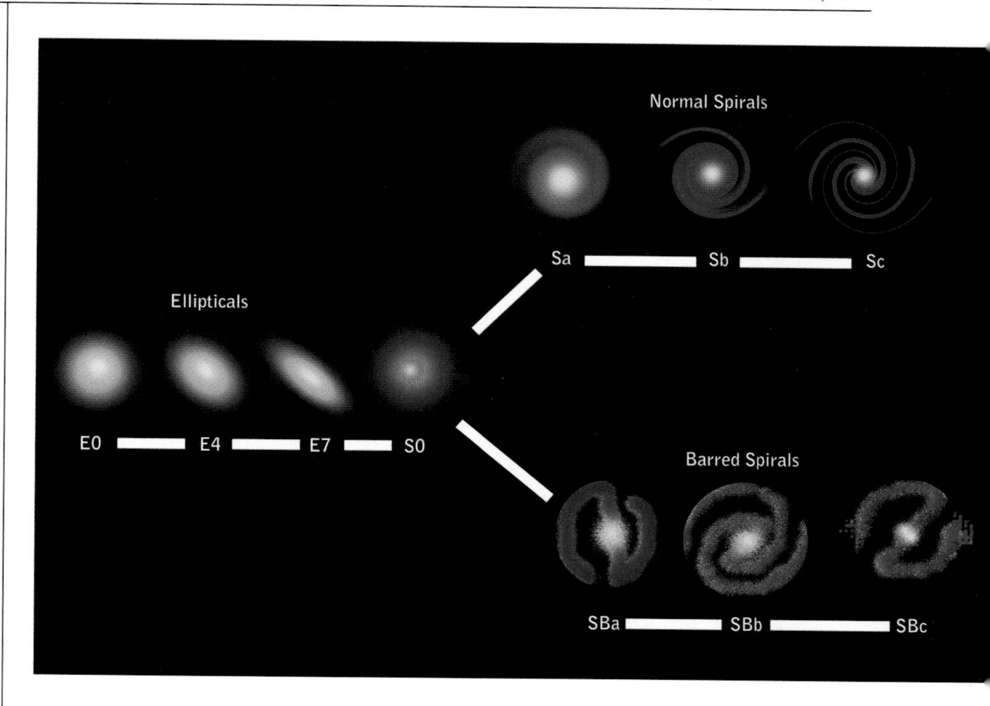

A forkful of galaxies, Edwin Hubble's "tuning-fork" diagram shows how these aggregations of stars, dust, and gas are classified by their shapes: spiral, elliptical, and barred spiral. The fork's handle holds different classes of elliptical galaxies, while the upper tine holds classes of spiral galaxies, and classes of barred spirals sit on the fork's lower tine.

Spiral Galaxies

Spiral galaxies, graceful stellar pinwheels, are some of the most stunning objects in the universe. There are two types of spiral galaxies, basic spirals (type S) and barred spirals (type SB). Both types are further subdivided into categories that relate to the brightness of the central bulge and the openness of their spiral arms. At one end are Sa galaxies. They have bright central bulges and a relatively closed spiral structure. At the other end are Sc spirals with their small central bulges and well-defined and open spiral arms. S0 spirals lie at the crossroads of Hubble's tuning-fork diagram, where ellipticals branch off into spiral and barred spirals. This type has a general spiral shape and central bulge, but lacks spiral arms. An S0 is also called a lenticular galaxy because its shape resembles a convex lens.

Barred spirals are classified in the same way as basic spirals. The difference is that barred spirals have an elongated nuclear bulge that looks like a bar of stars cutting through the **galaxy**'s nucleus. SBa spirals have a well-defined bar and spiral arms that begin curving outward from the ends of the bar. The other end of the spectrum holds SBc spirals,with just a hint of a bar and a very open spiral structure. About 20 percent of all spiral galaxies are barred spirals. The **Milky Way** is an SBc-type galaxy.

Spiral galaxies hold a mixture of old and young, hot and cool stars. They also hold gas and dust that promote star formation along their spiral arms. Dark dust lanes and bright **nebulae** lie along the inner curve of a spiral's arms. Spiral galaxies are generally fairly bright objects due to their hot, young O- and B-type stars.

Whirlpool Galaxy (M51, NGC 5194)

The Whirlpool galaxy is one of the most magnificent spiral galaxies in our sky. It is an 8.4-magnitude, Sc-type **spiral galaxy** some 25 million light-years away in the **constellation** Canes Venatici (the Hunting Dogs).

M51 is one of Charles Messier's original discoveries. He first spotted the galaxy on October 13, 1773, while observing a comet. In 1845 William Parsons (Third Earl of Rosse) made a detailed painting of M51, which was subsequently dubbed the Whirlpool. Lord Rosse observed the galaxy through a 1.8-meter **reflecting telescope** he built at Birr Castle, in Ireland.

The Whirlpool appears to be interacting with the small, **irregular galaxy** called NGC 5195 or M51B, which passed through massive M51 several million years ago. The encounter distorted smaller NGC 5195 from its original disk shape while it enhanced M51's spiral structure. Today, we see NGC 5195 linked to M51 by a long extended arm called a tidal bridge.

Irregular Galaxies

Like **spiral galaxies,** irregular galaxies hold a variety of star types and large clouds of gas and dust. But unlike spiral or **elliptical galaxies,** irregulars have no distinct structure, symmetry, or nucleus. Galaxies that don't fit into the other categories often end up in this classification. Irregular galaxies hold a fair number of ionized hydrogen (HII) regions, while the galaxy's stars tend to be hot, large O- and B-type stars.

M82

At magnitude 8.4, M82 is an easy object to see with a backyard **telescope.** This archetypical **irregular galaxy** is some 12 million light-years away in the **constellation** Ursa Major (the Great Bear).

Roughly 600 million years ago, M82 had a close encounter with its companion **spiral galaxy,** M81. The violent interaction lasted about a hundred million years and created more than a hundred bright, young, compact star clusters through M82's middle. These clusters are spherically shaped and hold up to one million stars each, leading astronomers to believe that they may be very young globular clusters. Before the encounter, M82 was probably a calm disk **galaxy.** Today, its explosive gas flow makes it a strong radio source (Ursa Major A). M82 is quite bright at infrared wavelengths, the brightest infrared galaxy in the sky.

Interacting Galaxies

Interacting galaxies are those caught in a gravitational embrace. The dance often results in galactic mergers or intense star formation. Such a gravitational waltz regularly distorts the component galaxies. A spiral's disk may be warped or its arms may be extended. Occasionally the gravitational dance creates long tidal tails of stars and gas that stream from one galaxy to the other, forming a bridge between the two—as is the case with **M51** and NGC 5195. Observations by the Infrared Astronomical Satellite (IRAS), a joint U.S.-Dutch-British satellite launched in 1983, found that many interacting galaxies are very bright in the far-infrared range of the spectrum. These IRAS galaxies are undergoing a period of star birth. They hold so many young, hot stars that many of them have luminosities as large as some **quasars.** Visually, most appear as intermingled pairs of **spiral galaxies** with extremely bright nuclei.

In some cases larger galaxies actually devour smaller ones. Supergiant **elliptical galaxies** have ten million times more mass than dwarf ellipticals. In close proximity, the supergiant elliptical easily has enough gravity to disrupt and rip apart the dwarf elliptical. The enormous galactic cannibal eventually pulls the pieces left behind into its own system.

The Antennae (NGC 4038 & 4039)

This merging pair of **interacting galaxies** does its gravitational dance some 60 million **light-years** away in the **constellation** Corvus (the Crow). Each **galaxy** has a bright nucleus and long curving tail of stars and gas ejected by the collision of the two. The system, also called the Ring-Tail, resembles the antennae of an insect. Through **ground-based telescopes,** the nuclei of the galaxies appear very bright without much detail.

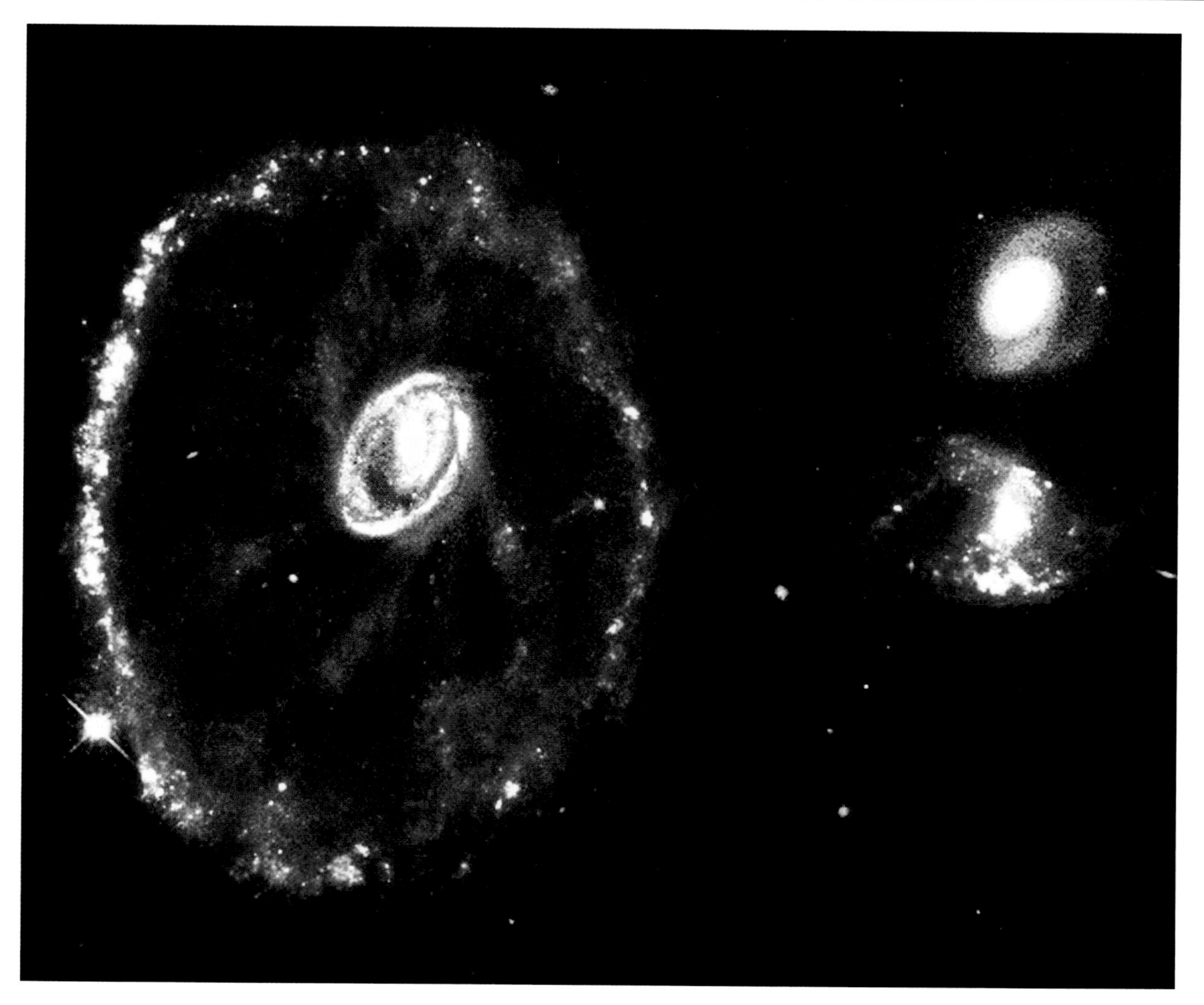

Survivor of a head-on collision with a neighbor, the Cartwheel galaxy shows evidence of the past crash in the cosmic wreckage of its outer ring. Like a rock tossed into water, the crash started a circular ripple of energy that plowed gas and dust before it.

Close inspection with the **Hubble Space Telescope** reveals that the collision has produced thick clouds of dust and rampant star formation. At least a thousand bright blue, young star clusters were uncovered.

The Cartwheel Galaxy

Once a normal **spiral galaxy** like our **Milky Way,** the Cartwheel galaxy underwent a head-on collision with one of the smaller galaxies nearby. Imagine a rock being tossed into a calm lake. As the rock enters the water, concentric ripples form and move outward. A similar process occurred when the smaller **galaxy** passed through the middle of the large spiral of the Cartwheel galaxy. The collision sent a great shock wave through the system. Moving at 322,000 kilometers an hour, the shock wave swept up gas and dust, creating a starburst ring around the remnants of the spiral's center. The ring holds several billion new stars and is 170,000 **light-years** across—the Milky Way would fit inside it. The Cartwheel is located some 500 million light-years away in the **constellation** Sculptor.

Radio Galaxies

Radio galaxies emit radiation at radio wavelengths one million times stronger than those of the **Milky Way.** Radio galaxies are the largest class of active galaxies and are typically **elliptical galaxies.** Very bright radio galaxies occasionally exhibit some of the high-energy characteristics of **quasars.** Some astronomers propose that they are the same type of object.

Extended radio galaxies have emissions that extend well past the outline we see in visible light. The emissions are concentrated in two huge radio lobes that can be as large as 200,000 **light-years.** Compact radio galaxies have radio emissions that are the same size or smaller than the galaxies' visible image. The first radio galaxy detected was Cygnus A. It emits a million times

Some 25 million light-years away, in the Hunting Dogs constellation, luminous young stars and glowing hydrogen swirl in the often-photographed Whirlpool galaxy. A spiral galaxy, this celestial beauty was first spotted in 1773 by French astronomer Charles Joseph Messier.

more radio energy than the Milky Way, yet it is about one billion light-years away from **Earth.**

Centaurus A (NGC 5128)

Centaurus A is one of the most unusual and dynamic **galaxies** known. A cross between a giant elliptical and a smaller disk spiral, Centaurus A is most likely the result of galactic cannibalism. In wide-field visible light images, Centaurus A shows a large, spherically symmetrical **elliptical galaxy** cut across its center by a dramatic disk-like dust lane—the possible remnants of a **spiral galaxy.** Hidden deep within the galaxy's center is a massive **black hole.** It has the size of a billion solar masses and takes up a region of space roughly the size of our **solar system.** The black hole may have grown to this size as it gobbled up material during the collision, or two massive black holes (one from each galaxy) may have merged.

Centaurus A is 11 million light-years away, making it the nearest **radio galaxy** to **Earth.**

Seyfert Galaxies

A Seyfert galaxy has a starlike nucleus that emits energy produced from processes not normally associated with stars and their evolution. Most of the energy is highly ionized and is characterized by strong, broad emission lines (bright points in a **spectrum**), which means the **galaxies** radiate enormous amounts of energy. These galaxies are called Type 1 Seyfert, and they act very similarly to **quasars.** Type 2 Seyferts have much narrower emission lines and are bright in the infrared wavelengths. Most Seyfert galaxies are spirals. They are named for astronomer Carl K. Seyfert, who first described them in 1943.

Hundreds of Seyfert galaxies have been identified, and roughly two percent of all **spiral galaxies** are Seyferts. They are three times more likely to be found as an interacting pair of galaxies than as a single galaxy alone. Roughly one quarter of Seyferts are oddly shaped. This suggests that Seyfert galaxies are the result of galactic collisions or gravitational interactions with their partners.

Seyfert galaxies can change their brightness by 50 percent in a period of just a few days to months. Astronomers believe these galaxies harbor supermassive **black holes** at their hearts. A sudden surge of material falling into a black hole would increase the galaxy's brightness rapidly. One of the most beautiful Seyfert galaxies is the face-on spiral NGC 7742, better known as the Fried Egg galaxy. Its yolk-yellow nucleus is roughly 3,000 **light-years** across. The Fried Egg is 72 million light-years away from **Earth** in the **constellation** Pegasus (the Winged Horse).

Quasars

Quasars are the most luminous kind of active galactic nuclei. They were discovered in the early 1960s and given the name "quasar" because of their starlike or quasi-stellar appearance. Photographs of these distant objects show no structure at all, only a bright, starlike point. Quasars are some of the most distant objects in the **universe,** yet they emit a tremendous amount of energy—typically 10,000 times the energy of an average **spiral galaxy.** To produce such large amounts of energy, a quasar must hold a **black hole** ten million to a billion times more massive than the **sun** in an area less than one **light-year** across.

Some quasars are remnants of the earliest days of the universe, when galaxies were just forming and still relatively close together. If some of these galaxies had supermassive black holes at their cores, then collisions would trigger quasar-like eruptions.

But if quasars were so common in the early universe, why don't we see them now? We do, if we assume that quasars are the cores of galaxies where large amounts of matter is flowing into a supermassive black hole. When the matter stops flowing in, the quasar settles down and we see a normal **galaxy.**

Galaxies such as superbright **radio galaxies** and **Seyfert galaxies** exhibit some of the same characteristics as quasars. The galaxies may have been even more active earlier in their lives.

Gravitational Lens

A gravitational lens is the bending of light by a gravitational body. As the light bends around the gravitational body (such as a **planet, galaxy,** or **black hole**), the image of the light source splits into multiple images that may ring the body. **Einstein**'s general theory of relativity predicted this phenomenon. The gravitational lens amplifies the light coming from the source, magnifying the image and making it appear brighter.

The first gravitational lens was discovered in 1979 when four **quasars** appeared in the shape of a cross on a photographic plate. This image of an "Einstein cross" actually showed one very distant quasar whose light had been distorted by a relatively nearby galaxy.

The **galaxy cluster** Abell 2218 is a stunning example of a gravitational lens. This rich cluster has a mass equivalent to 10,000 galaxies and at two billion **light-years** away it is relatively close by. The cluster's enormous gravity distorts distant galaxies into delicate arcs that seem to outline the gravitational lens. The galaxy cluster is so massive and compact that it produces multiple images of these distant galaxies.

Galaxy Clusters

Galaxies tend to occur in groups. The majority of them appear to occur in assemblages that hold two or three to as many as a few thousand members. Our own **galaxy** resides in a small, irregular community called the **Local Group.**

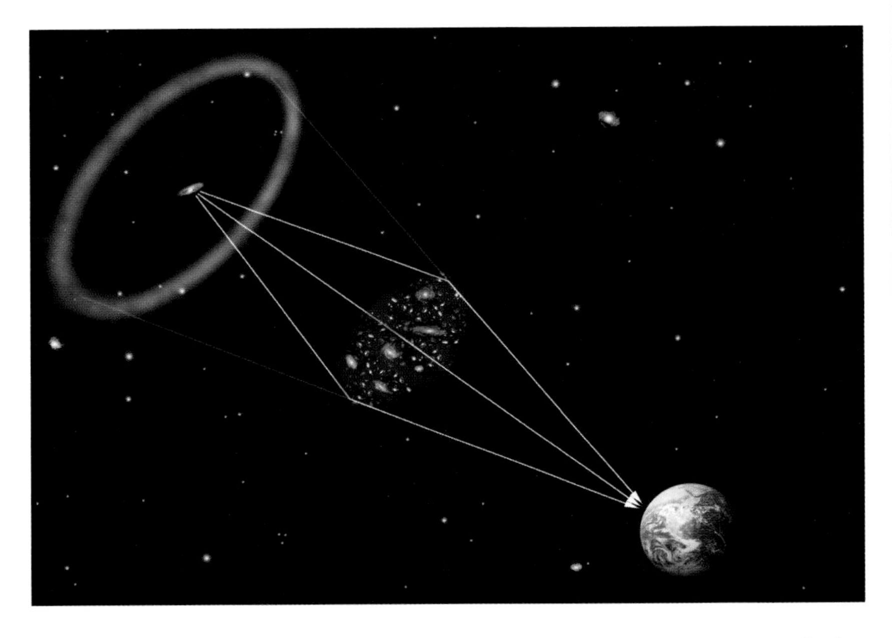

When light from a galaxy directly behind another hits the front galaxy or cluster, the gravity of the intervening cluster bends the light rays into multiple warped images. Viewed from Earth, the distant blue galaxy splits into images that appear at different places in the sky—a gravitational mirage.

In general, objects in the universe are spaced relatively far apart. **Planets** are spaced roughly 100,000 times their diameters and stars within a galaxy are spaced about one million times their diameters from one another. In a cluster of galaxies the spacing thins down to only about one hundred times a typical galaxy's diameter. Such massive objects, residing so close, sometimes invade each other's gravitational space. These **interacting galaxies** occasionally collide, merge, or become distorted.

The largest galaxy clusters are nearly spherical in shape and contain thousands of members that are typically massive elliptical and S0-type galaxies. The nearest large cluster to the Local Group is the Virgo cluster, some 55 million **light-years** away. This irregularly shaped, large grouping holds roughly 2,500 members, about 75 percent of which are spirals. The rest of the galaxies are ellipticals along with a few irregulars. The largest galaxy of the group is M87, a monster **elliptical galaxy** and strong radio source.

Some large clusters have an abnormally high number of galaxies at their centers. These so-called "rich clusters" typically hold about a thousand galaxies. The visible mass (stars, dust, and gas) of a rich cluster accounts for only about 10 percent of the cluster's total mass. The gravitational interactions between each galaxy with its cluster neighbors reveal that the cluster must hold more mass than we can see. At least 90 percent of a rich cluster's total mass must be dark matter.

Rich clusters are highly evolved groups that have had many interactions with one another. The intergalactic matter between individual members holds large amounts of highly ionized hydrogen and helium as well as heavier elements formed in **supernova** explosions. The galaxies in these clusters have interacted so often that gas has been stripped from the individual galaxies and pools between them. This enriched intergalactic medium is far different from the stuff between galaxies of irregular clusters. In irregular clusters most of the ionized gas is found in the galaxies themselves rather than between them.

One rich group is the Coma cluster. It holds more than 3,000 elliptical and S0-type galaxies roughly 300 million light-years away in the northeast corner of the **constellation** Coma Berenices. The cluster measures 20 million light-years across and possesses two equal clumps of galaxies, each centered by a large galaxy—one elliptical, and the other an S0. The Coma cluster is part of the Perseus-Pisces supercluster of galaxies.

The Perseus cluster is another rich cluster. It is 250 million light-years away in the direction of the constellation Perseus and is the brightest x-ray cluster we can see.

The Local Group

The **Milky Way** is part of a small group of galaxies called the Local Group. Our neighborhood holds three large spirals and more than 30 small galaxies in an area roughly three million light-years across. The three largest galaxies, in descending order, are the **Andromeda galaxy,** the Milky Way, and the Triangulum Spiral.

The smaller members of the Milky Way's group not only include the Large and Small Magellanic Clouds, but also the smaller and much closer Sagittarius dwarf elliptical galaxy (Sag DEG). Dwarf galaxies in the **constellations** of Ursa Minor, Draco, Fornax, Leo, Carina, Sextans, and Sculptor are also part of the group.

Recent studies speculate that the **Magellanic Clouds** may have once been part of the Andromeda galaxy. The Andromeda galaxy appears to have formed relatively close to the Milky Way, but it drifted away after a collision with a dwarf galaxy. This left the Magellanic Clouds and a few other dwarf galaxies to succumb to the Milky Way's gravitational pull. In the distant future, the Milky Way and the Andromeda galaxies may merge to form a giant **elliptical galaxy.**

The Local Group is part of a supercluster of galaxies centered on the Virgo cluster some 60 million light-years away. Our local supercluster is roughly 160 to 250 million light-years in diameter and appears to be disk-shaped, which may mean it is rotating. It is six times wider than it is thick, and 98 percent of the galaxies are confined to five percent of the volume. The Virgo cluster is at the center of our supercluster, and our own Local Group lies on its outskirts.

Andromeda Galaxy (M31, NGC 224)

At just 2.5 million light-years away, the Andromeda galaxy is the nearest large galaxy to the **Milky Way.** From a very dark-sky site, the galaxy is barely visible in the northern sky with the unaided eye. Through binoculars or a small **telescope,** Andromeda looks like an elongated, fuzzy cotton ball. Only long-exposure photographs taken through large telescopes reveal the galaxy's spiral structure.

Andromeda is an Sb-type **spiral galaxy** whose mass is slightly larger than that of the Milky Way. Andromeda has a large, bright nucleus that appears to have two components separated by five light-years. These characteristics may be the result of a collision with a dwarf galaxy some five to ten billion years ago.

Like the Milky Way, Andromeda holds a halo of **globular clusters.** But this halo is three times larger than the Milky Way's. The stars in Andromeda's globular clusters have more heavy elements than stars in the Milky Way's globular clusters, indicating that they are a bit younger than our globular cluster stars. Hovering about in the galaxy's gravitational field are at least four small elliptical companion galaxies. Two of them, designated M32 (NGC 221) and M110 (NGC 205), are the most prominent.

Large-scale Structures

Galaxies, galaxy clusters, and superclusters (large clusters of galaxy clusters) form intricate, spongelike structures, called large-scale structures, around enormous bubbles of empty space.

Once astronomers thought galaxies, galaxy clusters, and superclusters were scattered randomly across the **universe,** but in the 1980s astronomers Margaret Geller and John Huchra began to study the distribution of galaxies in three dimensions. The universe, in essence, appeared clumpy. Geller and Huchra found that galaxies are distributed in large-scale structures—long, narrow filaments that form thin walls around huge voids.

Most of the matter in the universe lies in the filaments or clumpy areas. But this is only 1 to 2 percent of the entire volume of space. Most of space is very empty.

One of these clumpy areas is some 300 million **light-years** away. It holds thousands of galaxies and spans an area of 250 by 700 million light-years but is only about 30 million light-years thick. Dubbed the "Great Wall," this formation is the largest known structure in the universe.

Astronomers have also found that our **Local Group** of galaxies is moving at roughly 600 kilometers a second toward a gigantic mass some 150 million light-years away. Called the "Great Attractor," this concentration of galaxies and dark matter is pulling all the galaxies in our part of the universe toward it.

15 NEAREST LOCAL GROUP GALAXIES

Galaxy	Distance (in ly)*	Diameter (in ly)*	Apparent Magnitude
Milky Way	0	100,000	–
Sagittarius	78,000	15,000	+7.7
Large Magellanic Cloud	160,000	20,000	+0.1
Small Magellanic Cloud	200,000	9,000	+2.3
Ursa Minor	225,000	1,000	+10.9
Draco	248,000	500	+9.9
Sculptor	250,000	1,000	+10.0
Carina	280,000	500	+20.9
Sextans	290,000	1,000	+10.3
Fornax	430,000	3,000	+8.1
Leo II	750,000	500	+11.5
Leo I	820,000	1,000	+10.4
Phoenix	1,270,000	1,000	+13.1
NGC 6822	1,750,000	8,000	+9.3
And II	1,910,000	2,000	+13.5

**All distances are approximate.*

Cosmic Mirages

J. Anthony Tyson

THE UNIVERSE IS CONNECTED ON ITS LARGEST AND SMALLEST SCALES. IN the earliest moments of the universe, tiny dark matter particles were created amongst temperatures and energies far higher than any imaginable on Earth. Today, that legacy is detectable in its cumulative gravitational effects on large-scale structures in the universe. The worlds of particle physics and astronomy are coming together in a transformed world view. Copernicus displaced Earth from a central position, and Harlow Shapley and Edwin Hubble removed our galaxy from any special location in space. Now, even the notion that the galaxies and stars comprise most of our universe is being abandoned. Emerging is a universe largely governed by particles of dark matter and, we are beginning to think, by an even stranger dominance of a smoothly distributed and pervasive dark energy.

How can cosmic mirages reveal "images" of dark matter? Imagine a room covered—walls, ceiling, and floor—with wallpaper. The wallpaper is decorated with a reliably repeating pattern of very close dots. Now imagine a magnifying glass suspended between you and one wall, a glass so clean and flawless the glass itself is invisible to you. Yet as you look in its direction, you'll notice that the pattern of dots behind it is distorted, as the light bends through the glass. By observing the pattern of distortion, you can get a very good idea of the size and shape of the magnifying glass, even though the glass itself is invisible to you. Similarly, the light bent by a clump of dark matter can help astronomers determine the shape of that clump, and map it.

To search for the "cosmic wallpaper," we have to look deep enough into the background universe so that there are thousands of galaxies projected near the foreground lens. And that begins with what astronomers have long known: If you build a bigger telescope and develop faster, more sensitive detectors of light, you will see deeper into the universe. There were always more galaxies to be seen and tricks to be learned to see them. Five decades ago, photographic emulsions typically recorded only about one quantum of light for every hundred hitting them. By the mid-1970s engineers at such places as Eastman Kodak, in collaboration with astronomers, had pushed the sensitivity of photographic plates to record-high levels, so that at least one of every 20 photons hitting a plate was recorded. Still only 5 percent efficient, this development was a great boost. In fact, it helped Richard Kron and me find the first hints of the cosmic wallpaper in the late 1970s.

Finding Radio Galaxies

As is often the case in research, the discovery of the wallpaper came as a by-product of unrelated projects. Working as a physicist at Bell Labs, I was naturally interested in pushing technology. I was also attracted to problems associated with seeing the faintest objects in the universe. I knew that so-called "radio galaxies" have microwave "hot spots" or "lobes" on either side of the parent galaxy, and that one peculiarity of this radio energy is that it is not caused by a thermal source; it is caused by electrons accelerated to very high velocities. Radio galaxies were known to emit "synchrotron radiation," at radio wavelengths when energetic electrons slammed into magnetic fields. But what caused the acceleration of the electrons in the first place?

To find out, Phil Crane, Bill Saslaw, and I applied for telescope time at the Kitt Peak National Observatory in Arizona, and eventually secured several observing runs on the Mayall four-meter reflector, the largest telescope at the observatory. We started the project by mounting a highly sensitive photomultiplier, an instrument used to detect and amplify light from faint sources, at the focus of the telescope, and scanned across the places where the radio lobes were. There were problems with using this technique because the instrument's photoelectric system and the Earth's atmosphere both changed from minute to minute. Occasionally, we saw evidence of a faint glow from the radio lobes, but there were too many other effects that we could not control. So we turned back to photography, but this time with a twist. First, with the help of Art Hoag at Kitt Peak, I learned how to "push" the efficiency of special photographic emulsions to the limit and exposed them at the focus of the Mayall Telescope for one hour. I took multiple exposures of several of the radio galaxies, each time using a special 20-by-25-centimeter Kodak plate. These plates were so sensitive that they had to be developed with a custom process at the end of each night. Of course, each plate covered an area of the sky far larger than we required for our project—and this turned out to be important.

Next, we converted the images on the plates to digital data by scanning each plate with a light beam. We used special software for faint light sources to bring out detail. This process produced "noisy" images peppered with false "galaxies" caused by chemical impurities in the emulsion. Astronomers at that time believed this effect set the faintness limit for galaxy detection. But we wanted to go fainter. Could we? And if so, how? Arizona's version of a monsoon provided an opportunity to find out.

Distant Galaxies

With our project half finished—with only a hint of light from those elusive radio lobes—I flew back to Tucson for another multiple-night run on the four-meter telescope. Then the rainstorm hit. Unable to use the telescope on Kitt Peak, I spent the week in the basement of the headquarters building in Tucson with our collection of plates. Frustrated by the lack of progress and with nothing else to occupy my time, I decided to scrutinize the "false" galaxies on various plates covering the same region of the sky. Most were clearly chemical events in the emulsion, accompanied by occasional pieces of dust and lint that did not appear on more than one plate. However, some of the fainter dots were present on multiple plates covering the same area of the sky. These barely perceptible black smudges on the digitally stretched negatives appeared at exactly the same sky coordinates each time. I first tried counting these "common" smudges, but soon realized that my learning curve was biasing the counts. In an attempt to get to even fainter light levels, I used a simple program to digitize images covering a common portion of the sky, and then added them together, generating a deeper image. At its faint limit, this deeper image revealed even more.

Obviously I was seeing out to great distances through a forest of galaxies spread over a huge volume of the universe. There were enough galaxies in every little patch of sky (about 6,000 in an area the size of the full moon) that it seemed possible that, within any narrow view of a foreground galaxy, you'd also get chance projected images of galaxies beyond them.

I was so absorbed in thought I was nearly hit by a car while walking back from the Kitt Peak Headquarters on the University of Arizona's Tucson campus. "What if each foreground galaxy had a large mass?" I wondered. Light rays from background galaxies projected within a few arc seconds would be bent by this mass. Each background galaxy would be moved to a new place on the sky, systematically warping its image. This might lead to a new and powerful means to statistically weigh the mass of a galaxy!

The blue, loop-shaped objects encircling galaxy cluster CL0024+1654 are actually multiple distorted images of a single blue galaxy lying behind the cluster. The gravity from this cluster along with its dark matter act as a gravitational lens.

It would take two decades and the development of a new imaging technology to fully exploit the cosmic mirage of distant galaxies, but later in the 1970s, Richard Kron discovered a hint of things to come. While studying the faint galaxies, he found that the galaxies at the faint limit of photographic plates were systematically bluer than brighter galaxies. The fainter and more numerous blue galaxies would eventually become useful tools for exploring mass. ■

The Universe

Cosmic Distance Ladder

Astronomers use overlapping indicators produced by several different methods—a cosmic distance ladder—to measure distances in space. For objects within our solar system, radar or laser ranging are used. Trigonometric **parallax** is used to determine distances out to about 1,600 **light-years** (500 **parsecs**). More distant measurements are made indirectly by calculating the difference in a star's intrinsic brightness compared to its apparent brightness. Distances in objects even farther away, such as **galaxies,** can be determined by the size of ionized hydrogen (H II) regions and the brightness of **globular clusters,** and **Type Ia supernovas.**

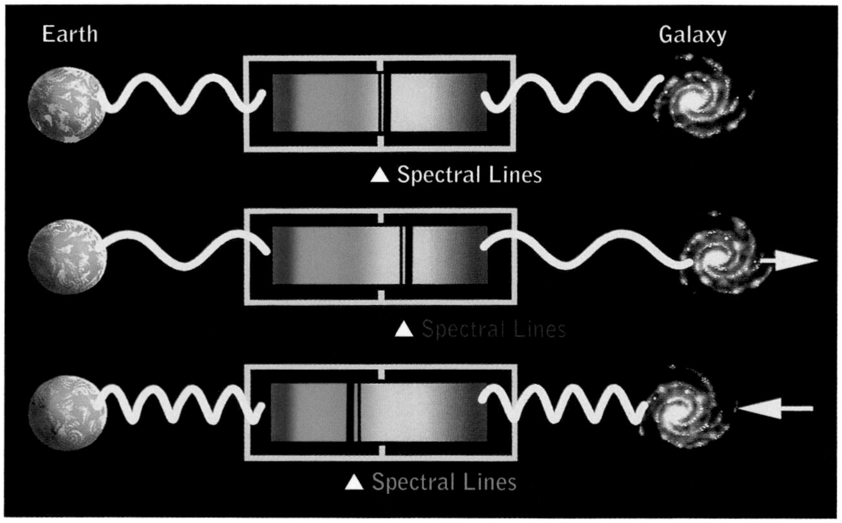

Which way is a galaxy moving? Scientists use changes in wavelengths of radiation to find out. The top illustration indicates that Earth and the galaxy are keeping a constant distance. In the middle, the galaxy is moving away from Earth (redshift); at the bottom, the galaxy is moving toward it (blueshift).

Redshift

Redshift is the displacement of electromagnetic radiation toward the longer wavelength red end of the **spectrum.** It is caused by the expansion of the **universe** and the curvature of **space-time.** As you sit at a railroad crossing waiting for an oncoming train, you'll find that the train's whistle has a higher pitch as it approaches and lower pitch as it passes by. The whistle is sending out a steady tone whose sound waves move out in a circle from the whistle. The change in pitch arises from the train's motion. As the train moves, the center of each new sound wave moves with it. The whistle's sound waves bunch up in the train's direction of motion (higher frequency) and spread out behind it (lower frequency). This phenomenon is called the Doppler effect after Austrian physicist Christiaan Doppler (1803-1853), who first described it in 1842. Six years later, French physicist Hippolyte Fizeau (1819-1896) suggested that light waves would behave the same way, and he turned out to be correct.

Objects within our **galaxy** may be moving toward or away from us. As an astronomical object moves away from us its spectral lines appear to shift toward the longer wavelength/red end of the spectrum (redshift). If it is approaching, its spectrum is shifted toward the shorter wavelength/blue end of the spectrum (blueshift).

Cosmological redshifts relate to the expansion of the universe. As the fabric of space-time expands, it carries all matter with it. Everything appears to be moving away from everything else, and the farther away a distant galaxy is, the greater its redshift, and thus the faster it appears to be moving away.

Gravitational redshift, also called Einstein shift, occurs when radiation must escape out of a gravitational field. **Albert Einstein** described the process in his general theory of relativity. As radiation escapes a gravitational force, it loses energy and its frequency decreases, moving it toward the red end of the spectrum. Such shifts are extremely small but have been measured for several **white dwarfs** as well as the **sun** and the **Earth.**

Light-year

A light-year is the distance that light travels in one year. Light travels at 300,000 kilometers a second, so it goes a distance of roughly nine trillion kilometers in a year.

Parallax

Parallax is the apparent change in position of an object due to a change in location of the observer. Hold your index finger straight up in front of your nose, then close your left eye, then your right. Your finger's position appears to move relative to objects that are much farther away. If you measure the angle of apparent movement of your finger, that is its parallax. In the same way, relatively nearby stars appear to shift as **Earth** makes its annual trip around the **sun.** The parallax of a nearby star can be used to determine its stellar distance, which is measured from the sun.

To calculate stellar distances, astronomers employ

trigonometric parallax. This is a method that uses observations made from either end of a known baseline: the distance between Earth's opposite orbital positions around the sun, which is two **astronomical units** (2 x 149 million km).

Imagine a large triangle formed by a star and Earth's orbital positions in January and in June. The star's parallax is half of its apparent angular displacement from the two Earth positions. Distance to the star from the sun is derived by dividing the distance between the Earth and the sun (1 AU) by the parallax angle.

Because most stars are too far away to measure their stellar parallax, spectroscopic parallax is used instead. Spectroscopic parallax estimates distance by comparing a star's spectral class (apparent magnitude) with its intrinsic brightness, or absolute magnitude.

This method depends on the **Hertzsprung-Russell (H-R) diagram.** Once we have a star's spectra, we can determine its spectral type and luminosity class. From the H-R diagram we can look up the star's intrinsic brightness and compare that to how bright it appears. The difference in the star's luminosity will tell us how far away it is.

Parsec

This unit of distance normally applies to objects well beyond the **solar system.** One parsec is the distance at which an object would have a parallax of one arc second, which is roughly equal to 31 trillion kilometers (30.86×10^{12} km) or 3.26 light-years.

Cepheid Variables

The time it takes a **Cepheid variable star** to brighten and dim and brighten again is directly related to its average brightness: the longer the period, the brighter the star. Because the star's period is easy to observe, we can figure out its intrinsic brightness (**absolute magnitude**). When we compare this to how bright the star appears in our sky, we can calculate its distance.

Hubble's Law

First proposed by astronomer **Edwin Hubble** in 1929, Hubble's law describes the relationship between distant **galaxies** and their movement away from us. The law states that the velocity at which a galaxy recedes is equal to its distance times the rate at which the universe is expanding. This expansion rate is called the Hubble constant.

The search for an accurate value of the Hubble constant has been one of the primary missions of the **Hubble Space Telescope.**

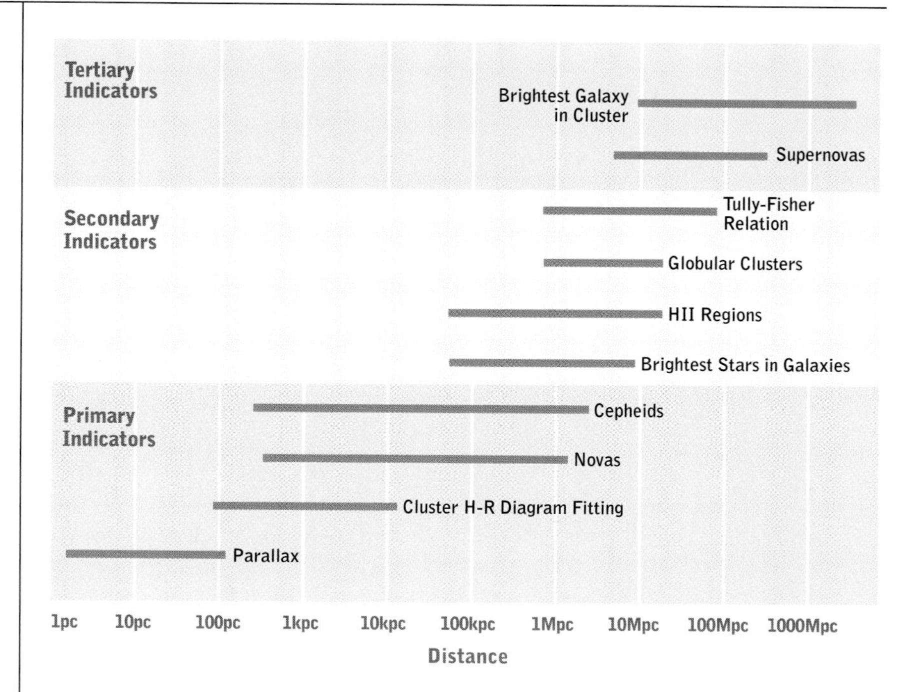

Overlapping indicators—a cosmological distance ladder—help astronomers gauge the distance of a galaxy. For closer galaxies, one way to calculate distance is to measure the time it takes a Cepheid variable star to brighten, dim, and brighten again. For more remote galaxies, scientists can assess the brightness of their Type Ia supernovas, or dwarf white stars.

Type Ia Supernovas

A Type Ia supernova occurs when a **white dwarf** star gathers enough matter from a companion star to begin an unrestrained gravitational collapse that results in a **supernova.** Because these white dwarfs should collapse and explode at the same mass limit, their explosions must all have the same brightness. Type Ia supernovas can be used to determine the distance to their host galaxies: The brighter the Type Ia, the closer the **galaxy.**

Expansion of the Universe

The idea that the universe is expanding stems from observations **Edwin Hubble** made in the 1920s. Hubble was the first to resolve individual stars in **galaxies,** proving that they were outside our own galaxy, the **Milky Way.** To learn more about their composition, Hubble began taking their spectra. When he broke a galaxy's light into its individual wavelengths, he found that everything was shifted toward the red end of the **spectrum.** This meant the galaxies were moving away from us. He also found that the farther away a galaxy was, the greater its **redshift.** Because the galaxy's spectrum is shifted toward the longer-wavelength/lower-frequency red end of the spectrum, it is moving away. The greater the redshift, the faster galaxies are receding. Everything is moving away from everything else. Hubble found this velocity-distance relationship in 1929 after taking the spectra of more than 20 galaxies.

Cosmological Assumptions

Cosmology is the study of the very nature and evolution of the universe. As we contemplate how the universe may have formed, we must make a few assumptions to fill gaps in our knowledge of the nature of the cosmos. First we assume that matter is uniformly distributed throughout the cosmos. Called homogeneity, this notion doesn't appear to hold true at first glance. A quick look through a pair of binoculars or a **telescope** finds lumps of matter in the form of stars, **planets,** and **galaxies.** But this distribution is localized and on a small scale. Even though a planetary system or galaxy is very large to us, it is only a small component in the larger scale of the universe. In fact, this **large-scale structure** finds galaxies spread evenly throughout space. Such even distribution means we can ignore individual galaxies and think of matter evenly spread across the cosmos.

DISTANCE INDICATORS

Object	Method of Determining Distance	Basis of Calibration
Nearby stars	Trigonometric parallax	Radar determination of an astronomical unit (AU)
Open clusters	Main-sequence fitting	Trigonometric parallax, moving clusters
A through M stars	Spectroscopic parallax	Trigonometric parallax, galactic cluster Color-Magnitude (C-M) diagram
O and B stars	Spectroscopic parallax	Galactic cluster C-M diagram, statistical parallaxes
Supergiant stars	Spectroscopic parallax	Galactic cluster C-M diagram
RR Lyrae stars	Period from light curve	Statistical parallaxes, globular clusters
Classic Cepheid variable stars	Period-luminosity relationship	Statistical parallaxes, galactic cluster C-M diagram
W Virginis stars (type II Cepheids)	Period-luminosity relationship	Statistical parallaxes, galactic cluster C-M diagram
Globular clusters	Integrated magnitude	RR Lyrae stars, C-M diagram
Novas	Maximum light	Expansion rate of shell
Ionized hydrogen regions	Angular size	Nearby galaxies
Supernovas	Maximum light	Nearby galaxies
Brightest galaxies in galaxy clusters	Integrated magnitude	Nearby galaxies
Galaxies	Hubble constant of redshift	Doppler shifts of nearest clusters of galaxies

The second assumption, called isotropy, says the universe looks the same in all directions. Again, this is not true on a small scale where we see individual galaxies and **galaxy clusters.** However, on the largest scales we see the same number of galaxies in all directions.

Next we must assume that the laws of physics we experience on **Earth** are the same everywhere else in the universe. This assumption is called universality. The farther out in space an object is, the farther back in time you are seeing it. It takes the light from a distant object time to reach Earth, so we don't see the objects in real time. The **sun**'s light takes just over eight minutes to reach Earth. Most galaxies are millions of **light-years** away, so it takes millions of years for their light to reach us. If the laws of physics change with time, there may be strange effects that occur when we look at these distant objects. Universality allows us to explain what we see using the laws of physics we know.

Finally, the assumptions of homogeneity and isotropy lead us to the cosmological principle. This fundamental rule says that observers in other galaxies see exactly the same thing that we see. Changes over time in a galaxy's evolution or a star's life cycle aren't included in the cosmological principle. As the universe expands, observers living in different galaxies at different times would see these objects at different stages in their evolution. Observers living at different times would have to correct for these differences.

Space-time

We live in three dimensions of space (up-down, left-right, forward-backward). In 1915 **Einstein** proposed that time was the fourth dimension—as you move through the three dimensions of space you also move through time. From that belief arises the four-dimensional continuum called space-time.

Einstein showed that gravity would bend the fabric of space-time. It is easier to visualize this concept in only two spatial dimensions. Imagine space-time as a giant rubber sheet stretched over a frame. Now place a bowling ball on the sheet. The sheet will bend around the bowling ball to accommodate it just as space-time bends around a massive object such as a star, **galaxy,** or **black hole.** The greater the mass, the more curved or warped space-time becomes.

Big Bang Theory

The big bang theory is often portrayed as the definitive model for the formation of the universe. In truth, the

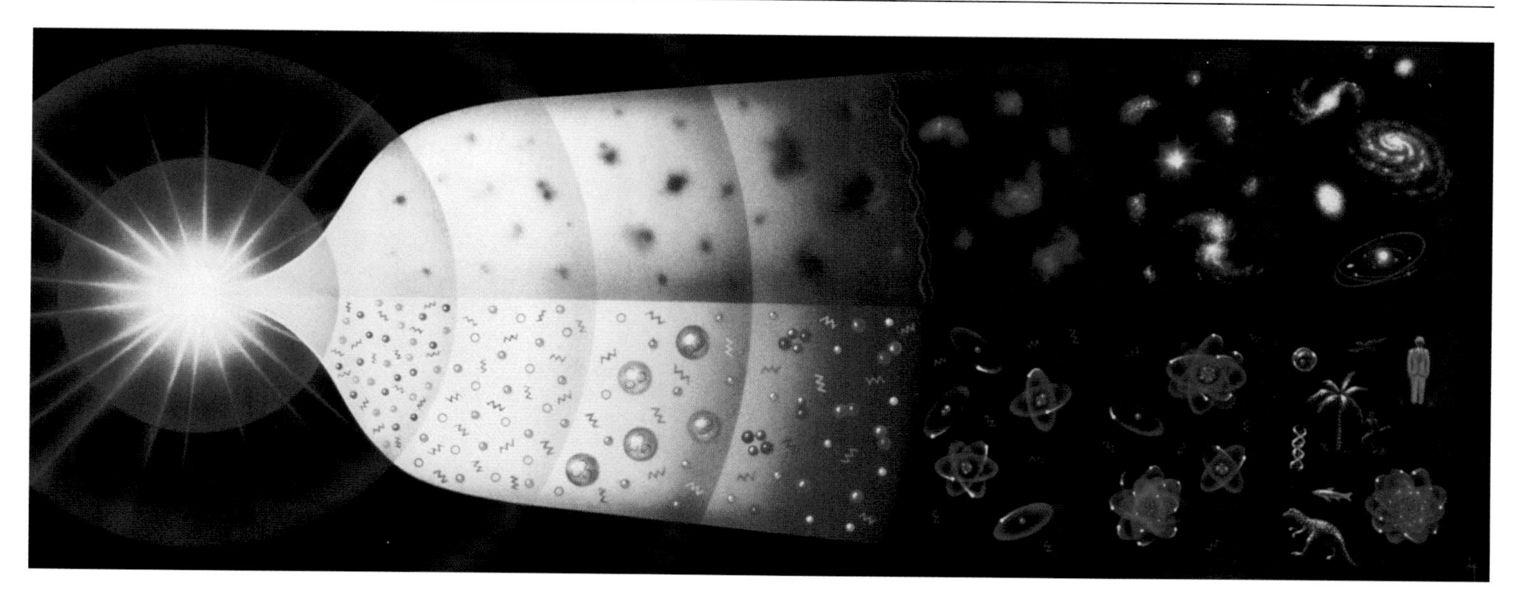

What happened after the big bang? The illustration shows what might have followed: At first, high temperatures preclude anything but particles and radiation (far left). Some 300,000 years later, atoms form (center). Some 500 million years after the bang, nuclear fusion reactions create stars and heavy elements (far right), the basic building blocks of life.

big bang is an assumption based on what scientists observe in the universe. It is a theory with deficiencies that have yet to be fully explained. In fact, the name big bang stems from a sarcastic comment made by astronomer Fred Hoyle in the 1950s. It was his way of conveying his doubts about the theory.

The big bang is thought to originate in a singularity. Singularities are theoretical points in space where density, the force of gravity, and the curvature of **space-time** are all infinite. The force of gravity in a singularity is so great that the fabric of space-time curves in on itself such that singularities have no radius or size. The laws of physics do not exist in a singularity. It is a complex, theoretical object whose behavior is best described mathematically.

Even though singularities are theoretical, there is some observational evidence for their existence. A singularity is thought to be the driving engine at the heart of a **black hole.**

Because we do not understand the physics inside a singularity, cosmologists do not reconstruct the history of the universe from time zero. Instead they speculate on what the universe was like a few ten-millionths of a second later. At this time, the universe was filled with high-energy photons with temperatures over one trillion kelvins and densities close to an atomic nucleus. These photons were gamma rays with very short wavelengths and very high energies.

As the universe expanded, wavelengths of the gamma rays grew longer, which lowered their energy, cooling the universe. As this ferment of hot gas and radiation continued to cool, atomic particles and subsequently atomic nuclei formed. The protons, neutrons, and electrons that form the structure of matter in our universe were formed during the first four seconds of the birth of the universe. At about 30 minutes, nuclear reactions had stopped and roughly one-quarter of the mass of the universe was helium with the other 75 percent hydrogen. This is the same hydrogen-helium ratio the oldest stars have today.

The universe was dominated by radiation for the first million years. At roughly the 300,000-year mark, the universe had cooled enough for nuclei and electrons to combine, forming neutral atoms. As free electrons were being snapped up by atomic nuclei, the universe became transparent and radiation was free to travel throughout the cosmos. Photons from this time are still spotted today as cosmic background radiation.

The name big bang and the idea that it originated in a singularity make it easy to imagine an explosion that happened at the "center" of the universe or at some point "over there," and then everything expanded to fill the space around it. That would be wrong. The theory involves the entire universe. It was a rapid expansion or inflation of all space and time in one instant, and it isn't over yet because the universe is still expanding.

Another way to imagine this expansion is by picturing a loaf of raisin bread. Think of the dough as the fabric of space-time. As the dough begins to rise, the raisins (**galaxies**) all appear to move away from each other in all directions.

Cosmic Background Radiation

The **big bang theory** predicted a residual background radiation from the early universe that should exist today. In 1965 two engineers with Bell Labs were

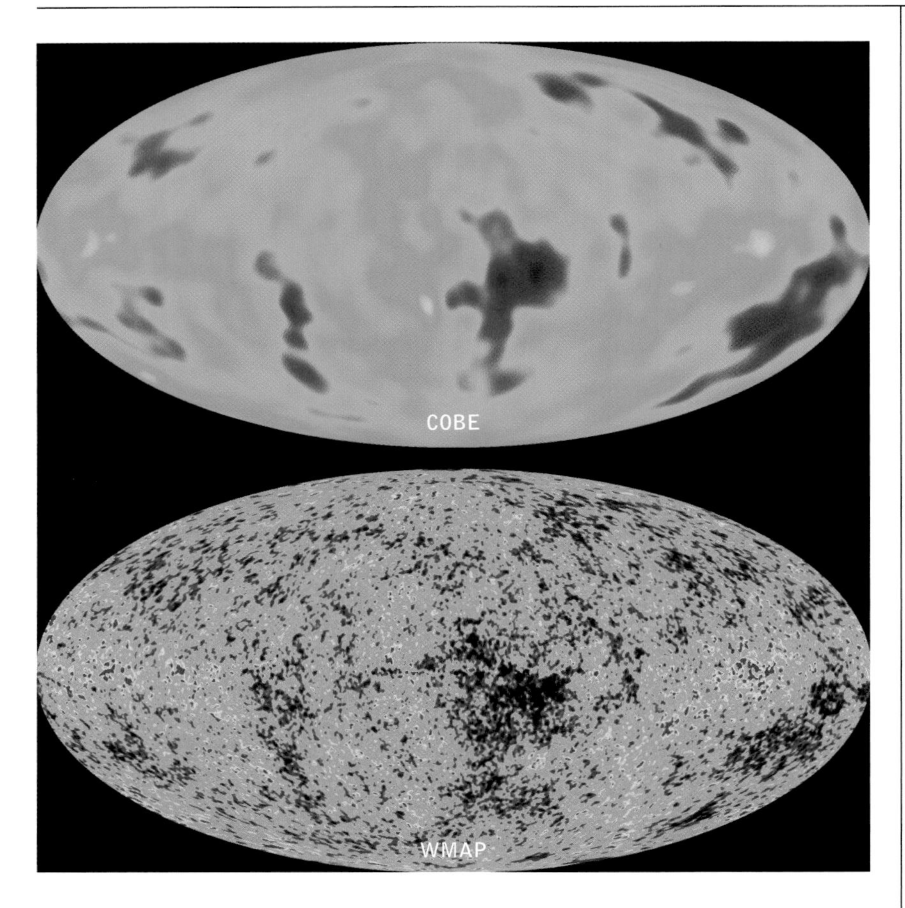

"Baby pictures" of the universe, these two different images reflect improvements in scientific tools. In 1992, COBE detected tiny temperature fluctuations, shown as color variations, in the infant cosmos. The image created by WMAP in 2003 brings increased focus and more details.

looking for the source of static that was interfering with satellite communications. What they found was a consistent signal emanating from every point in the sky at the wavelength predicted for this background radiation. In 1989 **NASA** launched a **satellite** (**Cosmic Background Explorer,** COBE) that mapped this radiation over the entire sky. By 1992 COBE had found tiny differences in the background radiation. These fluctuations were no more than 30-millionths of a degree warmer or cooler than the average temperature. These structures existed around the time when the first galaxies were forming.

Antimatter

During the very early universe, gamma ray photons had enough energy to decay and convert their energy into particle pairs of matter and antimatter. Antimatter particles are opposite counterparts to normal particles. Normal electrons have a negative charge. Antielectrons, called positrons, are electrons with a positive charge and a spin opposite that of electrons. When a normal particle and an antiparticle meet, they annihilate each other, converting their mass into energy or gamma rays.

Cosmic Strings

Cosmic strings are theoretical linelike imperfections in **space-time.** They are infinitely thin but very massive. Cosmic strings can take the shape of an infinite line or a closed loop. These regions of high mass are thought to have encouraged **galaxy** formation. Their high mass and subsequently large gravitational force would have gathered matter together, forming infant galaxies.

The Geometry of Space-time

Einstein's general theory of relativity proposed that gravity could bend or curve **space-time.** We experience this curvature as gravity. Einstein's theory predicts that the entire universe has a general curvature and that the geometric form that space-time is curved into directly relates to the fate of the universe.

There are three possible geometric shapes the universe might have. The first has a positive curvature and is called a closed universe. This configuration would look spherical and would have a finite volume with no edge. You could travel in a straight line and eventually end up where you started because the closed universe is finite. If the universe is closed and it holds enough mass, eventually the expansion would stop. It would then begin to gravitationally contract, ultimately shrinking down to a singularity, in what is called the big crunch (the opposite of the **big bang**). Some theories speculate that the universe oscillates between these expansions and contractions in a process dubbed the big bounce.

The second possible geometric shape of the universe has a negative curvature. Called an open universe, it appears saddle shaped. This type of universe would be infinite and gravity would be too weak to halt the expansion.

The final geometric shape is a zero curvature, flat universe. Here too the universe is infinite. The expansion would continue forever, but a bit slower than it would for an open universe.

Gravity Waves

In his general theory of relativity, **Einstein** predicted that the force of gravity exerts extremely weak, wavelike disturbances called gravity waves that propagate at the speed of light. Similar to waves in the ocean, gravity waves are ripples in the fabric of **space-time.** They are generated when some type of mass is accelerated, oscillated, or violently disturbed. Gravity waves are one of the weakest forces of nature and have not been observed directly. Only those produced by highly massive objects could possibly be detected. The most likely sources of gravity waves include close binary star systems that hold massive **neutron stars, black holes,** and **supernovas.**

Grand Unified Theory

The four primary forces of nature are the force that binds atomic nuclei (strong nuclear force), the force that holds atoms together (electromagnetic force), the force that controls radioactive decay (weak nuclear force), and gravity. The grand unified theory attempts to bring together the first three of the four forces.

In 1873 James Clerk Maxwell published his *Treatise on Electricity and Magnetism.* In it he suggested that electricity and magnetism were the same force. In 1967 the Weinberg-Salam-Glashow theory predicted that the weak nuclear force was combined with the electromagnetic force at high energies into an electroweak force. This electroweak force was confirmed in 1983 when massive elementary particles called W and Z bosons were detected. Today's theorists speculate that the electroweak force and the strong nuclear force could also have been combined at even higher energies during the early universe.

Age of the Universe

The age of the universe can be estimated by chemical elements, the age of the oldest star clusters, the age of the oldest **white dwarf stars,** and the expansion rate.

The lower limit of the age of the universe can be determined by studying the oldest globular star clusters. Observations suggest that the oldest **globular clusters** are between 13 and 14 billion years old. But because it is difficult to estimate the exact distance to globular clusters, it is hard to determine their intrinsic brightness and therefore their age.

Estimates for the upper limit relate to the universe's expansion rate or Hubble constant. The calculation is simple. We know the distance to a **galaxy** by its **redshift,** which also gives us the velocity at which it is receding. All we need to do is divide distance by velocity to find out how long the universe has been expanding. Estimates using this method put the age at roughly 15 billion years old.

In March 2004 the Hubble Space Telescope Science Institute released images of the most distant galaxies known. The image, called the Hubble Ultra Deep Field (HUDF), may reveal galaxies just 400 million years after the **big bang.** By studying such images we may further refine our estimates of the age of the universe.

Accelerating Universe

In the very first moments following the birth of the universe, the fabric of **space-time** expanded exponentially. In a mere fraction of a second the universe grew by a factor of 10^{50} (that's a 1 with 50 zeros after it!).

As time passed, the force of gravity began slowing the expansion down. With all the visible matter and **dark matter** in the universe, you would think that gravity is still applying the brakes.

In the mid-1990s, two separate research teams (the Supernova Cosmology Project and the High-Z Supernova Search Team) set out to measure this deceleration due to gravity. Both teams used very distant **Type Ia supernovas** to measure the expansion rate, and what they found was startling. The supernovas appear farther away than they should be. The universe is speeding up, not slowing down!

Long before Earth, the sun, and the Milky Way, the first stars may have burst into the universe in a celestial fireworks display, as depicted in this artist's rendering. Views from the Hubble Space Telescope hint that the universe made most of its stars in a torrential firestorm just a few hundred million years after the big bang.

Dark Energy

If the expansion of the universe is accelerating, some type of antigravity force has to be driving it apart. Dark energy radiating from deep space provides that repulsive force.

Albert Einstein first proposed this antigravity force in his general theory of relativity. Called the

DETECTING THE SIGNATURE OF THE BIG BANG

Robert W. Wilson

IN SCIENCE—JUST AS IN LIFE—THE PEOPLE WHO STUMBLE UPON DISCOVERIES ARE NOT always the people who are looking to make them. And that helps explain how two scientists working for the phone company discovered the first hard evidence of the big bang. Arno Penzias and I teamed up to make the Bell Labs six-meter-wide horn reflector useful for astronomical observations. With its set of electronic amplifiers and receivers, especially the ruby traveling wave maser, the reflector was potentially the most sensitive instrument in the world. Its well-shielded design also made it one of the most discriminating of antennas, able to sort out subtle differences in radio noise sources.

Arno was already well along in converting this equipment for astronomical measurement when I arrived at Bell Labs. He knew that the first step was to calibrate the receiver's temperature scale, and for that purpose he really overdid the job! He built what is known as a "cold load"—basically an extremely cold, and therefore low-noise, calibration source. It was a cylindrical nest of waveguides, gas baffles, nitrogen pre-coolers, and gaseous and liquid helium containers and absorbers. All this was wrapped in a "dewar"—essentially a highly efficient, well-insulated thermos bottle. We poured some 25 liters of liquid helium in it and calculated the radiation temperature at the top to be approximately 5 kelvins. That's an extremely low temperature, close to absolute zero, more than 268 degrees below the freezing point of water on the Celsius temperature scale.

Arno handled most of the cold load preparation while I set up our radiometer, the device that actually measures the temperature of things, like an extremely sensitive electronic thermometer does. Radio telescopes actually measure the temperature of radio sources. The sky has a temperature, the moon has a temperature, and so does anything that has any heat at all, even if that heat is what we would describe as "cold." The radio spectrum is simply an extension of the spectrum of visible light that we are familiar with. The electromagnetic spectrum, as it is known to scientists, has high-energy gamma rays and x-rays at one end, the visible region as a narrow wedge in the middle, and the microwave and radio spectrum stretching out at the low-energy end of the spectrum. Every physical body in the universe that has any temperature at all radiates some energy across this spectrum. The coolest objects will be "brightest" in the low-energy range—the radio and microwave. Radio astronomers and communications engineers call these thermal sources noise because they are incoherent and cause fluctuations (such as snow in a TV set) even if their average value is subtracted out. Our job was to account for all the noise sources so that their average value could be subtracted from our measurements of the horn to give the temperature of the cosmos beyond the Earth.

There were indications all along that sources of radio radiation were not being accounted for. After the six-meter horn was built and was being used with the Echo satellite, Ed Ohm, who was a very careful experimenter, added up all the components of the system and compared that figure to his measured total. He had predicted a total system temperature of 18.9 kelvins, but he found that he consistently measured 22.2 degrees, some 3.3 kelvins more than he had expected. However, that was within the measurement errors of his instrumentation, so he didn't consider the discrepancy to be significant.

I distinctly remember that, as Arno and I put the helium-cooled reference source on the six-meter reflector and got everything working, we knew we were going to be either happy or sad with the result. The antenna's measurement of the reference source's temperature would either be within what we considered an acceptable range—based on our calculations of the "noise" it would also register—or it wouldn't.

Our first observations were disappointing. We had hoped that the discrepancies could be explained by the known limits of our instrumentation. Our observing technique, based on one developed by Robert Dicke, was to compare the antenna's temperature to that of the the cold load. When the antenna was pointed straight up, the radiation temperature was about 7.5 kelvins. The problem was that we had expected 2.3 kelvins from the sky, and possibly 1 kelvin from the absorption in the walls of the antenna.We saw something that was considerably more than that—the antenna's reading was hotter than the helium in the cold load and it should have been colder. Clearly we had a problem, and it lay either in the antenna or beyond.

We knew that unless we could get to the bottom of this mystery we would not be able to do an experiment I had wanted to do since my days at Caltech: determine whether the galaxy was surrounded by a halo of radiation. Arno's immediate reaction was, "Well, I built a pretty good cold load, since imperfections in it would have the opposite effect," so we looked for other sources of error.

Maybe it was the Earth's atmosphere.
Maybe it was New York City.
Maybe it was the Milky Way.
Maybe it was many discrete sources.
Maybe it was the pigeons.
Maybe it was the Van Allen radiation belts.

We lived with this problem for about a year while understanding our system and completing an accurate measurement of the flux of Cassiopeia A. If we could not understand our system at our first frequency of 4 gigahertz where a galactic halo should be very weak, we had no hope of measuring it at 1.4 gigahertz where we hoped to detect it.

We were really scratching our heads about what to do until one day when Arno was talking on the phone with Bernie Burke about other matters, he happened to mention our results and our inability to determine what was going on.

Burke told Arno that a group at Princeton was working on something that might explain what we were seeing. Robert Dicke, the leader of Princeton's physics group, who had developed a terrific switching device for microwave radiometry and was now interested in cosmology, had made a fascinating prediction. He suggested that if we live in an oscillating universe—one that expands and then contracts back to another super-hot big bang—it will, at the time of each big bang, cleanse itself of all heavy elements and begin the expansion over again, repeating this cycle throughout eternity. One outcome of this cyclic behavior is that in each cycle the universe will eventually relax into a state of thermal equilibrium. The expansion rate of the universe, and its subsequent cooling, causes the thermal spectrum to be Doppler-shifted from the highest gamma-ray range all the way to the other end of the spectrum—into the microwave and radio range. Although this "reddening" will stretch out the spectrum, it will still look like a thermal spectrum.

Dicke realized that this radiation might be visible in the microwave range. His experience with radiometers had taught him that the thermal radiation left over from the big bang was something that he could look for, and the understanding that this was possible made an enormous contribution to astronomy. Prediction or not, this notion gave Dicke's graduate students a good reason to build a sensitive radiometer, which they were well along with doing in late 1964, just when we were struggling with our own radiation-detection equipment. There had been other predictions of the thermal signature of the big bang, and even some obscure astronomical measurements, but these were unknown to us, and to Dicke's group.

Arno Penzias (left) and Robert W. Wilson (right) stand in front of Bell Labs' horn-shaped antenna in Holmdel, New Jersey. Using the antenna in 1964 they inadvertently discovered remnants of cosmic microwave background radiation, effectively validating the big bang theory.

When Burke told Arno what the Princeton group was up to, Arno called Dicke. This led to visits back and forth, and eventually to our realization of what it was we had discovered—the telltale radiation from the big bang. Still, we made one final check. We took a signal generator, attached it to a small horn, and took it around the top of Crawford Hill to artificially increase the temperature of the ground and measure the characteristics of our antenna sensitivity in the backfield. We did not find anything unexpected. Only then did we send a notice to a journal to formally tell the world what we had observed. ■

cosmological constant, this repulsive force was a mathematical fix that helped to balance the universe against its own gravity. General relativity predicted that the universe must either expand or contract, but Einstein thought the universe was static. Adding this antigravity cosmological constant brought balance to the equation.

When **Edwin Hubble** discovered the expansion of the universe, Einstein got rid of the cosmological constant. He regretted adding the term to his original equation, thinking of it as his "greatest scientific blunder." Now it seems he may have been right after all.

Recent observations by the **Hubble Space Telescope** and the Wilkinson Microwave Anisotropy Probe (WMAP) have found that the bulk of the universe is made of dark energy. Current estimates of the mass/energy budget place dark energy at roughly 70 percent of the universe, while visible matter and **dark matter** make up less than 30 percent. Most of the universe is made of something we know virtually nothing about.

Is there life on other planets? Seeking an answer to that question, Stanley Miller (above) looked at the processes that could have led to life on Earth. In a closed container, Dr. Miller and a colleague mixed materials found in Earth's earliest atmosphere. The resulting "soup" held the building blocks of DNA.

The nature of dark energy is important to the fate of the universe. If dark energy is stable, the universe will continue to expand and accelerate forever. If dark energy is unstable, the universe could ultimately come apart. Dubbed the "big rip," this doomsday scenario has the universe accelerating to speeds that rip apart the fabric of **space-time** to a point were even atoms are torn apart. On the flip side, if dark energy is dynamic it could gradually decelerate and turn over to become an attractive force that contracts the universe into a "big crunch" implosion.

Life in the Universe

The study of the possibility of life elsewhere in the universe is called exobiology, astrobiology, or bioastronomy. Scientists studying the question look at conditions on Earth-like planets such as **Mars** and **Jupiter**'s moon **Europa.**

One of the first experiments aimed at discovering the processes that could have led to life on **Earth** was done in 1953. Stanley Miller and Harold Urey tried to reproduce the conditions on Earth at the time life is thought to have first begun. In a closed and sterilized glass container, they circulated gases (hydrogen, ammonia, and methane) of Earth's primitive atmosphere through water that represented Earth's oceans. To simulate lightning bolts, electric arcs were passed through the gases.

They let the experiment run for roughly one week, then analyzed the material in the flask. What Miller and Urey found was a primordial soup that held fatty acids, urea, and four amino acids, the building blocks of DNA. Amino acids can link together to form basic proteins. The experiment has also been done with hot silica to simulate hot lava flowing into the oceans and ultraviolet radiation to simulate the **sun**'s radiation upon an Earth with very little atmospheric protection. These experiments gave similar results.

In 1960 astronomer Frank Drake tried to detect signals from extraterrestrials with the 26-meter Green Bank radio antenna. Dubbed Project Ozyma after the Wizard of Oz, it was unsuccessful and abandoned after a few months.

More recently, NASA's Microwave Observing Program targeted roughly a thousand nearby stars with specially designed receivers. These receivers are being used in conjunction with radio telescopes at Arecibo, Parkes, and Green Bank. The highly sensitive equipment splits frequencies into two billion channels, each one hertz wide. Software then searches each channel for patterns that could be signals from intelligent life.

SETI Institute

The SETI (Search for Extraterrestrial Intelligence) Institute was founded in 1984. It is a private, nonprofit scientific research and educational organization dedicated to the exploration of the origin, nature, and prevalence of life in the universe.

The mission of the institute includes projects related to astronomy and planetary sciences, chemical evolution, the origins of life, biological evolution, cultural evolution, and, of course, the search for extraterrestrial intelligence.

Telescopes

Telescope

Although ancient Egyptians made glass around 3,000 B.C., and ancient Greeks knew the basics of refraction and reflection by about 300 B.C., the first telescope did not come into existence until roughly 1608, when a Dutch eyeglass maker, Hans Lippershey (1570-1619), applied for a patent for his spyglass. Lippershey's spyglass made distant objects appear closer. It was the first well-documented **refracting telescope.**

Early telescopes suffered from severe irregularities that distorted the objects viewed through them. One such irregularity is related to an object's color and is called chromatic aberration. Longer wavelength red light is refracted less than shorter wavelength blue light. As light made of many different wavelengths (white light) passes through a **lens,** each individual wavelength comes into focus at different distances from the lens. Chromatic aberrations can blur the image or make the object appear to have a false color.

To combat some of the problems, Niccolò Zucchi (1586-1670)—a mathematician in Rome—suggested using a mirror instead of a lens to focus light. In 1616 he created the first **reflecting telescope** but had poor results with it. Several others, including **Johannes Kepler,** Robert Hooke, and René Descartes, tried to produce workable instruments, but it was **Sir Isaac Newton** who first built a usable reflecting telescope in 1668.

The third type of telescope is a catadioptric. These telescopes combine properties of both refractors and reflectors. Catadioptric telescopes have a large lens or corrector plate at the front end of the telescope's tube. Light passes through the corrector plate toward the primary mirror at the back end of the tube. The primary mirror reflects the light back to the front of the telescope, where it is focused onto a convex-shaped secondary mirror. The secondary mirror sends the light to the back end of the telescope and out to an eyepiece through a hole in the center of the secondary mirror.

A telescope's light-gathering power depends on its aperture (the diameter of its primary lens or mirror). The larger the aperture, the dimmer the object it can see. The focal length is the distance from a lens or mirror to the point where light converged by it comes to a focus. Large catadioptrics are more portable and manageable than large reflecting telescopes or refractors because catadioptrics have the long focal length a large mirror would require but in a short, manageable space. This makes catadioptrics a good choice for backyard astronomers; the most common type is a Schmidt-Cassegrain telescope.

Refracting Telescope

A refracting telescope uses a series of **lenses** to bend or refract light to a focus forming an image. When most people imagine a telescope they picture the long, thin tube of a refracting telescope—the type of telescope first used by **Galileo Galilei** in 1609. Galileo's first telescope made objects appear roughly 30 times closer than they appeared with the unaided eye.

The simplest refractor has two lenses: an objective lens and an eyepiece. However, most refractors hold several lenses. The objective lens is at one end of the telescope, and it determines the light-gathering ability of the telescope. The larger the objective, the larger the telescope's aperture or opening and the more light it can gather. The eyepiece is at the other end, and it brings the image into focus. The objective is usually made up of several lenses. It is large and has a long focal length. Having a multi-lens objective allows several wavelengths of light to be focused on the same point, reducing chromatic aberrations.

The largest refractor in the world is at Yerkes Observatory in Williams Bay, Wisconsin. This 102-centimeter telescope was installed in 1897. The observatory was built after astronomer George Ellery Hale (1868-1938) convinced Chicago magnate Charles Tyson Yerkes (1837-1905) to fund the project. Dutch-American astronomer Gerard Peter Kuiper (1905-1973) used the Yerkes refractor to discover the Uranian **moon Miranda** in 1948 and **Neptune**'s moon Nereid a year later.

Lenses

There are two basic types of lenses, converging and diverging. Converging lenses, sometimes called convex lenses, are thicker in the middle than at the edges. These lenses bend light passing through the lens to a point or focus. A magnifying glass is an example of a converging lens.

Diverging lenses are thicker at the edges than at the center. They appear bowl-shaped or concave and are called concave lenses. Light entering a concave lens diverges or spreads out. The light appears to come from a point behind the lens.

The Hale Prime Focus Spectrograph

David DeVorkin

The Prime Focus Spectrograph of the five-meter Hale Telescope from the observatory on Mount Palomar in Southern California was for 40 years one of the most powerful tools of observational cosmology. When this spectrograph was put into operation in 1951 on the Hale Telescope, it created an unbeatable combination for the study of cosmologically interesting questions because, together, the two devices could image the spectra of faintest known galaxies. Using the spectrograms the Prime Focus Spectrograph produced, astronomers measured the expansion rate of the universe with a precision never before possible.

The spectrograph had to be handled in total darkness and cold, so it was sturdy, smooth-edged, and massive enough to respond slowly to temperature changes, like the presence of body heat. It was also very compact and could fit into the confined space of the observing cage at the top of the telescope tube. In practice, the spectrograph sat flatly on a column in the observer's cage. Light from the Palomar mirror would enter from below and strike a small polished slit on the bottom of the black metal housing. Some of the light would reflect off the jaws of the slit and be imaged by an eyepiece that the astronomer would use to make sure the instrument was properly lined up. The light that passed through the slit was then reflected onto a concave mirror, which collimated the beam and sent it to a diffraction grating, which in turn acted like a prism to break up the light into a spectrum. That spectrum was then photographed by a very fast "Schmidt" camera loaded with tiny glass plates of photographic emulsion.

It took many hours, sometimes many nights, to collect enough light from an extremely faint galaxy to record its spectrum. To record the spectrum of a galaxy, the light collected on the entire surface of the five-meter mirror was concentrated onto a tiny photographic plate inside the camera that was less than an inch across. The resulting spectrum was even smaller—a third of that width. Lenses made of diamond and sapphire were added to the cameras to keep the spectrum in focus across the entire field.

This spectrograph played a critical role in deciphering a wholly new class of extragalactic object. In the 1950s, as radio telescopes grew more sensitive and capable of pinpointing discrete sources of radio energy, dozens and then hundreds of sources remained unidentified by optical means. The Hale Telescope was called to action, and, by the 1960s, astronomers such as Maarten Schmidt and Jesse Greenstein were finding that the optical smudges that seemed to be linked to the radio noise had radial velocities, or redshifts, far greater than any normal galaxies. Schmidt and his colleagues found what are now called "quasi-stellar radio sources," "quasars," or simply "QSOs." Now known to be an extremely energetic (or luminous) form of galaxy, QSOs have greatly extended our baseline for determining the character of the Hubble constant. ■

Using evidence from this Prime Focus Spectrograph in conjunction with the Hale Telescope, astrophysicists Jesse Greenstein and Maarten Schmidt in the 1960s helped to establish the conception of quasars as extremely distant starlike bodies that give off intense amounts of energy.

To reduce aberrations in optical instruments such as telescopes, designers use several lenses with different refractive properties. The simplest refracting telescope may have four or five separate glass lenses.

Reflecting Telescopes

Reflecting telescopes collect and focus light using mirrors instead of **lenses. Sir Isaac Newton** first presented a working version of a reflecting telescope to the Royal Society in London in 1668. Reflectors avoid chromatic aberration by reflecting light rather than transmitting it through a lens. Reflectors can be made of material, lighter and thinner than glass, that is then coated with a reflective material. Early mirrors were made of polished metal. Late 19th-century reflectors had mirrors made of an alloy of copper and tin. French physicist Jean Bernard Léon Foucault (1819-1868) made the first glass mirror coated with fine layers of silver. Today's astronomical mirrors are generally made of glass coated with thin layers of aluminum.

By far the most common reflector design is a Newtonian telescope. Newtonians use a parabolic mirror to reflect incoming light to a point called the prime focus. A smaller, flat secondary mirror reflects the light at a 90-degree angle out through a hole in the side of the telescope's tube. Attached to this hole is an **eyepiece** that brings the light into focus for your eye.

Over the years Newton's design has been modified to create reflecting telescopes that have shorter focal lengths and therefore shorter, more manageable tubes. The Cassegrain telescope, developed in 1672, has a large primary mirror with a hole in its center. Incoming light is reflected to a point at the telescope's opening, where a convex-shaped secondary mirror reflects it back and out through the hole to an eyepiece at the back of the telescope. This configuration allows the telescope to have a long focal length in a relatively short tube. In 1930 Bernhard Voldemar Schmidt (1879-1935) added a thin lens to the front end of the tube. This corrector plate helps eliminate aberrations caused by the mirror. Schmidt-Cassegrain telescopes are popular types of backyard telescopes.

Eyepieces

The eyepiece—the lens or combination of lenses at the observer end of the **telescope**—magnifies and focuses an image to the point where an observer views it. There are many types of eyepieces, each with its own lens configuration.

One important aspect of an eyepiece is its exit pupil. The exit pupil is the diameter of the beam of light that leaves the eyepiece and reaches the observer's eye. The eyepiece's exit pupil should be smaller than the pupil of the observer's dilated eye. Otherwise the eye cannot take in all the light the eyepiece is transmitting. On average, a fully dilated human pupil spans seven millimeters, but it gets smaller as we age.

The exit pupil is calculated by dividing the focal length of the eyepiece by the focal ratio of the telescope, which is the telescope's focal length divided by its aperture. A telescope's magnification is calculated by dividing the telescope's focal length by the eyepiece's focal length. Eyepieces with a short focal length give a higher magnification than those with a longer focal length.

Telescope Mounts

There are several different types of supports, or mounts, for telescopes. They must be solid so that the telescope is protected from vibration, and they must compensate for **Earth**'s rotation. Most images taken through a telescope have a very long exposure, so any vibrations or glitches in the telescope's tracking system will yield a blurry image.

There are two main types of telescope mounts, equatorial and altazimuth. Equatorial mounts are aligned parallel to Earth's **axis** of rotation. As Earth spins on its axis, objects in our sky appear to rise and set like the sun. A motor drive compensates for this motion, allowing the telescope to track an object for long periods of time.

An altazimuth mount has one axis (altitude) perpendicular to the horizon and the other (azimuth) parallel to it. This type of mount is relatively cheap and easy to build, but it does not easily track the apparent motion of objects through our sky. To compensate for Earth's motion, high-speed computers must continually rotate both the altitude and azimuth axes as well as a drive on the optical axis to counter the rotation of the telescope's field of view.

Adaptive Optics

Large **telescopes** need methods called adaptive optics to counter distortions caused by **Earth's atmosphere.** Stars twinkle as a result of turbulence in our atmosphere. This turbulence distorts the light from distant stars, making them appear to bounce around or flicker. Large telescopes taking long exposures magnify that distortion, making a starlight point look like a blurry blob.

A technician inspects a huge mirror made for a reflecting telescope in Arizona. Reflecting telescopes collect and focus light using mirrors instead of lenses, eliminating chromatic aberrations that can blur an image or impart a false color to objects.

Turbulence in the atmosphere is detected by watching how the image of a bright star changes or by creating an artificial star with a high-intensity **laser.** A sensor determines how the atmosphere has distorted the incoming light. This information is immediately sent to mirror supports that make small adjustments—hundreds each second—to the mirror's shape to counteract distortions. The term "adaptive optics" should not be confused with the term active optics. Active optics is a technique that controls the shape of the primary mirror. To decrease a telescope's weight, large mirrors are cast very thinly. As the telescope tilts and moves to find objects, gravity distorts the mirror's shape. Active optics compensate for the mirror's flex with computerized mirror supports called actuators. They continually monitor the mirror's shape, making very minute adjustments.

CCD (Charge-coupled Device)

CCDs are small electronic imaging devices that have been widely used in astronomy for many years. These silicon chips hold a sensitive array of electrodes that efficiently respond to a broad range of wavelengths, which means they can detect very faint objects. Most of today's digital cameras have smaller and less sensitive versions of the larger astronomical CCD chips.

Ground-based Observatories

Ground-based Observatories

Most ground-based observatories are located on carefully chosen mountaintops. Our atmosphere interferes with light coming from distant astronomical objects. Placing an observatory at high altitude helps eliminate atmospheric distortions. The air is thinner and more transparent at high altitude, and it is also less turbulent. The mountaintops selected as observatory sites have smooth air flow around the mountain. This creates relatively stable skies that offer steady viewing. Ground-based observatories operate at visual and radio wavelengths. Those at very high altitudes can detect a portion of the infrared **spectrum** (short infrared wavelengths), but our atmosphere filters out most infrared, x-ray, gamma-ray, and ultraviolet radiation.

Mauna Kea Observatories

The 4,205-meter peak of Mauna Kea on the Big Island of Hawaii is the world's highest observatory. The site is above 40 percent of **Earth's atmosphere** and above 97 percent of the water vapor in the atmosphere. These conditions provide extraordinary transparency, dryness, and calm air in the skies above the site. Such extremely dry conditions make it ideal for infrared and submillimeter astronomy. The summit holds nine **telescopes** used for optical and **infrared astronomy** and three for submillimeter astronomy. The site also holds a radio antenna that is part of the **Very Long Baseline Array.** The observatory is managed by the Institute for Astronomy at the University of Hawaii.

W. M. Keck Observatory

High atop Mauna Kea, the twin Keck Telescopes hold the two largest single-mirror telescopes in the world. The ten-meter mirrors are composed of 36 hexagonal segments, each 1.8 meters wide and 75 millimeters thick. Each segment holds computer-controlled pistons that maintain the large mirror's precise figure. Adjustments to an accuracy of four nanometers (1,000 times thinner than a human hair) are made twice a second. These adjustments counteract the warping tug of gravity. Construction of the Keck I began in 1986; the telescope started operations in 1993. Keck II followed in 1996.

Kitt Peak National Observatory

In 1958, after a three-year survey of more than 150 mountain ranges across the United States, the National Science Foundation selected Kitt Peak, in Arizona, as the home of the National Observatory. The 81-hectare site is in the Quinlan Mountains southwest of Tucson. Located in the Sonoran Desert at an altitude of 2,095

NOTABLE NAKED-EYE OBSERVATORIES

Name and Location	Dates in Use	Significance in the History of Stargazing
Stonehenge, Salisbury Plain, England	~3000 to 1500 B.C.	Britons of the Neolithic Period may have used these megaliths to track solar and lunar alignments, especially during summer and winter solstices.
El Caracol, Chichén Itzá, Yucatán Peninsula, Mexico	800 to 1100	Mayan and Toltec inhabitants of Mesoamerica traced Venus at points along the horizon from this tower; useful for calendar accuracy.
Gaocheng Observatory, North-central China	Built 1279	Chinese astronomer Guo Shoujing, at behest of Mongol leader Kublai Khan, calculated the length of the year to within 26 seconds from here.
Samarkand Observatory, Uzbekistan	Built 1428	Islamic ruler of Timurid empire, Ulugh Beg, made a comprehensive catalog of 1,018 stars and their locations.
Observatory of Taqi al Din, Istanbul, Turkey	1577 to 1580	Astronomers of the Ottoman Empire tracked the passage of stars and configuration of constellations.
Bighorn Medicine Wheel, Wyoming, U.S.	1600 to 1800	Plains Indians determined date of summer solstice by the rising and setting of the sun and the appearance at horizon of stars Aldebaran, Rigel, and Sirius.
Jantar Mantar Observatories, India	Built 1724-35	Maharaja Sawai Jai Singh II established five observatories around India to chart and map movements of the stars, building on the work of Ulugh Beg.

One of the Keck Telescopes, the largest single-mirror telescopes in the world, peers at the night sky through an open dome on the summit of Mauna Kea, a 4,205-meter dormant volcanic peak on the Big Island of Hawaii. The twin Kecks have ten-meter mirrors composed of hexagonal segments computer-controlled to maintain precision.

meters, Kitt Peak has a high percentage of clear weather and a very steady atmosphere. The site holds the 4-meter Mayall Telescope as well as the 3.5-meter WIYN Telescope, named for a consortium that includes the University of Wisconsin, Indiana University, Yale University, and the National Optical Astronomy Observatories. The National Solar Observatory also has facilities on Kitt Peak: the McMath-Pierce Solar Telescope and the Solar Vacuum Tower. The McMath-Pierce Solar Telescope is the largest solar telescope in the world. Its tilted facade rises more than 30 meters above the mountain. This facade holds only a small portion of the telescope's 150-meter light shaft, the majority of which is underground.

The National Optical Astronomy Observatory (NOAO), headquartered in Tucson, operates Kitt Peak. NOAO also manages three other observatories, the National Solar Observatory with facilities at Kitt Peak and Sacramento Peak, New Mexico; and the **Cerro Tololo Inter-American Observatory** in Chile.

Cerro Tololo Inter-American Observatory

The Cerro Tololo Inter-American Observatory (CTIO) lies roughly 55 kilometers east of La Serena, Chile. Like its northern sister, **Kitt Peak National Observatory,** CTIO is jointly operated by the Association of Universities for Research in Astronomy and the National Optical Astronomy Observatory. CTIO sits at an elevation of 2,215 meters and holds the four-meter Victor M. Blanco Telescope. The Blanco Telescope was commissioned in 1974 as a southern compliment to the four-meter Mayall Telescope on Kitt Peak.

Mount Wilson Observatory

Above Pasadena, California, in the San Gabriel Mountains, Mount Wilson Observatory is one of the most historic observatories in the world, having been the home of such discoveries as the **sun**'s position in the **Milky Way,** and the expansion of the universe. At an altitude of 1,740 meters, the observatory was founded in 1904 by George E. Hale as a solar observatory.

Using funds from the Carnegie Institution of Washington, Hale installed the Snow Solar Telescope on loan from Yerkes Observatory in Williams Bay, Wisconsin. The site was further developed, and in 1917 the 2.5-meter Hooker Telescope was installed. The Hooker Telescope dominated astronomical research until the Hale Telescope at Palomar Observatory became operational some 32 years later. **Edwin**

Hubble made observations of **Cepheid variable stars** using the Hooker in 1919. His observations led him to the conclusion that the universe is expanding.

Palomar Observatory

Founded in 1934, Palomar Observatory holds the world-famous five-meter Hale Telescope. The telescope is named for astronomer George E. Hale who, in 1928, persuaded the Rockefeller Foundation to fund it. Corning Glass Works in New York State cast the Pyrex mirror of the Hale on December 2, 1934. The 20-ton disk took eight months to cool before it was shipped by rail to Pasadena, California, where it was ground and polished. Construction of the telescope's home, a thousand-ton rotating dome, began in the mid-1930s and was nearly complete by 1941, when the United States entered World War II. War delayed the completion of the 14.5-ton mirror until November 18, 1947. It then began a two-day trip to the observatory, 1,706 meters up in the San Jacinto Mountains northeast of San Diego. Though not fully operational, the telescope was dedicated on June 3, 1948. **Edwin Hubble** took the first image through the telescope in January 1949.

Palomar also holds the 1.2-meter Oschin Schmidt Telescope. Between 1948 and 1957, the telescope was used to systematically photograph the north **celestial pole** to declination minus 30 degrees. The Palomar Observatory Sky Survey (POSS), financed by the National Geographic Society, yielded 2,000 glass photographic plates holding images of stars down to magnitude 21. POSS was the first detailed photographic star atlas of the northern sky. In 1970 the European Southern Observatory's Schmidt Telescope in Chile and the United Kingdom's Schmidt Telescope in Australia completed the survey for the Southern Hemisphere. Today Palomar Observatory is increasingly plagued by light pollution from the rapid urbanization of southern California.

Apache Point Observatory

Located high above White Sands Air Force Base in the Sacramento Mountains of New Mexico, Apache Point Observatory holds a 3.5-meter **telescope** as well as the 2.5-meter telescope dedicated to the Sloan Digital Sky Survey. The Sloan Digital Sky Survey is a high-resolution photographic survey of one-quarter of the sky. Images are taken in five different colors, then processed so that the shape, brightness, and color of hundreds of millions of objects can be determined.

The sky survey will result in a three-dimensional map showing the distribution of matter and revealing the largest structures in our universe.

MAUNA KEA TELESCOPES

Name	Size	Wavelength	Operated by	Year
University of Hawaii 0.6-m telescope	0.6 m	Optical and infrared	University of Hawaii	1968
University of Hawaii 2.2-m telescope	2.2 m	Optical and infrared	University of Hawaii	1970
NASA Infrared Telescope Facility	3.0 m	Infrared	NASA	1979
Canada-France-Hawaii Telescope	3.6 m	Optical and infrared	Canada/France/University of Hawaii	1979
United Kingdom Infrared Telescope	3.8 m	Infrared	U.K.	1979
Caltech Submillimeter Observatory	10.4 m	Submillimeter	Caltech/NSF	1987
James Clerk Maxwell Telescope	15 m	Submillimeter	U.K./Canada/Netherlands	1987
W. M. Keck Observatory (Keck I)	10 m	Optical and infrared	Caltech/University of California	1993
Very Long Baseline Array	25 m	Radio	NRAO/AUI/NSF	1993
W. M. Keck Observatory (Keck II)	10 m	Optical and infrared	Caltech/University of California	1996
Subaru Telescope	8.3 m	Optical and infrared	Japan	1999
Gemini Northern Telescope	8.1 m	Optical and infrared	U.S./U.K./Canada/Argentina/ Australia/Brazil/Chile	1999
Submillimeter Array	8 x 6 m	Submillimeter	Smithsonian Astrophysical Observatory/Taiwan	2003

Space-based Observatories

Space-based Observatories

High above our filtering atmosphere, space-based observatories can see at all wavelengths of the **electromagnetic spectrum.** Although a small portion of the infrared spectrum does trickle through the atmosphere, most of the radiation is absorbed by water vapor, carbon dioxide, and oxygen. Far-infrared radiation—wavelengths longer than 40 micrometers—is absorbed by **Earth's atmosphere.** Radiation at far-infrared wavelengths provides clues to **planets, comets,** infant stars, and other relatively cool objects.

Ultraviolet radiation with wavelengths shorter than 290 nanometers is completely absorbed by Earth's ozone layer. Space-based observatories such as the International Ultraviolet Explorer, launched in 1978, and the Extreme Ultraviolet Explorer, launched in 1992, have mapped hot, excited regions of gas and massive stars. X-rays, gamma rays, and **cosmic rays** have wavelengths shorter than ultraviolet radiation. Observations at these wavelengths must also be done from space-based observatories.

Hubble Space Telescope

Since the early 1920s, astronomers have dreamed of a **telescope** in space that would see the cosmos at wavelengths unfiltered by **Earth's atmosphere.** That dream became a reality on April 25, 1990, when the **space shuttle** *Discovery* launched the Hubble Space Telescope (HST) into orbit. Since that time, HST has provided us with extraordinary images and has changed our understanding of the universe.

Designed and built in the 1970s and '80s, Hubble has 76 handholds and special grappling fixtures that allow the space shuttle's robotic arm to capture it. Hubble is the first orbiting observatory designed to accommodate regular space-based servicing missions. Roughly the size of a large bus and weighing about 11,100 kilograms, Hubble holds a 2.4-meter primary mirror that is extremely smooth. Orbiting some 612 kilometers above the **Earth,** Hubble travels at a speed of 28,000 kilometers an hour. It travels coast-to-coast in 10 minutes and orbits Earth once every 97 minutes.

Named for astronomer **Edwin Hubble,** HST holds an array of instrumentation that includes three cameras, two spectrographs, and fine guidance sensors used to point the telescope and measure an object's precise position. Because it observes objects from high above Earth's atmosphere, Hubble can produce very high-resolution images that are roughly ten times better than **ground-based observatories.** It can resolve features that are 0.1 arc seconds across.

Management of Hubble's science operations rests with the Space Telescope Science Institute (STScI) on the Johns Hopkins University Homewood Campus in Baltimore, Maryland. Data acquired by Hubble is sent to an orbiting tracking and relay satellite that conveys the information to a **ground station** in White Sands, New Mexico. The information is then transmitted to **NASA**'s Goddard Space Flight Center in Greenbelt, Maryland, which subsequently forwards the data to STScI in Baltimore.

On January 16, 2004, NASA Administrator Sean O'Keefe announced his decision to cancel all future shuttle servicing missions to Hubble. Safety guidelines following the ***Columbia*** tragedy were cited as the primary basis for this decision. A mission scheduled for 2006 would have replaced the Wide-Field Planetary Camera 2 (WFPC2) with a newer version that is two to three times more sensitive at infrared wavelengths than Hubble's Near Infrared Camera and Multi-Object Spectrometer (NICMOS). The Hubble's fine guidance sensors would have been replaced, and the Cosmic Origins Spectrograph that studies the chemical composition of far-distant interstellar gas would

HUBBLE & JAMES WEBB DATA

	Hubble ST	James Webb ST
Launched	25 Apr 1990	Projected August 2011
Main mirror	2.4-m diameter	6.5-m diameter
Light detected	Ultraviolet to infrared	Infrared
Primary mirror material	Ultra-low expansion glass	Beryllium
Optical resolution	0.1 arc seconds	~0.1 arc seconds
Wavelength coverage	115-2,500 nm	0.6-28 microns
Mass	11,100 kg	6,200 kg
Height of orbit	600 km	L2 orbit (1.5 million km from Earth)
Period of orbit	97 minutes around Earth	365 days around Sun
Intended lifetime	15 years	5-10 years

have been installed; astronauts also would have boosted Hubble's orbit to keep it from spiraling too close to Earth. With the cancellation of manned shuttle servicing missions, age, gravity, and wear will take their toll on Hubble's orbit and components. Eventually, the telescope's **orbit** will degrade, and the Hubble will fall to Earth.

James Webb Space Telescope

Currently scheduled for launch in August 2011, the James Webb Space Telescope (JWST) is the **Hubble Space Telescope**'s successor. The orbiting observatory will carry instrumentation for both visible and infrared astronomy. It will peer through dust and gas to unveil the birth of stars and planetary systems, shedding new light on how they form and interact. JWST will also study **galaxy** formation and evolution while probing the very nature and abundance of **dark matter.** These observations will help determine the shape and composition of the universe so that we may better understand its past, present, and future.

The observatory will hold a lightweight, deployable mirror roughly 6.5 meters across. Specifications call for the JWST to see objects 400 times fainter than those visible with ground-based infrared telescopes such as those in the **Keck Observatory,** as well as Hubble's NICMOS camera and the Space Infrared Telescope Facility. The JWST is named for NASA's second administrator. Under James Webb's direction, NASA undertook the challenge of landing a man on **Earth's moon.**

Cosmic Background Explorer

Launched on November 18, 1989, the Cosmic Background Explorer (COBE) **satellite** has mapped the microwave background radiation left across the entire universe by the **big bang.**

COBE's observations provided evidence that the early universe did indeed fit the prevailing big bang theory of how the universe formed. The satellite's differential microwave **radiometers** precisely mapped the background radiation while the far-infrared absolute **spectrometer** compared the **spectrum** of this background radiation to that of an ideal **blackbody** (an object that perfectly follows the laws of physics that dictate how matter should emit and absorb radiation).

COBE found an extremely smooth spectrum that deviates from a blackbody by less than 1 percent. These minute variations were early structures that later developed into **galaxies** and **galaxy clusters,** forming the **large-scale structure** of the universe that we see today.

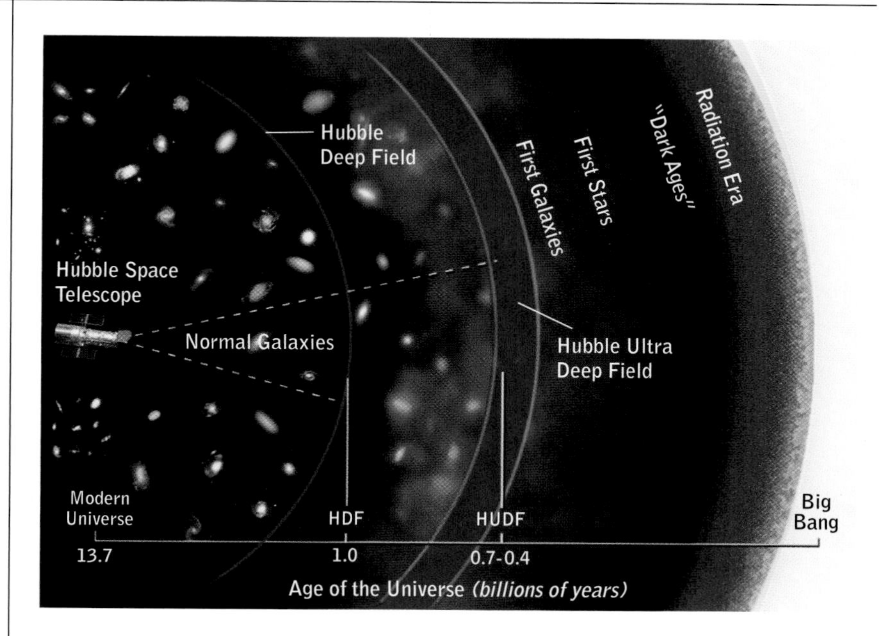

An instrument that shook previous perceptions, the orbiting Hubble Space Telescope (in the left center of the illustration) looked back through space and time 13 billion years to the age when the first galaxies were forming.

Deep Space Network

The Deep Space Network (DSN) is a system of three deep-space communications complexes used primarily to track **spacecraft.** The three facilities, all operated by **NASA,** are placed approximately 120 degrees apart around the world, which allows the DSN to remain in constant contact with spacecraft as **Earth** rotates.

In 1958 NASA recognized the need for a DSN to manage communications for all deep-space missions using robotic spacecraft. By centralizing its efforts, NASA eliminated the need for each individual mission to develop its own communications programs.

The DSN is the largest and most sensitive scientific telecommunications system in the world. The U.S. facility is in the Mojave Desert, outside Goldstone, California. The two others are at Tidbinbilla Nature Reserve southwest of Canberra, Australia, and at Robledo de Chavella,west of Madrid, Spain. Each complex holds a 70-meter-, 26-meter-, and 11-meter-diameter **antenna** as well as two 34-meter-diameter antennas. The Goldstone complex also includes two additional 34-meter beam waveguide antennas used primarily for research and development. The Network Operations Control Team at the Jet Propulsion Laboratory in Pasadena, California, controls the three centers. When not communicating with spacecraft, these radio dishes probe the heavens at radio wavelengths.

History of the Hubble Telescope

Robert W. Smith

The Hubble Space Telescope (HST) began life in 1946 as the gleam in the eye of Lyman Spitzer, Jr., then a professor of astronomy at Yale. Spitzer soon after moved to Princeton University where he had trained as a graduate student. One of a generation of U.S. scientists who learned the fundamentals of science politics during World War II, Spitzer had a courteous and modest manner that belied a great determination to further those scientific projects in which he believed, and for such projects he was prepared to fight tenaciously. Spitzer, along with many other astronomers, fought extremely hard, over a period of decades, to bring the HST into being. At times the project was close to being cancelled. It survived, but on the road to approval from the White House and Congress it was scaled back in size from the original plans. At first its main mirror was to be 3 meters in diameter, but the telescope as built had a primary of 2.44 meters in diameter. What was then called the Large Space Telescope (LST) was scaled back to become the Space Telescope, though more than one astronomer had quietly regarded the first name, LST, as recognition of the driving force behind the instrument—the Lyman Spitzer Telescope.

One of the HST's great attractions to astronomers such as Spitzer was that, by getting above the atmosphere, it would be able to detect a much wider range of wavelengths than an equivalent telescope on the ground. Thus, while the 2.5-meter Hooker Telescope at Mount Wilson was designed solely for optical wavelengths, the HST was fashioned as an optical, ultraviolet, and infrared telescope.

The HST is a striking and hugely impressive piece of technology. In the 1980s I followed the development of the telescope very closely as I researched its history. In 1985 I saw the telescope in a gigantic "clean room" at the Lockheed Missiles and Space Company in Sunnyvale, California. Held erect in a special mount, it was covered in silver-colored, multilayered insulation to combat the alternating swings from broiling heat to frigid cold as it slipped between the daytime and nighttime parts of its orbit. It seemed far bigger and more imposing than the 12-meter by 4.5-meter dimensions of its main structure would suggest. I was also struck by the HST as a marker of just how far telescope making had advanced during the 20th century. The Hooker Telescope had first been turned to the heavens in 1917. The HST's primary mirror is slightly smaller than the 2.5 meters of the Hooker, and so is quite small by the standards of the biggest contemporary ground-based telescopes. However, in the intervening 70 years between the completions of these two instruments, a remarkable series of technological advances, combined with the willingness of NASA and the European Space Agency to spend billions of dollars on the project, have led to a vastly more powerful scientific tool. The HST is entirely automated and is commanded from the ground, with astronauts periodically visiting it in orbit to make repairs and upgrades.

What I did not know at the time I first saw the telescope, and only one or two people suspected it, was that the HST's primary mirror had a flaw. It was a shade too flat at its edges by an amount less than a fraction of the width of a human hair. This nevertheless was a big error in the world of precision optics, and for its first three years in orbit following its launch in 1990, the HST was not nearly such a powerful telescope as its advocates had hoped. But a brilliantly successful repair mission by shuttle astronauts in 1993 restored the telescope to its full capabilities.

The Hubble Constant

If there was one scientific problem that became associated with the HST before it was launched, it was that of the determination of the distance scale of the universe and the determination of the Hubble constant. In 1977, when NASA issued a call for astronomers to become involved in the development of the HST, the determination of the Hubble constant was identified as the number one scientific problem. By 1984 this problem had become a "Key Project" for the HST. The goal was to come up with a value of the Hubble constant that was accurate to within 10 percent. Wendy L. Freedman of the Carnegie Observatories and her Key Project team painstakingly arrived at the value. The expansion of the universe, the "Hubble Flow," as it is called, increases by 70 kilometers per second for each megaparsec increase in relative distance. A megaparsec is equivalent to some $30{,}900 \times 10^{15}$ kilometers—a rather messy number. This implies that the universe is about 13 billion years old. However, while the current estimates of the Hubble constant now vary less than they did a decade or so ago, the problem has still not been solved in the sense of astronomers agreeing on a single value. Perhaps such a value will have to await future missions that will follow the HST into space.

The Hubble Deep Field

In late 1995 the HST spent ten days pointed toward a patch of the sky about a tenth of the diameter of the full moon, near the handle of the Big Dipper. The aim was to provide a "deep core sample of the universe" by securing an image of objects as faint as possible with the telescope. The result was an image of the "Deep Field"—more than 1,500 galaxies at the very farthest limits of the observable universe. In Edwin Hubble's most famous book, *The Realm of the Nebulae,* published in 1936, he wrote about great telescopes and their role in expanding our knowledge of the universe by pushing deeper into new regions. For the HST, the Hubble Deep Field is perhaps the most spectacular example of exactly this, and is perhaps the telescope's most scientifically significant image to date.* What one U.S. senator dubbed a "technoturkey," after the flawed mirror was first discovered, has proven to be the most famous and in some ways the most productive telescope ever built, one that has helped reshape and extend our views of the physical universe.

For a historian of astronomy, the launch of the HST stands as a watershed experience. For the first time I found myself seeing history made before my eyes. Surely the HST project was far too large for any one player to comprehend firsthand. But early on I gained access to masses of private and public papers that revealed the complex interrelationships of the players and the pressures that drove them to do what they did. Many of the principal figures in the history freely subjected themselves to highly structured, formal tape-recorded interview sessions, knowing that I and other colleagues would transcribe and edit these into a permanent record. These interviews underline that history is a dynamic process, our view of which can change with time, distance, and perspective. Much like the universe itself. ■

In 2004 the HST returned a deeper image: the Hubble Ultra Deep Field, revealing an estimated 10,000 galaxies.

Temporarily docked on the space shuttle *Discovery,* the Hubble Space Telescope underwent repairs and received new gyroscopes in December 1999.

2 | Our Solar System

The sixth planet from the sun, Saturn has 31 known satellites of differing sizes. In November 1980 the Voyager 1 spacecraft took a series of images as it passed through Saturn's system. In this artist's rendering, which is based on a montage of the images, Saturn's largest moon, Titan (upper right corner), appears smaller than Dione (forefront) due to each moon's orbital path and distance from the planet at the time the photographs were taken.

BY ASTRONOMICAL STANDARDS, OUR SOLAR SYSTEM IS A CLOSE AND FAMILiar neighborhood. The sun, the moon, and the nearer planets have been a visible part of human life and lore since the dawn of mankind. But the more we learn about these neighbors, the odder they seem. Each year brings surprising discoveries about the nine (or is it eight?) planets, their moons, asteroids, comets, and the sun. The very definitions for these heavenly bodies are now suspect: Pluto is smaller than several moons in the solar system and more similar to Neptune's satellite Triton than any other body. Pluto's moon, Charon, is big enough to be a companion planet. Comets fade into asteroids; asteroids are captured by planets to become tiny moons; large moons have atmospheres, volcanoes, and oceans, much like planets. And the neighborhood keeps growing. We have found so many new satellites around Jupiter that reference books cannot keep up. Billions of kilometers away, the icy worlds of the Kuiper belt are coming into view—distant solar system neighbors.

Astronomers are scrambling to reassess the solar system because of the wealth of new information gained from improved ground-based telescopes, new eyes in space such as the Hubble Space Telescope, and robotic exploration of the planets. Since the space age began with the launch of Sputnik 1 in 1957, only the moon has been visited by humans, but every planet except distant Pluto has been studied via planetary spacecraft, as have a number of other planets' moons, asteroids, comets, and the sun. Some of the most exciting missions involve the search for life. Missions to Mars have confirmed that it once had at least one key prerequisite for Earth-type life: liquid water. Data from missions to Jupiter's moon Europa indicate that the satellite harbors a vast, liquid ocean beneath a frozen surface. Could organic molecules or even larger life-forms exist within that water?

The quest to understand the solar neighborhood has implications for our own security. Huge, ancient craters on Earth show that it, like all the other planets, is vulnerable to impacts from asteroids and comets. Countries around the world have begun to cooperate on surveys of near-Earth objects. Some of the objects that pass close to Earth come from the solar system's farthest reaches: the Kuiper belt and the Oort cloud—vast, dark realms of scattered rock and ice that are the homes of comets and planet-like bodies.

The solar neighborhood has many dark byways, but thanks to human ingenuity, we are rapidly learning more about its inhabitants. The fact that every piece of knowledge comes with several questions attached only adds to the excitement of exploration.

Planets glow in the twilight above England's Stonehenge, which may have been an early astronomical observatory. Jupiter stands highest above the megaliths at upper left, while Saturn (left), Mars (top), and Venus (right) form a nearly perfect equilateral triangle just above the horizontal stone near the picture's center.

Solar System Basics

The Solar System

Our solar system consists of the **sun** and all the objects that orbit it. These bodies include nine major planets (although the status of **Pluto** as a planet is in debate); at least 138 moons; millions of **asteroids;** innumerable meteoroids; and trillions of **comets,** many of which originate in a far-distant region known as the **Oort cloud;** as well as the gas and dust of the **interplanetary medium.** The radius of the solar system—the distance from the sun's center to the outer reaches of the Oort cloud—is approximately 100,000 **astronomical units** (AU) or 100,000 times the distance from the sun to **Earth.**

The sun is by far the dominant object in the solar system, containing 99.8 percent of the system's mass. The four planets closest to the sun—**Mercury, Venus,** Earth, and **Mars**—are known as the inner, or terrestrial, planets. They are small, dense, and rocky. Beyond Mars is the **asteroid belt,** followed by the four large outer planets, the gas giants—**Jupiter, Saturn, Uranus,** and **Neptune**—made primarily of gases such as helium and hydrogen. Pluto, the farthest planet, is small and icy, and may simply be one of the larger bodies in an orbiting group of rocky objects called the **Kuiper belt.**

All the planets except Pluto circle the sun in **elliptical orbits** on approximately the same plane. Pluto's orbit is noticeably elongated and tilted with respect to that of the other planets, and at times passes within the orbit of Neptune.

Five planets (in addition to Earth) are visible to the naked eye and have been known since ancient times. Mercury, Venus, Mars, Jupiter, and Saturn have long been familiar as starlike beacons traveling across the seemingly fixed background of the stars. In 1781 astronomer **William Herschel** discovered the planet Uranus using his homemade **telescope;** in so doing, he doubled the size of the known universe. The difference between Uranus's predicted orbit and its actual path led astronomers to search for a more distant planet whose gravity might be tugging on Uranus. In the 19th century, both British mathematical astronomer John

The sun and some of its neighbors line up in this logarithmic interpretation of the solar system. The gauzy pink area that separates the sun and its neighbors from the largely unknown area beyond is called the heliopause. The Oort cloud, a spherical shell of comets, marks the outer edge of the sun's gravitational sphere.

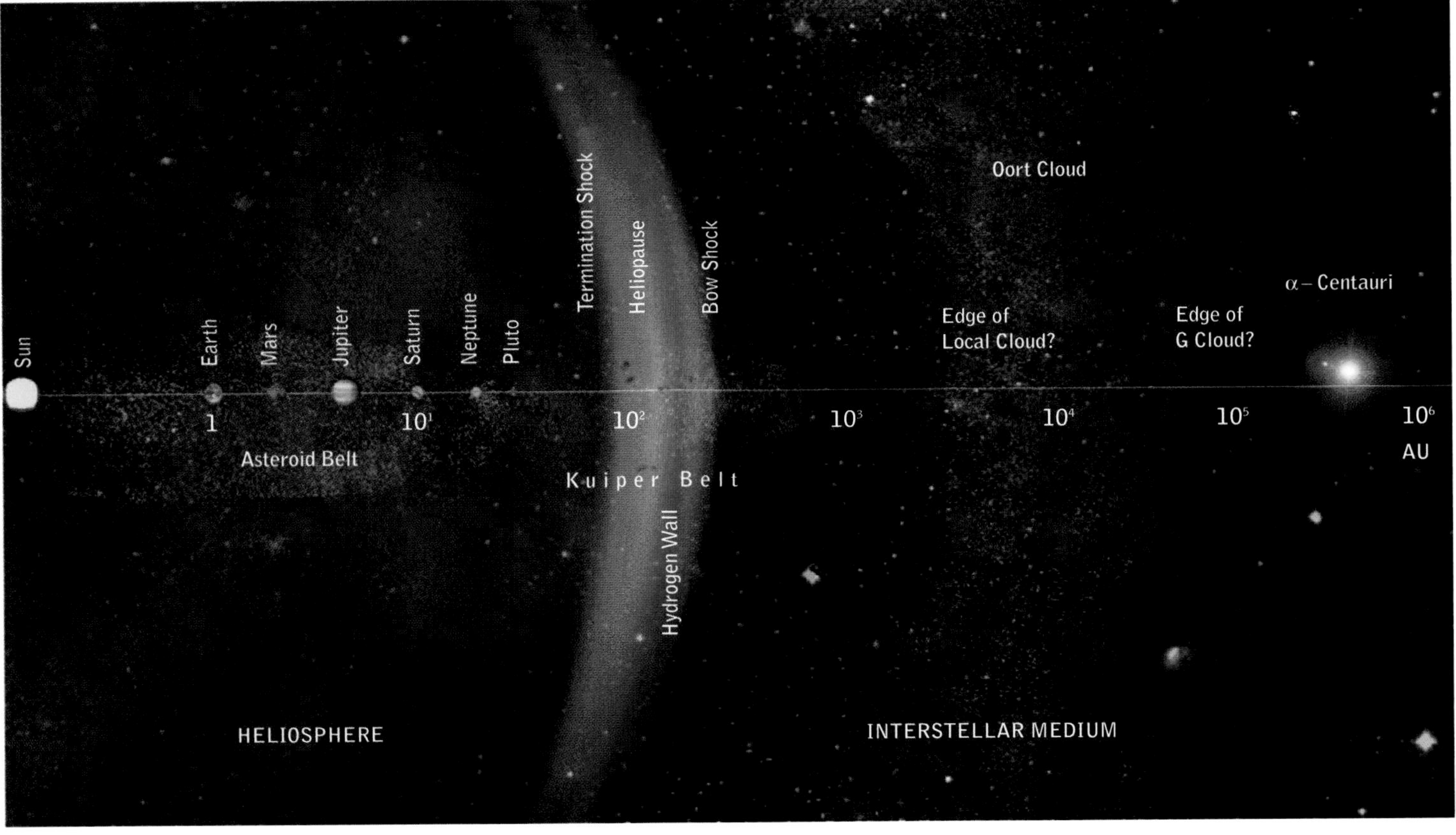

Couch Adams and French celestial mechanician Urbain Jean Joseph Leverrier determined mathematically where such a planet must be; their calculations were confirmed in 1846 when Neptune was discovered via telescope from the Berlin Observatory.

Because Neptune alone did not seem to account for all the gravitational discrepancies in Uranus's orbit, some observers kept searching for a more distant planet. In 1930 Clyde William Tombaugh used a photographic telescope to discover Pluto, a tiny speck that had moved 3.5 millimeters on two photographs taken six days apart. As it turned out, Pluto's mass was not enough to make a difference in the orbit of Uranus, so its discovery was serendipitous.

Sicilian astronomer Giuseppe Piazzi was the first to discover a large asteroid, which he named Ceres, in 1801. By the end of that century, several hundred asteroids had been spotted. Modern astronomers have added thousands upon thousands, with more to come. Most circle the sun in a wide belt between the orbits of Mars and Jupiter; a few hundred have also been found in Jupiter's orbit. In 1951 Gerard Kuiper predicted that another belt of rocky, orbiting bodies would be found just outside Neptune's orbit. His prediction was proved correct in the 1990s, as modern telescopes discovered hundreds of small, dim, rocky objects circling at the edge of the solar system, a group now known as the Kuiper belt. Another prediction made in 1951 seems to be true as well. Jan Hendrik Oort stated that a shell of rocky ice balls, the breeding ground of long-period comets, should orbit the sun at an extreme distance, up to 100,000 AU. Although the cloud has not been seen by any telescope, the trajectory of comets such as Hale-Bopp through the solar system backs up this claim.

Astronomers don't yet know whether Earth and its neighbors represent a typical solar system, a type that can be found around many other stars in many other galaxies. They may know soon, as more extrasolar (outside our solar system) planets are being discovered every year.

Formation of the Solar System

The **solar system** was born from a vast, cold, slowly rotating cloud of interstellar gas and dust—the solar **nebula**—almost five billion years ago. The process by which **planets** formed and the timeline of their formation are still not completely understood. However, most astronomers agree on a basic outline. Pulled by gravity, most of the solar nebula's gas and dust began to fall toward the center of the nebula, eventually forming a dense mass called the protosun. Meanwhile, the rest of the spinning nebula began to flatten into a disk. As the protosun continued to condense, its radiation began to vaporize the ice in the inner regions of the young system and push lighter gases, such as hydrogen and helium, toward the outer regions.

Tiny bits of rock and dust within the nebula began to stick together, at first chemically, and then gravitationally as they grew larger. Over many thousands of years, these rocks clumped together into **planetesimals,** mountain-sized boulders, that further accreted into orbiting protoplanets. In the inner regions of the young system, these planetesimals coalesced into the terrestrial planets. In the outer, cooler regions, they formed the rocky cores of the giant planets, which went on to accumulate thick, gaseous atmospheres from gas remaining in the nebula's disk. Meanwhile, the protosun was becoming more massive. Eventually pressures at its core became so high that fusion began, and the newborn **sun** started to release enormous amounts of radiation. This radiation scoured the young system, preventing new planetesimals from forming. In all, the planets took at least a hundred million years to form.

For close to a billion years more—until about 3.8 billion years ago—the solar system remained a rough neighborhood, with frequent collisions among the leftover, rocky debris of the solar nebula. The evidence of these collisions can be seen today on the scarred surfaces of our **moon** and other rocky planets.

Orbits of the Planets

The orbit of an object in the **solar system** can be described by six elements: its **semimajor axis** (distance from the **sun**), **eccentricity** (shape of the ellipse), inclination (the object's orbital plane compared to the solar system's orbital plane), the argument of the perihelion and the longitude of the **ascending node** (two elements that describe the orbit's orientation in space), and the longitude at the epoch (the object's position at a given time). Using these figures, an astronomer can calculate the orbit of a **planet, a moon, a comet,** or a **spacecraft** on its way to **Mars.**

Most ancient astronomers, though, observing the sun, moon, stars, and visible planets as they wheeled across the sky, made a logical assumption: The **Earth** was the stable center of the universe, and all celestial objects orbited it in perfect circles. Unfortunately for close observers of the five known planets, this simple assumption did not explain the planets' motions.

Astronomer Johannes Kepler reviews diagrams in front of a portrait of his mentor, Tycho Brahe. Kepler succeeded Brahe as imperial mathematician to the Holy Roman Emperor. On a quest to make mathematical sense of the universe, Kepler spelled out laws of planetary motion and measured the distance of the planets from the sun.

Moving more slowly than their starry background, the wandering worlds occasionally stopped in their progress and moved backward for a while in what is known as retrograde motion. Mars, for example, actually loops around in its path across Earth's skies. The Egyptian astronomer **Ptolemy** worked out an elaborate scheme in which the planets followed their own circular motions, called epicycles, while orbiting the Earth.

With the publication of *De revolutionibus orbium coelestium* in 1543, Polish astronomer **Nicolaus Copernicus** was able to restore some simplicity to celestial mechanics by showing that planetary orbits could easily be explained in a sun-centered system. Mars's loopy motion results from its position outside Earth's in the solar system and from its slower speed around the sun. When the Earth overtakes Mars in their respective orbits, the red planet seems to move backward relative to the distant stars. Copernicus's work was followed some 60 years later by that of **Johannes Kepler.**

Johannes Kepler (1571-1630)

Johannes Kepler was a German mathematician and astronomer whose three laws of planetary motion helped shape modern astronomy. He was born in Weil der Stadt, Germany, to a family of modest means, and received a scholarship to the Lutheran seminary at the University of Tübingen in 1587. There, a professor of mathematics introduced him to the heliocentric (having the sun as its center) theories of **Copernicus.** Kepler began to study planetary orbits, believing that they fit into a divine plan for the universe that could be explained geometrically.

Kepler's first book about the planets, *Mysterium cosmographicum* (1596), attracted the attention of the great observational astronomer **Tycho Brahe,** who invited Kepler to become his assistant in Prague, where Brahe was imperial mathematician to the Holy Roman Emperor. When Brahe died in 1601, Kepler succeeded him in the post and proceeded to write works on optics, astrology, and stars, as well as *Astronomia nova* (1609), in which he spelled out his first two laws of planetary motion. Kepler realized that the planets' orbits could not accurately be described by perfect circles. Instead, they are ellipses, elongated shapes with two focal points. Kepler's first two laws of planetary motion describe these orbits:

1. **The orbit of a planet about the sun is an ellipse with the sun at one focus.** There is no object at the second focal point. The longest diameter of the ellipse, passing through both focal points, is called the major axis, and half of that distance is the **semimajor axis,** which represents the planet's average distance from the sun.

2. **A line joining a planet and the sun sweeps out equal areas in equal intervals of time.** This means a planet's speed—its **orbital velocity**—increases as it moves toward perihelion, its closest point to the sun, and decreases as it moves toward aphelion, its farthest point.

In 1611 Kepler became the district mathematician of Linz and continued to publish prolifically. His *Harmonice mundi* (1619) outlined the distances of the planets from the sun and laid out his third law of planetary motion:

3. **The square of a planet's sidereal period around the sun is directly proportional to the cube of the length of its orbit's semimajor axis.** The farther a planet is from the sun, the longer its year (sidereal period), defined as one revolution around the sun.

Kepler's three laws acquired a solid mathematical underpinning later in the 17th century, when **Isaac Newton** formulated his laws of motion. Newton's brilliant explanations of gravity, force, and motion proved Kepler's laws so accurate, in fact, that 19th-century astronomers were able to use them to correctly predict the existence of an eighth planet, **Neptune,** from tiny perturbations in the orbit of **Uranus.**

He also published works supporting **Galileo**'s observations, highly accurate astronomical tables, treatises on **telescopes** and on the Christian calendar, and even (posthumously) a piece of science fiction in which the hero travels to the **moon.** For all of his work, but particularly for his brilliant exposition of planetary motion, Kepler is hailed as one of the greatest astronomers in history.

Astronomical Unit

An astronomical unit, abbreviated AU, is the average distance between the **Earth** and the **sun:** approximately 149,600,000 kilometers (1.49×10^8). Distances within the **solar system** are typically measured in AU.

Planet

The question "What is a planet?" became more difficult to answer as the 20th century ended. Taken from the Greek verb meaning "to wander," the word planet originally referred to the traveling lights in the sky that came to be known as **Mercury, Venus, Mars, Jupiter,** and **Saturn.** By the mid-20th century it was accepted that a planet was one of the nine bodies, larger than **asteroids,** that orbited the **sun.** No further definition was needed.

Several discoveries in the 1990s forced astronomers to reconsider. One was the first observation of **Kuiper belt** objects in 1992. These large, rocky bodies circle the sun in the same general region as **Pluto,** casting doubt on Pluto's status as a planet. In 1995 astronomers also confirmed the existence of **brown dwarf stars.** Cool, gaseous brown dwarfs are more massive than **gas giant** planets, such as Jupiter, but are otherwise similar: They are not massive enough to begin fusion, and they often orbit other stars. Adding to the excitement—and confusion—in the 1990s was the discovery of planets circling other stars. Most of the **extrasolar planets** found so far are very massive, close to the bottom limit for brown dwarfs. Astronomers have even found some "exoplanets," free-floating bodies that do not orbit a star; they are larger than Jupiter but smaller than brown dwarfs.

Therefore, any definition of planet based on mass, composition, or position in a system is becoming difficult. For now, most astronomers would agree that a planet is a spherical object, orbiting a star, that is larger than an asteroid but smaller than a brown dwarf. Further refinements of this definition will probably come as the 21st century advances.

DISTANCES WITHIN THE SOLAR SYSTEM

Light-time	Approx. Distance	Example
3 seconds	900,000 km	Earth–moon round trip
8.3 minutes	149,600,000 km	Sun to Earth (1 AU)
1 hour	1,000,000,000 km	1.4 x sun-Jupiter distance
1 year	63,000 AU	Light-year
4 years	253,000 AU	Next closest star

Gas Giant

A gas giant is a large planet composed mainly of gas swirling around a smaller rocky core. In the solar system, four planets are gas giants: **Jupiter**, **Saturn**, **Uranus**, and **Neptune.** Most **extrasolar planets** discovered so far are also gas giants, because only the most massive planets can be observed.

According to the prevailing theory of solar system formation, the outer planets picked up their thick coats of gas from the primordial gases of the solar system over millions of years. Unlike the inner planets, they were too far from the **sun** for these atmospheres—mostly hydrogen and helium—to be dispersed by solar radiation. Jupiter, the largest gas giant, has approximately 318 times the mass of the **Earth;** Saturn, 95; Neptune, 17; and Uranus, 14. However, these planets are not as dense as their terrestrial cousins; Saturn, for instance, has a density less than that of water.

In theory, gas giant planets could be as large as 13 times the mass of Jupiter. Above that limit, they would be massive enough to initiate fusion and would be classified as stars.

Planetesimal

Planetesimals are rocky, asteroidlike objects built from the coalescing dust, ice, and gas of a solar nebula in the early years of a **solar system.** According to most theories of **planet** formation, colliding planetesimals eventually combine to form full-fledged planets.

Barycenter

A barycenter is the center of mass in a system of orbiting bodies, and thus the point around which all those objects orbit. Although the **sun** contains most of the mass in the **solar system,** the system's center of mass lies just outside the sun's surface, not at its center. For this reason, the sun wobbles a little as it, too, orbits the barycenter.

Similarly, the barycenter of the **Earth-moon** system lies within the Earth, because the Earth's mass is so much greater than the moon's. In the case of binary stars, bodies of equal mass, the barycenter would lie at a point equidistant between the two stars.

Interplanetary Medium

The interplanetary medium is the mixture of dust, ionized gas (mostly hydrogen), charged particles of the **solar wind, electromagnetic radiation, cosmic rays,** and magnetic fields that pervades the space between the **planets** of the **solar system.**

The density of the medium is about five particles per cubic centimeter near the **Earth** and decreases steadily away from the **sun.** The interplanetary medium is dominated by the sun's magnetic field, which interacts with the solar wind and magnetic fields of the planets in complex ways.

Planet X

In the early part of the 20th century, before the discovery of Pluto, astronomers William Pickering (in 1909) and Percival Lowell (in 1915) theorized that another planet lay beyond the **orbit** of **Neptune.** They based their theory on perturbations in the orbits of **Uranus** and Neptune, seemingly caused by a more distant object. Lowell called this object "Planet X."

After astronomer Clyde William Tombaugh discovered Pluto in 1930, some astronomers who analyzed Uranus's motion speculated that still another planet remained to be found beyond Uranus, other than Neptune and Pluto. Voyager 2 flew by Neptune in 1989 and obtained very accurate readings of the planet's mass. These new figures showed that the orbits of the nine planets were all accounted for without the addition of another large body, and the notion of Planet X was laid to rest.

Moon

Moons are natural satellites, bodies that revolve around a larger body, such as a planet or large asteroid. Until the invention of the telescope, the only moon known to astronomers was Earth's. In 1610 **Galileo** used his new telescope to discover Jupiter's four largest moons, now called the Galilean satellites in his honor. Over the centuries, as ground-based observation improved, astronomers found moons around every planet except Mercury and Venus, ranging in size from Jupiter's giant **Ganymede** to **Phobos** and **Deimos**, the little, asteroid-like companions of Mars. In the 1970s and 80s, the Voyager 1 and 2 spacecraft detected many more tiny moons around the outer planets. In all, the solar system contains at least 137 moons; more may yet be discovered.

Analysis of moons and their **orbits** suggests that planets acquire moons in several ways. Satellites that orbit a planet closely, in the same direction as the planet's rotation and near the planet's equatorial plane, most likely formed at the same time and from the same rotating cloud of material as the planet itself. The Galilean moons are examples of this type. Others—**Neptune**'s **Triton**, for instance—have orbits that are retrograde to the planet's rotation. These moons appear to have been formed separately and captured by the gravity of the more massive planet. And **Earth's moon** may represent yet another way in which moons are born. Studies of the moon's geology and orbit support a theory that it was formed when a giant collision between the infant Earth and a huge, planet-size object blasted enough material into space to create the moon.

Life on the Moons?

A few of the solar system's natural satellites hold a particular interest for astronomers. For instance, **Triton** (**Neptune**), **Io** (**Jupiter**), and **Titan** (**Saturn**) all possess atmospheres. Triton has icy polar caps and geysers; Io, huge volcanic eruptions; and Titan may have hydrocarbon seas, similar to those that may have existed on the young **Earth.** Researchers are hoping to learn much more about these satellites as probes reach them in the early decades of the 21st century.

Syzygy

Syzygy (pronounced SIHZ-uh-jee) is the alignment of three bodies of the **solar system** in a straight line. The moon, when it is "new" or "full" is in syzygy with the **Earth** and **sun.** This alignment is also called opposition or conjunction when the Earth, sun, and another **planet** are involved.

MAJOR MOONS OF THE SOLAR SYSTEM

Planet	Total Found by 2004	Satellite	Year Discovered	Average Distance from Planet Center (in km)	Approx. Mass (x 10^{20} kg)	Approx. Diameter (in km)
Earth	1	Moon	—	384,400	734.9	3,476
Mars	2	Phobos	1877	9,380	0.001	27 x 22 x 18
		Deimos	1877	23,460	0.002	15 x 12 x 10
Jupiter	63	Metis	1979	128,000	0.001	40
		Adrastea	1979	129,000	0.0002	20 x 16 x 14
		Amalthea	1892	181,300	0.072	250 x 146 x 128
		Thebe	1979	221,900	0.008	116 x 98 x 84
		Io	1610	421,600	894	3,630
		Europa	1610	670,900	492	3,138
		Ganymede	1610	1,070,000	1480	5,262
		Callisto	1610	1,883,000	1080	4,800
		Himalia	1904	11,480,000	0.100	170
		54 others				
Saturn	31	Mimas	1789	185,520	0.375	418 x 392 x 383
		Enceladus	1789	238,020	0.65	513 x 495 x 489
		Tethys	1684	294,660	6.27	1,060
		Dione	1684	377,400	11	1,120
		Rhea	1672	527,040	23.1	1,528
		Titan	1655	1,221,830	1,345	5,150
		Hyperion	1848	1,481,000	0.20	328 x 260 x 214
		Iapetus	1671	3,561,300	15.9	1,436
		Phoebe	1898	12,952,000	0.072	220
		22 others				
Uranus	26*	Portia	1986	66,090	0.800	136
		Puck	1985	86,010	0.800	162
		Miranda	1948	129,390	0.66	481 x 468 x 466
		Ariel	1851	191, 020	13.5	1,158
		Umbriel	1851	266,300	11.7	1,169
		Titania	1787	435,910	35.2	1,580
		Oberon	1787	583,520	30.1	1,520
		Sycorax	1997	12,179,000	0.008	190
		18 others				
Neptune	13	Naiad	1989	48,230	0.002	58
		Thalassa	1989	50,080	0.0004	80
		Despina	1989	52,530	0.02	148
		Galatea	1989	61,950	0.04	158
		Larissa	1989	73,550	0.05	208 x 178
		Proteus	1989	117,650	0.5	436 x 416 x 402
		Triton	1846	354,760	214	2,706
		Nereid	1949	5,513,400	0.3	340
		5 others				
Pluto	1	Charon	1978	19,600	16.2	1,186

**Most sources list 27 moons for Uranus, however, according to NASA, "Status of the provisional satellite S/1986 U10 was revoked in December 2001."*

Anatomy of a Planet

Axis

An axis is an imaginary line, drawn through a body, around which it rotates. The alignment of a **planet**'s axis—its "tilt"—relative to the plane of its orbit around the sun determines whether the planet experiences seasons. **Mercury** is the only planet in the **solar system** without an appreciable tilt to its axis of rotation. **Uranus** is tilted by 98 degrees; compared to Mercury, it is spinning on its side.

Planetary Ring

Composed of particles of rock and ice that range in size from fine dust to large boulders, planetary rings may be the remains of **moons** or other large bodies that were captured and pulverized by the parent planet. Saturn's rings are among the most spectacular sights in the solar system, but all four giant planets—**Jupiter, Saturn, Uranus,** and **Neptune**—have rings. All rings are found within a certain distance from the planet, known as the Roche limit. Inside this limit, large satellites would be torn apart by the planet's gravitational forces.

Planetary Geology

Planetary geology is a wide-ranging field that studies the surface and internal processes of the **planets, moons, asteroids,** and other bodies in the **solar system**. Among other things, planetary geologists attempt to learn how the solar system was formed, why planets have evolved in such different ways, whether life might have evolved on other bodies, and what the geology of other planets can tell us about the Earth.

Except in the case of **Earth's moon,** from which the Apollo astronauts gathered samples, scientists learn about other planets from a distance. Ground-based and space-based **telescopes** yield data through **spectroscopy**; probes send back images and data from orbit

A swirl of colors, rings that may be the remains of pulverized moons surround Saturn. Voyager 2 shot this highly enhanced color view from a distance of 8.9 million kilometers. The color variations indicate differences in the chemical composition of the rings. Although Saturn's are the most colorful, all four giant planets have rings.

and from surface rovers. These data give geologists a surprising amount of information on the body: its size, shape, mass, density, and movement; the presence of a magnetic field; surface features such as **volcanoes, craters,** and canyons; composition of minerals on the surface, possible age of the surface, and evidence of water; atmospheric composition; and other data.

Tectonics

Tectonics is the study of changes to the structure of a **planet**'s or **moon**'s crust. Driven by internal heat, a planet's surface may crumple or crack into mountains and faults, and this form of tectonic activity is visible on all the terrestrial planets. However, for reasons that are still not clear, only **Earth** seems to have a crust broken into moving plates.

The Planetary Surface

Most of what scientists know about the geology of other worlds comes from remote observation of their surfaces. Geologists look for certain common processes: impact cratering; gradation, or flattening, caused by weathering forces; volcanism produced by internal heat; and tectonism—surface shifts that create mountains and faults. Using their knowledge of similar processes on the **Earth** and **moon**, researchers then attempt to link these features to the evolution and structure of the distant world. For instance, the fact that **Venus** shows relatively few impact **craters** implies that it has been resurfaced by lava since the early days of the **solar system,** when impacts were frequent. **Mars**'s large **volcanoes** indicate that the **planet** has little or no crustal plate motion so volcanoes remain over hot spots much longer than on Earth. On the other hand, some surface features seem to represent processes never seen on Earth: sulfur volcanoes on **Io**, for example, and nitrogen geysers on **Triton**.

Crater

A crater is a circular depression on the surface of a **planet, moon,** or **asteroid.** Some craters are caused by volcanic activity, but most are impact sites resulting from a collision with another body, such as a meteoroid or asteroid. Impact craters are visible on all the terrestrial planets and on most moons, and range in size from less than a millimeter to more than two thousand kilometers across. Craters are the result of extremely violent events, involving energies equivalent in some cases to that of thousands of hydrogen bombs. When a rocky object hits a rocky body, such as a moon, the motion of the moving body is converted into heat, sending out shock waves. The impacting object is typically vaporized, while the surface it hits is pulverized and ejected outward to create a circular rim. A typical, midsize impact crater might have a central peak, where the crater floor rebounded after impact; large craters may have several rings of ridges created from the repeated rebounding of the surface.

By observing the number and condition of impact craters on a moon's or planet's surface, scientists can determine the age of that surface and learn how much volcanism and weather erosion has taken place.

NAMES FOR COMMON PLANETARY FEATURES

Feature	Description
Catena	Crater chain
Chasma	Steep trough or canyon
Labes	Landslide structure
Lacus	Small plain
Macula	Dark spot
Mare	Large dark or low area
Mensa	Small, flat-topped prominence
Mons	Mountain or volcano
Patera	Shallow crater with complex edge
Planitia	Broad, low-lying plain
Planum	Large plateau
Regio	Large distinct area
Terra	Large highland region
Vallis	Twisting channel
Vastitas	Wide lowland plain

Basin

A basin is a large impact **crater.** Some of the largest surface features in the solar system are basins, including the South Pole–Aitken Basin, on the far side of **Earth's moon**, more than 2,092 kilometers across, and **Mercury**'s Caloris Basin, about 1,350 kilometers wide.

Ejecta

Ejecta is the pulverized and molten debris thrown from a volcanic eruption or from a **crater** during an impact. The carpet of ejected material that surrounds a crater is known as an ejecta blanket.

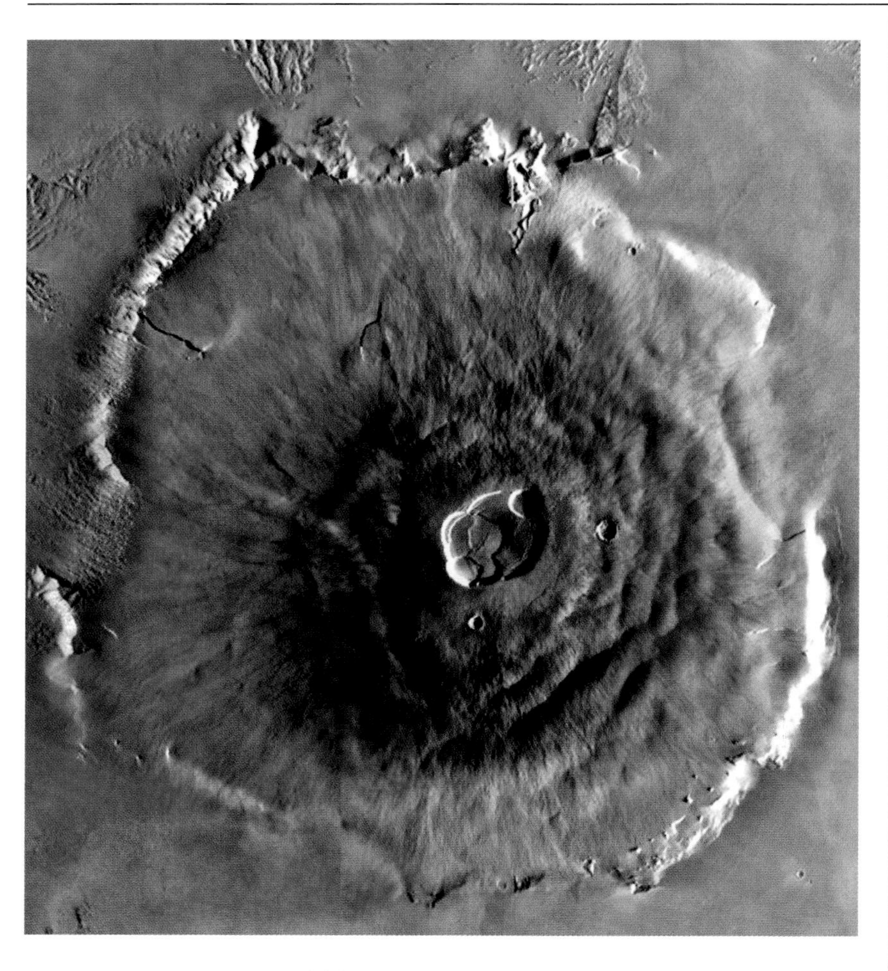

The caldera of Olympus Mons on Mars yawns wide in this mosaic created from images made by the Viking I orbiter. Because Mars has less gravity than other planets, its volcanoes balloon to great size. Olympus Mons climbs more than 26 kilometers high—making it the solar system's largest known volcano.

Volcano

In addition to **Earth,** several **planets** and **moons** in the **solar system** have volcanoes. This fact tells researchers about the body's geology. **Venus** and **Mars** are dotted with volcanoes built from magma deep inside. On Venus, extensive lava flows have given the planet thousands of volcanoes and a relatively smooth planetary surface. Mars, too, has volcanoes, but because Mars lacks Earth's moving plates and has less gravity than Earth or Venus, some have grown to great heights over their hot spots. Olympus Mons, for instance, is more than 26 kilometers high, about three times the height of Earth's Mauna Loa. Earth's moon also had active volcanism in its distant past, which resulted in vast lava flows across its impact **craters** rather than volcanic cones.

At least two other moons have dramatic volcanism. **Jupiter**'s satellite **Io** is the most volcanically active body in the solar system, erupting with more than a hundred times the lava of volcanoes on Earth. Io's magma comes not from internal radiation, but from friction created by the gravitational pull of Jupiter. **Neptune**'s **Triton** has active volcanoes that produce frigid plumes of nitrogen. The eruptions of these icy "cryovolcanoes" may begin when the **sun**'s faint heat warms pools of liquid nitrogen below the surface, making them boil.

Caldera

A caldera is a large depression at the summit or sides of a **volcano** caused by collapse or explosion. The caldera of **Mars**'s largest-known volcano, Olympus Mons, measures 90 kilometers by 60 kilometers.

Planetary Atmosphere

The atmosphere is the layer of gas that surrounds an astronomical body, such as a star, **planet**, or **moon**. Among the planets and moons of the **solar system**, **Venus**, **Earth**, **Mars**, **Jupiter**, **Saturn**, **Uranus**, **Neptune**, and **Saturn**'s moon **Titan** have significant atmospheres; **Pluto** has a very tenuous atmosphere, as do some large moons, and **Mercury** has almost none.

A body's mass and temperature determine whether it holds onto an atmosphere. Hot gases move around and are more likely to escape into space unless held by gravity. Small, hot Mercury cannot retain an atmosphere, but colder Titan, at about the same mass, can.

The atmospheres of the terrestrial planets differ. Earth's air, primarily nitrogen and oxygen, is enriched with oxygen by its plant life, which also removes carbon dioxide. The atmospheres of Venus and Mars consist mainly of carbon dioxide with small amounts of nitrogen. Venus's hot, heavy "air" presses on its surface at 91 times the surface pressure of Earth. Mars, less massive, has a thin atmosphere that increases in its southern summer as the southern polar ice cap vaporizes.

The atmospheres of the **gas giant** planets consist primarily of hydrogen and helium, with methane and ammonia also found on Jupiter and Saturn. Nitrogen and methane are the most abundant gases in Titan's smoggy, orange atmosphere, which may rain ethane and propane onto the moon's surface.

Magnetosphere

A magnetosphere is the area around a **planet** that is occupied by its magnetic field. The magnetosphere loops out from each magnetic pole, forming a shield between the planet and the **solar wind.** The side facing the **sun** is compressed by the solar wind, while the side away from the sun stretches out in an elongated magnetotail. Six planets in the **solar system**—**Mercury**, **Earth**, **Jupiter**, **Saturn**, **Uranus**, and **Neptune**—have significant magnetospheres.

The Sun

The Sun

The sun is a star and the center of our **solar system.** A G2-type star, a yellow dwarf, it falls into a common class, but it is brighter and more massive than about 90 percent of the stars in our **galaxy.** Like other stars, the sun is an enormous ball of hot gas—primarily hydrogen and helium—whose energy comes from nuclear fusion in its core. With a diameter of 1,392,000 kilometers, the sun could hold one million Earths if it were hollow. Its mass, 1.99 x 10^{30} kilograms, is about 333,000 times the mass of the **Earth** and represents more than 99 percent of the mass in the solar system. The temperature at the sun's visible surface, the **photosphere**, is 4,400 kelvins, but it reaches 16 million kelvins at its center and climbs again to 1 to 2 million kelvins in its **corona.** Like the **planets,** the sun rotates on its **axis.** Because the sun is not solid, the surface at its equator rotates once every 25.4 days, while the surface near the poles rotates about every 34 days.

The sun probably formed some 4.6 billion years ago from the condensing gases of a solar **nebula,** consisting mainly of hydrogen and helium with trace amounts of heavier elements such as calcium, sodium, and iron. It began as a protosun, a relatively cool, loose ball of gas that heated as its mass pulled in more gases and they collided. When temperatures and pressures had climbed high enough in the protosun's center, fusion began, releasing enormous amounts of energy. The sun is now about halfway through its life cycle. After another five billion years or so, it will run out of hydrogen to fuel its fusion; its core will collapse, and its outer layers will cool and expand, turning the sun into a red giant. Eventually, the outer layers will float away from the core, leaving behind a **white dwarf star.**

The sun's core is an unimaginably hot, high-pressure zone. At 16 million kelvins, with pressures 250 billion times higher than on the surface of the Earth, hydrogen nuclei fuse to create helium nuclei. The small amount of mass lost in this fusion process becomes huge amounts of energy in the form of photons, which radiate outward. The reaction also produces enormous numbers of the tiny, chargeless particles known as neutrinos. In all, the sun converts 600 million metric tons of hydrogen to helium every second, producing 400 trillion trillion watts of energy.

The radiation travels outward from the core through the radiative zone, its energy diminishing slightly as the photons collide with particles along the way. The last 20 percent of its trip takes the radiation through the convective zone, which transports the radiation via currents of gas that bubble to the surface. There, at the seething, grainy-looking photosphere, the radiation enters the sun's atmosphere as light. The solar atmosphere has two layers: the relatively narrow **chromosphere** and the extensive outer corona. Some of the gas in the corona moves fast enough to leave the sun's gravitational pull and fly into space as the **solar wind.** The charged particles of this wind spiral from the rotating sun at least as far as the orbit of **Neptune.**

The rotation of the sun's electrically charged gases creates a strong and complex magnetic field. Regions of the field rise through the photosphere into the corona, forming tangled loops that constantly break and reconnect. This magnetic field is probably responsible for many of the sun's most dramatic features. Where the field breaks through the photosphere, dark regions known as **sunspots** and bright active areas appear; enormous loops of gas called prominences and filaments, some of them larger than Earth, shoot forth; sometimes huge explosions, **solar flares,** erupt. The magnetic field's sphere of influence, known as the **heliosphere,** extends to the limits of the solar system.

Photosphere

The photosphere is the visible layer of the **sun.** Although the sun is not solid, the photosphere looks like a surface to human eyes because it is the region from which visible light is emitted. About 500 kilometers thick, it is cool relative to the rest of the sun, with a temperature of about 3,944 kelvins at its top.

THE SUN	
Average surface temperature	5505°C
Average core temperature	16,000,000°C
Rotation	25.4 days (at equator)
Equatorial diameter	1,392,000 km
Mass (Earth=1)	333,000
Density	1,408 kg/m^3
Surface gravity (Earth=1)	28.0

Active Region

An active region is an area of solar activity in the outer layers of the **sun**. Occurring where strong magnetic fields break through the **photosphere**, active regions emit intense radiation in a range of wavelengths. They may appear as **sunspots** in the photosphere, or as hot spots, called plages, in the **chromosphere. Solar flares**, prominences, and coronal mass ejections are other indications of active regions on the sun.

Sunspots

Sunspots have been known to observers for thousands of years. We now know that they are regions on the **sun**'s **photosphere** where a particularly strong portion of the sun's magnetic field slows gas rising to the surface. The center of a sunspot, depressed a little below the level of the surrounding gas, registers about 4,200 kelvins, as opposed to the photosphere's more usual 5,800 kelvins. Individual sunspots are typically twice the diameter of the **Earth.** They often appear in clusters, with the total number of sunspots increasing in an 11-year cycle. Early in each cycle, most of the sunspots appear around the 30° north and south latitudes on the sun; later in the cycle, they occur closer to the equator.

Chromosphere

The chromosphere is the layer of the **sun**'s atmosphere directly above the **photosphere.** A dim layer of gas about 2,000 kilometers high, it is seen as a pink glow around the outside of the **moon** during a **solar eclipse.** The chromosphere is hotter than the photosphere, reaching 10,000 kelvins in its upper layers. It is marked by spikes of gas called spicules that shoot up into the overlying **corona** and fade within a few minutes.

Corona

The corona is the extensive, wispy, hot upper layer of the **sun**'s atmosphere. During a **solar eclipse** it can be seen as huge, white sheets of gas streaming out from the body of the sun. Reaching out millions of kilometers from the sun's **chromosphere**, the corona is very tenuous—about ten trillion times less dense than air on Earth—and very hot, with temperatures reaching one to two million kelvins. This extreme heat may be carried by the sun's magnetic fields.

The corona changes constantly. Bright arches of hot plasma, known as coronal loops, at times extend from the sun's surface for thousands of kilometers.

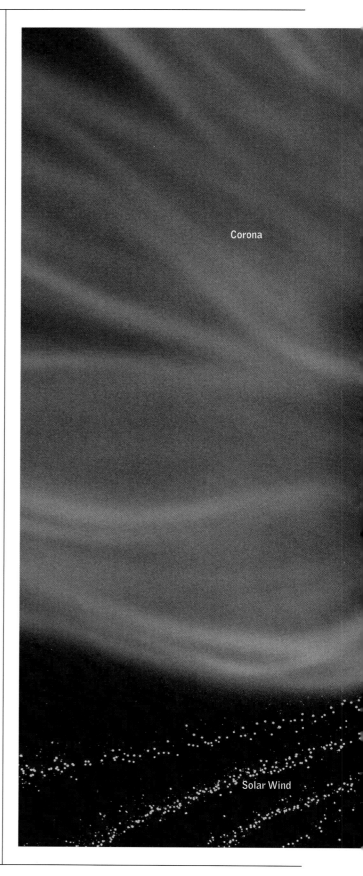

A cross section of our sun illustrates how it generates energy from nuclear fusion in its superheated, pressure cooker-like core. The energy travels outward through the radiative zone and the convective zone to the seething photosphere, the sun's visible layer, where it enters the sun's double-layered atmosphere—the chromosphere and the corona—as light.

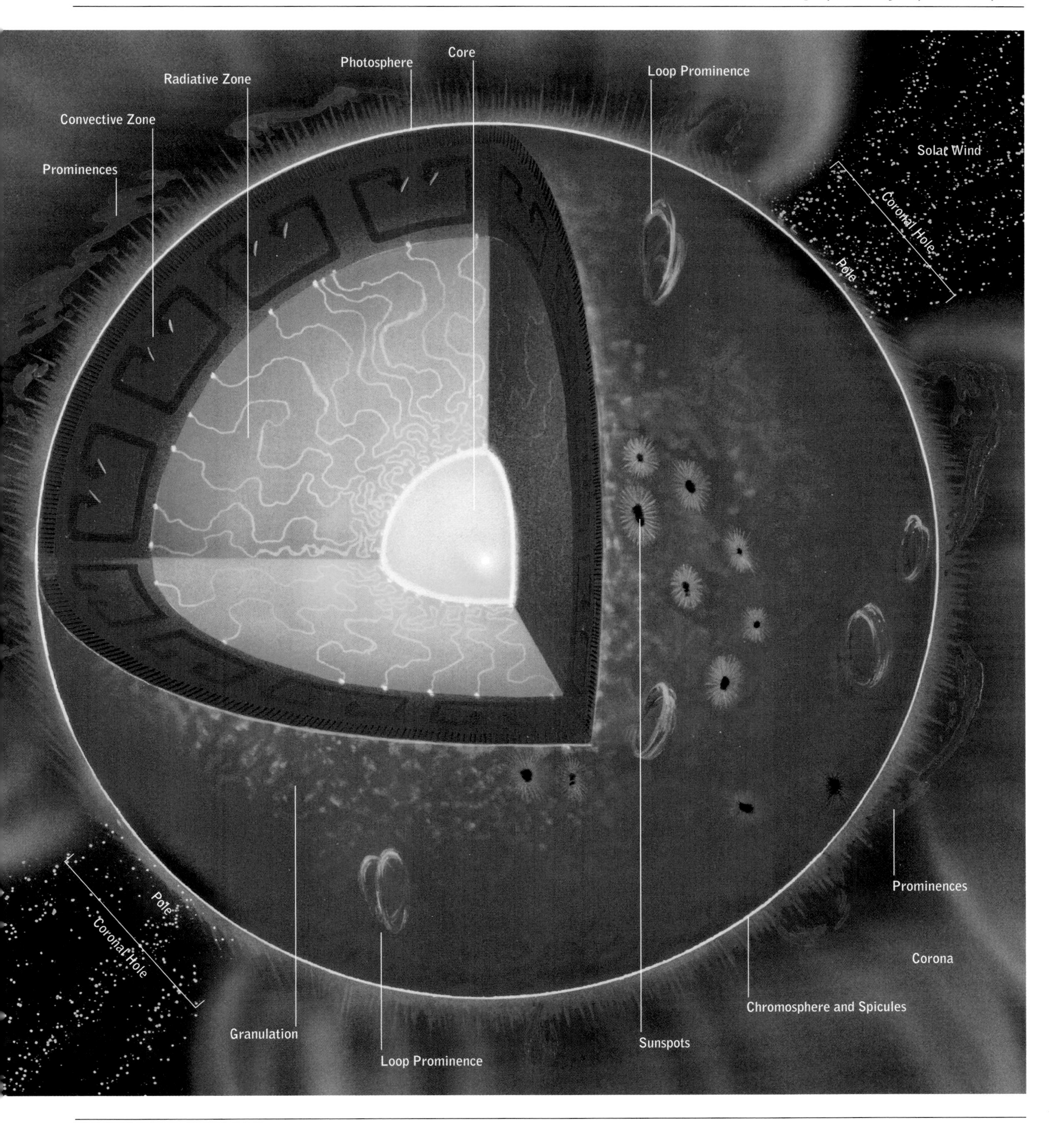
Core
Photosphere
Radiative Zone
Loop Prominence
Convective Zone
Solar Wind
Prominences
Coronal Hole
Pole
Prominences
Corona
Pole
Coronal Hole
Chromosphere and Spicules
Granulation
Sunspots
Loop Prominence

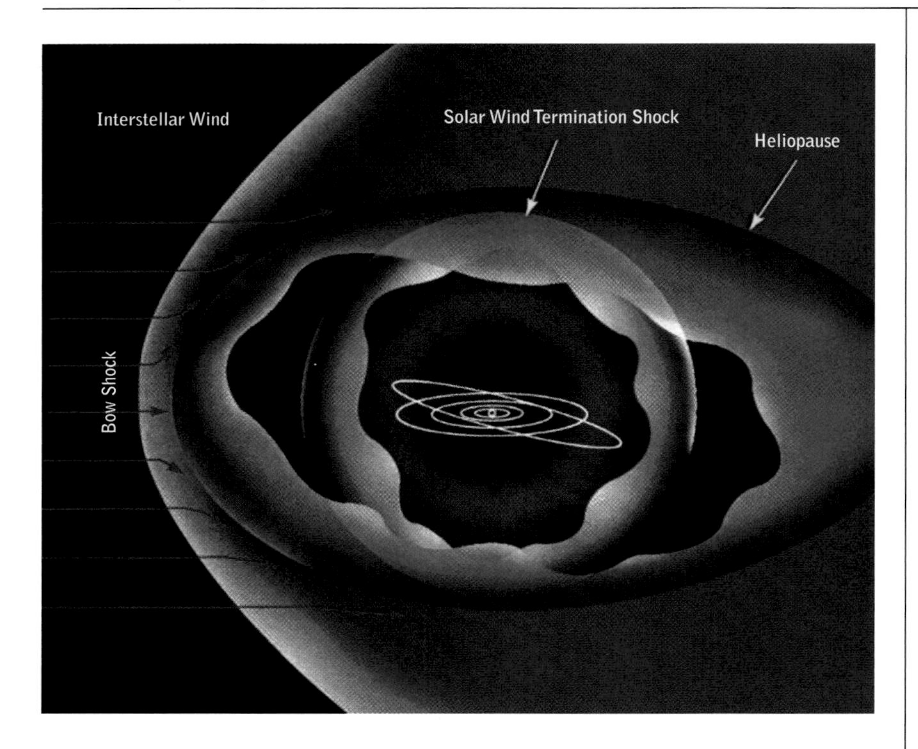

The heliosphere is a vast "bubble" of space created by the sun's solar wind and magnetic fields; it extends about 100 AU from the sun. At its outermost edge, called the heliopause, the solar wind meets the interstellar medium. If the interstellar wind is supersonic with respect to the solar wind, a shock occurs, and the heliopause bows.

Even more energetic are coronal mass ejections (CMEs), explosions of billions of kilograms of plasma thrown outward at high speeds. CMEs can result in solar storms. Coronal holes occur where magnetic field lines extend outward; these dark, low-density areas allow the **solar wind** to escape into space.

Solar Flare

Sudden eruptions on the **sun**'s surface, solar flares typically occur during the peak of the **sunspot** cycle. Violent releases of energy, they eject billions of tons of charged particles at more than 1,000 kilometers per second into space, as well as radiation ranging from radio waves to x-rays. In the few minutes that a flare usually lasts, its temperature can reach several million kelvins. When the charged particles of a solar flare reach **Earth**'s magnetic field, they can cause **auroras** and **geomagnetic storms**, disrupt **satellite** communications, and endanger **astronauts** in space.

Heliosphere

The heliosphere is the **sun**'s area of magnetic influence and the region in which the **solar wind** pushes back the **interstellar medium.** The heliosphere extends about 50 to 100 AU from the sun. At its boundary, called the heliopause, the pressure of the solar wind balances the pressure of the interstellar medium.

Solar Wind

The solar wind is an outpouring of atomic particles from the **sun**'s **corona.** Streaming out at one million tons of matter per second, the solar wind consists mostly of protons and electrons, with tiny amounts of silicon, sulfur, calcium, chromium, nickel, neon, and argon ions. This wind travels at up to 900 kilometers a second, with its fastest speeds occurring where it escapes from the sun through coronal holes.

Where the solar wind encounters planetary magnetic fields, it can cause **auroras** and press the field in toward the **planet.** It also makes **comets**' tails stream away from the sun. The solar wind can be felt as far as the heliopause at the limits of the **solar system.**

Solar Constant

The solar constant is the rate at which solar energy enters the top level of **Earth's atmosphere.** This amount actually varies depending on **Earth**'s distance from the **sun** and the amount of solar activity, but on average, at a distance of one **AU**, the solar constant is 1.35 kilowatts per meter squared. Other planets have solar constants measured in the same way.

Lunar Eclipse

A lunar eclipse occurs when the **moon** moves through **Earth**'s shadow; in other words, when the moon is on the opposite side of the Earth from the **sun** and is lined up along the plane of the **ecliptic.** Lunar eclipses always occur on or near the full moon.

In a total lunar eclipse, the moon enters the central, darker portion of the Earth's shadow, called the umbra. The moon turns a dark reddish color due to light scattered by **Earth's atmosphere.** The time period in which the moon is in complete shadow is called its totality. If it passes only through the outer, lighter portion of the shadow, called the penumbra, it dims to a light red color. In a partial eclipse, only part of the moon is cut off by Earth's shadow, and the moon looks as though it is passing through one of its phases. Total lunar eclipses typically occur once or twice a year. In contrast to **solar eclipses,** they can be seen from anywhere in the world the moon is visible.

Solar Eclipse

A solar eclipse occurs when the **Earth** passes through the shadow cast by the **moon.** This happens only during a new moon and when the moon is close to the plane of the **ecliptic.** Although the moon is much smaller than the **sun,** it is so much closer to the Earth that its apparent size in the sky is about the same, allowing its disk to cover the sun's completely during a total eclipse. The shadow cast by the moon onto the Earth's surface is no more than 270 kilometers wide and races across the Earth at more than 1,700 kilometers an hour. Therefore solar eclipses are visible to only a small number of people at any one time.

There are three kinds of solar eclipses. An observer standing within the darkest part of the moon's shadow, the umbra, witnesses a total eclipse as the disk of the moon covers the sun. Observers in the outer part of the shadow, the penumbra, see a partial eclipse, in which the sun is only partly obscured. And when the moon is farthest from the Earth in its **elliptical orbit,** its disk does not cover the sun completely, leaving a ring of bright light around its edges. This is called an annular eclipse.

Although a total eclipse lasts only a few minutes, it is spectacular. When the moon blocks the sun's surface, the sun's **corona** becomes visible, streaming into the darkened daytime sky. Observers should use proper filters or indirect viewing devices to prevent eye damage.

Solar Missions

In the 1960s the U.S. Pioneer missions 5 through 9 became the first probes to study the **sun,** collecting information on its magnetic fields, plasma, and surface. The German-American Helios probes 1 and 2 followed in the '70s. The closest craft to approach the sun, they swung as close as 45 million kilometers. The most fruitful era of solar observation began in the 1990s with the **launch** of several ambitious solar missions.

Ulysses

A joint project of **NASA** and the **European Space Agency** (ESA) launched in 1990, the Ulysses **spacecraft** was accelerated around **Jupiter** and sent into an orbit that allowed it to see the **sun** at polar latitudes during both the minimum and maximum of the solar cycle.

Yohkoh

Yohkoh—"sunbeam"—is a Japanese probe launched in 1991 with instruments from the U.S. and the U.K. These **detectors** study x-rays and gamma rays emitted from the **sun** during flares and other bursts of energy.

SOHO

Launched in 1995, the Solar and Heliospheric Observatory (SOHO) is a joint project of **NASA** and the **European Space Agency.** It was the first probe to be put into orbit around the **L1 Lagrangian point** between the **sun** and **Earth,** where its 12 instruments collect data on the sun's structure, atmosphere, and **solar wind.**

TOTAL LUNAR ECLIPSES 2005-2015

Date	Visible From
3 Mar 2007	Americas, Europe, Africa, Asia
28 Aug 2007	Americas, Eastern Asia, Australia, Pacific
21 Feb 2008	Americas, Europe, Africa, Central Pacific
21 Dec 2010	Americas, Europe, Eastern Asia, Australia, Pacific
15 Jun 2011	South America, Europe, Africa, Asia, Australia
10 Dec 2011	N. America, Europe, Eastern Africa, Asia, Australia, Pacific
15 Apr 2014	Americas, Australia, Pacific
8 Oct 2014	Americas, Asia, Australia, Pacific
4 Apr 2015	Americas, Asia, Australia, Pacific
28 Sep 2015	Americas, Europe, Africa, Western Asia, Eastern Pacific

TOTAL SOLAR ECLIPSES 2005-2015

Date	Visible From
29 Mar 2006	Europe, Africa, Western Asia
1 Aug 2008	Northeastern North America, Europe, Asia
22 Jul 2009	Eastern Asia, Pacific
11 Jul 2010	Southern South America
13 Nov 2012	Northern Australia, Southern Pacific
20 Mar 2015	North Atlantic, Faroe Islands, Svalbard

TRACE

NASA's Transition Region and Coronal Explorer (TRACE), launched in 1998, is collecting data on the **sun's** magnetic field so that scientists can study how the magnetic field affects the solar atmosphere and its activity. The data TRACE collects will complement the information gathered by **SOHO.**

Genesis

In 2001 **NASA** put another craft—Genesis—into orbit around the **L1 Lagrangian point.** Equipped with winglike collectors that pull in particles of the **solar wind,** Genesis will return to **Earth** in the fall of 2004. As it descends, it will drop a sample-return capsule that will be retrieved in midair by helicopter, and its tiny cargo of solar particles will be delivered to scientists for examination.

Laying Bare the Sun

William Harwood

In October 2003, one of the most powerful explosions ever recorded in the sun's outer atmosphere sent a vast cloud of electrically charged particles hurtling toward Earth like bird shot from a cosmic shotgun blast. Two days later, that blast slammed into our planet's protective magnetosphere, triggering spectacular auroral displays as showers of electrons and protons spiraled down magnetic field lines and crashed into the atmosphere above Earth's poles.

No one knows exactly what triggers the magnetic short circuits that apparently cause the sun's million-degree corona to suddenly generate such cataclysmic flares and even more powerful coronal mass ejections (CMEs)—mind-boggling explosions that periodically spew billions of tons of ionized gas into space. But a new generation of satellites is working to collect data needed to solve the sun's most puzzling riddles and to provide an early warning that would minimize the earthly effects of flares and CMEs.

"It's one thing to say a line of thunderstorms is coming," says George Withbroe, director of NASA's Sun-Earth Connection Division. "It's another thing to say, 'Oh, by the way, that line of thunderstorms includes tornadoes.' We want to improve the reliability of the predictions." For good reason. A single solar flare can unleash the energy equivalent of more than a billion thermonuclear bombs, accelerating charged particles to near the speed of light and increasing the temperature of an Earth-size region of the sun by tens of millions of degrees. Waxing and waning with the sun's 11-year sunspot cycle, flares may or may not be accompanied by CMEs. Along with disrupting power grids, both flares and CMEs can scramble communications, cause satellite malfunctions, and pose a radiation hazard to orbiting astronauts.

Even small variations in the sun's overall energy output can have a profound long-term effect on Earth. In the mid-17th century, for example, flares and CME activity virtually ceased for more than 50 years. The sun's total output dropped by only a quarter of one percent, but that was enough to produce the Little Ice Age, marked by especially harsh winters and a brief expansion of glaciers in the Swiss Alps. Understanding the sun's behavior will help us predict possible consequences to our own climate, which could go far beyond the onset of another ice age. "If there is long-term variability in the sun that can either increase or decrease global warming, that's important to understand," Withbroe says. "There are folks who study these issues who think maybe up to 30 percent of global warming in the past century is due to the sun getting brighter. Most of the warming appears to be caused by human activity, but if the sun has contributed 30 percent, that's important to know if you're trying to use the past in order to predict the future."

Studying the Sun with SOHO

Enter SOHO—the Solar and Heliospheric Observatory—a European Space Agency satellite launched by NASA in December 1995 that has quietly revolutionized knowledge about the sun's internal structure and how it interacts with the seething solar atmosphere. Equipped with a battery of light-splitting spectrometers, telescopes, and solar-wind monitors, SOHO orbits a point about 1.5 million kilometers from Earth, a gravitational eddy where it can remain with minimal effort. After eight years of operation, SOHO still has enough fuel for another seven years of observations.

SOHO was designed to study the entire sun, from its hidden core—where nuclear fusion reigns supreme—to its convective outer layers—where flares and CMEs are born and where the solar wind blows out into space. To probe the sun's interior, SOHO monitors ultralow-frequency sound waves that cause vast stretches of the visible solar surface to gently move up and down a few hundred meters every few hours. These sound waves, generated by surface convection, propagate throughout the sun's interior and return to the surface, causing the sun to vibrate like a gong.

How fast a given sound wave moves through the interior depends on its frequency as well as the temperature and composition of the material it passes through. By studying the precise nature of surface vibrations, helioseismologists can infer details about the sun's interior structure and gain insights about what drives the magnetic dynamo that presumably heats the corona and powers flares and CMEs.

Like Earth, the sun is layered; its thermonuclear core extends from the center to about 25 percent of the sun's radius. The so-called radiative zone, where energy is transported toward the surface by radiation, extends to 71 percent of the sun's radius. From there to the surface, energy is carried by turbulent convection. Data from SOHO indicate that the radiative zone rotates rigidly, as if it were solid. The convective zone, however, rotates differentially; areas near the equator come full circle every 25 days or so, while regions at higher latitudes spin more slowly.

"There's a lot of shear going on right there at the interface," Withbroe says. "That's a good place for magnetic fields in motion to wind up. So ultimately, the studies of the interior are going to give us clues as to how the whole dynamo works." Magnetic energy generated by that dynamo is believed responsible for heating the corona to its extreme temperatures. In a 100-kilometer-thick transitional zone between the sun's lower atmosphere and its corona, the temperature jumps from 10,000 kelvins to 1 million.

NASA's Transition Region and Coronal Explorer (TRACE) satellite, launched in April 1998, studies the sun's outer atmosphere with a high-resolution imaging system designed to complement SOHO's data. In its first year of operation, TRACE took more than 1.5 million images of the sun, providing new information on the mystery of coronal heating. Building on such successes, the United States, France, Germany, and the United Kingdom plan to launch STEREO—the Solar Terrestrial Relations Observatory—in 2006. It will consist of two identical spacecraft designed to study coronal mass ejections in three dimensions.

More Sun Observatories

NASA's next major sun spacecraft will be the Solar Dynamics Observatory (SDO), which will concentrate on high-resolution helioseismology from Earth orbit. "To really do helioseismology, you need a tremendous amount of data," says Withbroe. "We'll put SDO in geostationary orbit [35,786 kilometers above Earth's Equator] and have a dedicated antenna on the ground and take just huge quantities of data down." Scientists, he adds, will be able to "look below a sunspot, see its roots. You can see the roots of an active region, follow it as it crosses the surface of the sun. And we also have the capability to observe sound waves on the back side of the sun." NASA plans to launch SDO in April 2008.

Oscillating coronal loops—a rare phenomenon—occur when a filament eruption causes mayhem in the corona. TRACE, one of a new generation of instruments studying the sun, caught this early phase of the eruption. Such cataclysms affect Earth short-term; they also may contribute to long-range phenomena like global warming.

SDO will be the flagship mission in NASA's Living with a Star program, which Withbroe devised. In addition, NASA hopes to launch a small fleet of modest Solar Sentinel spacecraft in time for the next solar maximum, in 2011, to observe the far side of the sun, to characterize the evolution of CMEs and solar flares, and to improve the general forecasting of solar storms. Says Withbroe: "There are two places in the solar system we haven't been to—the sun and Pluto. And the sun is really unique. It's the only star we're going to be able to go visit for a very long period of time—unless we get some new physics." ■

The Mariner 10 spacecraft—the only mission to fly by this mysterious planet—captured this view of Mercury, which is seen by the naked eye only at twilight. The Mariner pictures show a dry landscape pockmarked by craters, something like the surface of the moon. The closest planet to the sun, Mercury remains largely unmapped and unexplored.

Mercury

Mercury

Mercury, the closest **planet** to the **sun,** has been known to astronomers since ancient times, despite the fact that the planet is visible to the naked eye only at twilight. At other times it is hidden within the sun's glare.

Most modern knowledge of Mercury is based on radar observations from **Earth** and on data received in 1974 and '75 from Mariner 10, the only mission to fly by the planet. Much about Mercury remains mysterious, and less than half its surface has been mapped.

Mercury has one of the most remarkable orbits and rotations in the **solar system.** Circling the sun at an average distance of only 57,910,000 kilometers, or 0.387 AU, its path is highly elongated, with its perihelion 23.8 million kilometers closer to the sun than its aphelion. Mercury pursues this orbit at a speedy 48 kilometers a second, completing its year in just 88 Earth days. Until the 1960s, astronomers believed that, like the **moon** circling the Earth, Mercury always presented the same face to the sun, so that its day was the same as its year. However, radar observations in 1965 showed Mercury's rotational period (the time it takes to make one revolution on itself) to be 58.6 Earth days. Interactions between the sun's pull and the planet's shape have locked its day into an exact three-to-two ratio with its year.

This results in some strange effects. Someone standing on the planet's surface (a very well-insulated someone) would see the sun rise every 176 days; at perihelion the observer would see the sun apparently stop in the sky and move backward for a while, before moving forward again as the planet's rotation caught up with its orbital speed.

In 1859 the French celestial mechanician Urbain Jean Joseph Leverrier reported a discrepancy between Mercury's predicted and actual orbits. Astronomers knew that the planet's perihelion would advance a little with each orbit—called **precession.** However, the amount of Mercury's precession was greater than Newtonian math could account for. Leverrier guessed that an unknown planet, which came to be known as Vulcan, was tugging on Mercury, but this planet was never found.

In 1915 **Albert Einstein** published a paper called "Explanation of the Perihelion Motion of Mercury by Means of the General Theory of Relativity." Einstein wrote that Mercury's orbit was affected by the enormous mass of the nearby sun, which distorted the fabric of **space-time** around it into a kind of gravity well. Einstein's formulas exactly accounted for Mercury's precession, helping to confirm the General Theory.

Mercury's diameter of 4,879 kilometers makes it the second smallest planet in the solar system, after **Pluto.** And yet Mercury is surprisingly dense for such a small body, which tells astronomers that it probably has a large, iron-rich core, accounting for perhaps 75 percent of the planet's radius.

Just how Mercury got this way is unclear. The planet might have formed from different materials than the other terrestrial planets due to its close-in orbit; or the young sun might have vaporized many of Mercury's lighter elements; or these lighter substances might have been knocked off the planet by a collision early in its formation.

Another unsolved mystery concerns Mercury's magnetic field. According to planetary science, it should not have one, yet Mariner 10 detected weak but significant magnetism. This implies that, like Earth, Mercury has a rapidly rotating, molten iron core—but Mercury spins very slowly, and the core of a planet this small should have solidified long ago. One theory speculates that the solid core is surrounded by a shell of iron mixed with sulfur, which would keep the iron liquid.

Hugging the sun as it does, Mercury's temperatures soar to 430°C on its dayside and drop below -183°C on its nightside. Pictures from Mariner 10 show an arid, cratered surface like the moon's. One huge crater, Caloris Basin, is 1,350 kilometers wide. The impact that made it was so powerful that its shock waves were felt on the opposite side of the planet, creating a hilly terrain.

Like the moon, Mercury has smooth plains probably formed by cooling lava earlier in the planet's history. Long, winding cliffs, or scarps, meander across the little planet. They may have formed as the young planet cooled and shrank.

Although Mercury is too hot to retain any significant atmosphere—the superheated gases would simply boil into space—instruments have detected the barest wisps of hydrogen, helium, sodium, and potassium. These elements may come to Mercury via the **solar wind** or may emanate from within the planet itself. In addition, radar astronomers have observed echoes from Mercury's poles that suggest the presence of water ice, of all things. Possibly, the permanently shaded craters at the poles are cold enough to hold on to ice that either remains from earlier ages or was delivered by colliding **comets.** Or the radar echoes may not indicate water ice, but some other reflective substance, such as sulfur. Like so many other aspects of Mercury, this one is still an enigma.

MERCURY	
Average distance from the sun	57,910,000 km
Revolution	88 days
Average orbital speed	47.9 km/s
Average temperature	167°C
Rotation	58.6 days
Equatorial diameter	4,879 km
Mass (Earth=1)	0.055
Density	5,427 kg/m^3
Surface gravity (Earth=1)	0.38
Known satellites	none

Missions to Mercury

Mercury has been visited by only one probe, Mariner 10. Launched by **NASA** in 1973, Mariner 10 used the gravity of **Venus** to whip itself around the **sun** three times, visiting Mercury once in each **orbit** during March and September of 1974 and March of 1975. The flyby photographed about half of the planet's surface, measured its mass and magnetic field, and registered trace amounts of an atmosphere.

In August 2004 NASA launched another gravity-assisted mission, called Messenger, to both Venus and Mercury. When the probe reaches Mercury in 2009, it will photograph almost all the previously unseen surface and investigate Mercury's many mysteries: its density, the structure of its core, its magnetic field, its tenuous atmosphere, and the reflective material at its poles. And in 2011, if all goes well, a joint project of the **European Space Agency** and Japan, known as BepiColombo, will send two orbiters to Mercury for further exploration.

Venus

VENUS

A brilliant beacon in the morning and evening, Venus is the brightest object in the night sky after the **moon.** This is partly because it is close to the **Earth:** Venus is the second **planet** from the **sun,** with a mean orbital distance of 108.2 million kilometers, or .723 **AU.** But

This false-color radar image—a mosaic of Magellan images—shows the topographical variation of Venus. Although the elevations appear distinct, their heights are very similar. The difference between the shallowest and tallest point on Venus is a matter of some 20 meters.

the brightness of Venus is also due to a blanket of reflective clouds, an indication of its intensely thick atmosphere. Because its orbit is closer to the sun than Earth's, Venus exhibits phases, like **Earth's moon.** In 1610 **Galileo** commented on this fact with the cryptic note: *"Cynthiae figuras aemulatur mater amorum,"* or, "the mother of love [Venus] imitates the figure of Cynthia [the moon]."

In size and mass, Venus is Earth's sister planet. Its radius of 6,052 kilometers is just short of Earth's, and its mass is 82 percent of Earth's. With an orbit of 224.7 Earth days, its year is also similar to our planet's, although its spin is retrograde, or west-to-east, unlike that of the other inner planets. The rotation of Venus is very slow: about 243 Earth days, making its day longer than its year.

Until the 1950s many people thought that beneath its clouds Venus would turn out to be a steamy, welcoming, tropical world. But microwave readings from the planet in 1958 indicated that it might be much hotter than tropical, and these readings were confirmed by data from a series of probes, including the Venera, Mariner, and Magellan missions. The picture they painted was not a pretty one. Venus is not only inhospitable, but also an arid inferno of crushing pressures and acid clouds.

In the first place, Venus is even more torrid than its proximity to the sun would indicate. Its mean surface temperature is a blistering 464°C , hot enough to melt lead, making Venus the hottest planet in the **solar system.** This heat is fostered by the now-infamous greenhouse effect. Sunlight that passes through Venus's clouds is absorbed by the planet's surface and released again as infrared radiation, which is trapped next to the planet by Venus's carbon dioxide–rich atmosphere. This greenhouse effect heats the planet by hundreds of extra degrees.

Although Earth also exhibits a greenhouse effect, it is much less extreme due to the differences between the two planets' atmospheres. Venus's atmosphere is

VENUS	
Average distance from the sun	108,200,000 km
Revolution	224.7 days
Average orbital speed	35 km/s
Average temperature	464°C
Rotation: retrograde	243 days
Equatorial diameter	12,104 km
Mass (Earth=1)	0.815
Density	5,243 kg/m^3
Surface gravity (Earth=1)	0.91
Known satellites	none

IMPORTANT MISSIONS TO VENUS				
Spacecraft	Type	Nationality	Launch Date	Encounter Date
Mariner 2	Flyby	U.S.	Aug 1962	Dec 1962
Venera 4	Lander	U.S.S.R.	Jun 1967	Oct 1967
Mariner 5	Flyby	U.S.	Jun 1967	Oct 1967
Venera 7	Lander	U.S.S.R.	Aug 1970	Dec 1970
Venera 9	Orbiter/lander	U.S.S.R.	Jun 1975	Oct 1975
Pioneer Venus 1 & 2	Orbiter/probe	U.S.	May 1978; Aug 1978	Dec 1978
Magellan	Orbiter	U.S.	May 1989	Aug 1990

96 percent carbon dioxide, and unlike Earth, the planet contains (apparently) no plant life and no liquid water to absorb the gas. The thick, dense atmosphere of Venus crushes down on the planet's surface at about 91 atmospheres, or 91 times the pressure of air at Earth's surface. A permanent, 19 kilometer-deep layer of yellowish clouds hangs 50 kilometers above Venus's surface. Composed of concentrated sulfuric acid, the clouds emit a drizzle of acid rain that evaporates before it reaches the perpetually dry terrain.

The air moves sluggishly at the surface, but 60 kilometers above it, winds whip around the planet at about 100 meters a second. That means Venus's upper atmosphere circles the entire planet every four Earth days, much faster than Venus's own rotation. In general, the atmosphere transports heat efficiently around the planet, so that the dayside and the nightside, and the northern and southern hemispheres, have much the same (high) temperature year-round.

In 1990 the Magellan orbiter thoroughly mapped the surface of Venus, yielding an extraordinarily detailed picture of the planet's terrain. The orbiter confirmed findings of earlier probes, which showed a relatively smooth surface with two highlands: Ishtar Terra in the north and Aphrodite Terra along the equator. Ishtar Terra is bounded on three sides by mountains, including the Maxwell Montes, which reach 11 kilometers above the surface, higher than Earth's Mount Everest.

Because the planet's surface is fairly smooth and flat and exhibits relatively few **craters**, scientists believe that it has been resurfaced by cooling lava. Thousands of **volcanoes** dot the plains of Venus, although it is not clear if any are still active. The large amounts of sulfur found in Venus's clouds were probably supplied by volcanoes at one point in its history. Heat from inside the planet has also fractured its surface, giving rise to odd features called coronae—circular structures pushed up by heat from below—and arachnoids—domes with weblike cracks.

The planet's size and shape, so similar to Earth's, as well as its evidence of volcanism, lead scientists to believe that Venus should have a liquid iron outer **core,** like Earth's. And yet instruments have detected no magnetic field around the planet. This may be because its spin is too slow to generate such a field, or its core may be significantly different from Earth's. Clearly, the planet, so similar and yet so different from our own, has much left to reveal.

Missions to Venus

After the Soviet missions Venera 1, 2, and 3 lost contact before reaching Venus, the U.S. probe Mariner 2 performed a successful flyby of the **planet** in 1962, picking up high temperature readings. This mission was followed by Venera 4, 5, and 6, which managed to penetrate the atmosphere and send back some data before losing contact—possibly after being crushed or melted by the extreme conditions on the planet. Venera 7 was the first probe to survive all the way to Venus's surface, followed over the years by a steady stream of Veneras up to number 16, as well as the Soviet craft Vega 1 and 2. These missions added to our knowledge of Venus's dense atmosphere as well as its volcanic surface.

Pioneer Venus 1 and 2 and Magellan were the most significant U.S. missions. The Pioneer Venus 1 orbiter radar-mapped much of the planet's surface and gathered more information about the atmosphere, which Pioneer Venus 2 supplemented with four atmospheric probes. In 1990 the Magellan orbiter mapped almost 98 percent of the planet's surface using radar, with a resolution better than 300 meters.

Earth

Earth

The third **planet** from the **sun** has several characteristics that set it apart from the other eight. The most obvious one is the presence of liquid water on almost 71 percent of its surface. The planet's nitrogen-oxygen atmosphere is unique as well, as are the constantly shifting plates of its crust. And, as far as we know, Earth is the only planet to harbor life.

The Earth is the largest and most massive of the rocky planets, with a mass of 597.4 x 10^{22} kilograms and a radius of 6,378 kilometers. It orbits an average of 149,597,900 kilometers from the sun, completing one orbit every 365.256 days. Its **axis** is tilted 23 degrees to the plane of the **ecliptic,** giving it distinct seasons in its northern and southern hemispheres. Studies of the Earth's mass, seismic patterns, and magnetic field indicate that the planet has a solid iron inner core heated to 5,000 kelvins, almost as hot as the surface of the sun. This core rotates even faster than the Earth's surface. Above it is a liquid iron outer core and then a mantle, less dense than the core. The mantle is hot enough to be stretchy and plastic, though not quite liquid. The thin, brittle crust rides on top of the mantle and forms the surface that holds the ocean and supports life.

The Earth seems to be the most geologically active planet in the **solar system.** Unlike the other planets, its crust is broken into plates that are in constant motion, borne along by currents of heat from below. Where plates collide, the crust is pushed up into mountains and earthquakes occur. Where they pull apart in mid-ocean, molten rock (magma) wells up from below to form new underwater mountain ranges. Magma also shoots upward through hot spots to give the planet active **volcanoes.** Where the Earth's crust rises above the ocean, it forms the landmasses known as islands and continents.

EARTH	
Average distance from the sun	149,600,000 km (1 AU)
Revolution	365.256 days
Average orbital speed	29.8 km/s
Average temperature	15°C
Rotation	23.9 hours
Equatorial diameter	12,756 km
Mass	5,974,000,000,000,000,000,000,000 kg
Density	5,515 kg/m^3
Surface gravity	9.81 m/s^2
Known satellites	1
Largest satellite	Earth's moon

Earth's most distinctive feature is its ocean. Water covers more than two-thirds of the planet's surface in the ocean, lakes, rivers, and polar and alpine ice. Water is also present as vapor in the atmosphere. In fact, unlike the other planets, the Earth receives just the right amount of heat to hold water in three forms: liquid, solid (ice), and gas (water vapor).

Where did this water come from? Probably most or all was incorporated into the Earth when it formed 4.6 billion years ago and then escaped into the atmosphere through fissures and volcanoes. Some may also have arrived via **comets,** which are essentially dirty snowballs from space. Just how much water came from which source is still a matter of debate. Once **Earth's atmosphere** was formed, however, the water was trapped on the planet, endlessly circulating via the water cycle. The water we drink today is billions of years old.

Earth's ocean holds most of its life. It serves as a planetary thermostat, storing heat in summer and releasing it slowly in winter, thereby keeping temperature swings moderate. It also absorbs carbon dioxide, some of which is taken up by plankton, removing it from the atmosphere.

The ocean is responsible for the atmosphere that Earth currently has. When the planet first formed, it had a thin hydrogen-helium atmosphere, like that of **Mercury.** In time, as volcanoes released gases from inside the Earth, a thick layer of water vapor and carbon formed. When liquid water began to cover the surface, that ocean absorbed much of the carbon dioxide, some of which was incorporated into the shells of sea creatures. As plants began to grow in the ocean and on land, they, too, took in carbon dioxide, releasing oxygen. Eventually the atmosphere came to be dominated by nitrogen, formerly just a trace element, and oxygen, along with small amounts of argon, neon, helium, hydrogen, and xenon.

Earth's atmosphere, essential to life, wraps the planet in a thin sheet of gases that merges into space at around 600 kilometers above the ground. The lowest

Seen from space, Earth exhibits a striking difference from the other eight planets of the solar system: more than two-thirds of its surface is covered with water. Possibly for this reason, Earth is the only planet in the solar system known to support life.

layer, the **troposphere,** is the densest, followed by the **stratosphere, mesosphere, thermosphere,** and **exosphere.** Temperatures steadily decline with height up to the thermosphere, and then climb again in this region where the stray molecules are heated by the sun's high-energy radiation up to 375 kelvins or more.

Both natural and human activities affect the atmosphere. In recent centuries the clearing of forests and the burning of fossil fuels have added significant amounts of carbon dioxide to the air, and most scientists now agree that this gas is contributing to a global warming trend. In addition, chlorofluorocarbons (CFCs), chemicals found in refrigeration and spray cans, entered the atmosphere in the later 20th century and ate holes into a layer of ozone (O^3) that protects life on the ground by absorbing ultraviolet radiation. CFC emissions have been greatly reduced, and countries are struggling to deal with global warming, but both continue to threaten the atmosphere.

Like some other planets, Earth has a magnetic field. Scientists believe it is generated by the heat of Earth's iron inner core, which causes convection currents in the liquid iron outer core. Magnetic field lines loop out from Earth's south magnetic pole and return at the north magnetic pole: These magnetic poles are about 11 degrees away from the planet's rotational poles and

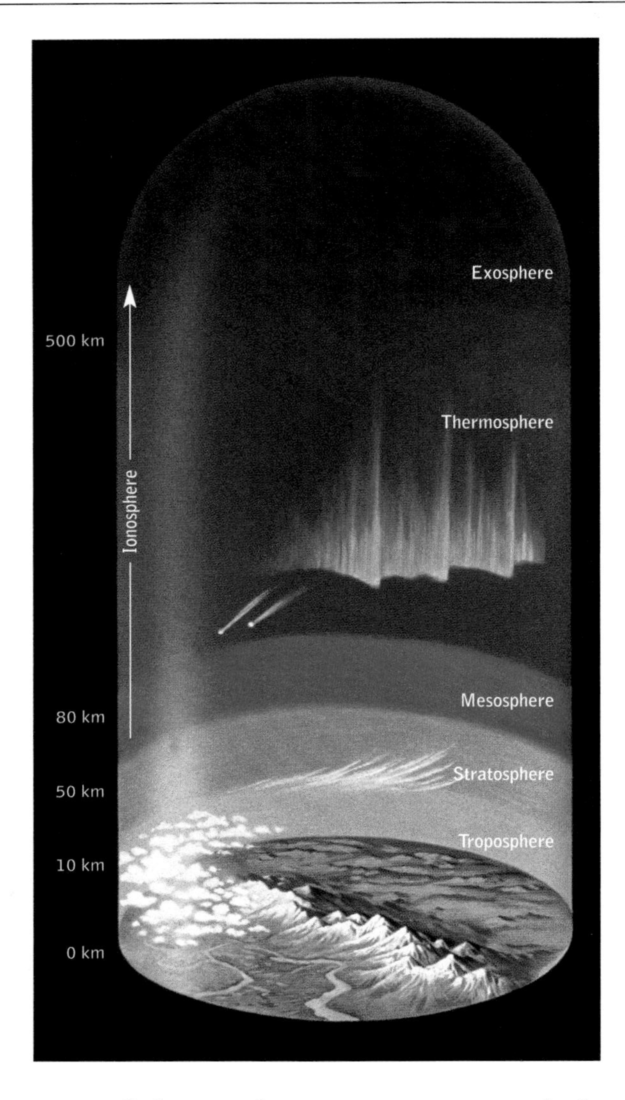

Pressing down on Earth, a layered atmosphere of gases about 600 kilometers high supports life, recycles water, and protects the planet from harmful radiation. Clouds form and weather occurs in the lowest layer, the troposphere, the densest part of the atmosphere.

move a little every day. For reasons not completely understood, the polarity of these magnetic fields reverses at irregular intervals ranging from tens of thousands to hundreds of thousands of years. The magnetic fields as a whole form a **magnetosphere,** a magnetic shell.

Biosphere

All the parts of the **Earth** that contain life, taken together, are known as the biosphere. Having begun in Earth's ocean more than three billion years ago, life in one form or another can now be found in a layer about 20 kilometers deep from the greatest depths of the ocean to the lower layers of the atmosphere; 97 percent of the Earth's biosphere falls in its ocean.

As far as we know now, Earth is the only **planet** that possesses life.

Earth's Atmosphere

Earth's atmosphere is the layer of gases, held close to the planet by gravity, that supports life, recycles water, and protects the surface from damaging radiation. It consists primarily of nitrogen (78.08 percent), oxygen (20.94 percent), argon (0.93 percent), and trace gases including carbon dioxide, helium, hydrogen, and xenon. The atmosphere as a whole is about 600 kilometers high, although its upper layers are very tenuous and merge gradually into space. It can be broken into five layers—**troposphere, stratosphere, mesosphere, thermosphere,** and **exosphere**—characterized by density, composition, and temperature. Almost all the atmosphere's mass, and all its weather, is concentrated in the troposphere. Air pressure at sea level is about 100 kilopascals (1000 millibars), decreasing approximately by half with every 5.5 kilometers of altitude.

Troposphere

Lowest level of Earth's atmosphere, the troposphere extends about ten kilometers from the planet's surface. It is the densest region of the atmosphere and, except for the **thermosphere,** the warmest, with the average temperature at **Earth**'s surface 15°C. Almost all the atmosphere's water vapor is in the troposphere, where most clouds form and where weather occurs.

Stratosphere

The stratosphere is the layer of the atmosphere that begins just above the **troposphere,** 10 kilometers high, and extends to 50 kilometers. Strong, steady winds blow in the thin, dry air of the stratosphere, making it the preferred level for jet travel. This is the region of the ozone layer that absorbs most of the **sun**'s ultraviolet radiation, preventing it from reaching the ground.

Mesosphere

The atmospheric layer just above the **stratosphere** is the mesosphere. In this zone, which extends from 50 to 85 kilometers above Earth's surface, temperatures drop sharply, bottoming out at about -93°C.

Thermosphere

The thermosphere is the atmospheric layer beginning 85 kilometers above **Earth** and soaring up to 600 kilometers. X-rays and other short-wave radiation from the **sun** energize the sparse molecules of the thermosphere so that temperatures rise sharply, climbing to 1727°C. Because these heated molecules are so few and far apart, however, the thermosphere would feel extremely cold to humans.

Exosphere

The exosphere marks the outer boundary of **Earth's atmosphere.** Beginning at the top of the **thermosphere,** at altitudes around 600 kilometers, the gases of the exosphere eventually merge into space. At these heights, they consist mostly of thick wisps of hydrogen and helium.

Ionosphere

The ionosphere is the region of **Earth's atmosphere** in which the **sun**'s radiation ionizes molecules so that they reflect high-frequency radio waves back to Earth. It begins in the upper parts of the **mesosphere** and extends throughout the **thermosphere** and into the **exosphere.** The ionosphere has proved very useful in promoting long-range radio transmission.

Magnetosphere

Earth's rotating iron core creates electrical currents that turn the Earth into a giant magnet. Looping out from Earth's south magnetic pole and returning at the north magnetic pole, magnetic field lines form a huge shield, or magnetosphere, around the planet.The magnetosphere protects the Earth from the particles of the **solar wind;** because they are electrically charged, the particles are trapped by the magnetic lines of force in two enormous rings called the **Van Allen belt.**

The magnetosphere is not, in fact, a sphere. On the dayside of the Earth, where it is blasted by the solar wind, it is pressed into a flattened arc extending about ten Earth radii into space. On the nightside it streams away from the solar wind for hundreds of Earth radii in a magnetotail. The boundary between the magnetosphere and the solar wind is called the magnetopause.

Van Allen Belts

Discovered in 1958 by the first U.S. artificial **satellite,** Explorer 1, and named after physicist James Van Allen, the Van Allen belts are two doughnut-shaped magnetic rings around the **Earth** that contain charged particles trapped from the **solar wind.**

The inner belt, roughly one Earth radius above the Equator, contains primarily high-energy protons and electrons. The outer belt, about three Earth radii above the Equator, holds mainly lower-energy electrons.

Aurora

Glowing, colorful, shifting lights in the high atmosphere, auroras are known as the aurora borealis, or northern lights, in the Northern Hemisphere, and the aurora australis, or southern lights, in the Southern Hemisphere. They are most commonly seen in regions around the 67° north and south latitudes near the magnetic poles.

Auroras are caused by showers of charged particles that escape from the **Van Allen belt** and cascade into the **ionosphere.** As they collide with atmospheric gases, the gases fluoresce, giving off light whose color depends on the gas emitting it. The most spectacular auroras occur during **geomagnetic storms,** such as those caused by coronal mass ejections.

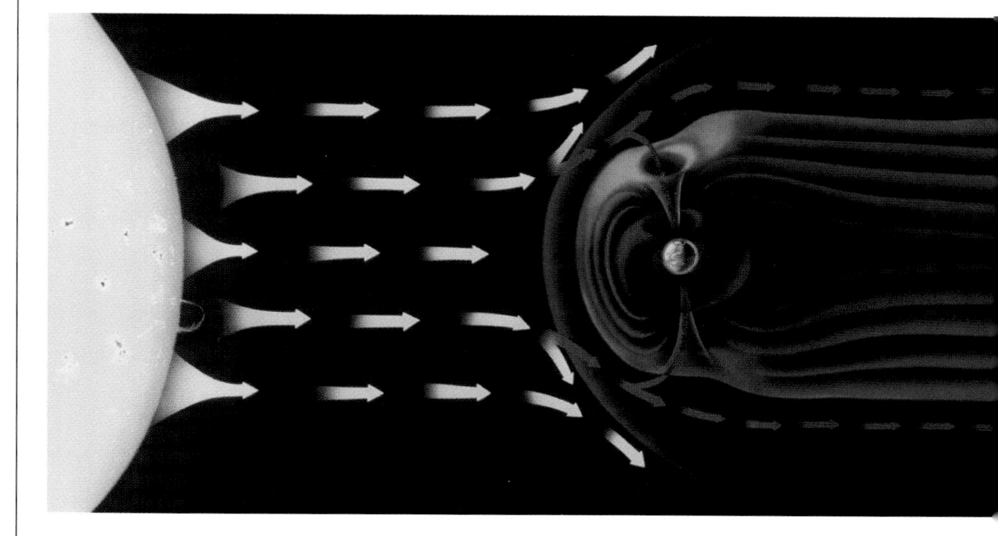

When the Earth's powerful magnetic field (called the magnetosphere and shown here in blue) meets the solar wind (the stream of magnetic particles pushed out by the sun, shown here in yellow), the clash traps some of the sun's particles in the doughnut-shaped Van Allen belts (in red).

Geomagnetic Storm

A geomagnetic storm is a major—often worldwide—disruption of **Earth**'s magnetic field caused by intense solar activity, such as a solar flare or a coronal mass ejection. Two or three days after the solar event, a blast of **solar wind** reaches the **magnetosphere,** compressing and intensifying it. Substorms lasting a half hour or so may occur during a geomagnetic storm, raining charged particles down through the upper atmosphere and causing brilliant **auroras.**

Earth Impact Crater

If **Earth**'s surface were not constantly being smoothed and shaped by plate tectonics, wind, water, and living organisms, **craters** would be a common feature of its landscape. Just like the other terrestrial planets, the Earth has been bombarded by meteorites, **asteroids,** and **comets** throughout its history, particularly in the early days of the **solar system.** Most of these craters have been erased, but about 150 remain to demonstrate just how devastating such an impact can be. Perhaps the most famous is the Chicxulub (pronounced CHEEK-sho-loob) Basin buried under sediment on Mexico's Yucatán Peninsula. At 170 kilometers across, it is evidence of an asteroid impact that may have contributed to mass extinctions at the end of the Cretaceous period, 65 million years ago. An even larger crater was discovered off the northwestern coast of Australia in 2004. Scientists speculate that this crater, dubbed Bedout, may be evidence of an even greater extinction at the end of the Permian period, 251 million years ago.

Tides

A tide, in its astronomical sense, is the movement or stretching of a large mass that is orbiting in a gravitation field. In its most familiar form, a tide is the regular variation in the surface of **Earth**'s ocean. Both the **sun** and the **moon** exert a gravitational pull on the Earth. The moon's influence is stronger, since it is so much closer. This pull is not equal over the entire mass of the Earth; it is stronger in parts of the Earth that are closer to the moon as Earth rotates. This uneven stress creates tidal forces that deform the Earth's shape, particularly the shape of its liquid ocean. As the Earth rotates, the ocean bulges outward both on the side facing the moon and on the side away from it. When the sun, moon, and Earth are aligned at the new and full moons—in **syzygy**—tides are highest and are called spring tides. When the sun and moon form a right angle from the Earth—called quadrature—at the moon's first and third quarters, tides are lowest and are called neap tides.

Solstice

Solstice refers to both a time of the year and a place in the sky. The summer solstice, usually June 21 in the Northern Hemisphere, is the day with the longest period of daylight and the point where the **sun** reaches its highest altitude in the sky as seen from this part of the world. The winter solstice, usually December 21 in the Northern Hemisphere, marks the day with the shortest period of daylight; it is also the point in the sky at which the sun reaches its lowest altitude in the year. The dates for the summer and winter solstices are the opposite in the Southern Hemisphere.

Equinox

An equinox is one of two points on the **celestial sphere** at which the celestial equator crosses the **ecliptic.** This event occurs when the sun, as viewed from **Earth,** appears to cross Earth's Equator, because Earth rotates on an **axis** tilted about 23 degrees from the plane of its orbit around the sun. When the sun seems to move north across the Equator (about March 21), this vernal equinox marks the beginning of spring. The sun seems to move south across the Equator (about September 22) on the autumnal equinox. On these days, day and night are of equal length.

Ecliptic

Viewed from Earth the ecliptic is the plane in which the sun appears to move; it is inclined at an angle of 23.27 degrees from the Earth's Equator. The sun thus appears to cross Earth's Equator twice each year with the seasons.

LARGEST EARTH IMPACT CRATERS

Name	Diameter (in km)	Age in Years	Location
Sudbury	200	1,850,000,000	Ontario, Canada
Bedout	200	251,000,000	Australia
Chicxulub	170	65,000,000	Mexico
Acraman	160	570,000,000	Australia
Vredefort	140	1,970,000,000	South Africa
Manicouagan	100	212,000,000	Quebec, Canada
Popigai	100	35,000,000	Russia
Puchezh-Katunki	80	220,000,000	Russia
Kara	65	73,000,000	Russia
Beaverhead	60	600,000,000	Montana, U.S.

Painting the sky with light, the aurora borealis, or northern lights, creates an Impressionist landscape in Alaska. When energy-charged particles escape from the Van Allen belt and collide with gas molecules in the Earth's upper atmosphere, the clash produces patterns of light, whose colors depend on the type of gas involved.

Earth's Moon

Earth's sole natural satellite, the moon is locked in synchronous orbit with Earth. The familiar near side (top) displays a rocky surface marked with light-colored craters and dark-shaded "seas" filled with lava rocks. The far side (bottom), riddled with craters, has a thicker crust.

Moon

The moon is **Earth**'s only natural satellite. With a diameter of 3,476 kilometers, it is one-quarter the size of the Earth and the fifth-largest satellite in the **solar system.** Some astronomers have even suggested that the Earth and moon are close enough in size that they should be considered a double-planet system.

In 1969 the moon became the first and, so far, the only extraterrestrial body visited by humans. The **Apollo** astronauts came home with 382 kilograms of lunar rocks that gave scientists insight into the age, composition, and origins of the moon.

The moon orbits the Earth at an average distance of 384,400 kilometers. It has a synchronous rotation, which means that it rotates on its **axis** at the same rate that it circles the Earth, which is why the same side always faces our planet. The moon's orbit and its day are therefore the same length: 27 days, 7 hours, and 43 minutes. The moon's gravitational pull causes tides in Earth's oceans. At the same time, high tides on the Earth exert a gravitational tug on the moon, giving its orbit a little extra energy, so that it is gradually moving away from the Earth at the rate of 3.8 centimeters a year. The moon's mass, 7.35 x 10^{22} kilograms, is only about 1/80th as large as Earth's, and its density is considerably less as well, 3,340 kilograms per cubic meter (compared to 5,515 kg/m^3 for the Earth). Scientists estimate that the moon has a small iron-rich core that occupies about 20 percent of the satellite's total radius.

The moon is an airless, dry, rocky world whose main surface features, its **craters** and maria ("seas"), are clearly visible to observers on Earth. Millions of impact craters mark the surface of the moon, ranging in size from the immense South Pole–Aitken Basin, more than 2,092 kilometers across, to microcraters less than one millimeter wide. Lunar mountain ranges up to 7,920 meters high rise around the biggest impact basins. Unlike mountains on Earth, they are the broken remains of the rims of impact basins rather than the result of tectonic movement. The dark gray, smooth-looking expanses that ancient observers called maria never held water. They are impact basins that were resurfaced in the distant past with cooling lava from the moon's interior. Almost all the moon's maria are on the side facing the Earth; the far side is

MOON	
Average distance from Earth	384,400 km
Average orbital speed around Earth	1.02 km/s
Average temperature	-20°C
Equatorial diameter	3,476 km
Mass (Earth=1)	0.01
Density	3,340kg/m^3
Surface gravity (Earth=1)	0.17

much more cratered and mountainous.

The moon seems to be geologically quiet, with no active **volcanoes** and no evidence of plate **tectonics.** It does, however, have moonquakes: mild tremors that do not register more than two on the Richter scale and typically originate deep within the mantle. Moonquakes take place when gravitational stresses from the alignment of the Earth and **sun** are strongest.

Although the moon apparently never had liquid water, astronomers received a surprise in 1994 when the Clementine spacecraft detected frozen water at the moon's poles. Hidden from the sun in shadowy craters, this ice probably piggybacked to the moon on **comets.** The moon's nightside is very cold, -169°C, but in sunlight temperatures shoot up to 117°C.

A fine-grained, fragmented material called regolith, made of pulverized rocks, covers the moon's surface. (It is not called soil because it does not contain any organic material.) In some places regolith is up to 20 meters thick. Analysis of moon rocks shows them to be basalts, made of solidified lava; anorthosites, ancient feldspar-containing rocks; and impact breccias, composite rocks formed from different materials welded together in the heat of meteorite impacts. The basalts are about three billion years old, dating back to a time when the moon's surface was molten. The highland rocks, about 4 to 4.5 billion years old, represent the moon's original crust.

Although the moon's rocks are approximately as old as the Earth's, they differ considerably in that they have lower amounts of the dense elements, such as iron, and show no evidence of volatile, easily evaporated substances such as water. They resemble the rocks of Earth's surface, but not those found deeper. These findings, and the moon's low density in general, support a dramatic theory for the moon's origin. Instead of being formed at the same time as Earth or being captured by Earth's gravity, as is the case with many other natural satellites, the moon was probably born in an enormous collision.

According to this theory, early in Earth's existence, about 4.5 billion years ago, an object the size of **Mars** struck our planet and blasted great chunks of the Earth's outer layers, as well as pieces of itself, into orbit. This orbiting matter eventually coalesced into the moon. In its early years the infant moon would have been hot, with a molten surface. After it cooled somewhat, it went through a period of intense bombardment by space debris, resulting in many of the craters we see today. Molten rock welled up from the interior to flood the impact basins, forming maria about three to four billion years ago. Eventually, the surface solidified and **meteor** bombardment died away, leaving the moon a quiet, dusty testament to its violent birth. In its early days the moon was much closer to the Earth, perhaps 213,000 kilometers away, its molten surface looming large in the sky. High tides on Earth exert a gravitational tug on the moon that gives its orbit a little extra energy, gradually sending it away.

As they circled the moon in 1969, the Apollo 11 astronauts saw something impossible to see from Earth: the far side of the moon. As the moon is in synchronous rotation, only one side is visible from Earth—the near, relatively smooth side. The far side is rougher and strewn with numerous craters.

Phases of the Moon

The apparently changing shape of **Earth**'s "inconstant moon" is simply a matter of lighting. The half of the moon that faces the **sun** is illuminated; the half facing away from the sun is dark. As the moon circles the Earth in its 29.5-day orbit, the portion of its sunlit hemisphere that faces Earth is constantly changing in a regular cycle of lunar phases.

When the moon is between the sun and the Earth, the side facing us is dark (though dimly lit by reflected sunlight from Earth, called earthshine); this phase is the new moon. During the next seven days, the illuminated side begins to come into view along the moon's eastern edge in a phase known as a "waxing crescent." By the first quarter moon, observers on Earth see half of the moon lit and half dark. The phases move on to a "waxing gibbous moon," until the moon reaches the opposite side of the Earth from the sun and is fully

MAKING THE MOON

J. Kelly Beatty

WHO HASN'T GAZED AT A LUNAR CRESCENT SUSPENDED IN THE EVENING sky, or watched with awe as a full moon rises majestically over a distant horizon? Each of us has surely pondered how Earth came to have such a stunning satellite. Astronomers have wondered, too. In fact, determining the moon's origin was a major reason for having Apollo astronauts trudge across the stark lunar landscape to scoop up dusty rocks.

Several theories have been proposed for the formation of the moon—some more plausible than others. The debate continues to this day; a current theory hypothesizes that the moon blasted away from Earth as the result of a huge collision.

Before the Apollo expeditions, theories of lunar origin were variations on three basic themes: the moon formed in Earth's general vicinity and became a captive body when it ventured too close to our planet; Earth and moon formed together as a kind of double planet; at some point the early Earth was spinning so rapidly that a large mass ripped away and into orbit.

None of these ideas survives scientific scrutiny. For example, the Earth-moon system has too much angular momentum—our planet spins too fast, and the moon circles it too quickly—for them to have formed side by side. Conversely, Earth would have had to spin incredibly fast, in just 2.6 hours, for the moon to have cleaved away, and the resulting pairing would have had four times too much angular momentum. Samples show that, unlike Earth, the moon is lacking in iron. Nor do lunar rocks contain any water. Yet they contain oxygen's three isotopes in ratios that are virtually identical to Earth's—in fact, the bulk lunar composition roughly matches that of our planet's mantle.

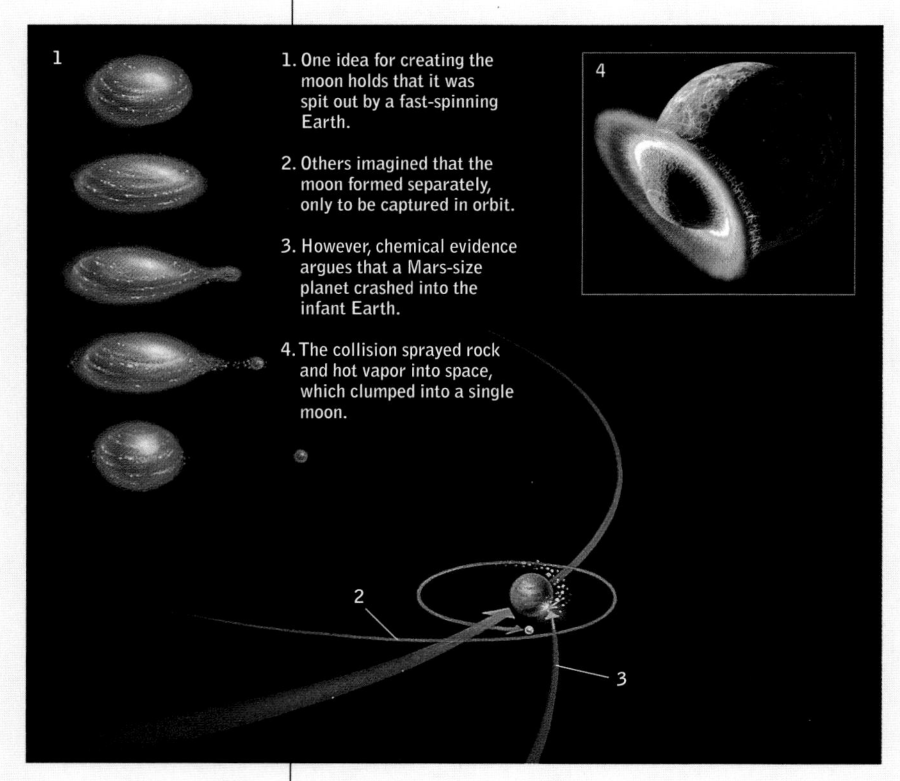

1. One idea for creating the moon holds that it was spit out by a fast-spinning Earth.
2. Others imagined that the moon formed separately, only to be captured in orbit.
3. However, chemical evidence argues that a Mars-size planet crashed into the infant Earth.
4. The collision sprayed rock and hot vapor into space, which clumped into a single moon.

These harsh post-Apollo realities left scientists bereft of good ideas, save for one. In the mid-1970s, two teams independently suggested that the moon could have been blasted away from the early Earth in a titanic collision. Computer modelers have simulated how such a collision would play out. They conclude that the impactor had to be at least as massive as present-day Mars, and that before striking it must have already segregated into an iron-rich core and a less-dense mantle. An off-center hit would have sprayed a huge jet of iron-poor mantle material into space as white-hot vapor, while the impactor's core remained behind and merged into Earth's. The vapor quickly smeared out into a doughnut-shaped disk, and as it cooled, the gas condensed into a chaotic swarm of tiny particles. Most of the disk cascaded back into Earth, while higher-flying matter collected into ever larger clumps. Gravitational perturbations between the nascent moon and the disk's remaining matter created a mutual repulsion, which pushed the still molten satellite outward and into an orbit tilted about ten degrees to Earth's Equator.

Impact specialists find this scenario quite appealing, and it neatly explains most of our satellite's major characteristics—its size, composition, and orbit. ■

illuminated as a full moon. Over the next week, the lighted portion shrinks in a "waning gibbous moon" until it is again half light, half dark—the last quarter. Finally, the waning crescent shrinks into darkness and another new moon.

Synodic Month

The synodic month, also called the lunar month, is the time it takes the **moon** to complete one full orbit of the **Earth** relative to the **sun.** In other words, it is the time that elapses between one new moon and the next, a period of 29.5 days. Because the Earth is orbiting the sun, the moon travels through more than a 360-degree circle before it completes a full cycle of phases. Thus, the synodic month is longer than the **sidereal month.**

Sidereal Month

The sidereal month is the time it takes the **moon** to complete one 360-degree orbit of the **Earth** relative to the background stars (and ignoring its position relative to the sun). The moon's sidereal orbit takes approximately 27.3 days. It is shorter than a **synodic month.**

Mascon

A mascon (short for mass concentration) is an area on the **moon** with a higher gravitational pull than the areas around it. Mascons are associated with the basalt-filled basins of lunar maria and may contain high-density rocks from the moon's mantle.

Lunar Missions (Unmanned)

The **moon** was the first extraterrestrial body visited by unpiloted **spacecraft** and remains the only one touched by humans. In 1959 just two years after **Sputnik,** the Soviet Union sent its Luna 3 probe around the moon, photographing the far side for the first time. From 1966 to 1976, 14 Luna missions successfully visited the moon. Some were orbiters; others sent probes to the surface, where they collected samples of rocks and regolith before returning to **Earth.** Two Luna missions sent out robotic Lunokhod vehicles that photographed and analyzed the surface. The Luna program contributed greatly to human understanding of the moon's composition, gravity, temperature, and other features. The U.S. sent out its own probes in the 1960s—the Ranger, Surveyor, and Lunar Orbiter missions. Rangers 7 through 9 sent back detailed images of the moon's terrain. Surveyor probes tested the feasibility of soft landings on the lunar surface in anticipation of the **Apollo** missions. The five Lunar Orbiter spacecraft also helped to prepare for the Apollo landings by photographing potential landing sites.

After the extraordinary success of the Apollo program, mission planners turned toward other targets until the 1990s. The most successful of the recent U.S. missions was Clementine, which orbited the lunar poles in 1994 and found evidence of water ice in their craters. Lunar Prospector also orbited the poles and was deliberately crashed into the moon in 1999, where it could not confirm the presence of ice. Launched in the fall of 2003, the **European Space Agency**'s Smart 1 spacecraft is slowly making its way to the moon using an ion engine; Japan is planning a lunar mission, Selene, scheduled for launch in 2005.

MAJOR UNMANNED LUNAR MISSIONS

Spacecraft	Type	Nationality	Launch Date
Luna 3	Flyby	U.S.S.R.	1959
Ranger 7	Planned impact	U.S.	1964
Ranger 8-9	Planned impact	U.S.	1965
Luna 10-13	Orbiter	U.S.S.R.	1966
Lunar Orbiter 1-2	Orbiter	U.S.	1966
Surveyor 1	Lander	U.S.	1966
Lunar Orbiter 3-5	Orbiter	U.S.	1967
Surveyor 3, 5, 6	Lander	U.S.	1967
Surveyor 7	Lander	U.S.	1968
Luna 14	Orbiter	U.S.S.R.	1968
Luna 16	Lander	U.S.S.R.	1970
Luna 17	Lander; released Lunokhod 1 rover	U.S.S.R.	1970
Luna 19	Orbiter	U.S.S.R.	1971
Luna 20	Lander	U.S.S.R.	1972
Luna 21	Lander; released Lunokhod 2 rover	U.S.S.R.	1973
Luna 22	Orbiter	U.S.S.R.	1974
Luna 24	Lander	U.S.S.R.	1976
Clementine	Polar orbiter	U.S.	1994
Lunar Prospector	Polar orbiter and planned impact	U.S.	1998

Mars

Mars

Of all the **planets,** Mars holds the most allure for humans. From the first telescopic observations in the 1600s to contemporary scientific missions, people on **Earth** have looked to the fourth planet as their best hope for learning that we are not alone in the **solar system.** Both the Greeks and Romans named the planet after their god of war—Ares and Mars, respectively—because its distinctive reddish color was associated with bloodshed. In 1659 Dutch physicist Christiaan Huygens turned a homemade **telescope** toward the planet and sketched its light and dark features. He noted that they appeared and disappeared approximately every 24 hours, leading him to suppose (correctly) that Mars had a rotation period about the same as Earth's.

Casting a pinkish glow, Mars, known since ancient times as the red planet, has several large geologic features, including the 5,000-kilometer-long Valles Marineris trough system and four Tharsis volcanoes (left center). Permanent ice caps at the poles wax and wane with the seasons.

In 1877 Italian astronomer Giovanni Schiaparelli mapped what he saw as a network of straight, dark lines on the surface of Mars. He called them *canali,* meaning "channels," but the word was soon mistranslated into "canals," fueling speculation that intelligent life existed on Mars. Among those who took up this position was American Percival Lowell, who built an observatory in Arizona in order to observe the Martian surface. He, too, thought he spotted canals. Other astronomers did not see these canals and were skeptical of the claims for intelligent life on Mars. When the first spacecraft visited the planet, in the 1970s, it became clear that there were no such structures. The canals that Lowell and others thought they saw are natural features that may, in fact, be dry riverbeds.

Mars is considerably smaller than Earth. Its radius of 3,397 kilometers is about half Earth's, and its mass is one-tenth of our planet's. Its highly elliptical orbit averages 227,920,000 kilometers from the sun, about one and a half times farther than Earth's. At certain times when the Earth moves between Mars and the **sun,** the red planet comes especially close to ours; in the summer of 2003, Mars mission planners took advantage of such an occasion to send several probes to the planet.

As Christiaan Huygens noted, Mars rotates on its **axis** at a speed similar to Earth's, making its day 24.62 hours long. Because it orbits farther from the sun, its year lasts 687 Earth days. The tilt of its axis is also similar to Earth's, averaging 25 degrees, although in the past that angle has been as great as 60 degrees. This tilt means that Mars, like Earth, has distinct seasons.

MARS	
Average distance from the sun	227,920,000 km
Revolution	687 days
Average orbital speed	24.1 km/s
Average temperature	-65°C
Rotation	24.6 hours
Equatorial diameter	6,794 km
Mass (Earth=1)	0.107
Density	3,933 kg/m^3
Surface gravity (Earth=1)	0.38
Known satellites	2
Largest satellite	Phobos

Microscopic Martian "blueberries," or sphere-like grains, fill a region of the planet nicknamed the "Berry Bowl." The triple blueberry (left of center) hints that these formations are not ejected volcanic material but, instead, grew in preexisting wet sediments, adding weight to the idea that Mars once held water.

Because Mars has an orbit more elliptical than circular, its southern summer is much hotter than its northern one.

Mars has no global magnetic field, but it does have local regions of magnetism. These are probably remnants of an earlier time when the planet had a hotter, more active core. This iron core, representing about 25 percent of the planet's mass, has probably cooled too much to support a full magnetic field now.

Consisting of 95.3 percent carbon dioxide, 2.7 percent nitrogen, and 1.6 percent argon, Mars's atmosphere does not resemble Earth's and would be poisonous to humans even if there were enough to breathe. It is 200 times thinner than the air on Earth, with air pressure at the surface only 0.01 percent that of Earth's. Mars's atmosphere does contain enough water vapor to form clouds over certain areas of the planet, and fine red dust from the surface frequently hangs in the air.

Mars is cold. Temperatures average -65°C, only occasionally reaching above freezing on the warmest days of summer. Variations in temperatures across the surface drive strong winds, sometimes raising immense summertime dust storms that can cover the entire planet.

Because the atmosphere is so thin and the planet has no protective magnetosphere, the sun's ultraviolet rays beat down on the surface with full force, virtually sterilizing the ground.

Mars can be divided into two mismatched halves. Its southern hemisphere is old, elevated, rough, and heavily cratered. Its northern hemisphere has smoother, younger terrain, dotted with **volcanoes.** Near the equator is a large bulge known as the Tharsis uplift, which supports enormous volcanoes, including the largest in the solar system, Olympus Mons. This gently sloping volcano, more than 26 kilometers high and 600 kilometers across at its base, probably attained its current size through repeated outpourings of lava over the eons. Because Mars does not have plate **tectonics,** Martian volcanoes can stay in place and grow indefinitely over the hot spots that feed them.

Running more than 4,000 kilometers around the equator, Valles Marineris is an immense system of connected canyons, in places deeper than Mount Everest is high. It probably formed billions of years ago when the planet's surface was more geologically active and split apart. Ice caps, made of water ice and carbon dioxide ice, cover each pole. These wax and wane dramatically over the seasons, releasing carbon dioxide gas into the atmosphere in the summer and gaining it again as ice that freezes out of the air in winter.

Thanks to several **missions to Mars**'s surface, we know that the planet is dry and rocky, covered with a fine red dust and sand made primarily of iron oxide, silicon, and sulfur. In places the sand is blown into dunes reminiscent of those in the Sahara.

Although the atmosphere of Mars is too thin and cold to support liquid water now, it seems clear that the planet had liquid water in the past. Winding across the dry Martian surface are channels that resemble dry riverbeds, complete with tributaries and deltas. These are the features that inspired some astronomers to conjure up canals. Some valleys show signs of powerful floods. And analysis of rocks by the Opportunity rover indicates that they were once saturated with water.

Scientists do not yet know where this water came from, how much existed, or where it all went. Some is frozen into ice on the polar caps and into frost on the Martian surface. More may lie frozen under the surface in regions around the poles. It is possible that water exists as a kind of permafrost throughout the planet's regolith, or even in liquid form under the ice. Because water is a prerequisite for the evolution of Earth-like life, the history and existence of water on Mars is a major area of study for planetary geologists.

Phobos

The larger of **Mars**'s two small, dark, and peculiar moons, Phobos was discovered in 1877 by Asaph Hall. It takes its name, which means "fear" in Greek, from one of the two attendants of Ares, the Greek god of war. Phobos is 27 by 22 by 18 kilometers in size, roughly cratered, and looks like an asteroid. It may, in fact, be an asteroid captured by the **planet**'s gravity, or it may be a piece of planetary debris left over from the earliest days of the formation of Mars.

Phobos orbits the Martian equator very close to the planet's surface—only 2.77 Mars radii away. With its rapid orbital period of 7 hours 39 minutes and a retrograde path (west to east), the little moon would appear to an observer on Mars's surface to rise in the

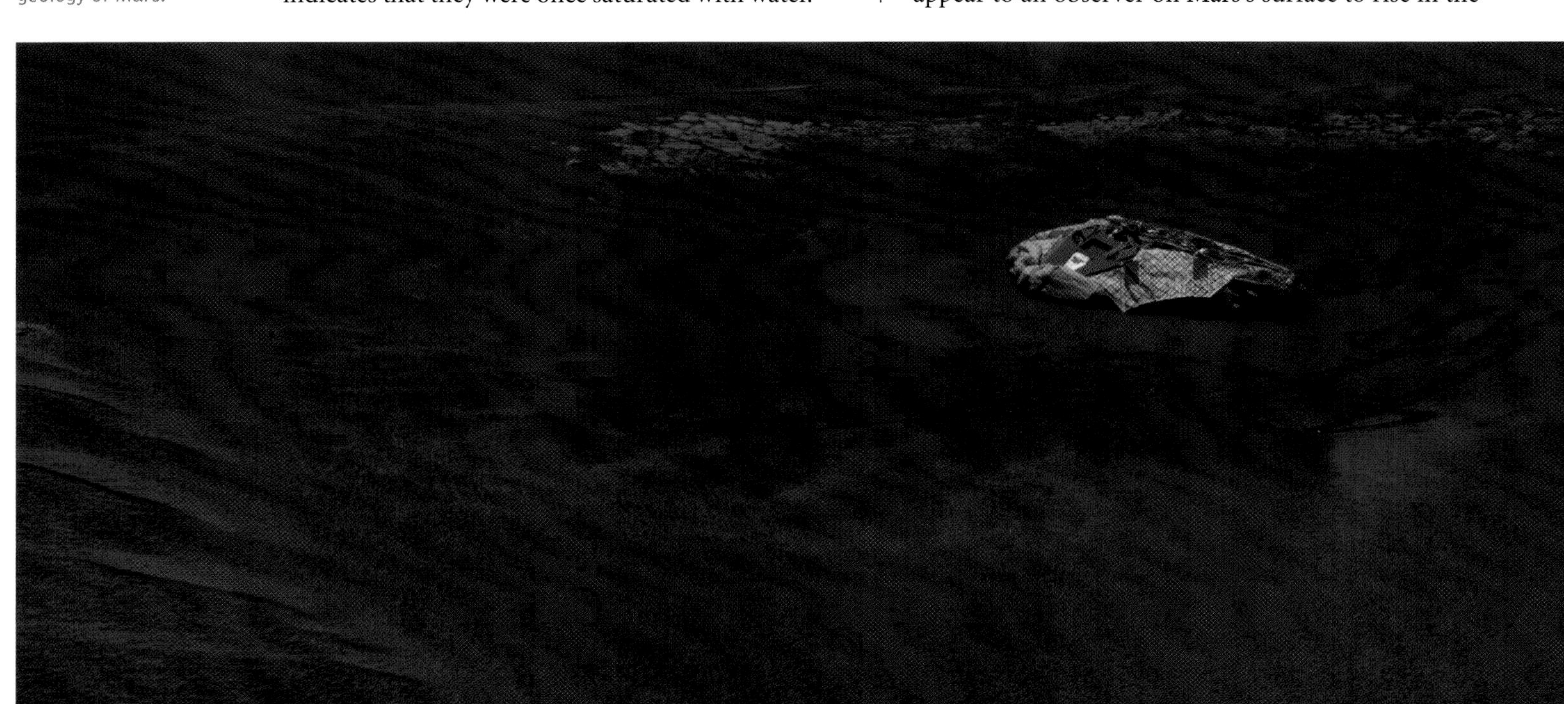

Opportunity, a rover vehicle, sent back this 2004 image of the lander that carried it to Mars and the deflated airbags that protected it. The rover explored the Meridiani Planum, while another rover, Spirit, landed on the opposite side of Mars. Pawing the Martian surface, the two rovers gathered valuable information about the geology of Mars.

west and set in the east three times a day. Like **Earth**'s moon, its orbit is synchronous, so that one side always faces the planet. Phobos is in the last years of its existence, astronomically speaking. Its orbit is slowly spiraling inward toward the Martian surface, and within 50 million years it will either be torn apart by Mars's gravitational forces or will crash into the planet.

Deimos

The smaller of **Mars**'s two tiny, irregular moons, Deimos is about 15 by 12 by 10 kilometers in size and may, like its partner, be a captured **asteroid** or a piece of early planetary debris. Deimos, like Phobos, is locked into a synchronous **orbit,** with the same side always facing the planet, and was discovered by Asaph Hall in 1877. Its name, "panic" in Greek, is taken from one of the two attendants of Ares, Greek god of war. Unlike Phobos, Deimos has a relatively slow orbital period of 30 hours 17 minutes and rises in the east. It is fairly far from the Martian surface, almost seven Mars radii, orbiting around the equator.

Life on Mars

Although early descriptions of canals on Mars turned out to be wishful thinking, Mars-watchers have not given up hope of finding evidence that life once existed—or still exists—on the red planet.

Most observers agree that the key to life is the presence of liquid water. Currently, the surface of Mars appears to be both dry and lifeless. It is frigid, has little atmosphere, and is bombarded by lethal ultraviolet radiation. Close-up studies show no traces of recognizable plant or animal life. However, images of Mars from orbiters such as the Mars Global Surveyor show fairly convincing scenes of old riverbeds, gullies, and floodplains. Rocks analyzed by the Opportunity rover in early 2004 appear to have been saturated with water at one point in their history. Frozen water at the poles and in the Martian regolith might have been liquid in a warmer, more humid past. Mars, it seems, was once a wet world.

A few scientists believe that proof of ancient life on Mars exists in a meteorite found in Antarctica in 1984. Analysis of this meteorite and a few others from different locations on **Earth** shows that they came from Mars, apparently blasted off the **planet** millions of years ago by a violent impact. The Antarctic meteorite, known as ALH 84001, appeared to contain organic compounds and tiny fossils that look like bacteria. These claims are controversial and are not upheld by a majority of scientists. Recent studies have shown that some of the evidence may not be biological in origin.

STORM OF THE CENTURY

J. Kelly Beatty

IN MAY 1971, AS MARS AND EARTH DREW CLOSE TOGETHER IN THEIR ORBITS, THE United States and the Soviet Union were ready to ratchet up the stakes in the exploration of the red planet. Soviet rocketeers dispatched two massive craft, named Mars 2 and 3, each designed to go into orbit around the planet and drop a camera-equipped lander onto the surface. The U.S. countered with two smaller Mariner spacecraft designed to orbit Mars for a comprehensive look at its surface and its seasonal climatic changes. Mariner 8 was lost during launch; its twin, Mariner 9, had to carry out all the mission's objectives. Four months later, as the spacecraft sailed outward, Mars was whipping itself into a frenzy. It was summer in the planet's southern hemisphere, and the sun's warmth had triggered strong winds and a growing dust storm. Astronomers watched as Mars's atmosphere grew choked with ocher dust within weeks. Never before had they seen the planet enveloped in such an intense, long-lasting dust storm, and it could not have come at a worse time.

Mariner 9 reached the red planet first, firing its braking rocket and slipping into orbit on November 19th. Mars 2 arrived about two weeks later, separating into its lander (which crashed due to a malfunction) and a successful orbiter. Mars 3's lander reached the surface intact, but its transmissions ceased abruptly seconds after touchdown. Many believe the raging global dust storm may have toppled or damaged it. Unfortunately, because the craft's activities were completely preprogrammed and irreversible, Soviet engineers could not delay the landing until after the winds subsided. For the same reason, the Mars 2 and 3 orbiters automatically relayed to Earth a steady stream of pictures showing Mars's virtually blank disk.

NASA engineers had the luxury of waiting out the storm, which at its peak extended a shroud of dust more than 60 kilometers high. Most of Mariner 9's scientific work was deferred until the atmosphere cleared, but the craft's early images showed a quartet of dark spots poking through the dusty pall. Disbelieving scientists finally realized these had to be enormous mountains—crater-topped volcanoes—jutting from the surface. As the dust settled, Mariner 9 began its mapping duties. The camera returned pictures of the Martian surface and produced the first full-disk photo mosaic of Mars. By the following October, its historic mission over, Mariner 9 had relayed more than 7,300 images to Earth. Even though a stuck filter prevented the camera from recording in color, the spacecraft had surveyed almost the entire planet for the first time. It showed us a Mars than no one had imagined, a planet of towering mountains, vast canyon systems, intricately layered polar regions, and a network of desiccated riverbeds suggesting that water once flowed across the surface. Our notions of Mars as a planet, and as a target for future exploration, were forever changed. ■

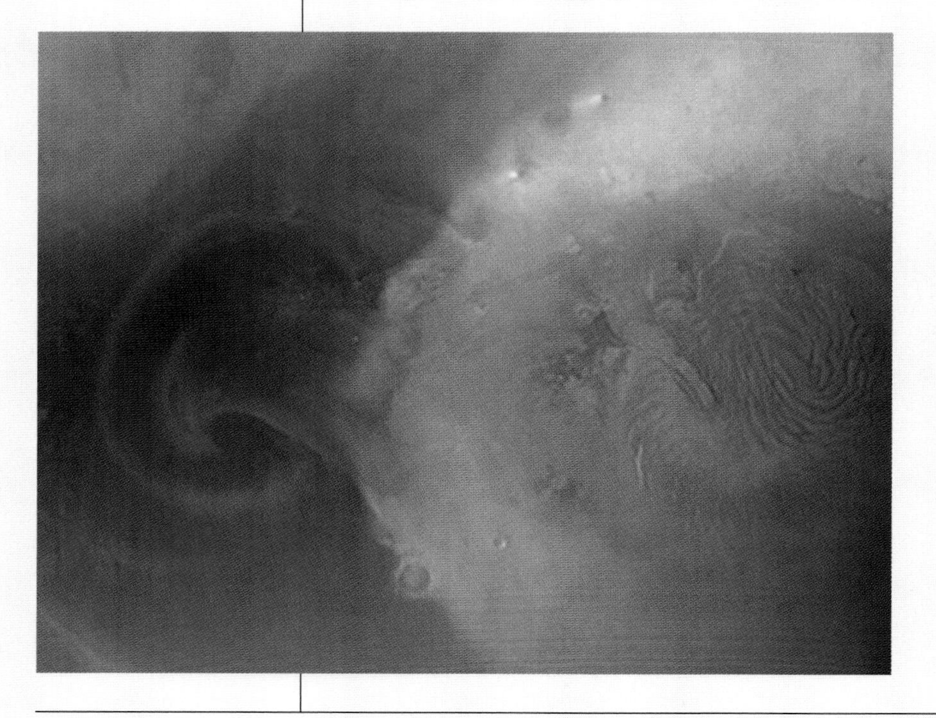

Due to its thin atmosphere, Mars routinely experiences strong winds during the summer. Frequently these winds create large dust storms. Mariner 9 returned this image of dust clouds from a polar storm in 1971.

No direct evidence yet exists for life on Mars, but the one-time presence of liquid water on its surface certainly keeps the door open to the possibility. Some scientists take heart from recently discovered microbes on Earth that can exist in extreme conditions in undersea hot-water vents and Antarctic ice. These "extremophiles" might provide a model for Martian life-forms.

Missions to Mars

The red planet has provided a rich field of study for robotic missions. Lured in part by visual evidence that the **planet** may once have had liquid water, the U.S., the Soviet Union/Russia, and other countries have launched, or attempted to launch, more than 30 missions to **Mars** since 1960. Most of these have been plagued by difficulties ranging from launch failure to communications breakdowns. Nevertheless, the successful missions have sent a wealth of information and images back to **Earth.** The first, Mariner 4, was a simple flyby that returned images of the surface. The next phase of exploration brought many successful orbiters. As early as 1971, the Mariner 9 orbiter was able to send back images that covered more than 90 percent of the planet's surface, as well as images of **Phobos** and **Deimos** and measurements of atmospheric temperatures and pressures.

Launched in 1975, the Viking 1 and 2 orbiters and landers mapped the planet even more completely from the sky and searched (unsuccessfully) for traces of life on the surface. After a hiatus, and then some costly failures for both the U.S. and the Soviet Union, the Mars Global Surveyor entered into a polar orbit around Mars in 1997, providing wide-angle images of the planet, laser mapping, data on heat emission, and information on the planet's magnetic field. The 2001 Mars Odyssey arrived that year and collected information on the changing Martian climate and the presence of water ice near the poles.

More ambitious missions explored the Martian surface with robotic vehicles. Mars Pathfinder, a probe, landed on the planet in 1997 and launched a six-wheeled rover called Sojourner that photographed the surface. The **European Space Agency**'s Mars Express orbiter reached the planet in 2003. Although the lander it released, Beagle 2, lost contact with the mother ship, the Mars Express orbiter is still studying the surface with cameras and **spectrometers.** Japan's Nozomi probe, designed for studying the Martian atmosphere, left Earth in 1998, but developed fuel problems and then was damaged by a **solar flare,** so the mission had to be abandoned.

Better news awaited Mars-watchers in early 2004, when two U.S. rovers, Spirit and Opportunity, landed successfully on the opposite sides of the planet—Spirit in Gusev Crater and Opportunity in Meridiani Planum. Trundling about on the Martian surface and scraping at local rocks, the rovers sent back valuable data on local geology. In one of the most exciting developments in Mars exploration, rocks examined by Opportunity showed strong evidence of having once been under water, supporting the idea that lakes or even oceans once covered portions of Mars.

SUCCESSFUL MISSIONS TO MARS

Spacecraft	Type	Nationality	Launch Date	Encounter Date
Mariner 4	Flyby	U.S.	Nov 1964	Jul 1965
Mariner 6	Flyby	U.S.	Feb 1969	Jul 1969
Mariner 7	Flyby	U.S.	Mar 1969	Aug 1969
Mariner 9	Orbiter	U.S.	May 1971	Nov 1971
Viking 1	Orbiter/lander	U.S.	Aug 1975	Jun 1976
Viking 2	Orbiter/lander	U.S.	Sep 1975	Aug 1976
Mars Pathfinder	Lander/rover	U.S.	Dec 1996	Jul 1997
Mars Global Surveyor	Orbiter	U.S.	Nov 1996	Sep 1997
2001 Mars Odyssey	Orbiter	U.S.	Apr 2001	Oct 2001
Mars Express	Orbiter	ESA	Jun 2003	Dec 2003
Spirit; Opportunity	Rovers	U.S.	Jun 2003; Jul 2003	Jan 2004; Jan 2004

Exploring the Outer Planets

Missions to the Outer Planets

Because of the enormous distances covered, missions to **Jupiter, Saturn,** and beyond are extremely challenging and expensive. In the 1970s mission planners at **NASA** organized a multiplanet mission that would take advantage of a rare alignment of Jupiter, Saturn, **Uranus,** and **Neptune.** Two spacecraft, Voyager 1 and Voyager 2, would use the gravity of the four planets to boost themselves past the gas giants.

Launched in 1977 and called the Grand Tour, these valuable missions collected a wealth of information about the giant planets, discovering **moons,** rings, atmospheric compositions, magnetic fields, and more. Beginning in 1979, Voyager 1 transmitted 19,000 images of Jupiter and its larger satellites and 17,000 images of Saturn and its large moons. Voyager 2 also sent back images of Jupiter, Saturn, and their moons, finding additional rings around Saturn and volcanic activity on Jupiter's moon **Io.** Voyager 2 then went on to fly by Uranus and Neptune in 1986 and 1989, respectively, sending back data on their moons, rings, magnetic fields, and atmospheres. To date Voyager 2 is the only spacecraft to have visited those planets.

Both spacecraft are now heading out of the **solar system,** still under power and transmitting data. By 2010 they will begin to reach the heliopause and should send back the first information ever received from that outer boundary of the solar neighborhood.

The farthest planet of all—**Pluto**—and its moon, **Charon,** are targeted for future study by the New Horizons mission. Scheduled for launch in 2006, the U.S. spacecraft, about the size of a small lifeboat, will swing through Jupiter's system at about 80,000 kilometers an hour in 2007 for a gravity boost before making the long journey to Pluto. Arriving in 2015, the craft will photograph and map Pluto and Charon in visible and near-infrared wavelengths. It will also collect data on Pluto's atmosphere and on the surface composition of both little worlds before, most likely, retargeting itself for a close encounter with a **Kuiper belt** object.

NASA has proposed a mission to explore Jupiter's icy moons, **Callisto, Ganymede,** and **Europa.** The nuclear-powered orbiter would be launched about 2015. It would orbit each moon and look for clues to its origin—and the possibility that it may sustain life in subsurface oceans.

Encountering the Outer Planets

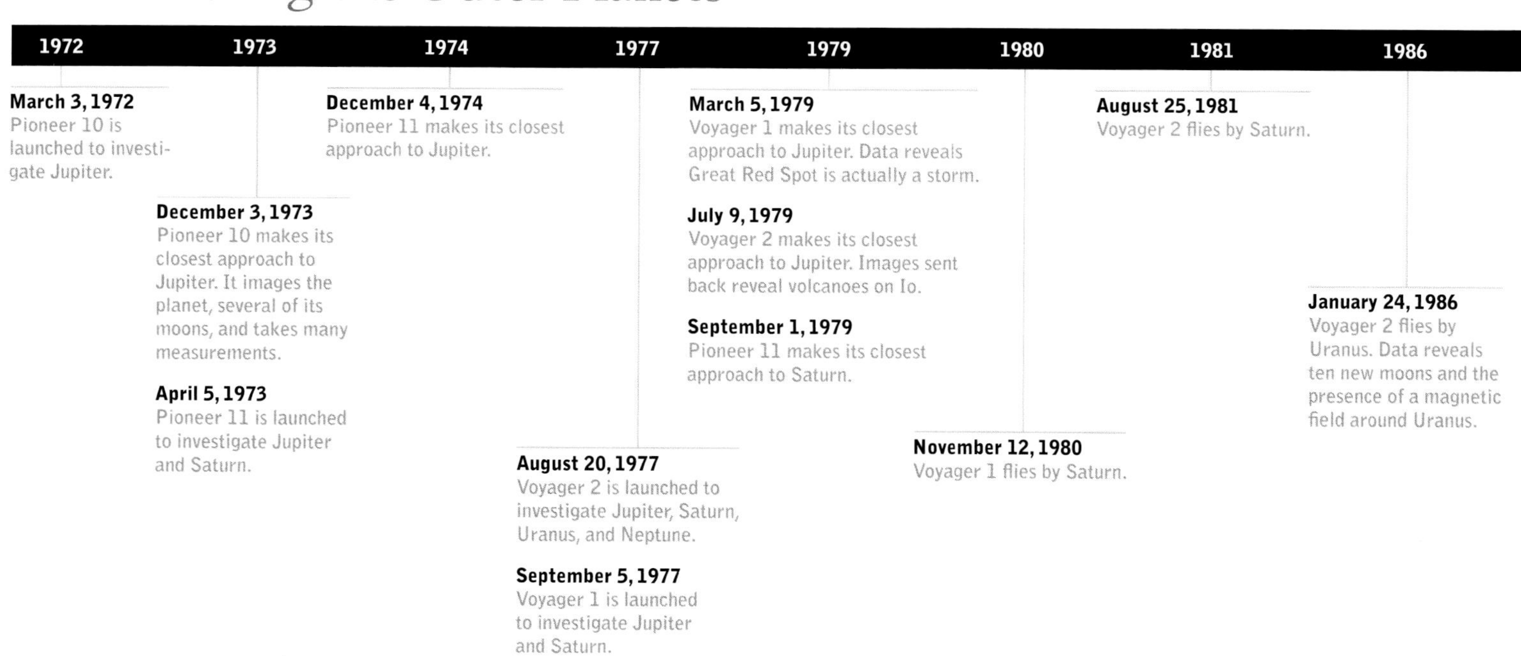

1972
1973
1974
1977
1979
1980
1981
1986
March 3, 1972
Pioneer 10 is launched to investigate Jupiter.
April 5, 1973
Pioneer 11 is launched to investigate Jupiter and Saturn.
December 3, 1973
Pioneer 10 makes its closest approach to Jupiter. It images the planet, several of its moons, and takes many measurements.
December 4, 1974
Pioneer 11 makes its closest approach to Jupiter.
August 20, 1977
Voyager 2 is launched to investigate Jupiter, Saturn, Uranus, and Neptune.
September 5, 1977
Voyager 1 is launched to investigate Jupiter and Saturn.
March 5, 1979
Voyager 1 makes its closest approach to Jupiter. Data reveals Great Red Spot is actually a storm.
July 9, 1979
Voyager 2 makes its closest approach to Jupiter. Images sent back reveal volcanoes on Io.
September 1, 1979
Pioneer 11 makes its closest approach to Saturn.
November 12, 1980
Voyager 1 flies by Saturn.
August 25, 1981
Voyager 2 flies by Saturn.
January 24, 1986
Voyager 2 flies by Uranus. Data reveals ten new moons and the presence of a magnetic field around Uranus.

New Horizons spacecraft approaches the farthest planet, Pluto, and its moon, Charon, in an artist's conception of an encounter planned for 2015. The spacecraft would map the planet and examine its atmosphere. The large dish antenna will help the spacecraft communicate with Earth.

1989 1990 1992 1994 1995 1997 1998 2003 2004

August 24, 1989
Voyager 2 flies by Neptune. It discovers six new moons. Data reveals that complete rings surround Neptune. The probe continues into interstellar space.

October 18, 1989
Galileo orbiter and probe are launched from space shuttle *Atlantis* to investigate Jupiter.

October 6, 1990
Ulysses (an ESA probe) is launched to study the sun. It will use Jupiter to launch itself into a heliospheric orbit.

February 8, 1992
Ulysses uses Jupiter's gravity to launch itself into a heliospheric orbit. Data reveals much about Jupiter's magnetosphere.

July 16-22, 1994
Galileo observes direct collision of comet Shoemaker-Levy 9 with Jupiter.

November 1995
Last contact with Pioneer 11.

December 7, 1995
Galileo probe, released from orbiter on July 13, 1995, enters Jupiter's atmosphere. It survives 58 minutes and descends some 200 kilometers before it disintegrates.

October 15, 1997
Cassini orbiter and Huygens probe are launched to investigate Saturn and its moon Titan.

February 17, 1998
Voyager 1 becomes the most distant human-made object in space. It continues to head into interstellar space.

January 23, 2003
The last time a signal is detected from Pioneer 10.

September 21, 2003
Galileo orbiter is deliberately crashed into Jupiter to avoid a possible impact with Europa.

February 4, 2004
Ulysses makes its second approach to Jupiter.

June 30, 2004
Cassini reaches Saturn.

Jupiter

Jupiter, the largest planet, displays a surface that looks like banded marble in a mosaic made from images snapped by the Cassini spacecraft in 2000. Powerful winds that lash the planet create dark and light bands, and streaks form as Jupiter's intense jet streams shear clouds apart.

Jupiter

Aptly named after the king of the Roman gods, Jupiter is the largest **planet** in the **solar system**—more than twice as massive as all the other planets and satellites combined. Called a **gas giant,** it is composed mainly of hydrogen and helium, like the **sun,** although most of the gases are liquefied under pressure. Fierce winds whip around Jupiter's surface, creating the light and dark bands that give the planet its distinctive appearance through a **telescope,** and enormous storms spiral across its cloud tops.

Formed from the same materials that made the original solar **nebula,** Jupiter is sometimes called a failed star, one that lacked sufficient mass to begin fusion in its core. However, Jupiter falls far short of the heft needed for even a small star. It would have to be about 80 times more massive to generate sunlike energy. Nevertheless, it is huge. Its radius at the equator is 71,492 kilometers, more than 11 times that of **Earth;** more than 1,321 Earths could fit inside its volume. Its mass, 18,986 x 10^{23} kilograms, is almost 318 times greater than Earth's, although far less than it would be if Jupiter were solid.

This fifth planet orbits the sun at a distance of 5.2 AU or 778,570,000 kilometers, more than three times as distant as **Mars.** Jupiter's orbit is 11.87 Earth years long, but its day, at only 9.9 hours, is short. Jupiter rotates more rapidly near its equator than near its poles, and its rapid spin causes it to bulge noticeably around the middle.

Jupiter has a thin, gaseous upper atmosphere overlying a liquid planet. The upper atmosphere, the colorful shell visible to telescopes, consists of three layers: an uppermost belt of ammonia-ice clouds, a middle band of ammonium hydrosulfide, and a lower layer of water ice. Strong, steady jet streams blow the light-colored zones and dark-colored belts in alternating directions at more than 640 meters each second. Scientists are not sure what causes their rich, reddish hues.

Intense storms—cyclones, anticyclones, and thunderstorms—disrupt Jupiter's high clouds. The most noticeable of these is the **Great Red Spot,** but other storms, almost as large, have been seen to form and move steadily around the planet, sometimes merging into larger tempests.

Visiting probes have also spotted gigantic flashes of lightning, possibly originating in the water-ice clouds below. Heat from within the planet, probably generated by its initial gravitational collapse, drives and sustains these storms and winds.

Jupiter's center is a relatively small iron-rich core. Compressed by the extraordinary bulk of the planet around it, the core experiences pressures 70 million times greater than those at sea level on Earth and temperatures of 20,000 kelvins, more than three times as hot as the surface of the sun. Surrounding the core is a 50,000-kilometer-thick layer of liquid helium and dense hydrogen. Pressures up to three million times the air pressure at Earth's surface make the hydrogen behave like a liquid metal. Heat-driven convection currents within this metallic ocean probably generate Jupiter's powerful magnetic field. A 20,000-kilometer-thick layer of liquid hydrogen surrounded by a thin layer of gaseous hydrogen and helium lies beneath the colorful shell of the upper atmosphere.

Jupiter's **magnetosphere** swells hundreds of millions of kilometers into the solar system, with a magnetotail extending past **Saturn.**

Not until Voyager 1 reached Jupiter in 1979 did scientists learn that Jupiter, like the other gas giants, has a ring. Thin and delicate, the three-part ring is probably composed of the remains of little **moons.**

Great Red Spot

First reported by German amateur astronomer S. H. Schwabe in 1831, the Great Red Spot is a reddish oval on **Jupiter** some 26,000 kilometers long and 14,000 kilometers wide. Two Earths could easily fit side by side within its confines.

Astronomers believe the Great Red Spot is an enormous storm, a high-pressure vortex known as an anticyclone, wedged in the upper atmosphere between two of Jupiter's jet streams. Its size has varied over the years, but heat from deep within the planet supplies a steady source of energy for this permanent storm.

Moons of Jupiter

Jupiter has more natural satellites than any other planet in the **solar system. Galileo** discovered the four largest—**Ganymede, Io, Europa,** and **Callisto**—in 1610. In the 19th and 20th centuries, astronomers found more, the Voyager and Galileo spacecraft added to the collection, and sensitive Earth-based **telescopes** in the early 21st century found others, for a total of 63 by early 2004. Ganymede is the largest moon in the solar system. It, as well as Europa and Callisto, might have oceans of water under its frozen surface.

Io

Although close in size to **Earth's moon, Jupiter**'s moon Io bears little resemblance to our own gray, cratered satellite. The innermost of Jupiter's four Galilean moons (discovered by **Galileo** in 1610), Io is a colorful ball of active volcanism. Passing **spacecraft** and the **Hubble Space Telescope** have detected at least a hundred active **volcanoes** on Io. One of them, Loki, is the most powerful volcano in the **solar system,** regularly emitting more heat than all Earth's volcanoes combined. Io's volcanoes produce enormous amounts of lava that resurface the moon, giving it a smooth, uncratered appearance. Sulfur compounds from the volcanoes stain the moon's surface red and yellow.

Io has so much internal heat because it is caught up in a tug-of-war between Jupiter's gravity and that of the other large satellites. Tidal forces stretch and squeeze its rocky body, generating the energy that powers its constant volcanic activity.

Europa

The second closest of **Jupiter**'s Galilean **moons,** Europa is one of the most interesting bodies in the **solar system** to those who look for extraterrestrial life. A little smaller than Earth's moon, Europa has a smooth, bright, shiny surface that might well be water ice. Crackled and crumpled like ice floes in the Arctic Ocean, this ice—if ice it is—appears to be pushed around by liquid underneath.

Europa, like **Io,** is close enough to Jupiter to be stressed by tidal forces from the **planet**'s gravity, which could generate enough heat to keep water liquid under the surface. If this possible sea is warm enough, it might support life. Europa also has a very thin oxygen atmosphere.

Ganymede

Giant Ganymede is the largest **moon** in the **solar system** with a radius of 2,631 kilometers, greater than that of **Mercury.** Ganymede's relatively low density

JUPITER	
Average distance from the sun	778,570,000 km
Revolution	11.9 years
Average orbital speed	13.1 km/s
Average temperature	-110°C
Rotation	9.9 hours
Equatorial diameter	142,984 km
Mass (Earth=1)	317.8
Density	1,326 kg/m^3
Surface gravity (Earth=1)	2.36
Known satellites	63
Largest satellites	Io, Europa, Ganymede, Callisto

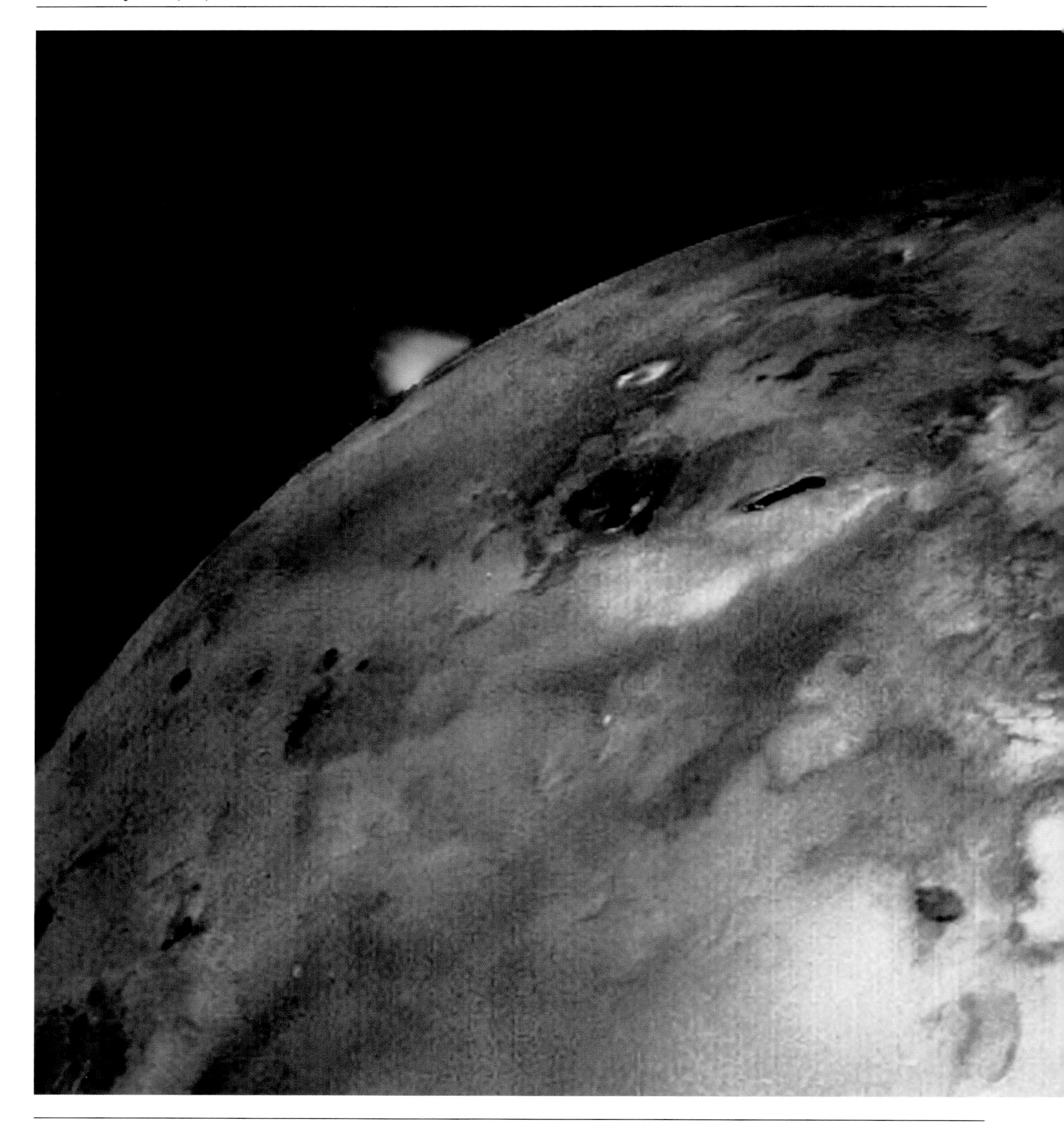

indicates that its upper layers may contain a great deal of ice. Ridged and cratered, Ganymede's rough surface points to a history of tectonic movement. One of its most interesting features is its pronounced magnetic field, detected by the **Galileo** mission. No other moon has a global **magnetosphere.** Scientists guess that this magnetism is a remnant of an earlier period of tidal flexing that generated electrical currents within the satellite. Like **Europa,** Ganymede appears to have a very tenuous oxygen atmosphere.

Callisto

The outermost of the four Galilean satellites of **Jupiter,** Callisto is also the third-largest moon in the **solar system,** after Ganymede and **Saturn**'s **Titan.** Unlike its more active brethren among the Galilean moons, Callisto is dark and tectonically quiet, with an ancient, heavily cratered surface. It does have a weak magnetic field, which may indicate that electrically conductive water beneath its surface is reacting to Jupiter's magnetic field.

Missions to Jupiter

Aside from the multiplanet Voyager missions, only three spacecraft have undertaken the long journey to Jupiter. From 1972 through 1974, two fairly simple probes, Pioneer 10 and 11 transmitted data about the Jovian atmosphere and images of some of the larger **moons,** and discovered three new satellites before continuing out of the **solar system,** sending their last transmission about 12 billion kilometers from the **sun.** The Pioneer missions were famous for carrying metal plaques that depicted the solar system, male and female figures, and other stylized drawings—greeting cards to any intelligent, extraterrestrial life-forms that might happen to come across the lonely probes.

In 1989 the Galileo orbiter took off from the **space shuttle** *Atlantis,* picked up speed in a slingshot trajectory around **Venus,** twice around **Earth,** and reached Jupiter in 1995. There it released a probe into the **planet**'s atmosphere. For 58 minutes the probe sent back data about Jupiter's stormy skies and atmospheric composition. Its readings showed a much clearer, drier atmosphere than expected, and scientists believe that the probe was simply dropped through a region of unusually good weather. The main Galileo **spacecraft** went on to orbit the planet, studying its magnetic fields and **Great Red Spot** as well as all four Galilean moons before burning up in Jupiter's atmosphere in 2003, when it was deliberately sent to do so.

Jupiter's moon Io erupts in volcanic fury in an image captured by Voyager 1. Silhouetted in the dark space above the planet at left, a plume of ejected material rises about 160 kilometers. Spacecraft and the Hubble Space Telescope have detected at least a hundred active volcanoes on Io.

Saturn, photographed by the Cassini spacecraft, appears to hang suspended from its rings, which Galileo called "cup handles." Made of ice particles ranging in size from specks to boulders, the rings—seven main ones and hundreds or thousands of ringlets—orbit Saturn at different speeds. Scientists think the rings may be the remains of pulverized moons, comets, or asteroids.

Saturn

SATURN

Graced with sweeping rings, Saturn is one of the most beautiful sights in the **solar system.** This sixth **planet** has been known since ancient times and often makes a brilliant appearance in the night sky, although it is 9.5 AU or 1,433,500,000 kilometers from the **sun.**

Saturn, the second-largest planet in the solar system, is a **gas giant** like **Jupiter,** consisting almost entirely of hydrogen (96 percent) and helium (3 percent). Although it is huge—with a radius of 60,268 kilometers—and 765 Earths could fit inside its volume, Saturn has a mass density less than that of water; it would float if a pool of water could be found large enough to support it. Saturn completes one orbit every 29.47 **Earth** years, but has a short day of only 10.65 hours. The planet's rapid rotation whirls its body into a flattened ball that bulges at the equator.

Like Jupiter, Saturn is essentially an enormous sphere of dense liquid surrounded by a thin, gaseous atmosphere. Its cold, hazy upper atmosphere lacks the sharply colorful bands seen on Jupiter, but Saturn, too, has powerful winds. Most blow eastward at more than 500 meters per second, five times as fast as the fastest winds on Earth. Vast storms, marked by white clouds of ammonia ice, sometimes rise into sight.

At about 400 kilometers below its cloud tops, Saturn's gases are under enough pressure to become a sea of liquid hydrogen some 30,000 kilometers deep. This liquid overlies an even denser ocean of electrically conductive liquid hydrogen, 14,000 kilometers in depth. Intense pressures keep Saturn's core of ice and rock molten.

The Voyager **spacecraft** found that Saturn gives off roughly two times the heat it takes in from the sun. The planet is not large enough to have retained that much heat from its formation; instead, scientists believe that the heat is a by-product of helium rain, which may be condensing and precipitating out of Saturn's outer regions into its metallic ocean.

Galileo was probably the first person to see Saturn's rings, which looked to him like "cup handles" on either side of the planet when viewed through his telescope in 1610. In 1659 Dutch astronomer Christiaan Huygens accurately described "a thin, flat ring," and Italian Giovanni Cassini discovered a dark gap between the two main sections in 1675.

Shining brilliantly in the sunlight, the rings are made of ice particles—ranging in size from dustlike specks to small boulders—with a small amount of rocky grit. Today, we know that there are not two, but in fact seven primary rings consisting of hundreds or even thousands of ringlets.

Each ring orbits Saturn at a different speed. The main rings are identified by letters; in order from closest to farthest from the planet, they are D, C, B, A, F, G, and E. Between the A and B rings is a 5,000 kilometer gap known as the Cassini division. A second, smaller gap called the Encke division, 270 kilometers wide, can be seen in the outer portion of the A ring. From their innermost to outermost edges, the rings stretch across about 282,000 kilometers in a sheet only 10 meters to less than a kilometer thick.

Five of Saturn's main rings fall within the Roche limit: the distance from a planet within which any satellite would eventually be torn apart by gravitational forces. This fact, combined with the discovery that all the rings seem to be younger than their parent planet, leads scientists to believe that the rings may be the remains of pulverized **moons, comets,** or **asteroids.**

SATURN	
Average distance from the sun	1,433,500,000 km
Revolution	29.5 years
Average orbital speed	9.7 km/s
Average temperature	-140°C
Rotation	10.7 hours
Equatorial diameter	120,536 km
Mass (Earth=1)	95.2
Density	687 kg/m³
Surface gravity (Earth=1)	0.92
Known satellites	31
Largest satellites	Titan, Rhea, Iapetus, Dione, Tethys

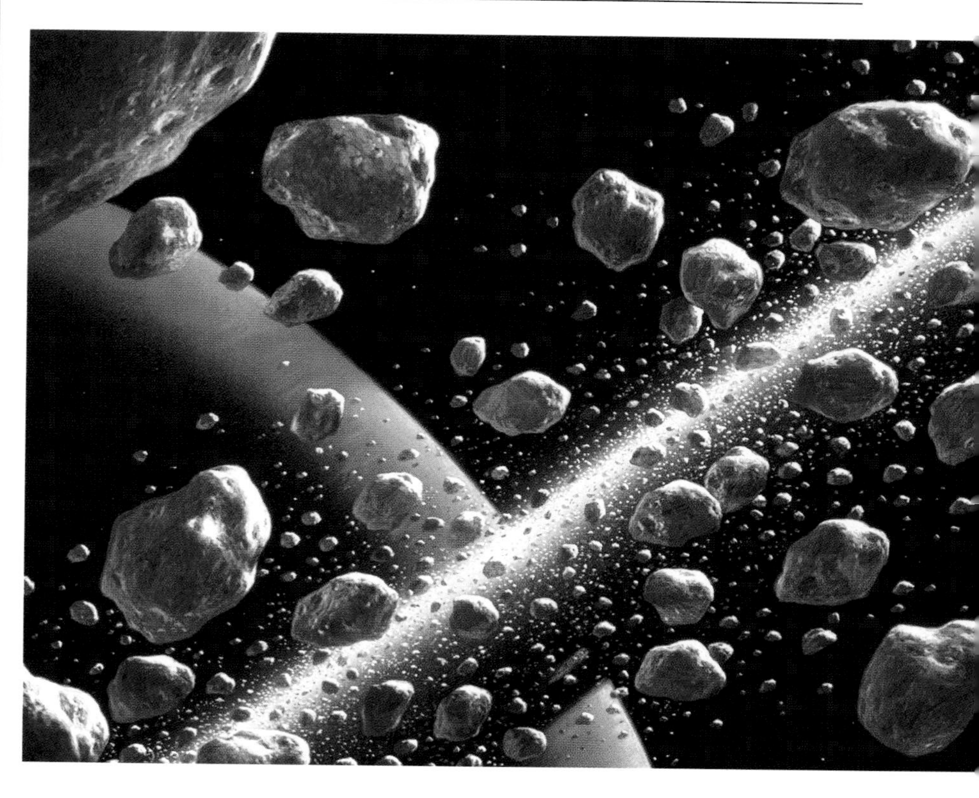

A closer view reveals that Saturn's seemingly smooth rings consist of rubble—icy hail-like particles, dust, and rocky grit that orbit the planet at blur-producing speeds. The rubble in one of the rings is so dense that no light passes through it.

The **Cassini mission,** which reached Saturn in June of 2004, reinforced this theory when it flew into orbit past the moon Phoebe and through the gap between Saturn's F and G rings, sending home detailed photographs of both the little satellite and the rings. The rings were not smooth, but rippled, with their icy particles clumping together like traffic jams on a highway. The Cassini division contained dirty material that looked like the dark substance seen on the surface of Phoebe, supporting the idea that the rings are created from the rubble of moons. And Phoebe itself was icy and heavily cratered, leading scientists to believe that it was a stray, captured by Saturn in the early years of the planet's formation. In its four-year mission, the Cassini orbiter will undoubtedly yield a wealth of information about the ringed planet and its intriguing moons.

Moons of Saturn

Saturn has a large and curious family of **moons.** As of April 2004, 31 had been discovered, ranging from giant **Titan** to tiny Suttung. Titan, with its thick atmosphere, is particularly interesting to *(continued on page 161)*

Destination Saturn

William Harwood

"Imagine a world that's smaller than Mars and bigger than Mercury, where the air is four times denser at its surface than the air in this room, and the surface pressure is about the same as you'd experience at the bottom of a neighborhood swimming pool." It was September 1997, six weeks before launch of the most sophisticated interplanetary spacecraft ever built, and Jonathan Lunine, a planetary scientist at the University of Arizona, had reporters enthralled.

"On that world," he continued, "the distant sun is never seen, and at high noon things are no brighter than a partly moonlit night on the Earth. Because of its great distance, the cold is so enormous that water is always frozen out of the atmosphere. Nitrogen is nearly so, but not quite. And the simplest organic molecule, methane, is there to take the place of water as a cloud-former, possibly a rainmaker, and maybe even the stuff of lakes or seas of hydrocarbons."

Hundreds of kilometers above the surface of a distant world, methane is broken down by cosmic rays and the distant sun's faint light, creating more complex organic compounds that rain down. Volcanism and meteoroid impacts shape the surface and provide the energy needed to create even more organic molecules.

The alien world described is not a planet but a moon: Titan, the largest satellite of Saturn and the target of a small probe hitchhiking to the ringed planet on NASA's nuclear-powered Cassini orbiter. While life almost certainly does not exist on Titan today, conditions there are thought to be similar to those on Earth shortly after the birth of the solar system, making Titan a high-priority target for both NASA and the European Space Agency (ESA).

"The organic chemical cycles that go on may constitute a chemical laboratory for replaying some of the steps that led to life on Earth," Lunine said. "Titan is in some ways the closest analogue we have to the Earth's environment before life began, and this makes Titan very important."

Cassini and ESA's Huygens probe were launched atop an Air Force Titan 4B rocket in October 1997. Costing $3.4 billion through the end of its primary mission in 2008, Cassini is one of the most expensive interplanetary spacecraft ever built, second only to NASA's Viking Mars project in the 1970s. (Viking's official cost—$2.73 billion—was figured in 1973 dollars; that's nearly $4 billion in today's dollars.)

In an era of faster, better, and cheaper spacecraft, where quick development, modest mission objectives, and tight budgets are the rule, Cassini stands out as a glaring exception. Yet it is expected to deliver a tremendous bang for all those bucks. For starters, its primary target is well over a trillion kilometers from Earth at its closest, about 21 times farther than Mars is at its nearest approach. Just getting there takes not six months but seven years, and involves a three-planet gravitational bank shot.

Once there, Cassini will spend at least another four years studying the Saturnian system, accomplishing in one mission what would take more modest spacecraft several flights to match.

"It will examine the planet's atmosphere, magnetosphere, and rings, as well as its many icy moons," said Wesley Huntress, NASA's director of space science at the time of Cassini's launch. "The mission represents a rare opportunity to gain significant insights into major scientific questions about the creation of the solar system, prelife conditions on early Earth, and just a host of questions about Saturn."

The Huygens Probe

Cassini entered into orbit around Saturn on June 30, 2004. If all goes well, Huygens will descend about six months later, spinning at seven rotations per minute for stability. It will hit the moon's upper atmosphere at an altitude of about 1,200 kilometers and a velocity of 22,088 kilometers an hour. In just three minutes, the

In this artist's rendition, the Huygens probe descends to Saturn's moon Titan as the Cassini spacecraft streaks past at upper right; the ringed planet looms in the distance. Huygens should land in January 2005, making more than 1,000 images during its two-and-a-half-hour descent.

probe will slow to less than 1,450 kilometers an hour, experiencing heat-shield temperatures up to 11,980°C and a braking force that is 16 times the pull of Earth's gravity.

A small pilot chute will deploy, pulling out Huygens's 8-meter-wide main chute at an altitude of about160 kilometers. For the next 15 minutes, the spacecraft will slowly descend while its instruments make initial observations of Titan's atmosphere. Then, at an altitude of about 112 kilometers, the main chute will be cut away and a 3-meter-wide stabilizer chute will open. Huygens should hit the surface at 24 kilometers an hour or so, two and a half hours after the descent begins. Data collected throughout the descent and landing will be beamed up to Cassini for relay to Earth.

The probe is expected to make more than a thousand images of Titan's surface and clouds, make spectroscopic measurements, and monitor the penetration of sunlight. A different instrument will measure wind velocity, while yet another will analyze atmospheric gases. An onboard weather station will not only measure temperature, pressure, and electrical activity but also look for lightning. Finally, a surface science package equipped with nine different sensors will try to determine the physical nature of whatever surface the probe lands on, liquid or solid, and should provide a point of reference that will help scientists interpret the radar imagery that will be carried out by Cassini.

"We hope, we have good confidence, the probe will survive landing," ESA project scientist Jean-Pierre Lebreton said before launch. "The landing speed is very low and...we have capability to do measurements for half an hour on the surface."

Cassini's Continuing Mission

Once the Huygens mission is complete, the science team will settle down for the most comprehensive study of Saturn and its moons ever attempted. Cassini, by using Titan's gravity to alter its trajectory, will make 70 ever changing orbits, accomplishing repeated close flybys of half a dozen moons.

Cassini's 18 instruments and 27 sensors include cameras to make close-up and wide-angle photographs of Saturn, its larger moons, and its glorious ring system. A visible and infrared mapping spectrometer will chart the distribution of specific minerals and chemicals in the atmospheres of both Saturn and Titan, as well as in the planet's rings and on the surfaces of Titan and other moons. It also will search for lightning and active volcanoes on Titan. A composite infrared spectrometer will measure infrared emissions from each target—the rings, the surfaces, and the atmospheres—to produce vertical temperature profiles and help to determine atmospheric composition. A third spectrograph will use ultraviolet imaging to determine how much hydrogen and deuterium are present. A remote-sensing microwave package includes a powerful radar system capable of seeing through Titan's cloud cover and imaging features as small as 350 meters across. Six other instruments will probe Saturn's magnetic field and assess how it interacts with the sun's solar wind, the energetic plasma associated with Saturn's magnetic field, and the particles trapped within it.

That's a lot of scientific horsepower to invest in a single spacecraft. But if any planet deserves red-carpet treatment, surely it is Saturn.

Magnificent Saturn

While all four of the gas giants—Jupiter, Saturn, Uranus, and Neptune—feature complex ring systems, Saturn's is by far the most spectacular. It is also the only one easily visible from Earth. A backyard telescope reveals two bright rings separated by a dark gap called the Cassini division. Large observatory instruments reveal a half-dozen clearly defined ring regions and four gaps. But their true nature was not revealed until NASA's Voyager probes flew past in 1980 and 1981. To the delight and amazement of astronomers, the Voyager cameras revealed more than a thousand individual rings whirling about the planet in a complex, ever changing dance orchestrated by the gravity of Saturn and a handful of small moons.

Composed of icy particles that range in size from tiny motes to boulders five meters or so across, Saturn's most clearly defined rings extend more than 97,000 kilometers from their innermost boundary to their outer reaches. Despite their large radial size, the rings are tissue-paper thin, relatively speaking, just a hundred meters or so thick at most.

The planet itself, second only to Jupiter in volume, is big enough to hold 765 Earths. Its density, however, is less than that of water and its overall mass is just 95 times that of Earth. Long-lasting hurricane-like storms and a 1,770-kilometer-per-hour equatorial jet stream roil its upper atmosphere. Saturn emits 87 percent more energy than it absorbs from the sun, possibly due to the friction of liquid helium that rains down through layers of hydrogen that compose its deep interior. All in all, it makes a beautiful, if inhospitable, place to visit. ■

astronomers. Mid-size moons—Mimas, Enceladus, Tethys, Dione, Rhea, and Iapetus—have been known since the late 18th century. All of them appear to be quite icy. One of them, Enceladus, is so highly reflective that scientists believe it is covered with recent water ice, possibly ejected by ice **volcanoes.** Iapetus is much brighter on one half than the other and may be collecting dark material on one side as it sweeps through its orbit.

Several of the moons reflect intense bombardment in their histories. Mimas has a **crater** one-third its size, the remains of an impact that would have nearly shattered the satellite.

Lumpy little Hyperion tumbles, slows down, and speeds up in its eccentric **orbit.** The only known body in the **solar system** that rotates chaotically, it is probably pulled about by tidal forces from Titan.

Titan

Titan, **Saturn**'s largest **moon,** is the second-largest natural satellite in the **solar system.** Discovered by Christiaan Huygens in 1655, it has a radius of 2,575 kilometers, making it larger than either **Mercury** or **Pluto.** And alone among the solar system's moons, Titan has a dense atmosphere, a fact that greatly intrigues scientists.

Images of Titan sent back by the Voyager missions revealed very little. The moon is covered by a thick orange haze, and its surface features are not visible. Its hazy, frigid atmosphere consists primarily of nitrogen, with minor amounts of methane and more complex compounds such as ethane and acetylene. These hydrocarbons may be joining together to add to Titan's smog. They may also be precipitating out of the atmosphere to form liquid lakes or oceans of ethane or methane on the moon's surface.

Although Titan is too cold to support Earth-like life, its chemistry resembles that of early **Earth.** Astronomers hope to learn much more in the following years, using data sent back by the **Cassini mission,** which will deploy the Huygens probe into Titan's atmosphere in early 2005.

TITAN	
Average distance from Saturn	1,221,830 km
Revolution	15.94 days
Average temperature	-178°C
Rotation	15.94 days
Equatorial diameter	5,150 km
Mass (Earth=1)	44.4
Density	1,881 kg/m^3
Surface gravity (Earth=1)	0.137

Shepherd Moons

Four of **Saturn**'s small **moons** play a key role in shaping the giant **planet**'s rings. Pan, Atlas, Pandora, and Prometheus are known as shepherd moons because their gravitational fields help to separate and confine the orbiting particles of Saturn's **rings.**

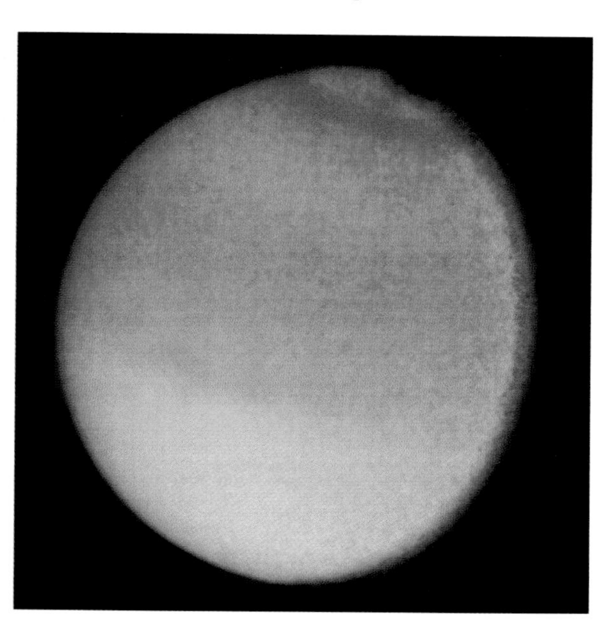

Titan, Saturn's largest moon, glows like a giant orange in this photograph by Voyager 2 in 1981. The southern half, below a well-defined equator, appears lighter in color than the northern hemisphere, and a dark collar rings the north pole. The different shades reflect cloud circulation in Titan's atmosphere.

Pandora and Prometheus hug either side of the thin F ring; Atlas shepherds the outer edges of the A ring, and Pan holds open the Encke gap in the A ring.

Cassini Mission

A joint project of **NASA,** the **European Space Agency,** and the Italian Space Agency launched in October of 1997, the Cassini **spacecraft** entered **Saturn**'s orbit on June 30, 2004. Plunging through a gap between the F and G rings, Cassini made its closest approach to the **planet:** 19,980 kilometers from the cloud tops. It will circle the giant planet for four years, studying Saturn's composition, atmosphere, **magnetosphere, rings,** and moons. A special imaging system will peer through **Titan**'s dense clouds to its surface for the first time. And in January 2005, the orbiter will deploy the Huygens probe into Titan's atmosphere. As it parachutes toward the surface, the probe will send back data on Titan's chemistry to **Earth** via the orbiter. If it survives its landing, the probe may also radio back information from the mystery moon's surface.

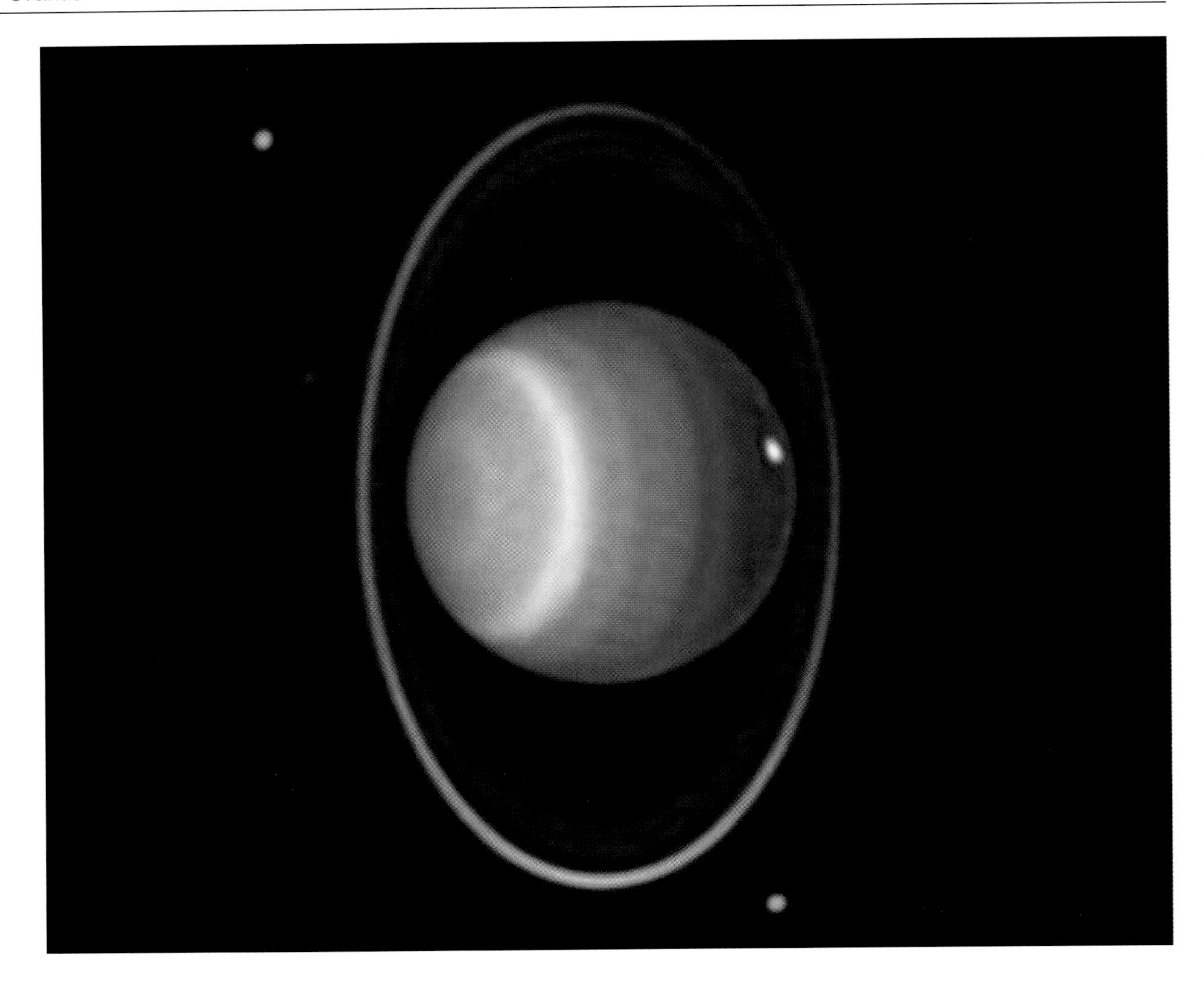

The Hubble Space Telescope captured this false-color view of Uranus, surrounded by its four narrow rings and 10 of its 26 known satellites. Blues and greens mark clear areas where sunlight can penetrate, while yellows and grays indicate a haze or cloud layer causing some light reflection. Oranges and reds mean very high cloud cover.

Uranus

Uranus

Uranus, the seventh **planet** from the **sun**, is just barely visible to the naked eye under good viewing conditions, but is so distant and dim that it was not known as a planet until 1781, when **William Herschel** discovered it using a **telescope.**

Fourteen times as massive as **Earth,** with an equatorial radius of 25,559 kilometers, Uranus is halfway between the giant planets **Jupiter** and **Saturn** and the smaller terrestrial planets in size. It is about twice as far from the sun as Saturn, in the distant, chilly regions of the solar system, and Uranus takes about 84 Earth years to complete one orbit.

A fairly rapid spin gives it a day 17.2 hours long. Unlike any other planet except **Pluto,** Uranus spins on its side—that is, its **axis** of rotation is almost perpendicular (inclined at 97.8 degrees) to the plane of its orbit around the sun.

The reason for this is unknown—astronomers speculate that the planet was knocked on its side by a huge collision early in its history. As a result, Uranus's seasons are long; the poles experience years of continuous sunlight in summer and continuous darkness in winter. However, temperatures do not vary greatly between the winter and summer sides of the planet, possibly because it is so far from the sun's heat to begin with, or because it has some way of distributing heat within its atmosphere. At the top of its clouds, the planet measures a frigid -215°C.

Through a telescope, Uranus looks like a smooth, blue-green ball. Its color comes from clouds of methane ice in its hydrogen-helium upper atmosphere. Unlike the other giant planets, Uranus has no strong source of internal heat to roil its atmosphere, so it does not exhibit the dramatic storms seen elsewhere in the **solar system.** It does, however, have jet streams, high-altitude winds that blow in the same direction as the planet's rotation.

Below the gaseous atmosphere is a liquid hydrogen shell, about 8,000 kilometers deep, which surrounds an ocean of compressed water, methane, and ammonia 10,000 kilometers in depth, kept liquid by pressure-induced high temperatures. Uranus's core is molten rock.

Just as Uranus's axis of rotation is skewed compared to the other planets, so too is its magnetosphere. Its magnetic field, similar to Earth's in strength, is tilted 59 degrees from its axis of rotation; the entire **magnetosphere** wobbles as the planet rotates.

Eleven thin **rings** orbit Uranus. Consisting of ball- to boulder-sized particles, they are less reflective than Saturn's rings and may be coated with sooty carbon dust. Because they orbit Uranus's equator, they are perpendicular to the plane of the solar system, framing the planet like rings around a bull's-eye.

Moons of Uranus

By 2004, 26 **moons** had been discovered circling Uranus. A 27th, S/1986 U10, had its moon status revoked in 2001. The five largest—**Miranda,** Ariel, Umbriel, Titania, and Oberon—have been known since before the space age. Titania, the biggest of the group, is about half the size of **Earth's moon.** Ice-covered and dark, the larger satellites, particularly Miranda, show signs of considerable melting and fracturing in their pasts.

URANUS

Average distance from the sun	2,872,500,000 km
Revolution	83.8 years
Average orbital speed	6.8 km/s
Average temperature	-215°C
Rotation: retrograde	17.2 hours
Equatorial diameter	51,118 km
Mass (Earth=1)	14.5
Density	1,270 kg/m^3
Surface gravity (Earth=1)	0.89
Known satellites	26
Largest satellites	Oberon, Titania, Umbriel, Ariel

MOONS OF URANUS

Name	Distance from Uranus (in km)	Diameter (in km)
Cordelia	49,770	40
Ophelia	53,790	42
Bianca	59,170	54
Cressida	61,780	80
Desdemona	62,680	64
Juliet	64,350	94
Portia	66,090	136
Rosalind	69,940	72
S/2003 U2	74,800	24
Belinda	75,260	80
Puck	86,010	162
S/2003 U1	97,700	32
Miranda	129,390	472
Ariel	191,020	1,158
Umbriel	266,300	1,169
Titania	435,910	1,580
Oberon	583,520	1,520
S/2001 U3	4,280,000	12
Caliban	7,230,000	96
Stephano	8,002,000	20
Trinculo	8,571,000	10
Sycorax	12,179,000	190
S/2003 U3	14,345,000	12
Prospero	16,418,000	30
Setebos	17,459,000	30
S/2001 U2	21,000,000	12

Miranda

Miranda, the innermost and smallest of **Uranus**'s five large **moons,** is a geologist's dream (or nightmare). Its crazy-quilt surface features **craters,** mountains, valleys, cliffs, fractures, faults, ridges, and grooves. Some areas appear to be much younger than others.

The satellite might have been blown apart in a massive collision and then reformed, or it may have cooled so quickly during its original formation that it solidified before its densest rocks could sink to its core.

Neptune

Neptune

Neptune was expected before it was discovered: Working independently of one another, English mathematician John Couch Adams and French astronomer Urbain Jean Joseph Leverrier used discrepancies in the orbit of **Uranus** to predict the existence of a more distant **planet** whose mass was pulling on Uranus. In 1846 German astronomer J. G. Galle found Neptune just where the two men had said it would be.

In size and composition Neptune is much like Uranus; it is a cold, **gas giant** about 17 times as massive as **Earth,** with an equatorial radius of 24,764 kilometers. At 30 AU from the sun, its **orbit** is far more distant than that of Uranus. At times this eighth planet becomes the ninth as **Pluto**'s eccentric orbit brings it closer to the sun than Neptune—which last occurred between 1979 and 1999. It takes Neptune 163.8 years to orbit the sun; it has not completed one full Neptunian year since it was discovered. Its day is 16.1 hours long.

In most ways, Neptune's atmosphere resembles that of Uranus. Icy methane clouds give the planet its dark blue appearance; below them, the gaseous atmosphere is composed mainly of hydrogen and helium. Again like Uranus, the body of the planet is liquid: a molten

Remote Neptune lies 30 times farther from the sun than Earth and receives only a miniscule fraction of the light that brightens our planet. A cold gas giant 17 times the size of Earth, Neptune appears dark blue beneath a veil of methane clouds. The Great Dark Spot and a slash of light (left) indicate storm activity on the planet.

NEPTUNE	
Average distance from the sun	4,495,100,000 km
Revolution	163.8 years
Average orbital speed	5.4 km/s
Average temperature	-214°C
Rotation	16.1 hours
Equatorial diameter	49,528 km
Mass (Earth=1)	17.1
Density	1,638 kg/m^3
Surface gravity (Earth=1)	1.12
Known satellites	13
Largest satellite	Triton

rock and ice core inside a shell of pressurized water, ammonia, and methane, which in turn is surrounded by a shell of liquid hydrogen. Unlike Uranus, Neptune is stormy. When Voyager 2 visited the planet, it sent back images of a huge, dark storm, much like **Jupiter's Great Red Spot,** about as wide as Earth's diameter. By 1994 the Great Dark Spot had disappeared, but other tempests, marked by light and dark clouds, come and go. Winds up to 2,200 kilometers an hour sweep the planet, westward at low latitudes and eastward at mid-latitudes. Neptune is so far from the sun that it does not receive enough solar heat to generate these winds; its heat source must be internal, though what powers it is unknown. The average temperature is -214°C, very chilly but slightly warmer than Uranus, although Uranus is closer to the sun.

Neptune's strong magnetic field, possibly generated by electrical currents within its swirling internal ocean, is tilted 47 degrees to its axis of rotation. Neptune has 13 **moons,** all but **Triton** are relatively small. Six thin, dim rings circle the planet. The outermost, Adams, contains three darker, clumpy sections. Astronomers speculate that these clumps are held together by the gravitational influence of Neptune's shepherd moon, Galatea.

Moons of Neptune

Only two of Neptune's 13 moons were discovered before the space age. Amateur British astronomer William Lassell spotted the largest, **Triton,** by **telescope** just 17 days after Neptune itself was discovered. American astronomer Gerard Kuiper found the third-largest, Nereid, in 1949. Nereid has the most highly elongated orbit of any satellite in the **solar system,** ranging from about 1,353,600 kilometers to 9,623,700 kilometers from Neptune. Dark, lumpy Proteus, the second largest moon, orbits too close to the planet to have been seen by early **spacecraft.** It and other, smaller moons were discovered either by Voyager 2 or by modern, ground-based telescopes.

Triton

Neptune's only large satellite, about three-quarters the size of **Earth's moon,** Triton stands out for several reasons. One is its unusual **retrograde orbit;** it is the only large moon to circle its parent **planet** in a direction opposite to the planet's rotation. This motion probably means that Triton was not formed at the same time as Neptune, but was captured later. Triton is the coldest moon in the **solar system.** Its icy surface reflects so much of the dim sunlight that reaches it that its surface temperature is an immobilizing -240°C. Despite its extreme cold, Triton evinces some activity. Erupting plumes of dark material have been seen near its south polar cap—possibly freezing nitrogen, methane, and dust ejected by ice **volcanoes** triggered by the sun's heat.

Because of its retrograde motion, Triton is doomed. Its orbit is slowly decaying, and in about 250 million years it will come close enough to Neptune to be pulled apart by the planet's gravity, possibly to form yet another **ring.**

MOONS OF NEPTUNE

Name	Distance from Neptune (in km)	Diameter (in km)
Naiad	48,230	58
Thalassa	50,080	80
Despina	52,530	148
Galatea	61,950	158
Larissa	73,550	208 x 178
Proteus	117,650	436 x 416 x 402
Triton	354,760	2,706
Nereid	5,513,400	340
S/2002 N1	15,686,000	48
S/2002 N2	22,452,000	48
S/2002 N3	22,580,000	48
S/2002 N4	46,570,000	60
S/2003 N1	46,738,000	28

Pluto

PLUTO

Pluto is a world with an identity crisis. Although it is generally considered to be the ninth (and final) **planet** of the **solar system,** some astronomers feel that Pluto should be demoted from planet status and considered just another member of the distant, orbiting band of minor worlds known as the **Kuiper belt.** Some believe that Pluto and its moon, **Charon,** are so similar in size and orbit that they form a double planet.

Pluto's discovery was a classic case of serendipity. After astronomers discovered **Neptune** by analyzing discrepancies in the orbit of **Uranus,** some believed that a ninth planet that was also affecting Uranus's motion would yet be discovered. After painstakingly studying photographs of the night sky taken several days apart for months, in 1930 American astronomer Clyde William Tombaugh discovered one tiny dot moving against the background of stars. This was Pluto. As it turned out, the little world was too small to have affected Uranus's orbit significantly.

The smallest of the nine planets, Pluto has a radius of 1,195 kilometers, making it about one-fifth the size of **Earth.** In many ways it is a twin of Neptune's moon **Triton:** small, icy, and frigid.

Orbiting the **sun** at an average distance of 39 AU, Pluto is so distant that, to an observer on its surface, the sun would look like a very bright star. Even the **Hubble Space Telescope** has been unable to see many details on Pluto's surface. Its orbit is highly elliptical, varying from 4,435,000,000 kilometers at perihelion to 7,304,300,000 kilometers at aphelion. For about 20 years out of every 248-year orbit, Pluto swings inside

Pluto's moon, Charon, seems to hover over the planet, but actually lies more than 19,000 kilometers away. Some astronomers believe the two should be considered a double planet because they are so similar in size and orbit. Astronomers first sighted Pluto in 1930, Charon in 1978.

the orbit of Neptune, making it the temporary eighth planet; the last time this occurred was between 1979 and 1999. Pluto will remain the farthest planet from the sun until April 5, 2231.

The orbit of the planet Pluto is more inclined to the **ecliptic** than that of any other planet. It is tilted at an angle of 17 degrees to the plane of the solar system, and its **axis** of rotation is skewed as well, leaning at 122 degrees to the ecliptic, so that, along with Uranus, Pluto spins almost on its side.

Pluto rotates slowly, giving it a day 6.4 Earth days long. Alone among the planets, it is locked into a mutually synchronous rotation with its moon, so that to an observer on Pluto's surface, Charon would stay in the same spot in the sky at all times, looming large overhead.

Pluto has never been visited by a **spacecraft,** so our knowledge of it comes from long-distance telescopic observation. It probably consists of a mantle of water ice around a rocky core. The temperature on Pluto, not surprisingly, averages an intensely cold -225°C.

When close to the sun, the planet possesses a very thin, extended atmosphere with a pressure only one one-millionth that of Earth's. Due to the planet's low gravity—about 6 percent of Earth's—Pluto's atmosphere leaks steadily into space from its upper regions as fast-moving molecules escape the planet's weak gravity.

The atmosphere's three main gases—nitrogen, methane, and carbon monoxide—undergo phase transitions, changing between solid and gas as the temperature warms or cools. When Pluto is closest to the sun, its atmosphere is gaseous; as it moves away from the sun in its long, **eccentric orbit,** those gases freeze and fall to the surface like snow in a very long, intensely frigid winter.

PLUTO	
Average distance from the sun	5,870,000,000 km
Revolution	248 years
Average orbital speed	4.7 km/s
Average temperature	-225°C
Rotation: retrograde	6.4 days
Equatorial diameter	2,390 km
Mass (Earth=1)	0.002
Density	1,750 kg/m³
Surface gravity (Earth=1)	0.06
Known satellites	1
Largest satellite	Charon

CHARON	
Average distance from Pluto	19,600 km
Revolution	6.4 days
Average temperature	-228.15°C
Rotation: retrograde	6.4 days
Equatorial diameter	1,186 km
Mass (Earth=1)	0.0003
Density	1,850 kg/m³
Surface gravity (Earth=1)	0.021

The Hubble Space Telescope has revealed bright and dark spots across Pluto's surface, perhaps icy and rocky areas with icy polar caps. Much more information should come our way if **NASA** succeeds in launching its New Horizons probe, which is scheduled to visit Pluto and Charon about ten years after launch in 2006.

After swinging by **Jupiter,** the New Horizons probe would photograph and map both Pluto and its moon, Charon, in detail, and it would also study the double system's weather, atmosphere, and temperature.

Charon

Astronomers at the U.S. Naval Observatory discovered **Pluto**'s moon Charon in 1978. They were studying photographs of Pluto and noticed an odd bulge to one side of the **planet.** This bulge turned out to be a large, close-in satellite. With a diameter of 1,190 kilometers, Charon is almost half Pluto's width, making it closer in size to its parent planet than any other moon in the **solar system.**

Covered with water ice, the satellite is locked into a remarkable, mutually synchronous orbit with Pluto, circling the planet every 6.4 **Earth** days. Because this is the same speed as Pluto's rotation, the moon always appears above the same spot on the planet.

Charon is also surprisingly close to Pluto, only 19,600 kilometers away. For these reasons, some astronomers call Pluto and Charon a binary, or double, planet. Charon may be a captured **Kuiper belt** object, or, like **Earth's moon,** may have formed from debris knocked loose by a collision between Pluto and some other large body.

An asteroid named Ida, a mini-planet of rock and metal, seems to dwarf its satellite, or moon, named Dactyl (at right). Like all asteroids, Ida spins on its axis as it circles the sun. Craters pockmark the surface of the asteroid.

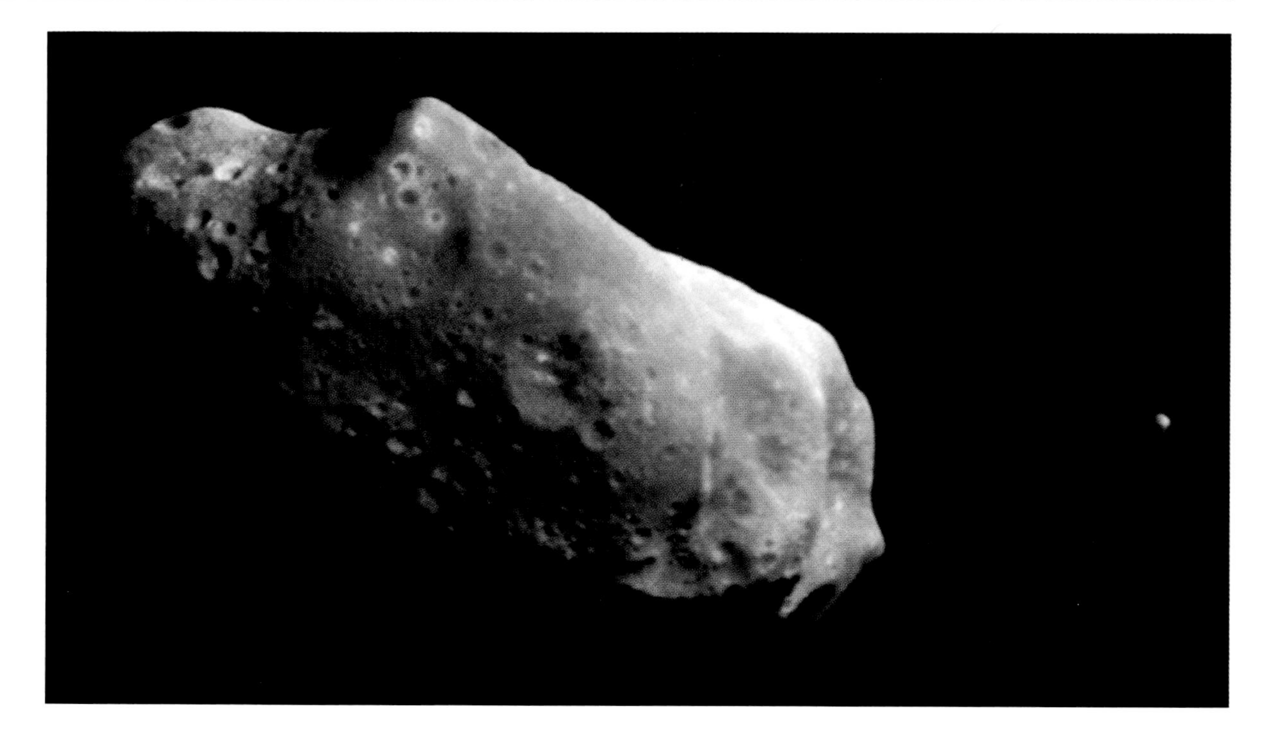

Asteroids

Asteroid Belt

Most of the **asteroids** in our **solar system** are confined to a wide belt between the orbits of **Mars** and **Jupiter,** known as the asteroid belt. Located between 2.0 and 3.3 AU from the sun, this belt may contain millions of asteroids, although they are spread out through such a large volume of space that a **spacecraft** traveling through the region would rarely encounter one. Within the belt, many asteroids collect in orbiting groups. Noticeable gaps between them, called Kirkwood gaps, are caused by the gravitational influence of Jupiter.

Asteroids

Asteroids, also called minor **planets,** are chunks of rock and metal left over from the early days of the **solar system.** Ranging in size from 100 meters to almost 1,000 kilometers across, millions of asteroids orbit the **sun,** most of them in a belt between the orbits of the planets **Mars** and **Jupiter.**

Some also orbit closer to the sun, while others are nearer to Jupiter; a few occasionally brush past the **Earth.** Despite their great numbers, the total mass of all asteroids in the solar system is only one-thousandth the mass of Earth.

Italian astronomer Giuseppe Piazzi discovered the largest known asteroid, Ceres, in 1801. Although it is spherical and orbits the sun, at 940 kilometers in diameter it is too small to be considered a planet. Only two other known asteroids, Pallas and Vesta, have diameters greater than 300 kilometers, although some, such as Ida, have managed to capture their own little **moons.** Most asteroids are lumpy, battered, irregularly shaped rocks less than one kilometer across. Almost all spin on their axes as they circle the sun; most have 6- to 20-hour rotations.

The majority of asteroids fall into one of three types: dark and carbonaceous C-types, gray and stony S-types, and metallic M-types. Astronomers identify them by analyzing their reflected light. Asteroids closest to the sun tend to be S-types, while those farthest away are more likely to be C-types. These three types are further divided into subclasses; B-type asteroids are a subclass of C-type. These differences probably reflect conditions in the solar **nebula** when the asteroids were forming. Although asteroids are spread through a large volume of space, occasionally they collide, knocking off smaller pieces. Scientists believe that this process gives birth to meteoroids, a few of which enter **Earth's atmosphere** as meteorites.

The nature and orbit of asteroids is of more than merely scientific interest to most people on Earth. An Earth-asteroid collision may have caused or contributed to the extinction of dinosaurs 65 million years ago. Any sizeable asteroid that hit the Earth now would cause widespread devastation. A number of governmental and private agencies are attempting to catalog and track any such potentially dangerous objects.

Trojan Asteroids

The Trojan asteroids are two asteroid groups that follow **Jupiter**'s orbit, one group preceding the giant planet around the sun and the other following it. The Trojan asteroids are centered on two **Lagrangian points,** places where the gravitational influences of Jupiter and the sun are balanced. The first to be discovered, in 1906, was named 588 Achilles; the others also take their names from heroes of the Trojan War.

Near-Earth Objects

Near-Earth objects (NEOs) are **asteroids** or dead **comets** that follow orbits close to **Earth**'s. Three groups of asteroids—Atens, Apollos, and Amors—are known to orbit in Earth's neighborhood. These are believed to be wanderers from the main **asteroid belt** that have been jostled out of their usual orbits by the gravitational forces of the **planets.**

Asteroids that might pose a danger to Earth are called Potentially Hazardous Asteroids, or PHAs. One such object probably created the enormous Chicxulub Crater on the Yucatán Peninsula some 65 million years ago; the dust it kicked up into the atmosphere may have changed the climate drastically enough to cause the extinction of the dinosaurs. Another, smaller asteroid exploding in **Earth's atmosphere** may have produced the massive Tunguska explosion in Siberia in 1908.

The Earth's atmosphere protects us from objects smaller than 50 meters in diameter. Impacting objects up to about one kilometer wide would create intense but localized damage. Astronomers calculate that one such object strikes the Earth every few centuries. Any object more than two kilometers in diameter could cause a worldwide "impact winter" by filling the atmosphere with dust; these impacts occur once or twice every million years.

Understandably concerned about future impacts, many public and private agencies have begun to catalog and assess NEOs. Teams in the U.S., France, Japan, and China are surveying the sky. **NASA**'s Spaceguard Survey, launched in 1998, intends to identify 90 percent of NEOs larger than one kilometer by 2008. To date, astronomers have cataloged more than 500 PHAs, none of them on a direct collision course.

NEAR Shoemaker Spacecraft

Built by **NASA** and launched in 1996, the NEAR (Near Earth Asteroid Rendezvous) Shoemaker spacecraft was the first to land on an **asteroid.** The probe reached 433 Eros on February 14, 2000, and collected information about the 33-kilometer-long asteroid's mass, composition, and gravity from orbit before finishing its mission by landing gently on Eros's surface.

WELL-KNOWN ASTEROIDS

Number and Name	Date Discovered	Category	Diameter (in km)*	Of Note
1 Ceres	1801	C-type	960 x 940	First asteroid discovered and the largest known
2 Pallas	1802	B-type	570 x 525 x 482	Second largest asteroid
4 Vesta	1807	Unknown	530	Third largest asteroid; the brightest main-belt asteroid
243 Ida	1884	S-type	58 x 23	First asteroid found to have a satellite asteroid (Dactyl—discovered 1993)
433 Eros	1898	S-type	33 x 13 x 13	First near-Earth asteroid discovered
951 Gaspra	1916	S-type	19 x 12 x 11	First asteroid to be imaged by a spacecraft (Galileo in 1991)
1036 Ganymed	1924	S-type	32	Largest near-Earth asteroid

**All distances are approximate.*

Comets

Comets

Comets are small, icy bodies left over from the formation of the **solar system** and believed to exist in large numbers at great distances from the **sun.** Most probably orbit in the **Kuiper belt,** 30 to 1,000 AU from the sun, or in the **Oort cloud,** located halfway to the nearest star at 30,000 to 100,000 AU. Only when these bodies are disturbed by the gravity of a passing star or dust cloud do they leave their distant homes and drop toward the sun and the inner solar system.

On average, comets are a few kilometers in diameter. Aptly described as icy snowballs, comets consist of small amounts of rock and dust embedded in frozen water, carbon dioxide, methane, carbon monoxide, ammonia, and small amounts of more complex organic molecules such as formaldehyde and hydrogen cyanide.

These compounds probably represent the composition of the **nebula** from which the solar system formed; comets in the Oort cloud may have accreted near the current orbits of **Jupiter** and **Saturn,** then been ejected into more distant orbits by the two giant planets' gravity.

Comets in their distant homes are invisible, but when their orbits are perturbed and they swing in toward the sun, they undergo dramatic and visible changes. Solar radiation begins to vaporize the comet's icy body, creating a glowing cloud of gas, called a coma, that can be up to one million kilometers across. Even larger than the coma is a thin cloud of hydrogen, detectable only at ultraviolet wavelengths, extending for millions of kilometers around the comet. The **solar wind** creates at least two tails: one, the ion tail, is made of electrically charged atoms; the other, the dust tail, is broader and flatter. Both tails extend for millions of kilometers and always point away from the sun. Buried inside the brilliant coma is the nucleus of the comet, a dark, icy rock.

Most observed comets are long-period visitors, meaning that they appear in **Earth**'s neighborhood no more than once every 200 years. Many of them probably start in the spherical shell of the Oort cloud, and

Illustrating with a model of a comet, Harvard Professor Fred Whipple tells students how these celestial bodies were born: The icy rock balls a few kilometers wide are left over from the formation of the solar system and live far from the sun, dropping into the inner solar system only when disturbed.

therefore can cross the plane of the solar system at any angle. Short-period comets, most of which may originate in the Kuiper belt, are those that have been caught up into tight orbits in the solar system and swing by more frequently, typically staying in the plane of the **ecliptic. Halley's comet,** which appears every 74 to 79 years, is a short-period comet; comet Hyakutake, a particularly bright spectacle in 1996, is a long-period comet. Comets do not live forever. Every time one approaches the sun, it loses mass as its ices vaporize. Close-up observations of Halley's comet by the spacecraft **Giotto** showed that it lost, on average, 20 tons of water per second as it neared the sun. Even so, Halley will be able to make a thousand more trips past the Earth before vanishing from sight.

Comet Shoemaker-Levy 9

From July 16 to 22, 1994, comet Shoemaker-Levy 9 provided a thrilling experience for Earth's astronomers as its fragments smashed into **Jupiter**'s atmosphere. After coming too close to Jupiter in 1992, the comet broke into at least 21 pieces, which were discovered by Carolyn and Eugene Shoemaker and David Levy in 1993. In July 1994, the debris crashed into Jupiter, creating fireballs with temperatures of 7,500 kelvins (hotter than the **sun**'s **photosphere**) and dust clouds larger than the continent of Asia.

Halley's Comet

Undoubtedly the most famous of all comets, Halley's comet has been observed and recorded since at least 240 B.C.; its appearance above the Battle of Hastings in 1066 is commemorated in the Bayeux Tapestry. The comet is named for astronomer Edmund Halley, who first recognized that certain bright comets observed throughout history were actually one object with a 74- to 79-year period. Halley's comet last swung by the **Earth** in 1986, when it was observed by no fewer than six spacecraft: Japan's Sakigake and Suisei, the Soviet Union's Vega 1 and 2, the U.S.'s **ICE,** and European Space Agency's **Giotto.** Halley's comet will next return to Earth's neighborhood in 2061.

Oort Cloud

The Oort cloud is a spherical shell of hundreds of billions of cometary nuclei that probably encircles the **solar system** at distances from 30,000 to 100,000 AU—halfway to the nearest star. Its existence was first postulated by Ernst Öpik in 1932; in the 1950s, Jan Oort further developed Öpik's theories. Astronomers believe that the icy objects of the Oort cloud originally formed in the solar **nebula** somewhere between the orbits of **Jupiter** and **Neptune** and were subsequently swept out of the solar system by interactions with the giant planets. The Oort cloud is the source of most long-period comets. Until the 21st century, the far-distant Oort cloud remained unseen, but in 2003, scientists at **Palomar Observatory** in California discovered a reddish planetoid, approximately three-quarters the size of **Pluto,** that may be a close-in member of the cloud. The distant body, christened Sedna, is as much as 130 billion kilometers (900 AU) from the **sun.**

SELECTED BRIGHT PERIODIC COMETS

Comet	Orbit Period (in years)	Next Perihelion
21P/Giacobini-Zinner	6.5	2005
9P/Tempel 1	5.5	2005
75P/Kahoutek	6.2	2005
2P/Encke	3.3	2007
6P/d'Arrest	6.5	2008
46P/Wirtanen	5.5	2008
8P/Tuttle	13.5	2008
7P/Pons-Winnecke	6.4	2008
10P/Tempel 2	5.5	2009
81P/Wild 2	6.4	2010
1P/Halley	76	2061
109P/Swift-Tuttle	130	2127
C/1995 01 Hale-Bopp	2,400	4397

Kuiper Belt

Located between 30 and 1,000 AU from the **sun**—from the orbit of **Neptune** to well past the orbit of **Pluto**—the Kuiper belt is a broad, flat region containing billions of icy objects that were formed in the early days of the **solar system.** Unlike the similar members of the **Oort cloud,** the Kuiper belt's ice balls were just far enough away from the giant planets' orbits to remain in their original locations.

Astronomers believe that the Kuiper belt is the source of most short-period **comets.** It contains thousands of bodies larger than 100 kilometers in diameter. The planet Pluto, its moon, **Charon,** and Neptune's moon **Triton** may be Kuiper belt objects. In recent years, sensitive Earth-based **telescopes** have discovered

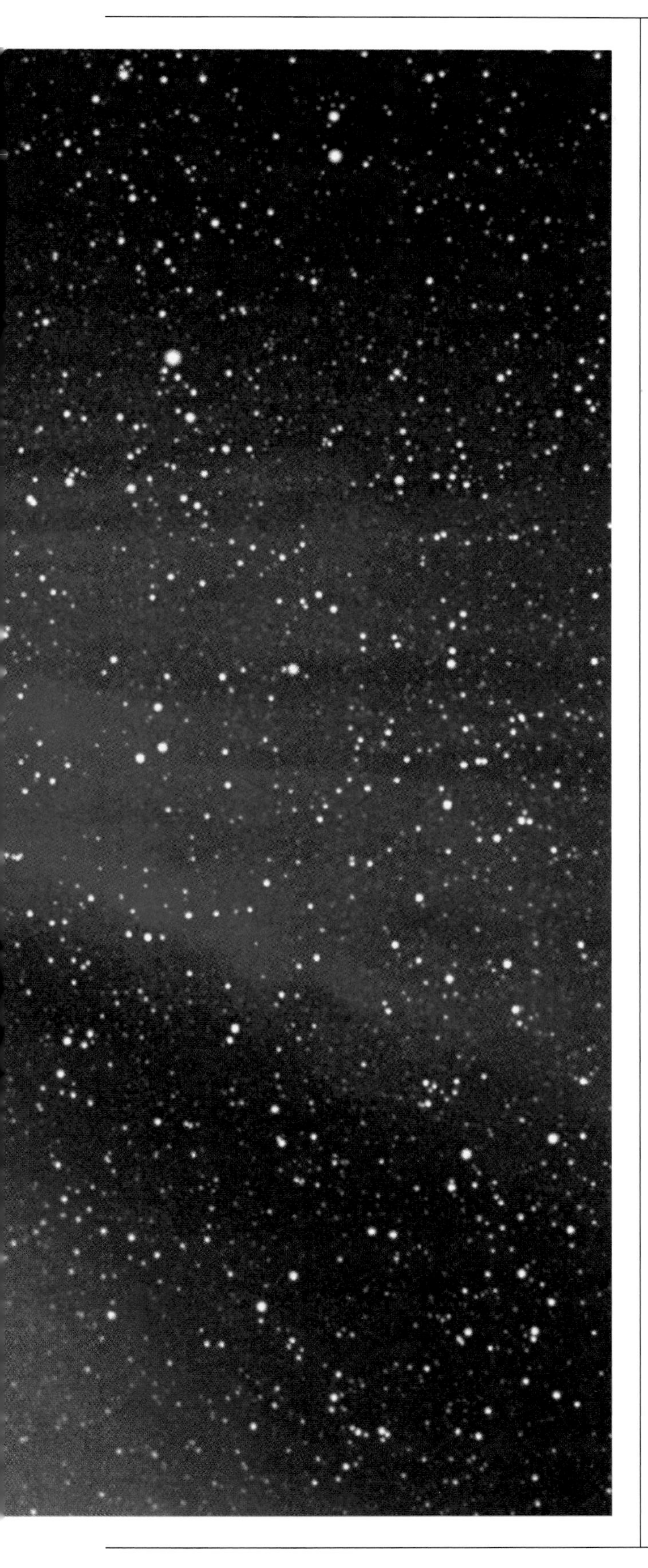

a number of sizeable Kuiper belt bodies, including Quaoar, about 1,300 kilometers wide, and an even larger planetoid temporarily called 2004 DW, perhaps 1,450 kilometers across.

Giotto Spacecraft

Built by the **European Space Agency** and launched in 1985, the Giotto spacecraft encountered and studied **Halley's comet** in March 1986 from a distance of 605 kilometers. Although the spacecraft was knocked off-kilter by a large particle 14 seconds before its closest approach, the robot managed to return a number of remarkable images of Halley's nucleus and determine that the dust and gas given off by the **comet** consisted of 45 percent water, 28 percent stony dust, and 27 percent organic compounds—complex materials with carbon in them. Giotto subsequently flew past comet Grigg-Skjellerup in 1992 and was able to send back information about that comet as well.

Stardust Spacecraft

In January 2004, **NASA**'s Stardust spacecraft flew through the dust and gas cloud surrounding **comet** Wild 2 and collected samples of its particles to bring back to **Earth.** When its reentry capsule parachutes to the ground in January 2006, Stardust will become the first robotic mission to return to Earth with physical samples from beyond the **moon.** The spacecraft also captured high-definition images of Wild 2's lumpy, irregular nucleus.

Deep Space 1

Deep Space 1 was an innovative robotic **spacecraft,** launched by NASA. In July 1999 it flew near **comet** Braille, and in September 2001 it encountered comet Borrelly, yielding close-up images of the comet's potato-like nucleus. The spacecraft tested 12 new technologies, including the (successful) use of ionized xenon gas as a **propellant.**

Rosetta

Launched on March 2, 2004, the **European Space Agency**'s Rosetta will be the first mission to orbit and land on a comet. Powered by two solar panels and **solar-cell** technology, Rosetta and its lander are scheduled for a November 2014 rendezvous with comet 67P/Churyumov-Gerasimento.

Every 2,400 years, a comet named Hale-Bopp orbits into the solar system. On its 1997 visit, Hale-Bopp showed what it was made of: water, ammonia, formaldehyde, and hydrogen cyanide—which can react to form amino acids, the building blocks of proteins needed for life.

THE GREAT CRASH OF 1994

J. Kelly Beatty

ONE OF THE GREATEST EVENTS IN THE HISTORY OF ASTRONOMY BEGAN WITH some bad film, a cloudy night, and four frustrated observers. On Palomar Mountain, in southern California, virtually in the shadow of the legendary five-meter Hale telescope, stands a small research telescope used almost exclusively for hunting comets and asteroids. The prospects for discovery seemed bright when the observing team of Eugene Shoemaker, his wife, Carolyn, and long-time collaborator David Levy, along with visiting French astronomer Philippe Bendjoya, arrived on March 21, 1993, to use the telescope. March 22 was their first night of viewing, but they discovered that their film had become fogged by accidental exposure to light. On the following night, March 23, they optimistically began their second night of viewing with good film.

Their optimism soured, however, soon after cracking the dome of the telescope open. Thin clouds overhead heralded advancing overcast. Resigned to the fact that any pictures they took would be compromised by the clouds, they decided not to waste any of their good film. Instead, using some of the film that was slightly fogged, they recorded an area of sky two times. Then the clouds thickened, and they called it a night.

Two days later, Carolyn examined the negatives closely. To one side, not far from Jupiter, was a fuzzy streak with a series of tails and thin lines at either end. "I don't know what this is," she said, bolting upright. "It looks like...like a squashed comet."

More powerful telescopes later revealed the blur to be a line of little comets arrayed like pearls on a string, each sporting its own tail. Based on other observations in the weeks thereafter, astronomers concluded that a single, giant ice ball had strayed to within 22,530 kilometers of Jupiter on July 8, 1992, and had been torn apart by the planet's gravity.

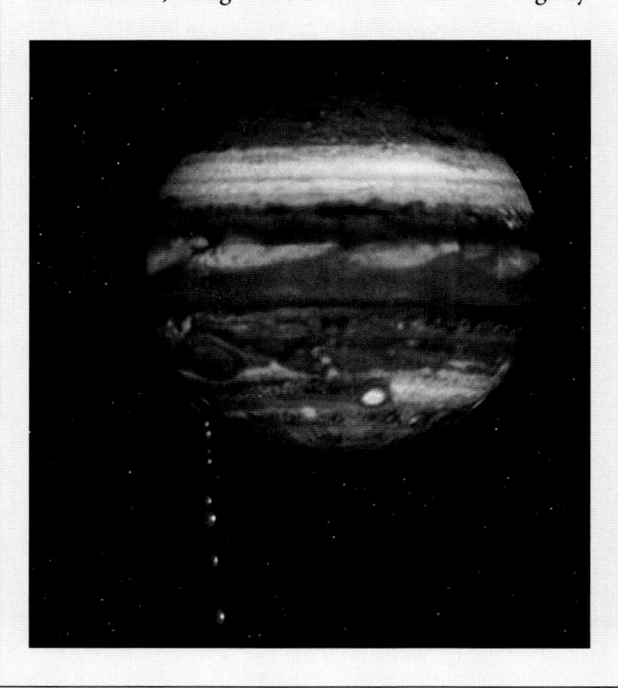

Even more stunning was the revelation that the shattered remains of comet Shoemaker-Levy 9, or S-L 9, were destined to strike Jupiter in July 1994. A cruel twist of geometric fate placed the target zone on Jupiter's far side, just out of sight from Earth. Computer-aided simulations tried to anticipate what would happen during each high-speed splash into Jupiter's atmosphere and what might be seen once the impact sites rotated into view. Some modelers suspected that the fragments were large, at least 1.5 kilometers across, and would strike with the kinetic-energy equivalent of a hundred billion tons of TNT—or more.

Others countered that Jupiter might well swallow the icy shards without a trace. But no one could predict the outcome with certainty. Never before had such an event been witnessed, and never before had so many of the world's telescopes—from mountaintop behemoths to humble backyard tubes on rickety tripods—turned their gaze to the same spot of sky. Even the orbiting Hubble Space Telescope and Jupiter-bound Galileo spacecraft were commandeered for the comet's deathwatch. Thanks to its location in interplanetary space, Galileo was able to view the target zone directly when the impacts commenced.

Fire in the Sky

Astronomers don't always fare well during highly publicized celestial events, so caution generally pervaded the pre-crash pronouncements. However, for once the celestial fireworks were spectacular. Over a period of six days beginning July 16, a score of cometary fragments bombarded Jupiter at 64 kilometers per second. Several created fireballs roughly 3,220 kilometers high, tall enough to peek around the planet's limb and be spotted by the Hubble Space Telescope.

Observers sat openmouthed as tremendous fireballs rotated into view and blazed into their telescopes' infrared-sensitive detectors. The superheated gas in the target zones had temperatures approaching 1110°C or more. As the conflagrations cooled, most left huge dark stains in the Jovian atmosphere. In some cases these splashes of sooty debris were larger than the entire Earth.

The black scars smeared out and faded after several months, though researchers continued to analyze the impact's consequences for years. Spectroscopists dissected the light from the superheated blast plumes in the hope of deducing something about the composition of the planet's atmosphere. But it proved difficult to disentangle which compounds had come in with the kamikaze fragments and which belonged to Jupiter.

Much effort went into pinning down the original size of Shoemaker-Levy 9 and of the pieces it became. No consensus was reached, though all the havoc wrought on Jupiter—the towering, incandescent plumes and the globe-girding tangle of dark atmospheric bruises—could easily have been the handiwork of an errant ice ball no larger than 1.5 kilometers across. An object of this size probably strikes Earth every 100,000 years on average, and, judging from the consequences seen on Jupiter, it would not be an event awaited with eager anticipation. ■

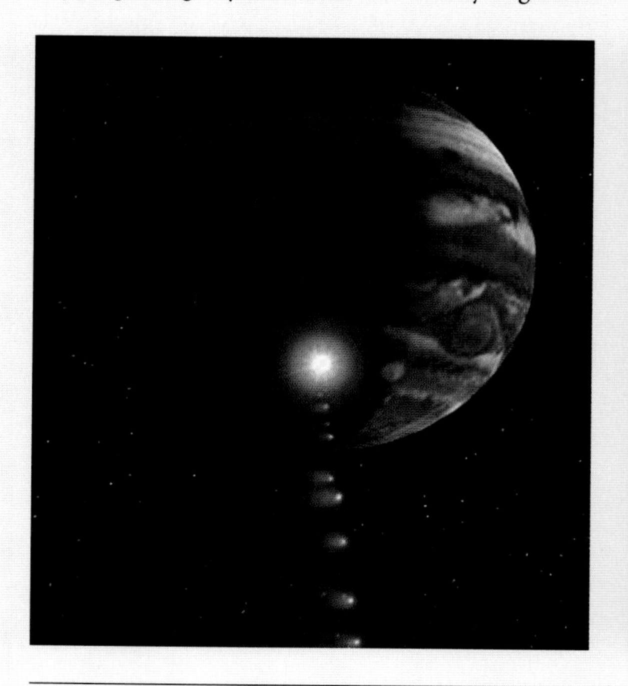

Showing four different perspectives, this artist's rendition of the collision of S-L 9's Fragment Q with Jupiter is superimposed on images taken of Jupiter by Voyager 2 in 1979.

Meteors

Meteors

Meteor, meteoroid, and meteorite are terms describing stages in the existence of interplanetary debris. Meteoroids are small chunks of rock and metal in space that range in size from less than a millimeter to tens of meters wide. Most often they are pieces of **asteroids;** sometimes they are bits of **comets;** rarely, they are pieces of the **moon** or **Mars** that were knocked off in a collision.

Meteors, particles of interplanetary debris that enter Earth's atmosphere and burn up, streak through the night sky near New Haven, Connecticut, in 2001. An annual November event, the Leonid meteor shower occurs as Earth passes through the dusty orbital path of comet Tempel-Tuttle.

As it moves through space, **Earth** sweeps up untold millions of little meteoroids, perhaps 200 million kilograms worth each year. Almost all of them vaporize in the atmosphere, never reaching the ground. As they shoot through the air, they heat up and leave a trail of glowing dust. This bright path is a meteor, or shooting star. At certain times each year, the Earth passes through a stream of dust left behind by a comet; as the dust burns in the atmosphere, it creates a veritable rain of shooting stars known as a **meteor shower.**

The more massive meteoroids that survive their trip to Earth's surface are known, once they are on the ground, as meteorites. The biggest of them may leave an **impact crater,** such as the Barringer Crater in Arizona, created by a meteoroid about 50 meters across. Meteorites fall into three main categories: stony, iron, and stony-iron.

Stony meteorites, the most common, are further subdivided into chondrite and achondrite objects. Chondrite meteorites are made of elements like those in the **sun,** and one group of chondrites, carbonaceous chondrites, contains complex carbon compounds and may represent the early composition of the **solar system.** Achondrites seem to be made of rock that was once molten and then crystallized.

Iron meteorites, comprising about 6 percent of all meteorites reaching Earth, are made mostly of iron and nickel; they are heavier than stony meteorites and less likely to break up in the atmosphere. Iron meteorites weighing up to 60,000 kilograms have been found around the world.

Not surprisingly, stony-iron meteorites are made of approximately equal amounts of stone and iron. These are less common than the other two types, representing about 2 percent of all known meteorites.

At least 30 stony meteorites found on Earth originated on Mars. Some of them are known as the SNC meteorites (the name is taken from their subgroups, the shergottites, nakhlites, and chassignites). These interesting basaltic stones hold trapped gases from the Martian atmosphere. Perhaps the most controversial Mars meteorite, however, is ALH 84001, found in Antarctica in 1984. This four-billion-year-old piece of Mars contains tiny, fossil-like structures that some scientists believe may be evidence of primitive bacterial life. Other scientists strongly disagree. The matter may not be settled until we have more evidence on the question from the surface of Mars itself.

Meteor Showers

On the average dark, clear night, an observer might see two or three **meteors** an hour. At certain times of the year, however, the **Earth** passes through a particularly dense trail of meteoroids, usually the debris left behind by a comet orbiting the **sun.** At these times, the dustlike particles burn visible trails through **Earth's atmosphere** at a rate of 15 to 100 sightings an hour in a spectacular meteor shower.

Each meteor shower seems to emanate from a particular point in the night sky, called the radiant, which is typically identified with its background **constellation.** Therefore, meteor showers are named after the constellations from which they seem to appear: the **Perseids** for Perseus, or the Ursids from Ursa Minor.

The Earth experiences about 30 of these regular meteor showers each year. Some occur during the daytime and thus are invisible to the naked eye. Others are sparse. A few, including the Perseids, **Quadrantids, Geminids,** and **Leonids,** are famous for their spectacular annual rain of light.

Perseids

One of the most prolific and reliable annual **meteor showers,** the Perseids appear every year between July 23 and August 20. Peak activity is on August 12. Consisting of debris from comet 109P/Swift-Tuttle, the shower has a maximum rate of about 80 meteors an hour. Records as far back as A.D. 36 mention the shower; in the 1860s, Giovanni Schiaparelli was the first to definitively associate the Perseids with the comet.

Quadrantids

The Quadrantid **meteor shower** appears to radiate from the **constellation** Boötes. The Quadrantids are visible between January 1 and 6, peaking from January 3 to 4 with a maximum rate of 110 an hour. They are tentatively connected to comet 96P/Macholz 1.

Geminids

The Geminid **meteor shower** occurs between December 7 and 16, with a peak on December 13. The meteors radiate from the **constellation** Gemini, near Gemini's twin stars Castor and Pollux, with a maximum rate from 60 to 100 meteors an hour. They are associated with asteroid 3200 Phaethon, which is probably the nucleus of a dead comet.

Leonids

The Leonid **meteor shower** did not come into its own until the early morning hours of November 13, 1833, when 150,000 "flaming stars" rained down upon the eastern half of the United States. After this spectacular storm, Denison Olmsted, Professor of Mathematics and Natural Philosophy at Yale University, studied every account he could find of the event. He concluded that **meteors** were particles that entered **Earth's atmosphere** at high velocities. Olmsted established their radiant in the **constellation** Leo.

Ernst Wilhelm Liebrecht Tempel discovered the Leonids' parent comet, 55P/Tempel-Tuttle, in the constellation Ursa Major on December 19, 1865. On January 6, Horace Parnell Tuttle independently spotted the comet. It was the 55th periodic (55P) comet to have its orbital path determined. Comet Tempel-Tuttle makes one revolution around the **sun** every 33.2 years. Its orbital path ranges from just inside Earth's orbit to beyond that of **Uranus.** Tempel-Tuttle's most recent journey by the sun (perihelion) was on February 28, 1998. This produced outstanding displays in November 1999 and 2001. The shower's maximum usually occurs around November 17, when an average of 10 to 15 meteors an hour can be seen in a dark sky.

WELL-KNOWN ANNUAL METEOR SHOWERS

Meteor Shower	Dates Active	Constellation	Comet Association
Quadrantids	Jan 1-6	Boötes	96P/Macholz 1 (suspected)
Lyrids	Apr 19-25	Lyra	C/1861 G1 Thatcher
Eta Aquarids	Apr 24-May 20	Aquarius	1P/Halley
Delta Aquarids	Jul 15-Aug 20	Aquarius	96P/Macholz 1 (suspected)
Alpha Capricornids	Jul 15-Aug 25	Capricorn	45P/Honda-Mrkós-Pajdusaková
Perseids	Jul 23-Aug 20	Perseus	109P/Swift-Tuttle
Orionids	Oct 16-27	Orion	1P/Halley
Taurids	Oct 20-Nov 30	Taurus	2P/Encke
Leonids	Nov 15-20	Leo	55P/Tempel-Tuttle
Geminids	Dec 7-16	Gemini	Asteroid 3200 Phaethon
Ursids	Dec 19-24	Ursa Minor	8P/Tuttle

3 | Reaching & Maneuvering in Space

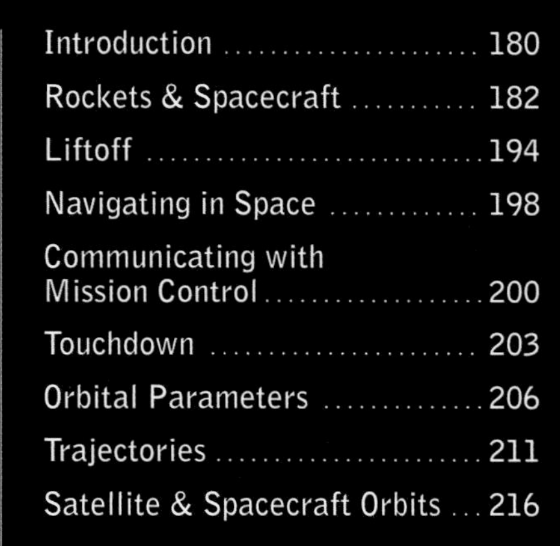

Roaring skyward, space shuttle *Atlantis* blasts off a launch pad at the Kennedy Space Center in Florida. To overcome gravity and reach orbit, spacecraft must attain breathtaking speeds. Huge quantities of rocket fuel are needed to propel the craft into space.

SPACEFLIGHT, THAT HALLMARK OF 20TH-CENTURY PROGRESS, HAS ITS ROOTS IN ancient achievements. The technology that put astronauts on the moon in 1969 was the product of centuries of scientific theory, military engineering, and backyard tinkering. As long ago as the 1200s, Chinese craftsmen understood the principles of action and reaction well enough to launch small rockets with some degree of accuracy. And the great thinkers of the scientific renaissance codified that understanding into the fundamental laws of physics that underlie all spaceflight. Galileo Galilei, Isaac Newton, and others laid out the rules of gravity and motion so clearly that it became possible to see how the forces that governed a falling object might also propel a spacecraft. Visionary writers such as Jules Verne captured that vision in fiction. Inspired by Verne, Russian mathematician Konstantin Tsiolkovsky wrote that humans were fated to "set foot on the soil of the asteroids, to lift by hand a rock from the Moon, to observe Mars...." Within a generation, Tsiolkovsky's designs for interplanetary craft took shape in the hands of Robert Goddard, Hermann Oberth, and Wernher von Braun.

The rapid progress in spaceflight seen in the 1960s owed as much to Cold War competition as it did to visions of Martian exploration. Thousands of satellites now circle the Earth, launched by complex, expensive rockets into a dense web of orbits. And past them, deeper into space, fly missions to the sun, Mars, Saturn, and beyond.

From the earliest days, almost all rockets were propelled by chemical fuels. That began to change at the turn of the 21st century. Ion and nuclear engines, solar sails, and laser propulsion are starting to replace chemical fuels, promising more efficient and faster ways to access the far reaches of the solar system and beyond. The energy requirements for reaching the nearest star system are extreme. Yet even the most astonishing methods of space propulsion are built around the basic principles that Newton gave us, and every voyage into space still obeys the rule of gravity.

The complex, computerized, ingenious machines that now fly into space might baffle an 18th-century mathematician, but the forces that move them and the graceful, arcing paths that they follow would be utterly familiar.

The physical laws of orbital motion are few and surprisingly simple. They govern home run baseballs, ballistic missiles, satellites, planets, moons, and interplanetary spacecraft. These laws, and the rockets, satellites, and orbits governed by them, are described in this chapter.

The first commercial spacecraft made a successful test flight on June 21, 2004, when the White Knight lifted the squid-shaped SpaceShipOne 15.24 kilometers above California. The smaller craft's pilot then fired a rocket that catapulted SpaceShipOne to a height of 100.12 kilometers, just 120 meters beyond the internationally recognized boundary of space.

Rockets & Spacecraft

Konstantin Tsiolkovsky

Born in the farming village of Izhevskoye, Russia, Konstantin Tsiolkovsky (1857-1935) was the father of modern spaceflight theory and aerodynamics. His was a largely solitary career. Partially deaf as a result of scarlet fever in his youth, Tsiolkovsky studied science and mathematics in Moscow before spending most of his life as a schoolteacher in the town of Borovsk. There, he built a wind tunnel and experimented with the effects of wind flow on models of dirigibles and other aircraft. At the same time, he began to develop theories about **rocket** propulsion, and in 1903 he published a paper called "Investigation of Outer Space by Reaction Devices" in a Russian scientific journal. The paper explored many fundamental issues of rocket science, including air friction, heat transfer, and fuel supply. In the next decades Tsiolkovsky published other groundbreaking theoretical papers on engine performance, **microgravity,** fuels, the use of booster rockets, and **space suits.** Ignored by the public during most of his life, in the 1920s Tsiolkovsky began to receive recognition from the Soviet Union for his work, and his theories greatly influenced the next generation of rocket scientists in Europe and the U.S.S.R.

Rocket

A rocket is a reaction engine; the term "rocket" can also mean a vehicle propelled by a rocket engine. Rockets work by exploiting **Newton**'s third law of motion, which states that for every action, there is an equal and opposite reaction. In the case of rockets, the action is the high-speed escape of gas through the rocket's nozzle. The reaction is the forward movement of the rocket.

Although the laws of motion were not clearly recognized until the 17th century, rockets have been used in war and as fireworks since ancient times. Having invented gunpowder in the first millennium A.D., the Chinese were, by 1232, launching military rockets, consisting of bamboo tubes stuffed with saltpeter and other compounds, at their enemies. The technology spread to Europe in medieval times, and until the 20th century rockets were used primarily as weapons, producing impressive explosions (as in the "rocket's red glare" seen at Fort McHenry in 1814) but little in the way of accurate damage.

The first modern scientist to outline the use of rockets for space exploration was **Konstantin Tsiolkovsky,** a Russian schoolteacher. Tsiolkovsky's 1903 paper describing the use of chemical **propellants** in rockets set the stage for 20th-century advances in rocketry. Tsiolkovsky's work was followed by that of American physicist **Robert H. Goddard.** Goddard built a series of rockets using both liquid and solid propellants that by 1937 achieved an **altitude** of about

Highlights of Modern Rocketry

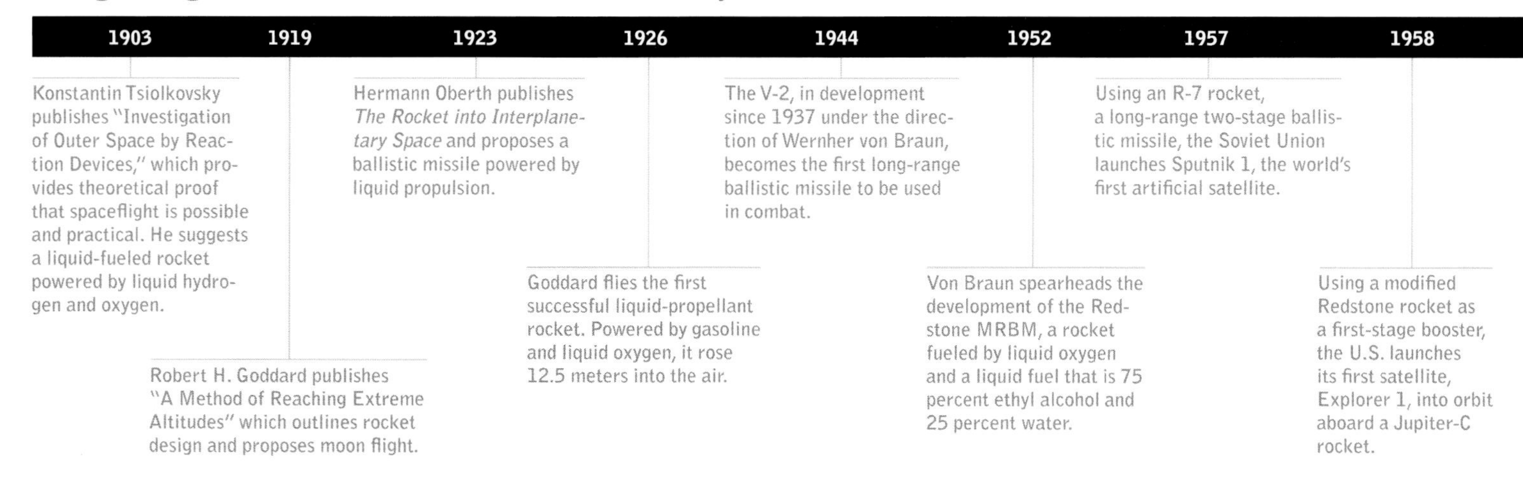

1903
Konstantin Tsiolkovsky publishes "Investigation of Outer Space by Reaction Devices," which provides theoretical proof that spaceflight is possible and practical. He suggests a liquid-fueled rocket powered by liquid hydrogen and oxygen.
1919
Robert H. Goddard publishes "A Method of Reaching Extreme Altitudes" which outlines rocket design and proposes moon flight.
1923
Hermann Oberth publishes *The Rocket into Interplanetary Space* and proposes a ballistic missile powered by liquid propulsion.
1926
Goddard flies the first successful liquid-propellant rocket. Powered by gasoline and liquid oxygen, it rose 12.5 meters into the air.
1944
The V-2, in development since 1937 under the direction of Wernher von Braun, becomes the first long-range ballistic missile to be used in combat.
1952
Von Braun spearheads the development of the Redstone MRBM, a rocket fueled by liquid oxygen and a liquid fuel that is 75 percent ethyl alcohol and 25 percent water.
1957
Using an R-7 rocket, a long-range two-stage ballistic missile, the Soviet Union launches Sputnik 1, the world's first artificial satellite.
1958
Using a modified Redstone rocket as a first-stage booster, the U.S. launches its first satellite, Explorer 1, into orbit aboard a Jupiter-C rocket.

2,740 meters. By this time, German scientists had entered the rocket-development business as well. Inspired by the work of Hermann Oberth, who published *The Rocket into Interplanetary Space* in 1923, and then by Oberth's apprentice **Wernher von Braun,** German researchers eventually created a devastating wartime weapon, the **V-2 rocket.** The heavy, 14-meter-high V-2s were launched in 1944 against England, killing and wounding thousands.

After World War II, Cold War rivalries spurred the development of larger and more accurate **intercontinental ballistic missiles** in both the U.S. and the Soviet Union. One such Soviet missile, the R-7, was modified to carry **Sputnik 1** into space in 1957.

The United States responded with the creation of the National Aeronautics and Space Administration (**NASA**). NASA presided over the construction of ever bigger and better **launch vehicles,** resulting in the massive **Saturn V** that ferried the **Apollo** astronauts to the **moon** and, later, in the combination of solid- and liquid-fuel rockets that propel the **space shuttle** into orbit.

Although a modern rocket is a complex instrument, it has the same basic structure as the homemade engine that Goddard launched in1926. Burning fuel within a combustion chamber produces hot, rapidly expanding gases. The gases push on all sides of the combustion chamber, but can escape only through a nozzle. As the gases stream out the nozzle, the pressure of the remaining gases inside the combustion chamber pushes the rocket forward. That reaction is called the rocket's thrust, and is measured in pounds (in the United States) or in newtons (in metric systems).

Rockets come in liquid-fuel or solid-fuel types. In both cases, the rocket propellant must combine a fuel with an oxidizer, which provides the oxygen needed for combustion in outer space. On the space shuttle, the solid rocket boosters that lift the orbiter from the ground contain powdered aluminum as their fuel mixed with ammonium perchlorate as an oxidizer. The external tank that takes over after the boosters are discarded contains liquid propellant: liquid hydrogen mixed with liquid oxygen (as the oxidizer). Solid-fuel rockets are relatively simple and safe, but their thrust cannot be controlled and they cannot be stopped and restarted. Liquid-fuel rockets are more volatile, but they can be stopped, restarted, and throttled to allow for maneuvering in flight.

The thrust of a rocket depends upon the mass of its propellant and the velocity at which its gases leave the engine. Rockets that carry **payloads** must provide enough thrust to lift the mass of the payload as well as the mass of the fuel and rocket itself.

Rockets today are used for launching probes and **satellites,** for atmospheric research, for space travel, and for military operations. They range from small, shoulder-fired bazookas to the massive engines that take the shuttle into space. Those that lift objects into orbit are typically multistage rocket combinations—launch vehicles—which discard each rocket as its fuel is used up.

1961
Launched into space by an R-7 rocket, Yuri Gagarin becomes the first man to fly in space.
1967
The first successful launch of the Saturn V, the first multiengine, three-stage launch vehicle capable of propelling spacecraft to the moon, is accomplished. The Saturn V would later be used to launch the manned missions to the moon. It was also used to launch Skylab.
1972
NASA shifts its focus to the development of a reusable rocket—the Space Shuttle Main Engine, which is fueled by liquid hydrogen and oxygen.
1978
The space shuttle orbiter prototype *Enterprise* performs the first successful test flights of Space Shuttle Main Engines in combination with solid rocket motors and boosters.
1981
The first successful launch of a space shuttle into low Earth orbit takes place.
1989
The first successful use of an Inertial Upper Stage rocket to deploy a spacecraft from the space shuttle occurs.
1992
The first successful use of a Transfer Orbit Stage rocket is employed to boost the Mars Observer into its Mars trajectory.

Almost all modern rockets use chemical fuels, but researchers are also designing alternative propulsion systems. Among them are highly efficient ion engines, which fire electrically charged atoms for thrust. **Deep Space 1,** the **spacecraft** that performed a

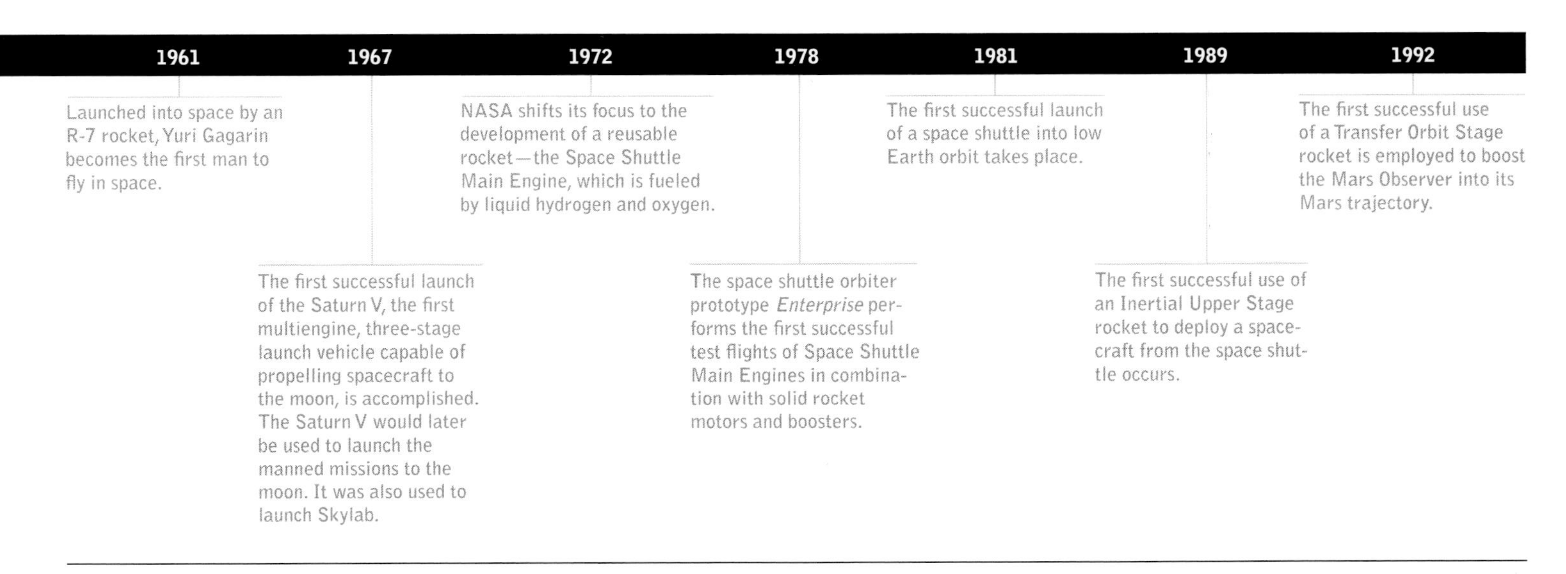

flyby examination of **comet** 19P/Borrelly in 2001, used an ion engine. Scientists are also investigating nuclear engines, which would be particularly powerful if they could harness a fusion reaction.

V-2 Rocket

The V-2 rocket, a **ballistic missile** used by Germany against European targets at the end of World War II, was originally designed by German scientist **Wernher von Braun** and his team as a **spacecraft** engine called the A-4. The German Army took over development of the rocket during the war and launched thousands of V-2s, armed with warheads, primarily toward Belgium and England. Although they were quite inaccurate, the missiles were capable of traveling hundreds of miles and causing severe damage.

After the war, the U.S. and the U.S.S.R. took V-2 parts and plans home from Germany and used them to develop their own increasingly powerful rocket engines. In 1949 one of these modified V-2s traveled a record 393 kilometers into the upper atmosphere from White Sands, New Mexico.

Propellants

Propellants are the chemicals used to create thrust in a **rocket.** They fall into two categories—solid and liquid—both of which work by combining a fuel with an oxidizer and burning it to produce hot gases.

Solid propellants are the oldest type, used since the days of ancient Chinese fireworks. Gunpowderlike mixtures of nitrates and carbon form a typical solid rocket propellant; the **space shuttle** uses ammonium perchlorate as an oxidizer and powdered aluminum as the fuel. Packed into a metal casing, these granular fuels burn from the inside out. They are relatively simple and safe, but cannot be throttled or shut down.

Liquid propellants are more volatile and dangerous than solid ones but are easier to control. They include petroleum fuels such as kerosene, supercooled gases such as liquid hydrogen and liquid oxygen, and highly toxic, hypergolic propellants such as hydrazine and nitric acid.

Robert H. Goddard

Robert Hutchings Goddard (1882-1945), an American physicist and inventor, was a pioneer of the science of **rocket** propulsion. Born in Worcester, Massachusetts, Goddard earned a science degree from Worcester Polytechnic Institute in 1908. He developed

After World War II, Londoners view a V-2 rocket displayed in Trafalgar Square. Originally designed as a spacecraft engine by German scientist Wernher von Braun, V-2 rockets became weapons of war—armed with warheads and launched toward England and Belgium during World War II.

tuberculosis after graduating and worked out several key ideas in rocket propulsion during his convalescence. By 1914 he had earned his first two U.S. patents: one for a liquid-fuel rocket, and the other for a two-stage solid-fuel rocket. In 1919 Goddard published a visionary paper—"A Method of Reaching Extreme Altitudes"—in the *Smithsonian Miscellaneous Collections.* In it he discussed the possibility of sending a rocket as far as the **moon.** Though ridiculed by the press for this science-fictional idea, Goddard continued to receive a modest grant from the Smithsonian Institution to continue his research. In 1926 he launched his first small rocket from an orchard in Auburn, Massachusetts. Powered by gasoline and liquid oxygen, it rose 12.5 meters into the air.

Physicist and space pioneer Robert Goddard poses with a rocket just before its successful launch 12.5 meters into the air in 1926. Although once ridiculed as "science fiction," Goddard's early work laid the basis for today's rocket-powered spaceflight.

Goddard's success attracted more attention and more funding, and in the next decades the scientist was able to build much more powerful rockets. Working primarily from a ranch in New Mexico, he designed the first gyroscopic guidance systems and **propellant** pumps and launched rockets that climbed to about 2,740 meters and exceeded the speed of sound. By the time of his death in 1945 Goddard had proved, both in theory and in practice, that rocket-powered spaceflight was not only possible—but also probable. Though he received little fame during his lifetime, after his death Goddard was awarded both the Congressional Gold Medal and the Langley Gold Medal.

Wernher von Braun

Wernher von Braun (1912-1977) was a German-born engineer and **rocket** scientist who became the father of the American space program. Born in Wirsitz, Germany, he studied mechanical engineering and aircraft construction at the Berlin Institute of Technology, later earning a Ph.D. in physics from the University of Berlin. While still an undergraduate, he joined the Society for Space Travel, headed by the visionary rocket scientist Hermann Oberth. By 1932 von Braun was known as a rising star and was hired by the German Army to develop missiles. At the village of Peenemünde, on the Baltic Sea, von Braun's team produced a series of successful liquid-fueled rockets, leading to the powerful **V-2 rockets** that bombarded Europe toward the end of World War II.

Von Braun and other German rocket scientists surrendered to U.S. troops in 1945 and were brought to the United States. Working at the Redstone Arsenal in Huntsville, Alabama, von Braun's team developed the Redstone, Jupiter-C, Juno, and Pershing missiles. In 1955, von Braun became a U.S. citizen, and in 1957 was finally able to return to his first love, spaceflight, after the Soviet Union launched **Sputnik 1.** Von Braun joined the newly created National Aeronautics and Space Administration (**NASA**), for which he created the massive launch vehicles, the Saturn I, IB, and V. In 1975, having resigned from NASA, he founded the private National Space Institute in order to promote public understanding of spaceflight.

Launch Vehicle

A launch vehicle is a system of **rockets** that boost a **spacecraft** away from the **Earth.** Most launch vehicles are used only once and are therefore known as expendable launch vehicles, or ELVs. The **space shuttle** system, which includes the orbiter spacecraft, is a reusable launch vehicle; most of its parts are recovered and restored.

A typical launch vehicle consists of two or more rocket stages that are discarded as their fuel is depleted. Rocket **propellants** may be solid or liquid: Europe's Ariane 5, for example, consists of a main-stage rocket propelled by liquid fuel and assisted by two or more strap-on boosters burning solid or liquid fuel. After the launch vehicle's rockets fall away, only the **payload**—such as a **satellite,** an interplanetary probe, or a manned spacecraft—remains.

The earliest launch vehicles were modified from **intercontinental ballistic missiles.** The R-7 that lifted **Sputnik** 1 into space in 1957 was one such redesigned missile, 29 meters long and weighing 267,000 kilograms fully fueled. As the race for space accelerated between the United States and the Soviet Union in the 1960s, both countries developed heavy-duty launchers to support manned spacecraft.

Saturn V

The three-stage Saturn V, which carried the **Apollo** missions, became **NASA**'s massive workhorse in the late 1960s and '70s. At 111 meters long, and weighing over 2,722,000 kilograms, it produced 33,375,000 newtons of thrust at **liftoff.** After taking **Apollo 8** around the **moon,** it powered all the subsequent moon landings, as well as **Skylab** 1. Over the course of 11 years, not one Saturn V ever malfunctioned.

Proton

The powerful, reliable Proton **rocket** has been the workhorse of the Soviet and Russian space program since the 1960s. Developed as an **intercontinental ballistic missile** and originally named the UR-500, the Proton launched probes toward the **moon, Mars,** and **Venus** in the 1970s and 1980s and placed satellites into **geostationary orbit.** A three-stage version of the Proton carried all the Soviet **space stations** into space. In the 21st century, the 40-meter-high Proton has served primarily as a commercial **launch vehicle,** taking satellites into orbit.

Modern Launch Vehicles

In the 21st century, a wide range of launch vehicles is lifting payloads into space between 50 and 100 times a year. **Spacecraft** owners choose a vehicle based on the weight of its **payload,** the payload's purpose, and its orbital destination.

Lightweight payloads going into **low Earth orbits** require much smaller rockets than massive payloads traveling into **geostationary orbit.** Titan, Delta, Ariane, and **Proton** ELVs, as well as the **space shuttle,** are some of today's heavy lifters. The Pegasus launch vehicle, designed for smaller payloads, is the first ELV to be launched from an airplane in flight.

Spacecraft

A spacecraft—sometimes known as a "spaceship"—is a vehicle designed to travel above the lower atmosphere. Spacecraft can be piloted or unpiloted, and include **satellites,** space probes, **space stations,** and the **space shuttle** orbiter. **Sputnik** 1, launched in 1957, was the first spacecraft. There are now thousands in orbit

MAJOR LAUNCH VEHICLES

Name	Country	Length	Payload
Ariane 5	European consortium	52 m	6 metric tons to GTO
Atlas/Centaur	U.S.	33 to 43 m	4,000 kg to GTO
Delta	U.S.	39 m	3,810 kg to GTO
Pegasus	U.S.	15.2 to 17 m	ranges from 410 to 500 kg to LEO
Proton	U.S.S.R./Russia	40 to 60 m	20,860 kg to LEO
R-7 Vostok	U.S.S.R./Russia	38.4 m	4,730 kg to LEO
Saturn V	U.S.	111 m	285,000 kg to GTO
Soyuz	U.S.S.R./Russia	49 m	5,500 kg to LEO
Space shuttle (STS)	U.S.	70 m (external tank)	30,000 kg to LEO
Taurus	U.S.	27.5 m	1,000 kg to LEO
Titan	U.S.	65 m	6,000 kg to GTO

GTO: geosynchronous transfer orbit; LEO: low Earth orbit

around the **Earth** as satellites, as well as others on trajectories toward other **planets.** Most spacecraft are propelled into space by launch vehicles that drop away as their fuel is exhausted. The majority have their own onboard power systems as well, and sometimes additional **rockets** for maneuvering.

Satellite

A satellite, in its broadest definition, is any object that orbits another object. **Planets, asteroids,** and **comets** that circle the **sun,** as well as **moons** that orbit planets, are examples of natural satellites. Since the space age began, the term "satellite" has increasingly become associated with artificial satellites: automated, orbiting **spacecraft. Sputnik** 1, launched in 1957, was the first artificial satellite. In recent years the number of artificial satellites placed into orbit has multiplied, and there are now more than 2,700 orbiting the **Earth** as well as a few orbiting the sun.

Modern satellites are expensive, custom-designed machines that serve a variety of military and civilian purposes, but they typically perform one of five basic functions. Communications satellites relay television, telephone, facsimile, and Internet signals; weather satellites collect data about weather patterns on Earth; global positioning satellites send radio signals to receivers on Earth to help airplanes, ships, and cars navigate; **remote-sensing** satellites study Earth's surface—its oceans, plant cover, chemical composition, and more; and research satellites, such as the **Hubble Space Telescope,** collect scientific data.

Satellites are sent into space aboard **launch vehicles** such as multistage **rockets** or the **space shuttle** orbiter. They usually consist of a bus—the metal or composite body; communications, propulsion, power systems; and computer—and a **payload,** the specialized equipment, such as cameras and antennas, dedicated to that satellite's function. **Solar cells, fuel cells,** and batteries provide onboard power, and an **attitude** control system keeps the satellite oriented. Although these complex devices are typically bulky and can weigh thousands of kilograms, engineers are also designing nanosatellites that weigh under ten kilograms and even picosatellites that weigh less than one kilogram.

A satellite's intended function determines its orbit. Spacecraft that need a close-up view of the Earth's surface, such as some remote-sensing and weather

With 33,362,000 newtons of thrust, Saturn V hurls the Apollo 11 spacecraft off the launchpad in a plume of flames, smoke, and liquid oxygen vapor. A workhorse of a rocket, the three-stage, 36-story high Saturn V did the heavy lifting that propelled the Apollo 11 astronauts to the moon in 1969.

satellites, often fly in **low Earth orbits** (LEOs), to 1,000 kilometers above the ground.

The higher the orbit, the slower the velocity. For instance, a satellite in LEO at 320 kilometers travels at 27,782 kilometers an hour and orbits Earth in 90.9 minutes. A satellite in orbit at 800 kilometers travels at 26,837 kilometers an hour and orbits Earth in 100.9 minutes.

Polar orbits, north to south, are one type of LEO; they allow for a wider coverage of Earth's surface. **Geosynchronous orbits** are much higher; at about 35,900 kilometers, they allow a satellite to travel at 11,300 kilometers an hour, completing one orbit every 24 hours and thus staying above the same spot on Earth's surface. These more distant craft have a broad view of much of the planet's surface and include many communications satellites.

Many countries, including the United States, Russia, Canada, Japan, China, and many western European nations, have satellites currently in orbit. Military and telecommunications satellites currently dominate the skies, to the extent that some orbits, once lonely outposts in space, are becoming as crowded as a popular beach on a sunny day.

Sputnik 1

The Soviet **satellite** Sputnik 1, launched into space on October 4, 1957, was the spark that ignited the space age. A Soviet team headed by Sergei Korolev built the 84-kilogram, beach-ball-size satellite and launched it into orbit aboard an R-7 **launch vehicle.** The simple craft circled the **planet** in **low Earth orbit** until early 1958. For the first 21 days of Sputnik's existence, it returned data about its onboard temperature and density readings of the atmosphere. The steady *beep-beep-beep* of its transmissions galvanized the United States into accelerating its own space program and strengthening science education so the U.S. would not lose the "space race" with the U.S.S.R.

Sputnik 1 was followed by several more Sputnik missions, some of which carried animals in order to test the feasibility of human spaceflight.

Retro-rocket

A retro-rocket is a **rocket** engine used to slow, stop, or reverse the motion of an aircraft or **spacecraft.** Piloted spacecraft may typically have a package of retro-rockets to slow them down for **reentry** into the atmosphere.

EUROPE'S WORKHORSE BOOSTER

Leonard David

EUROPE'S POWERFUL ARIANE 5 ROCKET IS BUILT, OPERATED, AND MARKETED BY the commercial launch firm Arianespace, headquartered southeast of Paris, France. This booster is capable of hurling into space heavy payloads, be they telecommunications, space science, military, or other type of spacecraft. The Ariane 5 has evolved as part of a family of Ariane boosters, with the inaugural flight of the European-designed Ariane 1 space transportation system having taken place in December 1979. Over the following decades, Ariane-class rockets have placed in orbit satellites belonging to a variety of nations, giving Arianespace a commanding global lead in the profitable booster-for-hire business. At present the company holds more than 50 percent of the international market for satellites launched to geosynchronous transfer orbit (GTO).

Since 1979 more than 155 Ariane flights have been rocketed into the sky from the sprawling Guiana Space Center, in French Guiana. French Guiana's near-equatorial location reduces a rocket's fuel needs by obtaining a free boost due to Earth's rotational velocity. That enables an Ariane 5 to launch heavier satellites than it could from a launch site farther from the Equator.

The space booster business is a highly competitive industry, with both the U.S. aerospace firms Lockheed Martin and Boeing also offering prospective satellite customers launch services.

Arianespace is backed by a combination of 41 aerospace manufacturers and engineering companies from 12 European countries, along with 11 banks, and the European Space Agency (ESA). It is from this technical, financial, and political melding of resources that Arianespace draws to maintain a competitive edge in the world launch market. The Ariane 5 now operates as Arianespace's workhorse launch vehicle. Arianespace's ELA-3 launch complex at the French Guiana spaceport was built to serve the heavy-lift rocket. This modern complex can handle a launch rate of up to ten Ariane 5 missions per year.

There are several key elements that comprise the Ariane 5 space transportation system: A cryogenic-fueled main stage, two solid propellant strap-on boosters, and an upper stage to place cargo into the proper target orbit. The Vulcain engine used in Ariane 5's main stage is one of the main technological developments of the Ariane 5 launcher program.

For an Ariane 5 to roar skyward, the main stage Vulcain engine comes to life seven seconds before the launcher lifts off. The two side-mounted boosters then ignite, delivering more than 90 percent of the total rocket's thrust at the start of flight. High above Earth, the booster's upper stage propels the payload to its final orbit, injecting a spacecraft into an accurate spot in space.

A Flightworthy Vehicle

The road to success for Ariane 5 has not been easy. On its first flight in June 1996, just 40 seconds after liftoff the rocket veered off course and broke apart. That failure resulted in the loss of European pride, not to mention a cluster of scientific satellites onboard designed to study sun-Earth interactions.

Subsequent flights of the Ariane 5, however, have given Europe added confidence that the rocket is a reliable, flightworthy vehicle. While Ariane 5 has experienced some glitches, it has delivered into orbit several telecommunications satellites. Additionally, the launcher orbited the ESA's XMM-Newton x-ray observatory satellite in December 1999; the Envisat Earth-monitoring satellite in February 2002; and launched in March 2004 ESA's Rosetta space probe to comet 67P/Churyumov-Gerasimenko—a decade-long mission that will reach that celestial target in 2014, then deposit a small probe onto the comet. The ESA comet chaser is well on its way. In late April 2004, in fact, the en route Rosetta space probe successfully performed its first scientific activity by imaging the newly found comet Linear.

A unique virtue of the Ariane 5 is that it is "human rated"—that is, it was originally developed by ESA to be topped by a manned mini-shuttle called Hermes. The piloted Hermes project, however, was eventually canceled.

The Ariane Family

Today, the Ariane 5 is a family of heavy-lift launchers. These rocket derivatives are step-by-step more powerful versions of the first Ariane 5–class vehicle that flew in 1996. New upper stages—modifications to strap-on rocket motors—along with other innovations, have augmented the launcher to meet ever changing needs of customers. For example, Arianespace will soon be offering the Ariane 5 ESC-A, the Ariane 5 ESC-B, and the Ariane 5 Versatile launchers.

Ariane 5 Versatile features an improved version of the booster's current upper stage, increasing the ability to place into geosynchronous transfer orbit a payload weighing up to 8,000 kilograms. The ESC-A version now slated for use in 2004 utilizes an upper stage capable of hauling into GTO payloads of up to 10,500 kilograms.

Then there's the Ariane 5 ESC-B that, once in operation in 2006, will be the most powerful of the Ariane 5 class launchers. A cryogenic upper-stage engine now under development, called the Vinci, would make possible pushing payloads into GTO that tip the scales at 12,000 kilograms.

And just like any competitive "trucking" line, Ariane 5 can accommodate a wide array of payloads, offering a full range of payload adapters, fairings, and dispenser equipment to eject satellites into space at the right locations.

For the 21st century, one new use of the Ariane 5 is launching ESA's Automated Transfer Vehicle, or ATV for short. This unpiloted supply vessel will dock with the International Space Station (ISS), hauling to the orbiting outpost quantities of water, air, nitrogen, and oxygen, as well as attitude-control propellant. Furthermore, the ATV is built to remove waste from the station and to nudge the space station to a higher altitude, thereby ensuring the laboratory's longevity in Earth orbit.

The first Automated Transfer Vehicle is labeled Jules Verne after the popular French science-fiction author of the 19th and early 20th centuries. Depending on the operational lifetime of the ISS, at least eight ATVs will be built by ESA and lofted by the venerable Ariane 5 booster.

Ariane 5 is the product of an aggressive European industrial team keen on responding to satellite customers' wishes, as well as dedicated to the objective of sustaining Arianespace as a world leader in the commercial launch business. ■

An Ariane 5 rocket, with the aid of two solid rocket boosters, rises from the Guiana Space Center. Another 30 Ariane 5s are in production, ensuring the model's competitiveness and longevity.

Payload

The payload is the cargo carried by a **spacecraft** or **rocket.** Typical payloads include **satellites** lifted into space on **launch vehicles** as well as the instrumentation of the satellite itself.

Heat Shield

A heat shield is a device or layer on a **spacecraft** or **missile** that protects it from the high heat—in the range of 1000°C—that the craft experiences when moving rapidly through the atmosphere. Heat shields typically come into play when a spacecraft reenters the atmosphere from space.

Most fall into one of two categories. Ablative heat shields are expendable devices, frequently nose cones, that absorb and dissipate heat as their outer layers vaporize. Reusable heat shields, seen on the **space shuttle** orbiter, consist of silica tiles over insulating material that are designed to absorb heat. The safety of reusable heat shields came into question after the **space shuttle** *Columbia* was destroyed following the loss of some tiles during flight.

Fuel Cell

Fuel cells are efficient battery-like devices that produce electricity while converting hydrogen and oxygen into water. Unlike batteries, fuel cells do not store energy; they produce it as long as they are supplied with their chemical reactants.

The **Apollo** and **Gemini** spacecraft used fuel cells to power their electrical systems and to produce water. These alkaline fuel cells carried chilled, pressurized oxygen, which was heated to a gas before entering the cell. The **space shuttle** also employs fuel cells; **NASA** is contemplating switching the shuttle from alkaline versions to newer proton-exchange membrane cells, which are lighter and less expensive.

Solar Cell

Widely used on **satellites, space stations,** and other **spacecraft,** solar cells are devices that convert sunlight into electricity. The cells are made from semiconductors, typically silicon crystal. When light strikes the surface of the semiconductor, it knocks loose electrons that are then channeled into an electric current.

Each cell produces a modest amount of current, so spacecraft usually have large arrays of the cells arranged in winglike panels. The **International Space Station,** for example, has eight arrays covering more than 2,900 square meters.

Advanced Propulsion Ideas

Traditional chemical **rockets** have worked well for decades, taking **satellites** into orbit, humans to the moon, and probes as far as the outer **solar system.** However, they are too inefficient, heavy, and slow to carry humans to other **planets** or to other stars. A voyage to **Mars** using chemical rockets, for instance, would take seven and a half months, all the while exposing the **astronauts** to dangerous radiation and the debilitating effects of prolonged **weightlessness.**

Space scientists have studied a number of alternatives to traditional chemical rockets. Some, such as **ion propulsion,** are already at work, while others require scientific breakthroughs before they become feasible.

Ion Propulsion

Ion engines produce thrust by ionizing, or giving an electrical charge to, the gas xenon. Accelerated to 30 kilometers a second, the ionized gas eventually propels a **spacecraft** about ten times as fast as standard chemical propellants. However, these efficient engines build up speed slowly and are not practical for rapid acceleration and missions to nearby objects.

Deep Space 1, a probe that performed a successful flyby of comet 19P/Borrelly in 2001, was propelled by an ion engine, and Japanese and U.S. scientists have already developed more efficient, High Power Electric Propulsion (HiPEP) engines that will be used to reach asteroids and the outer solar system.

Nuclear Power

Various kinds of nuclear-powered **rocket** engines have been proposed over the years, ranging from fission-based nuclear thermal rockets to fusion and matter-antimatter engines. By the 1960s both the U.S. and the U.S.S.R. were experimenting with nuclear thermal rockets that would pump hydrogen through a reactor core, heat it to 2500°C, and expel it for thrust. Although such an engine is much more efficient than a chemical rocket, the mass of the reactor and the shielding necessary reduce its effectiveness, and in general the idea of sending a nuclear reactor into space has been hard to sell to the general public. To date such an engine has not been used.

Fusion engines, not yet feasible, would be far more powerful. Hydrogen atoms entering a reactor would fuse to produce a larger, helium-4 atom and energy.

Heading toward a rendezvous in space, the Deep Space 1 craft approaches the comet 19P/Borrelly in 2001. The spacecraft was powered by an ion engine, which produces thrust by giving an electrical charge to the gas xenon. More advanced ion engines may help spacecraft reach asteroids and the outer solar system.

The extremely hot plasma that this would create, at temperatures of millions of degrees, would be expelled from the rocket via magnetic fields. Fusion-powered **spacecraft** could make the journey to **Mars** in perhaps three months. Even further away is the intriguing prospect of matter-antimatter annihilation as a source of power. If scientists could develop an engine that tapped the immense power released when protons and antiprotons collide, the resulting energy and charged particles would have exhaust velocities that approach the speed of light.

Solar Sailing

Spacecraft powered by solar sails take a completely different approach to propulsion and eliminate the use of **rockets.** Instead, such craft would be carried into space by traditional methods and then would unfurl a thin, lightweight, reflective mylar-film sail, perhaps half a kilometer wide. The very gentle but unfailing force of electromagnetic radiation from the sun pressing on the mirrorlike sail would propel the craft forward, accelerating it steadily until it flew at 90 kilometers a second, ten times faster than the **space shuttle.** Solar sailing craft are now under development at the U.S. Jet Propulsion Laboratory, with possible robotic missions to **Mercury,** outer **planets,** or other stars in mind.

Laser Propulsion

Theorists have proposed several ways of using the highly concentrated power of **laser** beams in **spacecraft** propulsion. One method is similar to nuclear propulsion. An onboard laser would heat the hydrogen **propellant** to thousands of degrees kelvin before the gas would be expelled at high velocities, but the high temperatures involved with this method make it challenging. However, the laser does not need to be onboard; engineers at **NASA** have already launched a tiny experimental craft by focusing a separate laser on the craft's propellant, heating it to extreme temperatures. A different use of lasers would be similar to **solar sailing.** Powerful, orbiting, solar-powered laser beams directed at a reflective sail could send a craft through space. At speeds of up to one-tenth the speed of light, such a craft could be practical for interstellar travel.

Liftoff

Launch

Launching a **spacecraft** from the Earth is a complex and hazardous operation, but its aim is simple—to overcome gravity and propel the craft into a desired orbit or trajectory in space. Almost all spacecraft are carried into space by **launch vehicles,** multistage **rockets** that drop away in flight as they exhaust their fuel. In order to reach orbit, spacecraft must achieve orbital velocity, the speed at which they will stay in orbit. To leave **Earth**'s pull altogether, they need to reach **escape velocity.** A spacecraft above the Equator is already in motion, traveling at more than 1,650 kilometers an hour as the planet rotates, and many launches take advantage of this motion to help propel the craft into space. A craft on an interplanetary trip may also make use of Earth's motion as it orbits the **sun.** However, it takes huge quantities of rocket fuel to lift a heavy spacecraft. On the **space shuttle,** for instance, the propellants weigh 20 times as much as the orbiter itself; the solid rocket boosters must generate 14,685,000 newtons of thrust each to lift the shuttle from the ground.

Most launches must occur during a particular time period, called a **launch window,** an interval of time that takes into account safety, Earth's location in space, and other factors. Once the craft and launch vehicle have been tested, they are taken to the **launch site** where engineers test them again, load **propellants,** and mate the craft to the launch vehicle. Controllers establish telecommunications with the craft and load command sequences into the computers. When every component has the all-clear the launch team gives the go-ahead for **countdown** and **liftoff.**

Coriolis Effect

The Coriolis effect (named after French physicist Gaspard-Gustave de Coriolis) is the apparent deflection in the path of an object traveling over a rotating

Riding the rails to space, the Soyuz TMA-2 rocket heads toward its launchpad at Baikonur, Kazakhstan. Russia and the United States are the oldest entrants in the space race, and Baikonur, far from major population centers, serves as Russia's equivalent to Cape Canaveral.

surface. In its most familiar form, the Coriolis effect means that air moving over the **Earth**'s surface veers to the right in the Northern Hemisphere and to the left in the Southern Hemisphere. The Coriolis effect applies to **spacecraft** launching from the surface of the rotating Earth. A craft launched directly eastward, in the direction of the planet's spin, would travel in a straight line relative to the Earth's surface, but when launched toward the north the ship would appear to veer toward the east because the planet would be spinning beneath it.

Launch Window

A launch window is the time period during which a **spacecraft** must launch in order to meet its safety and mission objectives. For example, the **space shuttle** must time its launch so that it has a daylight landing in case of emergency. To reduce the spacecraft's travel distance, interplanetary missions typically launch when **Earth** and the objective **planet** or other destination are aligned.

Launch Sites

Multiple countries and agencies operate at least 22 launch sites—bases from which **spacecraft** may be launched—around the world. Other military bases exist in secret locations. Launch sites must meet a number of criteria. Unless they are specifically used for high-inclination launches, they are best located near the Equator; they must provide flight paths clear of large population centers; and they need an extensive infrastructure for transportation, assembly, fueling, and downrange tracking. The U.S. and Russia have the world's largest space programs. The U.S. launches primarily from Cape Canaveral and Vandenberg Air Force Base, and Russia from the Baikonur Cosmodrome. The **European Space Agency,** France, Japan, and China also maintain space bases.

Gantry

A gantry is a large, movable frame, with platforms at different levels, that is used for holding and servicing **spacecraft** before a **launch.**

Countdown

A countdown is the process of marking off each step leading to the **launch** of a **rocket.** The count proceeds in inverse order, ending with T-time, expressed as T minus the number of minutes or seconds until launch.

ESTABLISHED LAUNCH SITES

Country	Launch Site
Australia	Woomera, Australia
Brazil	Alcântara, Brazil
China	Jiuquan, China
China	Taiyuan/Wuzhai, China
China	Xichang, China
ESA	Kourou, French Guiana
France	Hamaguir, Algeria
India	Sriharikota Island, India
Israel	Palmachim Air Base, Israel
Italy	San Marco Platform, near Kenya
Japan	Kagoshima Space Center, Japan
Japan	Tanegashima Space Center, Japan
Russia	Baikonur Cosmodrome, Kazakhstan
Russia	Kapustin Yar, Russia
Russia	Plesetsk Cosmodrome, Russia
Russia	Svobodnyy, Russia
U.S.	Edwards Air Force Base, California
U.S.	Kennedy Space Center, Florida
U.S.	Kodiak Launch Complex, Kodiak Island
U.S.	Reagan Test Site, Marshall Islands
U.S.	Vandenberg Air Force Base, California
U.S.	Wallops Flight Facility, Wallops Island, Virginia

Liftoff

Liftoff is the movement of a rocket-powered vehicle as it leaves the launchpad, ascending vertically. Liftoff applies only to craft moving vertically.

Escape Velocity

Escape velocity is the speed an object needs to achieve in order to break free from another object's gravity. An object attempting to leave **Earth**'s atmospheric grip must travel at 11.2 kilometers a second to reach escape velocity, which translates to 40,200 kilometers an hour.

THE SPACE ELEVATOR

Christopher Wanjek

ELEVATOR GOING UP...AND UP...AND UP A LITTLE MORE. SCIENTISTS AND ENGINEERS contemplating inexpensive and reliable access to space have set their sights on a modern-day version of Jack's beanstalk: an elevator reaching 100,000 kilometers, far beyond the International Space Station's 355-kilometer-high loft. This is no fairy tale. The space elevator, as it is known, would be a ribbon or cable tethered to Earth and rising to an orbiting platform a quarter of the way to the moon. Earth's gravity and the platform's centrifugal force, acting in opposite directions, would keep the cable taut. A cargo box containing a satellite could rise, or astronauts could even shimmy up the cable at a fraction of the cost of a rocket launch—a steady weeklong climb. The feat may be less challenging and expensive than other engineering projects under consideration, such as the proposed bridge over the Strait of Gibraltar connecting Spain to Morocco, or past accomplishments such as the transatlantic telegraph cable. The estimated price tag is ten billion dollars. The elevator would quickly pay for itself, though, lowering the cost of placing a satellite into space from $20,000 to about $200 a kilogram.

The concept of a space elevator dates back to 1895. Konstantin Tsiolkovsky, a Russian astronautics pioneer, envisioned a "celestial castle" sitting atop a thin tower, held up by centrifugal force like a rock swinging high at the end of a rope. Science fiction writer Arthur C. Clarke featured the space elevator in his 1979 novel *The Fountains of Paradise.* The elevator remained fundamentally impossible to build, however, because no material known could withstand the expected forces. The building material requires a tensile strength of over 100 gigaPascals. This is a measure of the material's resistance to snapping or deforming. Steel has a tensile strength of about 1 gigaPascal; quartz and diamond fibers can support about 20 gigaPascals.

The 1991 discovery of carbon nanotubes escalated the space elevator from the realm of science fiction into science reality. Nanotubes are cylindrical molecules of carbon stronger than diamond and steel, theoretically beyond 100 gigaPascals. With fiber in hand, the space elevator will be a challenge but not impossible to build.

Perfecting nanotube production is the first task. The longest fibers today are only about a meter long, with 63-gigaPascal tensile strength. Clearly much more is needed—produced inexpensively—to create what engineers foresee as a meter-wide, paper-thin ribbon made up of hundreds of fibers, each 100,000 kilometers long. Parts of the ribbon would need an aluminum coating to protect them from oxidation. The elevator's base would be a moveable ocean platform in the equatorial Pacific, far from air traffic and in a region with little lightning activity or severe weather.

Construction would begin with a rocket launch to geosynchronous orbit, about 35,900 kilometers high. This is the point at which a satellite takes exactly one day to orbit Earth and thus maintains a hovering position. The satellite would snake a cable back to Earth and gradually climb to 100,000 kilometers as more and more cable is released. Once the first cable was secured to Earth, engineers would send up robotic "climbers" that would sew new cable onto existing cable, creating a ribbon. This process would take about two years. That first satellite, now at 100,000 kilometers, would act as the necessary counterweight to hold the ribbon tight. Elevator operators would power the climb from Earth with lasers. Cargo could be released at any point after several hundred kilometers. Cargo let loose at 100,000 kilometers, whirling around at more than 11 kilometers a second, would have enough tangential velocity to escape Earth's gravitation field and fly to Saturn. Several payloads could climb the elevator at once.

With directed resources, the elevator could be in place by 2020. A second generation of faster elevators could halve the trip into space, sparing would-be travelers from an overload of elevator music. ■

Advances in nanotechnology, tether technology, electromagnetic propulsion, and space infrastructure could actually lead to the development of a space elevator 100,000 kilometers in length. Artist Pat Rowling conceived this view, looking down an immensely long cable toward Earth.

The "glass" cockpit of the space shuttle *Atlantis* features flat-panel screens that give the crew instant access to vital information. Built in 1984, *Atlantis* was totally refurbished with state-of-the-art computer technology in 1998.

Navigating in Space

Tracking Station

A tracking station, or ground station, is a facility set up on **Earth** to track an object through space. Tracking stations contain electronically controlled **antennas** that transfer signals to and from a **spacecraft** to track its position, collect data, and transmit instructions.

Onboard Computing

Modern **spacecraft** typically contain a number of small computer systems that control various functions of the craft, including telecommunications, data collection, timekeeping, and navigation. The systems are connected to a central computer that "oversees" all the other activities. **Astronauts** aboard piloted craft may use laptop computers for their own work.

Celestial Navigation

Celestial navigation is a method of directing a craft toward a desired destination by referring to known celestial bodies. Just as sailing ships used to navigate using stars such as **Polaris,** modern **spacecraft** may set or correct their courses by comparing star sightings at known coordinates with other celestial landmarks, such as the **moon** or **Earth.** The whole process can be done automatically using imaging instruments on board that locate stars and transmit data to navigational computers containing detailed star catalogs.

Altitude

Altitude, in astronomy, refers to the angular distance of a celestial body above or below the horizon. In spaceflight, it means the distance of a craft from the surface of the **Earth** or another celestial body.

Gyroscope

A gyroscope is a device with a spinning mass that can turn freely in any direction on its axis, mounted on a stable, immovable base (an inertia platform).

A **spacecraft** often has three gyroscopes, with each measuring its **altitude** on a different axis: pitch, yaw, or roll.

Attitude

For **spacecraft,** attitude means position or orientation in space as determined by the relationship among a craft's three axes and some fixed reference point. The three axes, related to the **space shuttle,** are pitch—movement of the nose up or down relative to the tail; yaw—swinging of the nose from side to side; and roll—literally rolling over to the right or left side. Attitude is monitored by inertial systems using **gyroscopes.**

Docking

Docking is the act of connecting two objects, such as a **space shuttle** and a **space station,** in space. It is a delicate procedure in which the first craft must align its three axes and make a final approach to the target at a rate of a few centimeters per second before engaging with a docking apparatus. The dangers of this maneuver were made apparent in 1997, when an automated Progress supply module flew off course during docking with the **Mir** space station and tore a small hole in its hull, depressurizing the station briefly before it was repaired.

Space Debris

A growing cloud of debris orbits the **Earth,** endangering working spacecraft and satellites. By the early years of the 21st century, agencies on Earth were tracking approximately 9,000 pieces of space junk and estimated that at least 110,000 objects larger than one centimeter were circling the planet. About half the objects are fragments from **launch vehicles,** discarded during separation of stages or created as a result of accidental or deliberate (in the case of military **satellites**) explosions. Other objects are inactive satellites or other spacecraft, dating from Vanguard 1, launched in 1958. Space junk also includes **astronaut** Edward White's glove, lost in 1965, and garbage bags discarded by the **Mir space station.**

Moving at high speeds, even tiny pieces of debris can damage a craft; **NASA** has replaced more than 80 **space shuttle** windows marked by impacts. Scientists have not yet arrived at a solution to the problem of space debris. Some have proposed setting up a system of lasers for vaporizing the junk. For now, they are tracking the larger pieces and hoping to avoid them.

TOP SOURCES OF SPACE JUNK AS OF JUNE 2000

Country or Organization	Satellites	Space Probes	Other Debris	Total
U.S.S.R./Russia	1,335	35	2,571	3,941
U.S.	741	46	2,971	3,758
P.R.C.	27	0	324	351
European Space Agency	24	2	233	259
Japan	66	4	49	119
Iridium	88	0	0	88
Intelsat	56	0	0	56
Globalstar	52	0	0	52
France	31	0	17	48
Orbcomm	35	0	0	35
India	20	0	4	24
United Kingdom	17	0	1	18
Eur. Telecom Sat. Org.	17	0	0	17
Canada	16	0	0	16
Germany	13	2	1	16
Italy	8	0	3	11
Indonesia	10	0	0	10
Australia	7	0	2	9
Brazil	9	0	0	9
Inmarsat	9	0	0	9
Luxembourg	9	0	0	9
NATO	8	0	0	8
Sweden	8	0	0	8
Arab Satellite Comm. Org.	7	0	0	7
South Korea	7	0	0	7
Mexico	6	0	0	6
Spain	6	0	0	6
Argentina	4	0	0	4
Czech Republic	4	0	0	4
Thailand	4	0	0	4
ASIASAT Corp.	3	0	0	3
FGER	3	0	0	3
Israel	3	0	0	3
Norway	3	0	0	3

Communicating with Mission Control

Telemetry

Telemetry is the transmission of measurements and information at a distance. When **Sputnik** 1 transmitted temperature and density readings to **Earth** via radio in 1957, this basic form of telemetry showed the world that spacecraft could supply valuable information from space. Today, telemetry is central to all satellite and space missions. It is used to track the position and status of **spacecraft,** collect scientific information, and monitor the physical condition of **astronauts.**

Telemetry systems have three parts: the measuring instruments or sensors aboard the spacecraft, whose data are converted to electrical signals via a transducer; a transmitter, which broadcasts those signals, typically via radio waves; and a receiving station, usually an antenna or network of antennas on Earth.

The measurements collected depend upon the mission of the spacecraft. They range from readings of sea-surface temperatures to the status of the plumbing aboard the **space shuttle.** Radio signals travel along prearranged frequencies to and from tracking antennas on Earth, such as those of the **Deep Space Network.** A signal sent from Earth to the spacecraft is known as an **uplink;** a signal from the spacecraft to Earth is a **downlink.** Instruments on the ground record downlink signals and convert them into readable data. For instance, a photograph of **Jupiter** taken by the camera aboard the **Galileo spacecraft** was converted into digital code, compressed, and transmitted at 120 bits a second. One of the three large receivers of the Deep Space Network then picked up the bit stream and transmitted it to the Jet Propulsion Laboratory, in Pasadena, California. There, computers reformatted it into an image.

Ground Track

A ground track is the imaginary path traced by a **satellite**'s movement over the surface of the rotating **Earth.** It can be plotted on a map in order to visualize the satellite's position relative to the ground. Ground tracks take various woven or looping shapes depending on the satellite's orbit. Satellites that exceed Earth's rate of rotation form direct tracks that progress eastward; those that lag behind Earth's rate of rotation form westward-moving tracks. Tracks are ascending when they

Advances in Communications

1957 | 1958 | 1959 | 1960 | 1961 | 1963 | 1969

October 4, 1957
The Soviet Union launches the first artificial satellite, Sputnik 1, into space. Communications with ground station are transmitted via shortwave radio.

January 31, 1958
The U.S. launches its first satellite of the space race, Explorer I. Data is stored in a miniature tape recorder and "dumped" on command by way of radio waves as the satellite passes over tracking stations on Earth.

October 18, 1959
Soviet satellite Lunik 3 returns images of the far side of the moon to Earth by developing and scanning photographs on board and radio transmitting them to a ground station in facsimile form.

August 12, 1960
NASA launches Echo 1A, the first passive communications satellite to reflect radio waves from ground-based transmitters to a receiving ground station thousands of miles away.

October 4, 1960
The U.S. Army launches the first active communications satellite, Courier 1B, which records signals from ground stations and retransmits them on command.

April 12, 1961
Soviet Lt. Yuri Gagarin becomes the first man to fly in space; his reentry is controlled by a computer program sending commands via radio signals to his shuttle.

December 24, 1963
The Deep Space Network is established. It creates the first integrated global communications capability to deep space through antennas and data-delivery systems set up in Goldstone, California; Canberra, Australia; and Johannesburg, South Africa (later moved to Madrid, Spain).

July 20, 1969
The Apollo 11 mission uses television transmissions between base and shuttle extensively for inspection, preparation, and documentation purposes of the first lunar landing. It results in one of the most enduring images in television history.

First man-made satellite to orbit the Earth, Sputnik 1 sent information about extraterrestrial temperatures to Earth via radio—an early example of telemetry, the science of measuring at a distance. Today, more advanced telemetry helps scientists track spacecraft, collect scientific information, and monitor the physical condition of astronauts.

1970 | 1975 | 1983 | 1998 | 2005 | 2010

November 17, 1970
The Soviets land the first remote-controlled moon rover, Lunokhod 1. It is steered via a TV-radio link from a control station in the U.S.S.R.

July 1975
The ATS-6, the first direct TV broadcasting satellite, plays a major role in the joint U.S.-Soviet Apollo-Soyuz shuttle-docking program, the first of its kind as well, as it relays TV-radio signals to the Houston Control Center.

April 4, 1983
The first Tracking and Data Relay Satellite System (TDRSS)—a telecommunications satellite—is sent into orbit to provide tracking, data, voice, and video services for NASA spacecraft on a globally integrated level with an associated ground station in White Sands, New Mexico. The system now consists of nine satellites.

October 1998
NASA implements the Small Deep Space Transponder (SDST) to provide direct-to-Earth communications and navigation services. It is the first transponder to use digital receiver technology to combine separate functions of spacecraft telecommunication systems into one integrated unit with less mass.

September 2005
ESTRACK, the European Space Agency's network of tracking stations that links satellites in orbit with the agency's Operations Control Center, will expand its reach with its tenth ground station in Cebreros, Spain.

2010
The anticipated development of the Advanced Deep Space Transponder (ADST) will provide direct-to-Earth and spacecraft-to-spacecraft communications and navigation services with digital signal-processing capabilities.

"*Endeavour*, this is Mission Control." Flight controllers in Houston's Mission Control Center get a big-screen view of space shuttle *Endeavour* poised on the launchpad at the Kennedy Space Center, in Florida. Once a spacecraft leaves the launchpad, the Houston-based team takes over ground control of the flight.

cross the Equator from south to north and descending when they cross from north to south.

Tracking

Tracking is the process of locating and following the path of a **satellite** or other **spacecraft.** Mission scientists on the ground measure the distance to a craft and its velocity by sending radio signals from **antennas** to the craft, which are then retransmitted by the craft's transponder; the difference between the two tones tells ground control where the craft is and what its velocity is. Dozens of **tracking stations,** typically containing large dish antennas, are found in various locations around the world. One famous set of tracking facilities is the **Deep Space Network,** administered by **NASA**'s Jet Propulsion Laboratory. The **European Space Agency** has a network of stations, ranging from Sweden to Kenya, known as ESTRACK.

Spacecraft Antennas

The antenna is the centerpiece of any **spacecraft** communications system. A device for sending and receiving radio waves, at its simplest an antenna can be a straight piece of wire. However, most **satellites** and other spacecraft have one or more high-gain dish, or parabolic, antenna that focus their signals so that they either receive or transmit radio waves along particular, narrow frequencies. Spacecraft may also carry low-gain, simpler antennas as backups. These have weaker signals than high-gain antennas but broadcast those signals over a wider frequency range. Spacecraft antennas are typically designed to receive or transmit only over certain frequencies: VHF (very high frequency) bands, with frequencies from 30 to 300 megahertz (MHz); UHF (ultrahigh frequency) bands from 300 to 3,000 MHz as well as other bands such as L band, S band, C band, X band, or K band.

Transponder

A transponder is an electronic device that receives radio signals on a particular frequency, amplifies them, and then retransmits them on another frequency. Powered by **solar cells** and batteries, transponders are critical to **spacecraft tracking** and communications. Large **communications satellites** can have 90 or more of them.

Downlink

Outgoing radio signals transmitted from a **satellite** or other **spacecraft** to ground stations or receivers are called the downlink. Downlink signals have different frequencies from the incoming, or **uplink,** signal.

Uplink

Incoming radio signals sent from ground stations to **satellites** or **spacecraft** are called the uplink. Uplink signals travel along different frequencies from **downlink** signals.

Touchdown

Atmospheric Drag

Atmospheric drag is a slowing force exerted by the atmosphere on moving objects. In spaceflight, it plays a large role in retarding the movement of **spacecraft** being launched from **Earth,** in eroding the orbit of **satellites** within the atmosphere, and in resisting a spacecraft's **reentry** to the atmosphere. The **Hubble Space Telescope,** for example, has no onboard propulsion capability at all and is slowly being brought back to Earth by atmospheric drag.

This resistance during reentry is both a hazard and a boon. The **space shuttle,** for instance, moves at 28,500 kilometers an hour as it enters the outer atmosphere, about 120 kilometers above the Earth's surface. If any spacecraft enters the atmosphere at too steep an angle, it will burn up in the friction- and pressure-generated heat. However, at the right angle, the atmosphere exerts just enough resistance to provide aerobraking, so that the craft, protected by a **heat shield,** can slow down without using a large quantity of **propellants.** The absence of an atmosphere on the **moon** makes it difficult to slow down enough to land there.

Deorbit

To deorbit is, simply, to leave orbit. The term is typically used in connection with **satellites** and other **spacecraft** orbiting the **Earth** and describes the process of leaving a stable orbit in order to return and land on Earth's surface—or, in the case of some obsolete satellites, to break up and burn in the atmosphere. Deorbit usually begins with a "deorbit burn," a braking maneuver in which rockets are fired to slow the craft and begin its descent into the atmosphere.

On September 17, 2003, the Jet Propulsion Laboratory in Pasadena, California, sent orders to the **Galileo spacecraft** to deorbit **Jupiter** and begin a steep descent into the planet's dense atmosphere. The resulting disintegration of the spacecraft ended Galileo's 14-year mission and 4.6-billion-kilometer journey into space.

Reentry

Reentry, for a **spacecraft,** is the time when the craft falls back into the atmosphere for its return to **Earth.** Reentry begins with a natural orbit decay or a **deorbit** maneuver that slows the vehicle enough that it descends into the outer reaches of the atmosphere, about 120 kilometers above Earth's surface. Traveling at speeds up to 39,000 kilometers an hour, spacecraft must enter the atmosphere at a precise angle: too shallow, and the craft will skip off the atmosphere like a stone from a pond's surface; too deep, and it will heat up too quickly and burn. In addition, a manned spacecraft must keep its negative acceleration (deceleration)

After a splashdown in the Atlantic, Alan Shepard—first American in space—climbs out of his Mercury spacecraft. Early space flights—like this one in 1961—used splashdowns to ensure soft landings. Today's space shuttles land on runways.

Endeavour
United States
6

below 8g's, the limit the crew can endure. Heat from friction and compression of air in front of the fast-moving craft brings temperatures up to 5500°C.

Glide Path

A **spacecraft**'s glide path, in its most general sense, is the flight path it follows as it approaches its landing. The path can be characterized by its length and angle of descent as well as by changes in direction. The **space shuttle,** for instance, follows a shallow, S-shaped glide path as it returns to **Earth,** banking in a series of gentle curves before flying a final spiral approach to the landing site.

Terminal Velocity

An object in free fall through a gas or liquid reaches terminal velocity when it ceases to accelerate downward and begins to fall at a constant speed. For a body falling through the atmosphere, this state occurs when the speed at which the gravitational force is pulling the object downward is both equal and opposite to the drag from the atmosphere (also called the air resistance) pushing it downward.

Coasting

A coasting **spacecraft** is one that is no longer under power, but is moving at a constant speed in a single direction due to inertia. Interplanetary craft coast in order to save fuel.

Touchdown

Touchdown is the landing of a **spacecraft.** Early **Apollo** missions actually experienced splashdowns by parachuting into the sea. Piloted craft from the Soviet Union and China have made soft landings on land, as have the U.S. **space shuttles,** which land on a runway like an airplane, aided by drag parachutes to slow their speed. The **Mars** exploration rovers Spirit and Opportunity, which descended to the Martian surface using parachutes, were cushioned by airbags upon touchdown.

Sites for spacecraft touchdown on **Earth** include the **Kennedy Space Center** at Cape Canaveral, Florida; Edwards Air Force Base in California; the Baikonur Cosmodrome in rural Kazakhstan; and the Jiuquan space facility at the southern edge of the Gobi, in China.

Touchdown! With a drag parachute reducing its speed, *Endeavour* lands at Edwards Air Force Base in California in 2002. After the space shuttle completed a smooth journey of 9.3 million kilometers and 217 orbits, bad weather at Cape Canaveral prompted ground control to switch its touchdown to the California site.

Orbital Parameters

Orbital Basics

Several terms must be defined or reviewed to describe the elements of an artificial **satellite**'s orbit around **Earth.**

Ellipse

An ellipse is a closed oval shape. The ellipse has two focal points, and its shape is defined such that from any point on its curve, the combined distance to the two focal points is equal. Most **satellite** orbits are ellipses, with one of the focal points being the center of mass of the **Earth** or whatever body the satellite is orbiting. A circle is an ellipse in which the distance between the two focal points is zero. In a circular orbit, the **perigee** and **apogee altitudes** are the same.

Perigee

In an elliptical orbit, the perigee is the point at which the orbiting **spacecraft** is closest to the **Earth.** The same term is used to describe the closest point of a satellite orbiting another planet or celestial body, including natural **moons** orbiting their **planets.**

Apogee

Apogee is the point in an elliptical orbit where the **spacecraft** or other object is at the farthest distance from the body it is orbiting.

Plane of the Equator

This is an imaginary flat surface through the Earth (or another celestial body) at its equator that extends into space in all directions. A satellite in orbit does not really "cross the equator," it crosses the equatorial plane.

Ascending Node

The ascending node is the point at which an orbiting **satellite** crosses the **plane of the equator** in a south-to-north direction.

Descending Node

The descending node is the point at which an orbiting **satellite** crosses the **plane of the equator** in a north-to-south direction.

Line of Nodes

The **ascending** and **descending nodes** lie on opposite sides of a celestial body on an imaginary line through the center of the body, called the "line of nodes."

Orbital Elements

There are six elements (or measures) that together will completely describe an orbit. The largest perturbation in the orbital elements is caused by the oblateness of the **Earth,** meaning its flatness at the poles. This causes the entire orbit to rotate with respect to the Earth's surface. Both the **argument of perigee** and the **right ascension of the ascending node** rotate several degrees per day, unless the orbit is specifically configured, or "frozen," to avoid this.

Semimajor Axis

This measure describes the size of the orbit. For a circular orbit, the semimajor axis is its radius. Remembering that the center of the orbit is at the center of the **Earth,** the semimajor axis equals the radius of the Earth plus the altitude of the satellite above the Earth's surface. For **elliptical orbits,** the major axis is an imaginary line through the center of the Earth (one of the **ellipse**'s focal points) that stretches from the **apogee** to the **perigee** of the orbit on opposite sides of the Earth. The semimajor axis is half of this length, and it is commonly used for ease of computations. In equations of orbital motion, the semimajor axis is expressed as "a."

Eccentricity

This measure describes the shape of the orbit. Eccentricity is the elongation of an **ellipse** away from the shape of a circle. It is calculated by comparing the distance between the two focal points to the length of the major axis. The measure can also be derived by comparing the radius of the **perigee** versus the radius of the **apogee.** A **Molniya orbit,** with a perigee of about 400 kilometers and an apogee of about 40,000 kilometers, is considered a highly elliptical or highly **eccentric orbit.** Because there is only one focal point for a circular orbit, the ratio works out to zero, or no eccentricity. Eccentricity is designated in orbital equations as "e."

Inclination

This measure describes the orientation of the orbit with respect to the **Earth**'s equatorial plane; the offset between the orbit's plane and the equatorial plane is measured in degrees from 0 to 180, and is expressed in orbital equations as "i." A **geostationary Earth orbit** in which a **satellite** orbits directly over the Equator has an inclination of 0 degrees. A **polar orbit,** which carries a satellite over the North and South Poles, has an inclination of 90 degrees. An inclined orbit between 0 and 90 degrees, in which the satellite is traveling in the same direction as the Earth's rotation, from west-to-east, is called prograde. A **retrograde orbit** circles the Earth from east-to-west, opposite to the Earth's rotation, and its inclination will measure between 90 and 180 degrees.

Argument of Perigee

This measure describes where the **perigee,** or low point, of the orbit is with respect to the Earth's surface. It is the angle, measured at the center of the **Earth,** between a line to the orbit's perigee and the **line of nodes.** It is measured from the **ascending node** toward the perigee, in the same direction as the **satellite**'s orbit; it can measure anywhere from 0 to 360 degrees, and is designated in orbital equations by "ω," the lower-case Greek letter omega.

Right Ascension of the Ascending Node

This measure describes the location of the **ascending** and **descending nodes** with respect to the **Earth**'s equatorial plane. It uses the vernal equinox—the location of the sun in the sky on the first day of spring in the Northern Hemisphere—as a reference point or zero point from which right ascension is measured. Right ascension in the sky is measured as an angle in the equatorial plane eastward—that is, in the direction of the Earth's rotation—from the vernal equinox. Thus the location of the node is expressed as the "right ascension of the ascending node." It will measure between 0 and 360 degrees, and is expressed in equations as "Ω," the upper-case Greek letter omega.

True (Mean) Anomaly

This measure describes where a **satellite** is within its orbit with respect to the **perigee.** It is expressed as the angle, measured from the center of the **Earth,** between the satellite's location in its orbit and the orbit's perigee. It is measured in the direction of motion of the satellite, from the perigee to the satellite, and is expressed in orbital motion equations as "ν," the lower-case Greek letter nu. A satellite's position is sometimes expressed as the time elapsed between its current location in its orbit and the time it passed through perigee.

Orbital Directions

Launching in the direction of the **Earth**'s rotation helps deal with the **rocket equation.** Although the Earth's rotation velocity at the Equator (0.465 km/s) is very small compared to the velocity required to get into space (11.2 km/s), launching in the direction of the Earth's rotation adds just enough to the velocity increment to allow added weight to the launch **payload,** so this approach is often used.

Prograde Orbit

This term is used for a **satellite** launched to the east to

Orbital Parameters

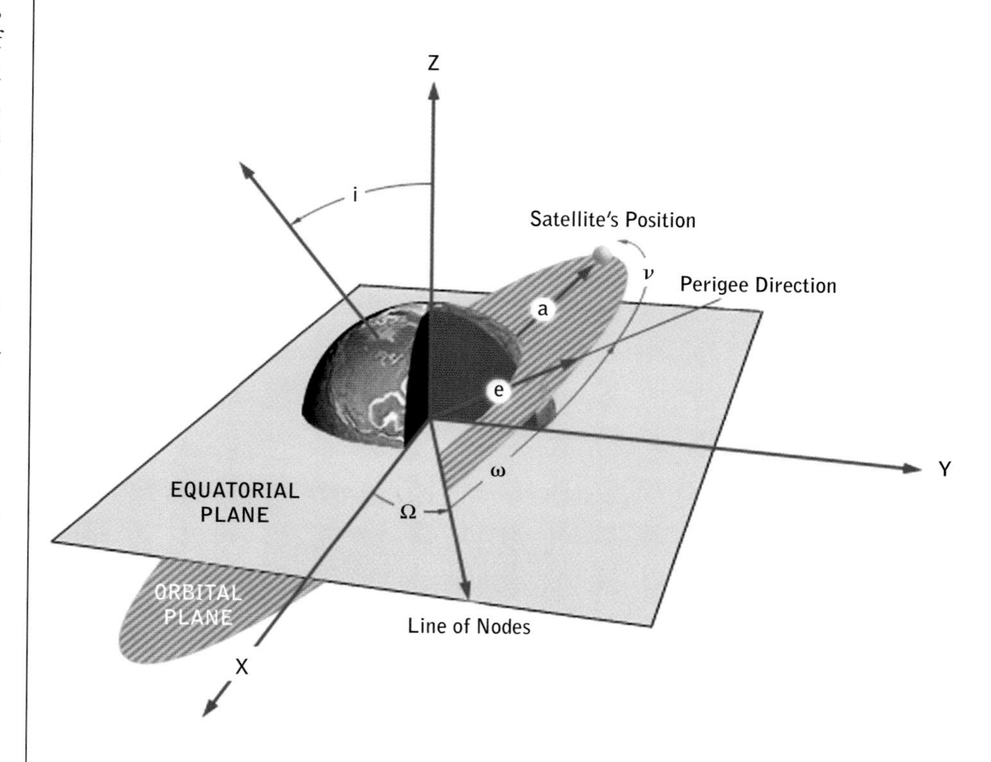

a	Semimajor axis	The size of the orbit
e	Eccentricity	The shape of the orbit
i	Inclination	Orientation of the orbit with respect to Earth's Equator
ω	Argument of perigee	Where the perigee, low point, of the orbit is with respect to Earth's surface
Ω	Right ascension of the ascending node	The location of the ascending and descending orbit locations with respect to Earth's equatorial plane
ν	True (mean) anomaly	Where the satellite is within the orbit with respect to perigee

take advantage of the **Earth**'s rotation. "Direct orbit" is another term used to describe the same thing.

Retrograde Orbit

This is a **satellite** orbit moving in a direction opposite to the **Earth**'s rotation. Used less commonly than **prograde orbits,** a retrograde orbit is appropriate for specialized concerns, such as staying always over areas of the Earth that are in sunlight.

The Rocket Equation

The recognized boundary of space is only 100 kilometers away from the **Earth**'s surface, but it takes a lot of energy to get there, and to stay there. How much energy was calculated in the 1890s by Russian mathematics teacher **Konstantin Tsiolkovsky** in his "rocket equation." He showed that the only information needed to predict rocket motion is the initial mass (weight) of the rocket and its fuel, how that mass changes as the fuel is burned, and the exhaust velocity of the burning fuel as it exits the rocket motor. Tsiolkovsky's simple equation is still the fundamental guide for all **satellite launches.**

Exhaust Velocity

This is the speed at which burning fuel exits nozzles of the **rocket** motor to produce the thrust for lifting a rocket and its **payload.** Nineteenth-century rocket propellants like gunpowder created an exhaust velocity of about 2,000 meters per second. Modern liquid-chemical fuels provide an exhaust velocity of up to 4,500 meters per second.

Mass Ratio

This is the ratio between the weight of a **rocket** plus its fuel at ignition versus its weight at any later time. So much energy is required to overcome **Earth**'s gravity and reach orbital height that at least 80 percent of a rocket's initial mass is fuel. A number of early satellite launches failed because the mass ratio was too near to one. The rockets lifted the vehicle off the launchpad a limited distance, kept it aloft until the fuel was expended, and then the entire **multistage launch vehicle** fell back to Earth. For the **space shuttle,** which must control its burn rate to keep launch velocity low enough to be tolerated by its crew, fuel makes up 94 percent of its weight before **launch.**

Velocity Increment

If the **mass ratio** and the **exhaust velocity** are known, the **rocket equation** will indicate the velocities needed for **launch,** attainment of orbit, and escape from the **Earth**'s gravitational pull. There are several basic relationships between **spacecraft** velocity and orbit:

- The higher the orbit altitude, the lower the velocity required to maintain orbit. The velocity of a spacecraft in circular orbit decreases with the square root of the orbital radius.
- The velocity of a spacecraft in an **elliptical orbit** is greater than one in a circular orbit at its **perigee** altitude.
- The velocity of a spacecraft in an elliptical orbit is less than one in a circular orbit at its **apogee altitude.**
- An increase in velocity to a spacecraft in a circular orbit will inject it into an elliptical orbit. The perigee will be at the original altitude—and at the point where the additional velocity is applied; the apogee will depend on the increment of velocity applied to the spacecraft.

Because of the Earth's rotation, an object on the surface of the Earth is already moving perpendicular to the Earth's gravitational pull at a velocity of 7.909 meters per second (or about 285 km/h). The velocity necessary to stay in space in an orbit at 500 kilometers altitude is 7.6 kilometers per second (27,360 km/h). The major concern at the vertical launch of a spacecraft is overcoming the Earth's gravitational pull quickly enough to get above the area of strong atmospheric drag to an altitude where an orbit can be maintained—above at least 200 kilometers. Attaining a circular orbit at 500 kilometers altitude requires a velocity increment of about 8.7 kilometers per second (or 31,320 km/h).

Multistage Launch Vehicles

Tsiolkovsky's **rocket equation** also told us that a single rocket, using any known **propellants,** could not possibly reach orbit. In a 1924 paper titled "Cosmic Rocket Trains," he proposed multistage launch vehicles with the mass of each stage being jettisoned when its fuel was expended. This was the major breakthrough needed to develop the first orbiting **satellites,** and it is still used for all satellite **launches** today.

Single-Stage to Orbit (SSTO)

The goal of launching a **satellite** with a single **rocket** stage is still being pursued, but will require the development of new, more efficient (more thrust per weight) propulsion techniques and materials.

Getting into Orbit

Tsiolkovsky's **rocket equation** also showed that the

An ash plume rises from Pagan volcano in the Mariana Islands of the Pacific Ocean in a view taken September 4, 1984, from the space shuttle *Discovery*—in low Earth orbit at 340 kilometers altitude. Part of the shuttle's vertical stabilizer is visible in the top of the image.

final speed of a **spacecraft** is dependent on only two things: the final **mass ratio** and the **exhaust velocity.** Thus attaining any orbit around the **Earth** for a **satellite** requires preplanning and many decisions that are constrained at **launch.** The **launch constraints** are even more confining for interplanetary launches.

Launch Constraints

The choice of the launch site, time, and direction depends on cost, downrange safety consideration, and the mission of the **spacecraft.** To protect people on the ground from the sometimes catastrophic results of **launch** failure, launches are directed over open ocean areas or large internal deserts, such as are available in Russia, China, and Australia.

For **polar orbits,** the latitude of the launch site is not constrained, because the **satellite** will be crossing all latitudes in its orbit anyway. But for **geostationary orbits,** the **launch site** should be as close as possible to the Equator to minimize the fuel and maneuver requirements to get the orbit aligned over the Equator.

The location of launch can also be constrained by the fact that the point where the rocket burn ceases is the **injection point** of the orbit, and the velocity, altitude, and location of this injection point define the orbit. For missions planning to use **parking orbits,** the time of launch is often less important, since the time of the initial "burn" to move the spacecraft to a **transfer** orbit can be chosen anywhere in the orbit. But for some satellites in parking orbits, there is a great need for maximum solar power for deploying (often a significant unfolding or opening up of systems once beyond **atmospheric drag**) and testing spacecraft systems. For the launches of interplanetary missions, the relative positions of the **planets** at time of launch, time of intersection, and, in some cases, recovery, are key to choosing launch times. These concerns can result in a **launch window** of just a few days over several years. Thus the impacts of launch location and time must be individually considered for each satellite launch.

Injection Point

This is the point in a **launch** when **rocket** propulsion ceases, requiring the **spacecraft**'s momentum to overcome gravity, and "injecting" it into orbit. The injection point becomes the **perigee** of the satellite's orbit.

Transfer

A transfer is any maneuver that changes the orbit of a **satellite** or **spacecraft.** One common example is the transfer approach typically used to attain the altitude of a **geostationary Earth orbit** (GEO). A satellite is launched with just enough velocity increment to inject it into a **low Earth orbit** used as a **parking orbit.** A controlled "burn" is then initiated to add just the right velocity to the spacecraft to inject it into a **geosynchronous transfer orbit,** also called a **Hohmann transfer trajectory**. This highly **elliptical orbit** will have a **perigee** at the **altitude** of the parking orbit and an **apogee** at the GEO orbit altitude. At apogee, another controlled burn puts the satellite into its GEO orbit. Because geostationary satellites stay over the same point on **Earth**'s surface, it is essential that the perigee burn in the parking orbit be initiated at the right location to ensure that the apogee occurs at the desired longitude over Earth's Equator. This two-step approach was used in early geostationary launches.

For today's launches into geostationary orbit, a different approach is used. A civil communications satellite may be launched directly into a geosynchronous transfer orbit from a location near the Equator like French Guiana (at 7° south latitude), but then require four different apogee "burns" to attain a series of less-than-geostationary-altitude elliptical transfer orbits for deploying and testing the new satellite's systems in an area of space that will avoid interfering with all the satellites already operating at GEO altitude. Even more complicated transfer maneuvers are used to conserve **propellant** aboard spacecraft moving into and out of orbits or supporting flybys of remote **planets** and **moons** in interplanetary spacecraft trajectories.

Getting Out of Orbit

Although gravity is always pulling a **satellite** toward the **Earth,** a satellite in a high enough orbit can be hard to get back down. In fact it takes the same amount of energy to **deorbit** a satellite as it does to get it into a high orbit in the first place, and the maneuvers can be equally complicated.

Stuck in Orbit

For satellites in high-altitude **geostationary Earth orbit** (GEO), it is not really feasible to bring them back to Earth. The **orbital decay** is very slow, and prohibitive amounts of fuel or other energy would be required to speed up the decay, or transfer them into lower orbits. So these satellites just do not come back. They are used until their onboard power is almost expended, and then are given a boost into a circular **graveyard orbit** above the GEO altitude, where they will stay out of the way for at least 100,000 years.

Deorbiting

Satellites in near **low Earth orbits** will return to the surface on their own in relatively short periods of time due to natural **orbital decay.** But for some spacecraft **payloads**—like the historical **Corona** spy satellite film canisters and the modern **space shuttle** with its crew—an organized deorbit is executed to control the location of **reentry,** and **touchdown,** splashdown, or, in the case of the shuttle, landing on a predetermined runway.

There are two ways to deorbit a satellite in low Earth orbit. One requires a "burn" to slow down the satellite's velocity and lower its **perigee** to less than 75 kilometers altitude above the **Earth**'s surface; at this height the orbit will decay and reenter the Earth's atmosphere almost immediately. The other is to apply a velocity-reducing burn to bring the perigee down to an altitude at which **atmospheric drag** will cause gradual orbital decay. The problem with this approach is the difficulty in predicting the time and place of reentry.

To safely deorbit and land the space shuttle, **astronauts** aboard start the process on the opposite side of the Earth from their runway at the Kennedy Space Center, in Florida. They fire the **rockets** into the direction they are moving to slow the shuttle's velocity a precise amount. This makes their near **circular orbit** impossible to maintain, and actually puts the shuttle into a new **elliptical orbit** with the **apogee** at the point of the "burn" and the perigee essentially at the runway on the other side of Earth's surface. The exact path must also take into account other variables, such as the slowing from atmospheric drag that helps lower the shuttle's speed from its orbital velocity of 29,000 kilometers per hour to its landing speed of 360 kilometers per hour.

Trajectories

Trajectory

A trajectory is the path of a projectile or other moving body through space. The term includes both closed, or bound, trajectories called orbits, and open trajectories, which typically take the form of a hyperbola, an unbound path which results in an object never returning to its **launch** or **injection point** unless complex maneuvers are applied to change its trajectory.

Ballistic Trajectory

A ballistic trajectory is one whose path depends only on the initial velocity and angle of projection; its downward path has no thrust, but assumes a **free-fall trajectory,** controlled only by momentum and gravity.

Consider the trajectory of a ball thrown into the air. Everyone knows it will fall back to **Earth** because of the pull of the Earth's gravity. But what is not widely known is that the ball's trajectory follows the same laws of motion and mathematical equations as **spacecraft** orbits and trajectories. The center of the Earth is the focal point for Earth orbits, so a ball held on the surface of the Earth can be considered to be in a circular orbit at the "altitude" of the Earth's radius. By throwing a ball into the air, you "inject" it into an **elliptical orbit.** If the Earth's surface did not get in the way, it would follow an elliptical orbit around the center of the Earth and come back to the place where you threw it. A **ballistic missile** follows the same laws of math and physics, but because the Earth's surface does get in the way, both the ball and the missile have an Earth impact point. The longest horizontal travel for a ballistic trajectory is reached if the original launch angle is 45 degrees. At both higher angles (a pop-up hit in baseball) and lower angles, gravitational force diminishes the horizontal travel distance.

Free-fall Trajectory

A free-fall trajectory is a downward **trajectory,** not controlled by any thrust, but only by a balance between remaining momentum of the object and gravity.

Suborbital Trajectory

This spacecraft trajectory is one with insufficient velocity to enter orbit—it is the **spacecraft** equivalent to the baseball analogy above. The **launch** is vertical to overcome gravity and **atmospheric drag** quickly, and the path is high enough to add horizontal travel based on the **Earth**'s rotation under the spacecraft, but the projectile has insufficient velocity to enter an "orbit" high enough to avoid an encounter with the Earth's surface.

Free-return Trajectory

This is an orbit between two celestial bodies that will naturally bring a **spacecraft** back to the vicinity of its origin. Used in the **Apollo** program as a safety factor for the manned missions to the moon, it clearly saved the lives of the **Apollo 13** crew. When an explosion destroyed their command-module propulsion system, the free-return trajectory automatically brought them back close enough to use emergency power from their lunar module to **deorbit** and return to Earth.

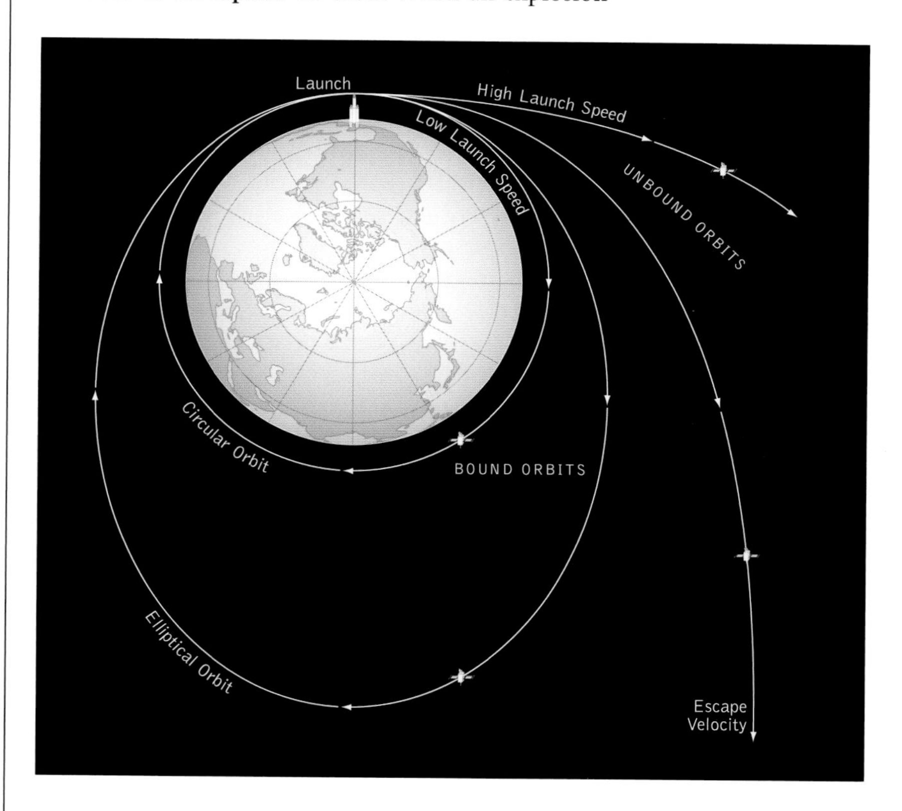

Trajectories depend on initial velocities. Ballistic trajectories do not obtain orbit because they collide with the Earth's surface. In a certain velocity range, closed trajectories attain Earth orbit. At greater velocities, unbound trajectories escape Earth's gravity entirely.

Cosmic Velocities

There are three cosmic velocities associated with four different types of **trajectories** for **spacecraft.** The first cosmic velocity is **orbital velocity.** For Earth it is about 27,800 kilometers per hour. Below orbital velocity, trajectories are ballistic. Between orbital and escape velocities is the realm of elliptical Earth orbits. The

VOYAGER: HELP FROM GRAVITY

Leonard David

TWO IDENTICAL ROBOTIC PROBES SENT OUT FROM EARTH IN THE LATE 1970S were destined to become the greatest space adventurers of our solar system in the 20th and 21st centuries. NASA's Voyager 1 was launched on September 5, 1977, joining its sister ship—Voyager 2—which had rocketed away earlier on August 20, 1977. It was a tour de force of space exploration—a grand tour of the gaseous planets in our solar system's family of worlds.

Between them, Voyager 1 and 2 explored all the giant planets in the outer solar system—Jupiter, Saturn, Uranus, and Neptune; 48 of their moons; and the distinctive system of rings and magnetic fields those planets retain. To accomplish their unprecedented treks, the Voyager spacecrafts utilized gravity-assist trajectories to reach ever more distant locations within the solar system. This gravity-assist technique can add or subtract momentum to increase or decrease the energy of a spacecraft's orbit. Using a planet's gravitational field, spacecraft can be deflected and propelled onto a different path. Such was the case for Voyager 1 and 2—each flew by massive Jupiter (Voyager 1 in March 1979 and Voyager 2 in July 1979) exploring that baffling globe of gas with onboard cameras and science instruments. Both then made use of Jupiter's intense gravity field to alter their trajectories toward Saturn.

At Saturn in November 1980, thanks to the planet's gravitational force, Voyager 1 was hurled out toward deep space. Still chalking up space mileage today, Voyager 1 is the most distant object from Earth built by humans. On its Saturn flyby in August 1981, Voyager 2 received a gravity assist from that ringed world, once again retargeting itself and setting out on another long-distance hop to Uranus. Arriving at that distant globe in January 1986, Voyager 2 used gravity assist to strike out toward Neptune. The probe reached that planet in August 1989, then sped onward and outward.

During their planetary "stopovers," the Voyagers relayed a bounty of scientific findings. Spacecraft observations found that Jupiter's Great Red Spot was a complex storm system that swirled in a counterclockwise direction. One of Jupiter's entourage of moons, Io, was pockmarked by active volcanoes—a sight not seen on any other body in the solar system. A puzzling surprise from Voyager involved another of Jupiter's many moons, Europa. Below this icy mini-world's topside, an ocean may exist—where there is water, the prospect that Europa might harbor life is a reasoned possibility.

Close scrutiny of Saturn by Voyager revealed long-lived ovals and other atmospheric features on the planet that were generally smaller than those on Jupiter. Saturn's elaborate ring system was also surveyed by the Voyagers, revealing among other finds that small moons "shepherd" ring material. The rings themselves consist of minuscule to house-size particles, believed to be leftovers of larger moons that were splintered by impacting comets and meteoroids.

Voyager 2's mission of discovery continued. Among many findings, it detected a magnetic field at Uranus, unknown until Voyager's arrival. Ten new moons were found orbiting the planet, with the spacecraft revealing that a previously known moon, Miranda, was rife with huge canyons and a mix of old- and young-looking geological features. Sojourning to Neptune, Voyager 2 images showed active geyserlike eruptions spewing on Triton, largest moon of that planet. Scientists were surprised to see several large, dark spots—reminiscent of Jupiter's hurricane-like storms—dotting Neptune.

The long-distance exploration of the dual spacecraft continues. As intrepid robotic explorers, and after having completing their primary studies, they have been recast as the Voyager Interstellar Mission (VIM) by the Jet Propulsion Laboratory (JPL) in Pasadena, California. JPL specialists built and operate the robotic probes. Instruments aboard the Voyagers are now searching for the boundary between the sun's influence and interstellar space, a region called the heliopause. ■

Gravitational slingshot: In this depiction of a gravity-assist flyby of Jupiter, the Voyager 2's trajectory is traced behind the planet (relative to the sun) as its velocity increases toward Saturn.

second cosmic velocity is **escape velocity.** The speed needed to break free of gravity and enter orbit is about 40,200 kilometers per hour. This is only about one-and-a-half times the orbital velocity, so the velocity range for Earth orbit is fairly narrow. Between escape-from-Earth and escape-from-the-sun velocities is the realm of hyperbolic trajectory flights in the **solar system.** The third cosmic velocity, escape velocity to exit the solar system from Earth, is roughly 150,000 kilometers per hour; beyond this is the realm of interstellar trajectories.

Earth-orbit Trajectories

There are several approaches for moving among orbits. Basic facts that guide these **trajectories** are: once a **spacecraft** has left the **Earth**'s surface, velocities are added parallel, not perpendicular, to the direction of motion; and adding velocity to a spacecraft in orbit does not increase its speed in that orbit, but increases the **apogee** of its orbit.

Hohmann Transfer Trajectory

This is the most economical and most common approach used for injecting **satellites** into **geostationary Earth orbit** (GEO). It was developed and published in 1925 by Walter Hohmann, city architect of Essen, Germany, but was first used 38 years later when the United States launched the first GEO satellite in 1963.

A satellite is launched with just enough velocity increment to inject it into a **low Earth orbit,** used as a **parking orbit.** A controlled "burn" is then initiated to add just the right velocity to the spacecraft to inject it into a **geosynchronous transfer orbit,** also called a Hohmann transfer orbit. This highly **elliptical orbit** will have a **perigee** (or low point, closest to Earth) at the **altitude** of the parking orbit where the maneuver was commenced, but the "burn" must add just the right amount of energy to make the **apogee** of this transfer orbit occur at the 35,900-kilometer altitude required for **geosynchronous orbits.** When the spacecraft reaches apogee, another controlled burn speeds it up just enough to allow it to escape its transfer orbit, and "circularize" the orbit at the geosynchronous altitude. Because geosynchronous satellites stay over the same point on the Earth's surface, it is also essential that the perigee burn in the parking orbit be initiated at the right location to ensure that the apogee ends at the desired longitude over the Earth's Equator. This two-step approach, while actually part of an elliptical orbit, is often called a Hohmann transfer trajectory because only half of the orbit is used to ascend to a new altitude before adjusting the orbit again. This maneuver can be used to reach a GEO orbit within hours.

High-energy Transfer Trajectory

A faster, steeper path up to a desired orbital **altitude,** the high-energy **transfer** requires a higher velocity than the **Hohmann transfer trajectory** and is appropriate only when time is of the essence. It is unsuitable for human crews, because the acceleration against gravity (g-force) is about ten g's, much more than a human can tolerate. One application is for antisatellite weapons, when reaching orbit altitude quickly is required to reach a target but there are no fuel requirements after hitting the target.

Other Transfer Trajectories

There are two other **transfer** approaches—low thrust chemical and electric propulsion—that require much less energy, but much more time to acquire orbital altitude. Using the very low-energy electric propulsion transfer approach to gradually spiral up to reach a **medium** or **high Earth orbit** would take months rather than the few hours required for a **Hohmann transfer.**

Interplanetary Trajectories

There are two basic approaches for reaching other planets from an Earth launch.

Direct Trajectory

The approach used most often to date for reaching other **planets** is a direct, high-energy trajectory for departure from **Earth** with sufficient thrust to exceed **escape velocity.** At about 900,000 kilometers, the spacecraft escapes the Earth's "sphere of influence"—the area where Earth's gravity is stronger than the **sun**'s—and enters a heliocentric **transfer** phase with a trajectory controlled by its momentum and the sun's gravity.

Planetary Gravity-assist Trajectory

In recent planetary missions, initial **launch** energy requirements have been greatly reduced by using non-direct trajectories with planetary gravity assists. One or more close encounters with the gravity fields of **planets,** or flybys, can give an energy boost that increases the speed of a **spacecraft** without the use of onboard **propellants.** Variants of this approach, increasingly being employed for interplanetary missions, can use a planet's gravitational field to speed up, slow down, or change the direction of spacecraft.

Lagrangian Points & Their Uses

Initially described in the late 1700s by Italian-born French mathematician Joseph Louis de Lagrange, these are unique points of gravitational equilibrium between two celestial bodies that are exercising gravitational force on each other; they are also called libration points. Looking at the Earth-moon system, they are located as follows:

- L1 and L2 are points of equal gravitational pull from the **Earth** and **moon.** Because the moon's gravitation is less than Earth's, these points are closer to the moon. L1 lies on a line between the Earth and the moon; L2 is on the same line but on the other side of the moon.
- L3 is on the same line, but on the other side of the Earth from the moon and at a distance above the Earth equal to that of the moon's orbit.
- L4 and L5 are on opposite sides of the Earth-moon line; both are along the path of the moon's orbit at points equidistant from the Earth and moon.

L1, L2, & L3 Uses

These three points in any two-body system, for example Earth-moon or Earth-sun, are naturally unstable, so a spacecraft at any of them would require very little fuel to cause a significant change in its **trajectory.** This characteristic of the L1 through L3 points was used as an opportunity to develop new fuel-efficient techniques for interplanetary spacecraft travel, such as chaotic control and the OGY technique.

Chaotic Control

A recently developed interplanetary spaceflight technique that requires less fuel for long extraterrestrial missions (outside the **Earth**'s gravitational field), this approach uses the instability of the L1, L2, or L3 **Lagrangian points** as a way point. Assuming it takes less fuel to reach one of these points than to reach the **spacecraft**'s ultimate goal, overall fuel consumption can be decreased by traveling first to a Lagrangian point and then doing a very small "burn" to boost the trajectory from there.

NASA first explored the use of this technique in a program named the International Cometary Explorer (ICE) initiated in December 1983. The experiment used a space probe at the end of its useful life after four years, the International Sun-Earth Explorer 3 (ISEE-3), and investigated the notion—very successfully—that using Lagrangian points and very short fuel "burns" could substantially extend the useful life of the spacecraft. Flying ICE by the Earth-moon L1 five times and giving a tiny boost on each flyby allowed it to travel on to two **comets** and collect valuable data.

OGY Technique

Named for three University of Maryland mathematicians—Ott, Gregobi, and Yorke—the OGY technique is a refinement of the chaotic-control method. First used in **NASA**'s **Genesis** mission that collected samples of the **solar wind,** a complicated OGY detour was designed to bring samples back to **Earth** in the most fuel-efficient manner possible for laboratory study. Genesis was moved to the Earth-sun L2 **Lagrangian point,** from which it could economically be moved to the Earth-moon L1. From there it flew a few "chaotic" orbits around the moon, and then moved into a stable orbit around the Earth from which the cargo capsule with its samples could be parachuted to Earth for recovery. This was a long, but fuel-efficient route home.

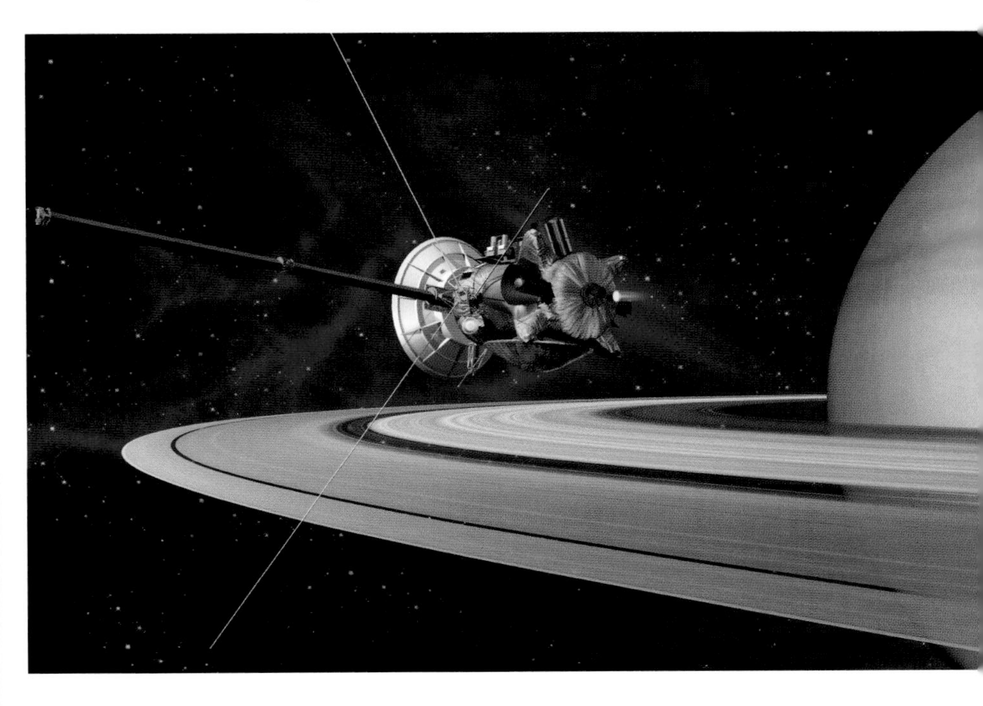

Reverse burn slows the Cassini spacecraft enough for capture into orbit around Saturn in a computer-simulated image. Cassini reached Saturn at the end of June 2004 after a gravity-assisted trip that used four planetary flybys (Venus, Venus, Earth, and Jupiter) to speed up and change direction.

L4 & L5 Uses

The L4 and L5 Lagrangian points are very stable; an object placed at one of these points, or in orbit around it, would stay indefinitely with no need to use propellant for maintaining the orbit. L5 in particular, through an L5 Society, an organization dedicated to the colonization of space founded in 1975, has long been imagined as an ideal place for a long-term space base for exploration, research, manufacturing, and human colonization.

Satellite & Spacecraft Orbits

Satellite Orbits

Satellite orbits are closed **trajectories,** which cause them to travel in a circular or elliptical path around one celestial body. Most man-made **satellites** are in orbit around the **Earth,** and their orbits are characterized in several ways—by shape, altitude, inclination, period, and use. Common examples from all of these categories are presented, defined, and explained below.

Reentry into the Earth's atmosphere is always a fiery affair. The Russian space station Mir, after 15 years of service in low Earth orbit, burns up on reentry over Nadi, Fiji, on March 23, 2001.

Circular Orbit

A circular orbit is a closed path, or orbit, for a **spacecraft** or **satellite,** that is a circle, with its center at the center of the Earth. An exact circular orbit is somewhat unstable because the Earth and its gravity field are not quite spherical, so considerable **propellant** for corrective **rocket** "burns" is required to maintain the circular path.

Elliptical Orbit

Any noncircular, closed path of a **spacecraft** follows the shape of an **ellipse,** and is an elliptical orbit. The center of the Earth is one of the two focal points, or foci, of the ellipse; the other focal point is at an empty location in space. The distance between the two focal points determines the **eccentricity** of the elliptical orbit.

Eccentric Orbit

This term refers to any **elliptical orbit** in which the ellipse is greatly elongated—that is, the major axis is much greater in distance than the minor axis.

Low Earth Orbit (LEO)

Definitions vary, but low Earth orbits are usually considered to be those at altitudes above the **Earth**'s surface ranging from 100 to 1,000 kilometers. These low orbits are the easiest and least expensive to achieve, and by far the most **satellites** are launched into LEO orbits. They also fall just below the base of the inner **Van Allen** radiation belt and thus avoid interference from highly charged electrons and protons from the Earth's **ionosphere** and from the **solar wind.** Another great advantage of this type of orbit is the high resolution of data, including imagery that can be obtained from sensors this close to the Earth's surface. One drawback is that all LEO orbits up to 1,000 kilometers altitude encounter **atmospheric drag.** In the lower altitudes the drag is highest, and these orbits are thus limited in their duration.

Polar Orbit

Polar orbits, typically at **low Earth orbit** altitudes, have an **inclination** at or near 90 degrees, thus perpendicular to the **plane of the Equator** and parallel to a line through the Earth's North and South Poles. **Satellites** in orbits with inclinations less than this can "see" to the poles if they have instruments aboard that image or sense a wide swath on either side of their **ground track.** This is one of only two orbits (the other being **Molniya**

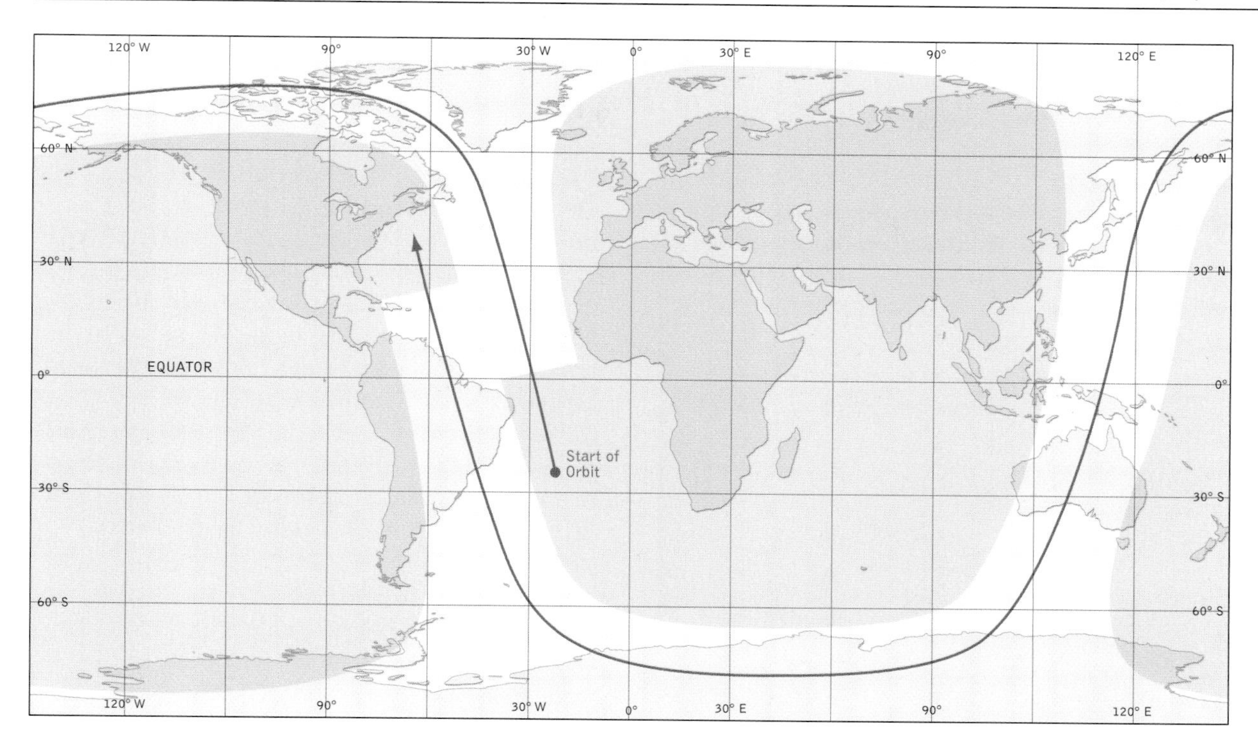

A satellite in polar orbit sees successive north-south swatches as the Earth revolves beneath it. The whole Earth surface will be covered, but how frequently depends on the orbit of the satellite.

orbit) that can look straight down on the poles and their ice sheets.

The greatest advantage of the polar orbit is that it"flies" over every point on the Earth's surface, and is thus optimal for global Earth-sensing missions. A disadvantage is that the Earth is constantly rotating below the satellite's track, so no spot on the Earth's surface can be sensed continuously. The orbital altitude and period must be designed to provide the desired frequency of crossing any point on Earth. This frequency of sensing a given location is often referred to as the "refresh rate."

Repeating Ground Track Orbit

This is an inclined orbit with a period designed so that it will make a certain number of complete orbits in a fixed number of whole days. Thus a **satellite** will return over the same spot on the **Earth** after that fixed number of days, and its **ground track** will overlay exactly. This type of orbit is very useful for sensing Earth phenomena for which a time series is desired, so changes or variations can be measured at fixed time intervals.

This orbit is complicated, however, by the oblateness of the Earth—the fact that the planet is not spherical and its rotation causes a bulging of mass along the Equator and a flattening at the poles—which must be taken into account.

Geosynchronous Orbit

At a particular **altitude** above the **Earth**'s surface—about 35,900 kilometers—and low **inclination, satellites** can travel west-to-east around the **planet** at the same speed as the Earth's rotation. Such an orbit is geosynchronous—"in synch" with the planet. Satellites in geosynchronous orbit are thus able to "look at" or provide communications for roughly the same area of the Earth at all times.

Similar orbits around other planets are called simply synchronous orbits, but that term is also sometimes used to mean the geosynchronous orbit.

Geostationary Earth Orbit (GEO)

The geostationary orbit is a variant of **geosynchronous orbit** in which the **inclination** is zero—the orbit remains directly over the **Earth**'s Equator—and the period is the same as the Earth's rotation. Thus, a **satellite** in this orbit will complete one revolution around the Earth in one day, traveling at a speed of 11,300 kilometers an hour. This GEO orbit requires a satellite to be at 35,786 kilometers above the Earth's surface. It is a very stable orbit, with incredibly slow **orbital decay,** allowing satellites to remain in orbit almost indefinitely, though it requires power to keep the orbital inclination at zero. GEO is sometimes called the "Clarke orbit," because it was first described by the popular science and science fiction writer Arthur C. Clarke in the October 1945 *Wireless World* magazine.

Injecting a satellite into this orbit requires multiple maneuvers, however, and it was not successfully achieved until the United States civil satellite program at **NASA** launched the first satellite into GEO orbit in 1963. The Soviets launched their first geostationary satellite in 1974, and the Europeans in 1981. In this orbit, a satellite appears fixed above a certain point on the Earth's Equator, and can "see" fully 42 percent of the Earth's surface, between 81.2 degrees north and south latitude and an equivalent width of 81.2 degrees of longitude to both the east and west. So three satellites in GEO orbit can completely cover the Earth in the equatorial and mid-latitudes, but not above 80 degrees latitude in the polar regions.

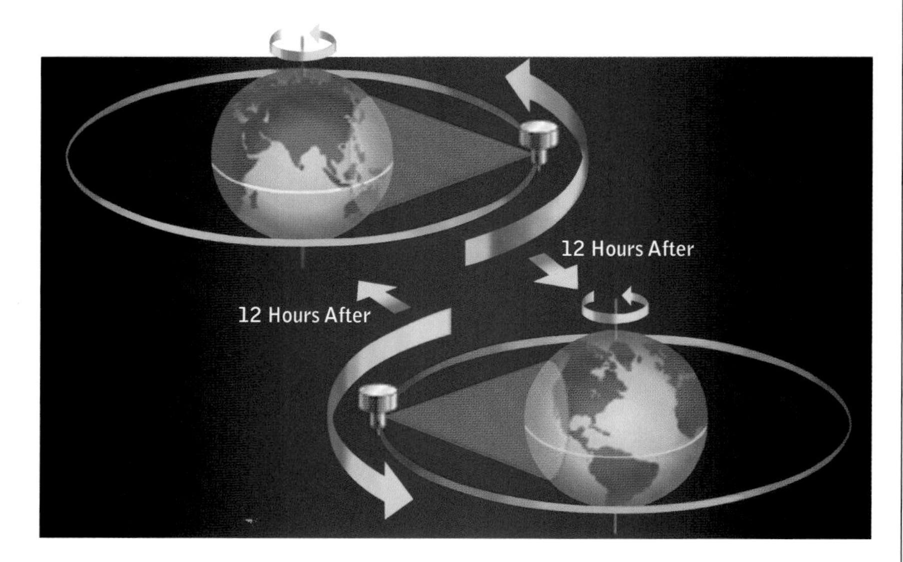

A satellite in geostationary orbit (GEO) at 35,786 kilometers altitude and zero inclination (directly over the Equator) will orbit at the same speed the Earth rotates and will stay fixed over the same point on the planet.

The geostationary orbit is the most widely used single orbit for Earth satellites, with close to 600 satellites launched into GEO orbits since 1963. It is ideal for stable, uninterrupted coverage for missions such as communications and weather sensing. Because orbital locations at this particular altitude are coveted, the longitudes over the Equator are monitored and meted out carefully by the **International Telecommunications Union** of the United Nations.

Semisynchronous Orbit

An orbit at an **altitude** of 20,330 kilometers, in which a satellite revolves around the **Earth** every 12 hours exactly, is called a semisynchronous orbit.

Medium Earth Orbit (MEO)

A term not used by all space scientists, medium Earth orbits are generally agreed to fall between about 1,000 kilometer **altitudes** at the base of the **Van Allen belt** and range up to an altitude just below **GEO** orbits.

Drifting Orbit

One **medium Earth orbit** is the "drifting orbit" used by some military communications **satellites.** The drifting orbit is designed with an **inclination** of zero degrees, but an **altitude** of 35,780 kilometers, just below that required for **geostationary orbit.** The satellite thus orbits the **Earth** somewhat more slowly than the Earth rotates, so the satellites seem to fall behind or drift with respect to the Earth's surface. This orbit can reduce dependency on a **GEO** satellite in a fixed position, as the drifting satellites are constantly moving and can fill in coverage gaps quickly if one of them fails.

Sun-synchronous Orbit

A sun-synchronous orbit is an orbit in which a **satellite**'s orbital plane maintains the same orientation to the sun at all times.

To achieve this orbital character, the satellite must be in a **low Earth orbit** with an **inclination** of about 98 degrees to the Earth's Equator. This is slightly more inclined than the 90 degrees of a classic **polar orbit,** and is thus a slightly **retrograde orbit,** meaning the satellite's movement across the surface of the Earth is tilted slightly in the direction opposite the west-to-east direction in which the Earth rotates.

Molniya Orbit

This is an **Earth** orbit described by a very elongated **ellipse,** thus an **eccentric orbit,** having a low **perigee** of about 400 kilometers over the southern pole and very high **apogee** of 40,000 kilometers over the Earth's north polar regions. Named after the Soviet scientist who developed the orbit, it is specifically designed to "linger" over the Arctic polar areas of the Earth.

The Molniya orbit takes advantage of **Kepler**'s second law. A **satellite**'s closeness to the Earth at its perigee over the South Pole requires it to move faster with respect to the Earth's surface during that part of the orbit, and its great distance from the Earth over the north polar regions, and its slower speed near apogee, allow the satellite to spend most of its time in this part of its orbit. The Soviets originally designed this orbit to support military communications with their remote northern military bases. But during the height of the Cold War, they presumably also used it for surveillance of American operations in the far northern latitudes.

High Earth Orbit or Highly Elliptical Orbit (HEO)
A closed flight path around the Earth at altitudes above that of **geosynchronous orbit** (about 35,900 kilometers) and up to about 96,500 kilometers above the Earth's surface is sometimes called a high Earth orbit (HEO). Many space scientists, however, use HEO to designate a highly elliptical orbit, such as the **Molniya orbit.**

Supersynchronous Orbit
This is another name for **high Earth orbit,** indicating that the **altitude** above the Earth's surface in this type of orbit is greater than the altitude of synchronous orbits.

Parking Orbit
A temporary **low Earth orbit** (LEO) used for initial injection into orbit of a **spacecraft** while waiting for a transfer maneuver to move it to a higher orbit, the parking orbit is also often used for check-out and testing of a **satellite**'s onboard systems.

A parking orbit is also used as a "storage orbit," at LEO altitudes sufficient to reduce **atmospheric drag,** for on-orbit spare satellites in constellations that have high-priority missions that cannot wait for **launch** of a replacement satellite if one fails.

Geosynchronous Transfer Orbit (GTO)
This orbit is a preferred transfer step for injecting a **satellite** into **geosynchronous** or **geostationary orbit.** It is an elliptical orbit, with a **perigee** at a low Earth **parking orbit** altitude, and its **apogee** at the altitude above the Earth's surface required for geosynchronous orbit—about 35,900 kilometers. Use of this approach requires two rocket "burns": one to get into this orbit from an **LEO** circular parking orbit, and the second at apogee to speed the spacecraft and "circularize" its orbit at the **GEO** altitude.

Graveyard Orbit
When sensors fail on an operational satellite and it can no longer perform its mission, the **satellite** is moved into a graveyard orbit where it will not collide or interfere with satellites that are still operating. It is particularly used for **spacecraft** in **geosynchronous orbits,** where **deorbiting** from that **altitude** would require too much fuel. The satellite is given a small boost just before its power dies to elevate it into a **high Earth orbit** several hundred kilometers above the **geostationary Earth orbit** altitude. There it can stay without further fuel or controls, and without interfering with geostationary satellites for thousands to hundreds of thousands of years. The decision to move a satellite to this orbit must be made while there is still sufficient energy aboard to execute the maneuvers required to change the orbit. Satellite systems that fail unexpectedly and catastrophically sometimes cannot be moved into their graveyard orbits, and must be tracked carefully to ensure they do not damage other working space systems.

Orbital Decay

All **elliptical orbits** tend to adjust over time to a **circular orbit.** Since orbits are a balance between gravitational pull and the momentum of the orbiting object, orbital decay is the result of one of these parameters changing. The gravitational pull is constant, but the momentum of the **spacecraft** decreases gradually because it is not traveling in a complete vacuum. In **low Earth orbit,** the satellite is slowed by **atmospheric drag,** or friction with the very thin Earth atmosphere; in higher orbits, the drag from solar radiation pressure gradually decreases a **satellite**'s momentum. These effects both lower the **apogee** of the orbit over time without affecting the **perigee.** Thus the eccentricity of the orbit is reduced, and all elliptical orbits eventually tend to become circular orbits at the altitude of the perigee. And once circularized, the circular orbits tend over time to decrease in **altitude.**

Interestingly, this effect of solar radiation is seen clearly in the life of satellite orbits, corresponding to the 11-year cycle of **sunspot** and solar radiation levels. Satellite orbits last considerably longer between the periods of high solar activity; satellites launched at the solar maximum display much more rapid orbital decay. The expected life of a satellite orbit depends largely on its altitude above the surface of the object it is orbiting. For typical Earth orbits, the range is from days to a million years.

EXPECTED LIFE OF A SATELLITE ORBIT

Altitude	Orbit Type	Duration*
200 km	Low Earth orbit	Few days
500 km	Low Earth orbit	Few weeks
600 km	Low Earth orbit	Few years
800 km	Low Earth orbit	Few centuries
1,000 km	Top of low Earth orbit	Several centuries
30,000 km	Near geostationary orbit	A million years

**Duration based on circular orbit at these altitudes.*

4 | Human Spaceflight

Floating free, astronaut Bruce McCandless test-flies the Manned Maneuvering Unit (MMU) on February 7, 1984. Powered by jets of compressed nitrogen, the MMU was created to assist spacewalking shuttle astronauts in efforts to retrieve and repair satellites. During the MMU's space debut, McCandless ventured up to 99 meters from the shuttle *Challenger*.

Friendship 7

LOOK UP INTO A STARRY SKY SOME NIGHT, AND YOU'LL SEE THE SAME VIEW witnessed by countless people since humanity began. If you happen to be in the right place at the right time, you can glimpse a bright, unblinking point of light that moves swiftly across your view, then disappears. It's not a star, but a home: the International Space Station, circling the globe at more than 28,160 kilometers an hour. Its presence in our skies signals that we live in a momentous time, one in which human beings are moving outward from their home planet to explore an endless frontier.

The age of human spaceflight began more than 40 years ago when Soviet pilot Yuri Gagarin's single-orbit voyage shook the world, as much for its Cold War political implications as for its historic significance, and it intensified a superpower space race that culminated with the first footsteps on the moon by Apollo 11 astronauts Neil Armstrong and Buzz Aldrin in 1969. Space pioneers like Wernher von Braun envisioned that the Apollo moon landings would be a jumping-off point for even more ambitious efforts—an orbiting space station, a base on the moon, and human expeditions to Mars.

But those dreams have proved elusive. Since Apollo, the course of human spaceflight has been shaped not only by technology but also by down-to-earth considerations such as budgets and shifting national priorities. Progress has been largely incremental, although there have been many dramatic moments, including the first flight of the space shuttle in 1981, the missions aboard Russia's Mir space station in the 1980s and '90s, and the repair of the Hubble Space Telescope by shuttle astronauts in 1993. And there have been moments of tragedy, most recently the loss of the shuttle *Columbia* and its seven-member crew in February 2003. The *Columbia* tragedy is a lingering reminder of the difficulties, and the costs, of the quest to explore space. At the same time, the spaceflight "club" is expanding, with China becoming the third nation to achieve a manned space mission, in October 2003. And even space tourism, another dream that dates back at least to the Apollo era, is slowly becoming a reality.

Today, spaceflight is still on the cutting edge of human experience. Only a few hundred people have ventured beyond the atmosphere, and they speak of an extraordinary adventure. But human spaceflight is more than an experience; it is a turning point in human evolution, one that was anticipated generations before it happened. As the Russian schoolteacher and spaceflight visionary Konstantin Tsiolkovsky wrote almost a century ago, "The Earth is the cradle of the mind, but one cannot live in a cradle forever."

John Glenn climbs into his Mercury spacecraft, Friendship 7, before one of several attempts to become the first American to orbit the Earth. After postponements of his launch due to mechanical problems or bad weather, Glenn finally succeeded on February 20, 1962. His three-orbit flight was a major milestone for NASA.

The Early Years

Animals in Space

The visionaries who created the first artificial **satellites** had more in mind than sending machines into space. Sergei Korolev, mastermind behind the Soviet space program and its **Sputniks,** long dreamed of the time when human beings would ride **rockets** into **Earth** orbit, and beyond. In the United States, German rocket pioneer **Wernher von Braun** also set his sights on manned space travel. Both longed to send explorers to the **moon** and more distant worlds, especially **Mars.** But no one could be certain whether people would tolerate the stresses of spaceflight—the accelerations of **launch** and **reentry,** and the alien experience of **weightlessness** in space. Before even one human space traveler could leave Earth, animals would have to pave the way.

So it was that on November 3, 1957, Sputnik 2 lifted off carrying the world's first space traveler, a dog named Laika. Sealed in a tiny capsule, Laika reached orbit some 1,507 kilometers above the Earth. Data from sensors on Laika's body, measuring her heart rate, blood pressure, and respiration, were transmitted to Earth. The craft contained enough food, water, and oxygen for ten days. But Sputnik 2 was not designed to be recovered; Laika would not return to Earth.

Immediately after the flight, the Soviet Union announced that Laika had adapted well to spaceflight and had survived for a week. But in 1999 veterans of the Soviet space program revealed that the capsule's environmental control system had malfunctioned; the cabin overheated, killing Laika. In 2002 another former Soviet scientist claimed that this had actually happened within four to seven hours of Sputnik 2's launch.

That Laika survived her launch, and her exposure to weightlessness—however briefly—gave encouragement to Korolev and others seeking to make human spaceflight a reality. During the next 28 months at least 13 dogs followed Laika into orbit; most fared much better, returning safely to Earth.

In the United States a variety of animals were enlisted as space travelers. Several rockets launched on suborbital flights in 1958 included mice in their **payloads,** and a squirrel monkey made the trip in December

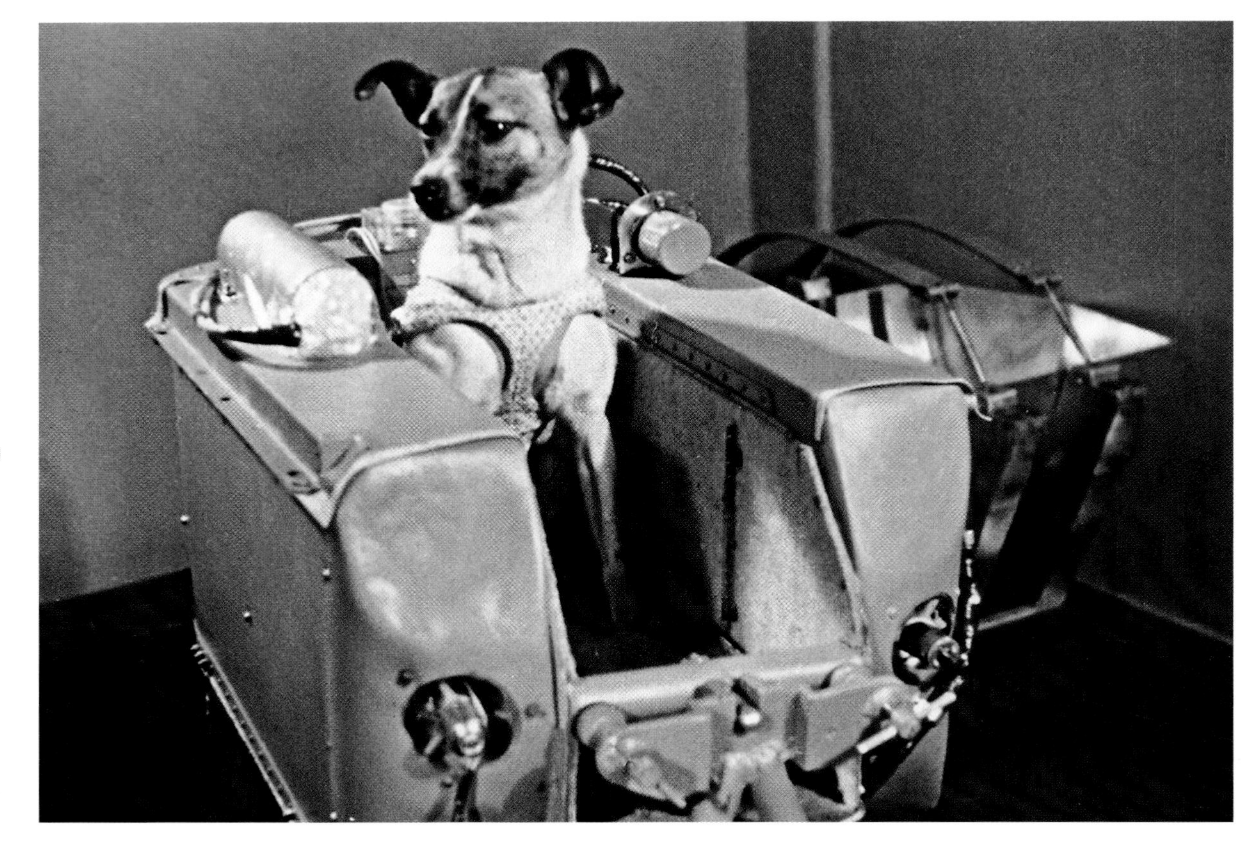

First space traveler, a dog named Laika, awaits her flight. Laika was a stray dog that was picked up on the streets of Moscow and prepared for spaceflight. Launched aboard Sputnik 2 on November 3, 1957, she survived a few hours in her small capsule before succumbing to overheating. Sputnik 2 burned up during its reentry into the atmosphere on April 4, 1958.

1958. By January 1961, with **NASA** working to send **astronauts** aloft for **Project Mercury,** a chimpanzee named Ham became Mercury's first passenger. His suborbital flight included some unplanned stresses, from higher-than-expected g-forces during launch and reentry to rough seas after splashdown, but Ham came through his ordeal in good condition.

Human spaceflight was closer than most people would have guessed just a few years earlier. NASA had selected seven pilots as the nation's first astronauts; each hoped to be the first man in space. A young Russian pilot named **Yuri Gagarin** beat them.

Yuri Gagarin, the First Man in Space

Yuri Alexeyevich Gagarin was 27 years old when he became the first man in space, on April 12, 1961. He had been an active-duty air force pilot when he was selected as one of 20 cosmonauts in 1959.

Gagarin's **spacecraft,** called **Vostok,** was designed to operate under automatic control or by commands radioed from **Earth,** so he would do no real piloting during his single-orbit mission. Propelled by a 30-meter booster, the same type of **rocket** used to **launch** the **Sputniks,** Vostok lifted off at 9:07 a.m. Moscow time. Sealed within its spherical cabin, Gagarin felt an upward surge and radioed, "*Poyekhali!* Here we go!" Eleven minutes and 16 seconds later, he was in orbit, following a path that ranged from 175 kilometers to 302 kilometers above the Earth.

Although some doctors had feared **weightlessness** might sicken Gagarin, he came through the flight with no ill effects. He was able to eat and drink normally, consuming foods packed in special squeeze-tubes. He saw the Earth passing beneath him and reported on his condition to mission controllers over a high-frequency radio, and by using a telegraph key.

Only one malfunction marred the otherwise successful flight, when a set of cables linking the spherical descent cabin with the cone-shaped retro module failed to disconnect as planned. The two pieces of the craft remained linked for several minutes, tumbling as they descended, until the cables finally broke free.

As Vostok continued its descent, friction with the atmosphere created an unearthly, purple glow surrounding the craft. Inside the spacecraft, Gagarin endured g-forces up to ten times normal. During the final descent, 108 minutes after liftoff, Gagarin ejected from the craft and parachuted to a safe landing near the Volga River.

Gagarin was anticipating an assignment to a second space mission when he was killed in the crash of a MiG-15 jet trainer on March 27, 1968.

Valentina Tereshkova, the First Woman in Space

A former textile factory worker and amateur parachutist named Valentina Tereshkova was one of five women selected to train for spaceflights in 1962. Tereshkova, whose radio call sign was *Chaika* (Seagull), logged almost three days in orbit aboard **Vostok** 6 beginning on June 16, 1963.

Tereshkova's flight overlapped with the five-day mission of Vostok 5, which was occupied by Valery Bykovsky and launched three days earlier. The two craft came within about five kilometers of each other during Tereshkova's first orbit. Although Tereshkova came through her flight in good condition, it was not until 1982 that another woman cosmonaut, Svetlana Savitskaya, followed Tereshkova into space.

Vostok & Voskhod

The first person to travel in space, Soviet cosmonaut **Yuri Gagarin,** got there in a **spacecraft** called Vostok (Russian for "east"). It was composed of two separate sections, or modules: A spherical descent module 2.3 meters in diameter, in which the cosmonaut rode; and the cone-shaped instrument module, containing the craft's retrorocket as well as small orientation rockets and other components needed for orbital flight. The entire assembly measured about 4.8 meters in length. Six piloted Vostok missions took place between 1961 and 1963, the longest being Valery Bykovsky's five-day flight in June 1963.

Although Vostok achieved many pioneering successes, its capabilities were limited. The Soviets' chief spacecraft designer, Sergei Korolev, was anxious to produce a next-generation spacecraft that could be used to send cosmonauts around the **moon.**

But Soviet leaders wanted more space successes, so Korolev created a modified version of Vostok, called Voskhod ("sunrise"). In October 1964 Voskhod 1 became the first multiperson spacecraft by carrying three cosmonauts on a day-long orbital mission. The flight was a risky one—to squeeze three men into a cabin designed for one, the cosmonauts had no space suits or ejection seats—but Voskhod 1 flew without mishap.

In March 1965 Voskhod 2 carried two space-suited cosmonauts into orbit; one of them, Alexei Leonov, emerged through a collapsible airlock and floated in the void for ten minutes, becoming the first human being to walk in space. The 1965 flight proved to be the final Voskhod mission; plans for more flights were abandoned as the Soviets focused on the moon race with the Americans.

A Vostok spacecraft is mated to its protective launch shroud, at left. The silvery sphere is the Vostok's descent module, in which the cosmonaut rode during the flight. To the right of the descent module, part of the cone-shaped instrument module is visible. The cylinder at right is the final stage of the rocket used to loft Vostok into Earth orbit.

NASA

By 1958 putting a man in space became the focus of the so-called space race between the United States and the Soviet Union. Although many believed such a task should be handled by the U.S. Air Force, President **Dwight D. Eisenhower** believed the human spaceflight program should be a civilian one. The National Aeronautics and Space Administration (NASA), created by Congress at Eisenhower's request, began operating on October 1, 1958.

Many of NASA's key personnel were highly experienced engineers, veterans of aeronautical research and development. Now, joined by a growing cadre of specialists, they applied their technical prowess to the new challenges of exploring space. Although NASA's most important goal was putting a man in space, the agency took on other challenges, including missions by space probes to explore the Earth's space environment, the **moon,** and the **planets.** Other **satellites** were created to study the **sun** and stars, and the **Earth** itself. But it was NASA's human spaceflight activities that gave the agency its greatest challenges—especially the call to land humans on the moon, made in 1961 by President John F. Kennedy. Achieving that goal with the missions of **Project Apollo** transformed NASA into an organization in which anything seemed possible.

But NASA's fortunes changed after the first Apollo lunar landings in 1969. Space budgets, which had risen to meet the awesome challenges of Project Apollo, were in decline. Faced with dwindling resources, the agency set its sights closer to home, focusing on the scientific and technological potential of **space stations** in Earth orbit. The major human spaceflight projects in the 1970s—a space station called **Skylab** and development of a reusable **space shuttle**—reflected that new direction. The '70s were also a time when unmanned explorers made history, including the twin Viking Mars landers in 1976, and the launch in 1977 of the "grand tour" of the outer planets by the Voyager 1 and 2 probes.

The shuttle era, which began in 1981, proved to be a time of triumphs and tragedies for NASA. The space shuttle proved to be amazingly versatile, and with it NASA rescued errant satellites and repaired the **Hubble Space Telescope.**

But the loss of the shuttle orbiters ***Challenger,*** in 1986, and ***Columbia,*** in 2003, underscored the difficulties and risks of human spaceflight and plunged the agency into some of its darkest periods. And NASA's robotic missions brought mixed results, with successes like Mars Pathfinder in 1997 and defeats such as the loss of the Mars Polar Lander in 1999.

Today, NASA is at a crossroads. With the **International Space Station** now permanently occupied, the space agency has been given a presidential directive to send humans back to the moon and on to Mars. But critics have questioned NASA's ability to meet those objectives. And no one can say whether the resources to accomplish those goals will become available. Still, NASA remains an organization capable of amazing achievements, as evidenced by the successes of the twin Mars exploration rovers, Spirit and Opportunity, in 2004.

Kennedy Space Center

Located next to Cape Canaveral on Florida's Atlantic coast, the Kennedy Space Center—named in memory of the slain President Kennedy—is the place where most of NASA's space missions begin. It has been the site for hundreds of manned and unmanned **launches,** including the **Apollo** lunar landings, **space shuttle** missions, and a host of robotic probes to study the **Earth,** the **sun,** the **solar system,** and the universe. Space vehicles are assembled and tested there, and it is also the place where the shuttle orbiter often lands and is refurbished after each flight.

Astronauts

Literally, it means "star sailor." But since the dawn of the space age, the word "astronaut" has conjured images of Americans voyaging into space. (Members of Russian space crews are referred to as cosmonauts, and Chinese astronauts are called Taikonauts.) They are the ones who walk in space, fly the **space shuttle,** and construct the **International Space Station.** They conduct scientific experiments, maintain and repair equipment, and make observations from their lofty perch hundreds of kilometers above the **Earth.** And they contribute to the design and testing of new space vehicles. Astronauts have lived in space for months on end, and they have walked on the **moon.** In short, astronauts do things most people can only dream about.

The first U.S. astronauts were the seven pilots chosen in 1959 for **Project Mercury.** These men, selected as much for their physical perfection as their superb piloting skills, became national heroes overnight, even before they flew in space. And so it was as the astronaut corps grew during the early 1960s—with a few gradual changes. Until 1963 all astronauts had to have test-pilot experience, but that requirement was dropped for the third astronaut selection. Then, in 1965, a handful of professional scientists was added, the first scientist-astronauts. And beginning in 1978, when **NASA** made the first selection of astronauts specifically for the space shuttle program, women and minorities were included for the first time.

Today, NASA has more than 100 active astronauts. They are no longer as famous as they were during the early days of space exploration. But there is nothing routine about what they do, and astronauts still represent the cutting edge of human experience.

Project Mercury

Created in 1958, Project Mercury's purpose was simply stated: Send Americans on the first flights into space. In 1959 seven young pilots—the nation's first **astronauts**—were chosen to fly the Mercury missions. Their small **spacecraft**—sometimes called a "capsule"—was shaped like a Styrofoam coffee cup measuring about 2 meters in length and about 1.8 meters across at its base. The capsule's exterior was covered in corrugated nickel-alloy plates for strength and heat resistance. To protect the astronaut from the fiery heat of **reentry** into the atmosphere, **NASA** spacecraft designer Max Faget gave the craft a blunt, gently curved **heat shield** coated with a resin material designed to char and burn off, carrying heat away from the craft. In case of an emergency during launch, Faget put a 4.5-meter escape **rocket** atop the nose of the craft to carry the capsule away from the booster. Attached to the heat shield was a package of solid-fuel **retro-rockets** that would slow the orbiting craft for its descent to Earth.

Inside the capsule the astronaut sat in a special formfitting couch designed to help him tolerate the g-forces of **launch** and reentry. In front of him was an instrument panel on which he could monitor the craft's systems and a viewing port for the craft's periscope. Although Mercury was designed to be flown by autopilot, it could also be controlled by the astronaut, using a control stick similar to a jet fighter's, to fire small jets on the capsule's exterior. The tiny cabin, which also contained systems for communications, environmental

PROJECT MERCURY

Spacecraft	Launch Date	Number of Earth Orbits	Flight Length	Notable Flight Results	Crew
MR3	5 May 1961	0	15 min 28 s	1st manned suborbital flight	Shepard
MR4	21 Jul 1961	0	15 min 37 s	2nd manned suborbital flight; spacecraft lost in recovery	Grissom
MA6	20 Feb 1962	3	4 h 55 min 23 s	1st manned orbital flight	Glenn
MA7	24 May 1962	3	4 h 56 min 5 s	2nd manned orbital flight	Carpenter
MA8	3 Oct 1962	6	9 h 13 min 11 s	Reaction control and frequency antenna modifications	Schirra
MA9	15 May 1963	22	34 h 19 min 49 s	1st launch of satellite while in orbit; 1st exclusively manual reentry	Cooper

U.S

control, and other functions, had barely enough room for its single occupant. Indeed, one astronaut quipped, "You don't get into Mercury; you put it on."

Two different rockets, each created for the nation's **ballistic missile** program, were used for Mercury missions—the Redstone rocket, employed for the initial, suborbital flights, and the more powerful Atlas, used for orbital missions.

Although NASA had hoped one of the seven Mercury astronauts would became the first human to fly in space, that did not happen. **Alan Shepard** made the first Mercury flight in May 1961, just a few weeks after Soviet cosmonaut **Yuri Gagarin** had already claimed that honor. Shepard's 15-minute suborbital flight was followed in July 1961 by a second suborbital mission, piloted by Gus Grissom. Then, in February 1962, it was time for Mercury's main goal: an orbital flight. After several delays, **John Glenn** rode his Mercury spacecraft into orbit atop an Atlas booster on February 20, 1962. Three more astronauts followed Glenn into orbit: Scott Carpenter made a three-orbit flight in May 1962. Wally Schirra orbited six times in October 1962. And Gordon Cooper stayed aloft for more than a full day in May 1963, making 22 orbits for the final Mercury mission.

To be sure, Mercury's successes did not eclipse the gains made by the Soviet Union, whose cosmonauts made much longer flights. But Mercury showed NASA that its astronauts could survive the rigors of flying in space and that its engineers could successfully tackle the challenges of human spaceflight. Even as Mercury ended, NASA had its sights on far more ambitious goals in space, including sending humans to the **moon.**

Alan Shepard

America's first man in space was a lanky New Englander named Alan Bartlett Shepard, Jr. Born on November 18, 1923, in East Derry, New Hampshire, Shepard grew up with a love of flying, and became a Navy test pilot. In April 1959 he was selected as one of the original seven **astronauts** for **Project Mercury.** At **NASA** he was known as much for his arrogant self-confidence as for his piloting skills. As pilot of the first Mercury mission, on May 5, 1961, Shepard made a 15-minute suborbital flight in the **spacecraft** Freedom 7, propelled by a Redstone booster. Freedom 7 reached an **altitude** of 187 kilometers before splashing down in the Atlantic. Grounded by an inner ear ailment from 1963 to 1969, Shepard waited a decade for his second spaceflight. In 1971, as commander of the **Apollo 14** mission, he piloted the third lunar landing and became the fifth man to walk on the moon. After retiring from NASA and the Navy in 1974,

The nation's first astronauts, selected for Project Mercury and dubbed the "Original 7," pose next to one of the F-106 Delta Dart jets they flew during training. From left: Scott Carpenter, Gordon Cooper, John Glenn, Virgil "Gus" Grissom, Walter Schirra, Alan Shepard, and Donald "Deke" Slayton. Of the seven, only Slayton did not fly a Mercury mission; grounded by doctors due to an irregular heartbeat, he made his only spaceflight in 1975, some 16 years after his selection as an astronaut.

Shepard was successful in a number of private business activities and created the Mercury Seven Foundation to provide scholarships for college science students. He died of leukemia on July 21, 1998.

John Glenn

John Herschel Glenn, Jr., the first American to orbit the **Earth,** was born July 18, 1921, in Cambridge, Ohio. After serving as a Marine fighter pilot in World War II and Korea, Glenn became a test pilot. Among his accomplishments was setting a transcontinental speed record in 1957. In 1959 he was selected as one of the seven Mercury **astronauts.** He was known not only for his technical skills but also for his charismatic charm.

On February 20, 1962, John Glenn and his Friendship 7 **spacecraft** were launched by an Atlas booster on a three-orbit mission. The flight had its harrowing moments. Before Glenn made his **reentry** into the **Earth's atmosphere,** ground controllers received an indication by **telemetry** that Friendship 7's **heat shield** might be loose. Fortunately, the signal turned out to be erroneous, and Glenn returned safely from his five-hour voyage.

After Glenn's astronaut career ended in 1964, he went on to become a corporate executive, then was elected to the U.S. Senate in 1974; he served there for 25 years. In 1998, at the age of 77, Glenn made a second spaceflight as a **payload** specialist on the **space shuttle** *Discovery.* Glenn's activities during the nine-day Earth orbit mission, designated STS-95, centered on experiments that were designed to investigate how the aging process affected his body's responses to flying in space.

Project Gemini

President Kennedy's 1961 call to land a man on the **moon** before the end of the decade confronted **NASA** with some enormous challenges. It did not take long for planners to realize that they could not move directly from the relatively simple **Project Mercury** missions to the complexities of a lunar landing.

No one could be certain **astronauts** could survive a two-week lunar voyage. Walking on the moon meant working in the vacuum of space, something no one had done before. And lunar crews would need to rendezvous with another **spacecraft** in orbit, which meant mastering the intricacies of orbital mechanics. To solve these unknowns, NASA created Project Gemini, named for the **constellation** representing the mythical twins Castor and Pollux.

The two-man Gemini spacecraft consisted of two components: a **reentry** module housing the astronauts, and an adapter module. The adapter contained a **retrorocket** section and an equipment section. The latter housed the electrical system, a set of maneuvering thrusters, and other components for use in orbit. On the outside, Gemini looked like a scaled-up version of Mercury, but it was actually far more sophisticated than its predecessor. For the first time Gemini astronauts were able to change their own orbit using the thrusters, aided in their maneuvers by an onboard computer. And when equipped with power-producing fuel cells instead of batteries, Gemini was capable of remaining in space for up to two weeks. To propel Gemini into space, NASA chose the Air Force's Titan 2 **ballistic missile.**

PROJECT GEMINI

Spacecraft	Launch Date	Number of Earth Orbits	Flight Length	Notable Flight Results	Crew
GT3	23 Mar 1965	3	4 h 52 min 31 s	1st 2-man orbital	Grissom, Young
GT4	3 Jun 1965	62	97 h 56 min 12 s	1st EVA	McDivitt, White
GT5	21 Aug 1965	120	190 h 55 min 14 s	Long flight duration	Cooper, Conrad
GT7	4 Dec 1965	206	330 h 35 min 01 s	Long duration; successful rendezvous	Borman, Lovell
GT6-A	15 Dec 1965	16	25 h 51 min 24 s	1st rendezvous in space (with GT7)	Schirra, Stafford
GT8	16 Mar 1966	7	10 h 41 min 26 s	Rendezvous and 1st docking	Armstrong, Scott
GT9-A	3 Jun 1966	45	72 h 20 min 50 s	Rendezvous; EVA	Stafford, Cernan
GT10	18 Jul 1966	43	70 h 46 min 39 s	Rendezvous; 2 EVA periods; docking	Young, Collins
GT11	12 Sep 1966	44	71 h 17 min 08 s	Record Gemini altitude of 1,190 km	Conrad, Gordon
GT12	11 Nov 1966	59	94 h 34 min 31 s	Aldrin performs total of 5.5 hours on EVA	Lovell, Aldrin

EVA: extravehicular activity

NOTABLE EARLY SOVIET SPACE FLIGHTS

Spacecraft	Launch Date	Number of Earth Orbits	Flight Length	Notable Flight Results	Crew
Vostok 1	12 Apr 1961	1	1 h 48 min 00 s	1st manned spaceflight; first orbital flight	Gagarin
Vostok 2	6 Aug 1961	17.5	25 h 18 min 00 s	Observations of the effects of weightlessness on crew	Titov
Vostok 6	16 Jun 1963	48	70 h 50 min 00 s	1st woman in space; successful tandem with Vostok 5	Tereshkova
Voskhod 1	12 Oct 1964	16	24 h 17 min 03 s	Medical observations made on effects of prolonged time in space on crew	Komarov, Yegorov, Feoktistov
Voskhod 2	18 Mar 1965	17	26 h 02 min 17 s	Successful EVA; miscalculated landing but crew was rescued	Belyayev, Leonov
Soyuz 1	23 Apr 1967	18	26 h 46 min 00 s	1st manned Soyuz flight; crashed upon reentry	Komarov
Soyuz 3	26 Oct 1968	64	94 h 51 min 00 s	Successful automatic and manual rendezvous with Soyuz 2 (unmanned)	Beregovoy

EVA: extravehicular activity

After two unmanned test flights, Gemini's piloted debut, Gemini 3, came in March 1965. Gus Grissom and John Young's three-orbit mission proved the craft spaceworthy, and racked up an important first, when the astronauts changed Gemini 3's orbit. During the four-day Gemini 4 mission, in June 1965, astronaut Ed White became the first American to walk in space. In August 1965 Gemini 5's Gordon Cooper and Pete Conrad spent a record-breaking eight days in space—a feat made more difficult due to the cramped conditions in the Gemini cabin. Then, in December 1965, Gemini 6's Wally Schirra and Tom Stafford accomplished Gemini's prime objective: They conducted the first space rendezvous, steering their craft to within 0.3 meters of Gemini 7. Meanwhile, aboard Gemini 7, Frank Borman and Jim Lovell chalked up a record 14 days in space, proving that astronauts could tolerate a round-trip lunar voyage.

The remaining Gemini missions had their share of crises and triumphs. In March 1966, after accomplishing the first space docking, Gemini 8's Neil Armstrong and Dave Scott narrowly escaped disaster when a faulty thruster on their spacecraft sent them tumbling wildly through space. On Gemini 9, 10, and 11, spacewalking proved the most difficult hurdle, as astronauts struggled to cope with the difficulties of working in a stiff, pressurized **space suit** in weightless conditions.

The Gemini program ended on a successful note. In September 1966 Gemini 11 accomplished the first rendezvous in a single orbit, and it made the first computer-controlled reentry. And in November 1966, Gemini 12's Buzz Aldrin showed that the spacewalking gremlins could be tamed during three separate excursions. Gemini gave NASA the experience—and the confidence—to reach for the moon.

Soyuz

A workhorse of the Soviet and Russian space programs, the Soyuz ("union") **spacecraft** was first flown in 1967. Although that first mission ended in disaster—a fouled parachute doomed cosmonaut Vladimir Komarov—the capabilities of the new craft were soon established. Roomier and more sophisticated than previous Soviet manned craft, Soyuz consists of a bullet-shaped **reentry** module in which as many as three cosmonauts ride during **launch** and landing, a cylindrical propulsion module, and an egg-shaped orbital module that provides a bit of additional space for activities in space.

During the late 1960s, Soviet space planners hoped to send cosmonauts to the **moon** using a Soyuz derivative called Zond. As it turned out, with the cancellation of the Soviet manned lunar effort, the primary role of Soyuz has been as a ferry craft to **space stations** in **Earth** orbit. A number of different versions of the Soyuz have been created, including the Progress unpiloted cargo ship used to transport supplies to the **International Space Station.** The current manned version of the craft, designated Soyuz TMA, can carry up to three cosmonauts, and can remain docked to the space station for more than six months.

Missions to the Moon

The Apollo Program

Project Apollo was created in response to President Kennedy's May 1961 challenge to land a man on the **moon** before the end of the decade. Although Apollo was a creation of the Cold War, its ultimate goal was peaceful: to expand the sphere of human exploration to **Earth**'s nearest celestial neighbor. At its peak, Apollo comprised some 400,000 people working at **NASA,** at aerospace contractors, and in laboratories around the nation and overseas.

Buzz Aldrin, who made history's first moonwalk with Neil Armstrong on July 20, 1969, photographed the imprint of his own boot on the powdery surface of the moon's Sea of Tranquillity. While this image has become a visual icon of the space age, Aldrin took the photo to provide scientists with information about the mechanical properties of lunar soil.

One of Apollo's first and most important challenges was choosing the best method to get to the moon. After much debate, engineers decided on a technique called lunar orbit rendezvous, which involved the creation of two separate **spacecraft,** of which one would ferry three **astronauts** to and from the moon while the other landed two of them on the lunar surface. To launch these two craft and send them on a path to the moon, NASA built the **Saturn V,** the largest rocket ever successfully flown. The Saturn booster plus the Apollo spacecraft contained millions of components, each of which required extensive work on design, fabrication, and preflight testing. Apollo was a massive engineering effort on an unprecedented scale.

During a total of 11 manned Apollo missions between 1968 and 1972, astronauts surpassed meeting Kennedy's challenge. Not one, but six teams of astronauts landed successfully on the moon, carrying out scientific explorations of their landing sites, setting up instruments that sent data back to Earth, and collecting hundreds of pounds of samples of lunar dust and rocks. Their orbiting companions circled the moon, took thousands of photographs of its surface, and conducted additional observations and experiments. The final three lunar landing missions were canceled as NASA struggled to cope with declining budgets and shifting priorities.

Apollo 1

On January 27, 1967, the crew intended for the first manned Apollo flight—veteran **astronauts** Gus Grissom and Ed White and rookie Roger Chaffee—took part in a prelaunch rehearsal for their two-week mission, which was slated for mid-February. The three astronauts were sealed inside their command module, which was pressurized with pure oxygen at 1.125 kilograms a square centimeter, slightly above atmospheric pressure. At about 6:30 p.m. EST, fire suddenly erupted inside Apollo 1. Within 20 seconds it spread throughout the cabin, creating such intense pressure that the spacecraft's hull ruptured. Ground crews struggled to open the craft's three-piece hatch. By the time they succeeded, many minutes had passed, and the astronauts were dead: They had been asphyxiated by toxic gases when their air hoses burned through. In the months that followed, **NASA** investigated the causes of the fire and redesigned the command module with fireproof materials and a new quick-opening hatch; high-pressure pure oxygen was no longer used during ground tests.

Apollo 8

In the summer of 1968, **NASA** made the bold decision to send the second manned Apollo flight, Apollo 8, on a voyage around the **moon.** The mission's crew—veterans Frank Borman and Jim Lovell and rookie Bill Anders—became the first astronauts to ride the **Saturn V** booster on December 21, 1968. About three hours after launch the men reignited the Saturn's third-stage engine to send Apollo 8 on a 66-hour voyage to the moon. Only Borman's brief bout with stomach flu marred the otherwise smooth moonward voyage. The view was awe-inspiring: As the first humans to leave **Earth**'s orbit, the astronauts saw their home planet shrink until they could cover it with an outstretched thumb.

In the early morning hours of December 24, the astronauts fired their service module's main **rocket** engine to go into lunar orbit. Borman, Lovell, and Anders made ten orbits, observing and photographing the moon's barren, pockmarked surface and testing navigation techniques that would be used on lunar landing missions.

During their 20 hours circling the moon, they also made two broadcasts to Earth using an onboard TV camera, including a reading from the book of Genesis that accompanied images of the stark moonscape. The astronauts fired their service module engine to return to Earth after midnight on Christmas Day, and splashed down in the Pacific on the morning of December 27.

Apollo 8's success gave NASA much needed experience to move ahead with the lunar landing attempt, and scored a historic first in the history of exploration.

Apollo 11

A dream of centuries came true on July 20, 1969, when Apollo 11 **astronauts** Neil Armstrong and Buzz Aldrin, piloting the **lunar module** *Eagle,* made history's first landing on the **moon.** The feat was the culmination of a decade of planning and preparation by hundreds of thousands of people working on Project Apollo. Apollo 11 met the challenge set forth by President John F. Kennedy in 1961—the goal "of landing a man on the moon and returning him safely to the Earth" before the end of the decade.

APOLLO MISSIONS

Mission	Launch Date	Flight Length	Notable Flight Results	Crew
Apollo 1	27 Jan 1967	no flight	Fire kills crew in preflight test	Grissom, White, Chaffee
Apollo 7	11 Oct 1968	260 h 09 min 03 s	1st manned Apollo flight; successful rendezvous with Saturn IVB	Schirra, Eisele, Cunningham
Apollo 8	21 Dec 1968	147 h 00 min 42 s	1st manned lunar orbit and launch of Saturn V; communications/tracking demonstrations	Borman, Lovell, Anders
Apollo 9	3 Mar 1969	241 h 00 min 53 s	Intervehicular and EVA crew transfer; rendezvous, docking, and propulsion demonstrations (LM)	McDivitt, Scott, Schweickart
Apollo 10	18 May 1969	192 h 03 min 23 s	1st lunar orbital mission with complete Apollo craft; LM rendezvous and CM docking in lunar gravitational field	Stafford, Young, Cernan
Apollo 11	16 Jul 1969	195 h 18 min 35 s	1st manned lunar landing	Armstrong, Collins, Aldrin
Apollo 12	14 Nov 1969	244 h 36 min 4 s	Successful lunar landing, exploration, and experiments in Ocean of Storms area	Conrad, Gordon, Bean
Apollo 13	11 Apr 1970	142 h 54 min 41 s	Mission aborted due to fire causing oxygen loss; transfer to CM for reentry	Lovell, Swigert, Haise
Apollo 14	31 Jan 1971	216 h 01 min 58 s	Successful lunar landing; general success with photography, experiments, and exploration with only minor problems	Shepard, Roosa, Mitchell
Apollo 15	26 Jul 1971	295 h 11 min 53 s	Use of lunar rover, three lunar EVA periods, one transearth coast EVA; successful surveying and sampling of Hadley-Apennine region	Scott, Worden, Irwin
Apollo 16	16 Apr 1972	265 h 51 min 05 s	Successful surveying and sampling of Descartes region; photography and experiments on moon and in flight	Young, Mattingly, Duke
Apollo 17	7 Dec 1972	301 h 51 min 59 s	Last lunar Apollo flight; successful surveying and sampling of Taurus-Littrow region; experiments and photography both on moon and in flight	Cernan, Evans, Schmitt

EVA: extravehicular activity; LM: lunar module; CM: command module

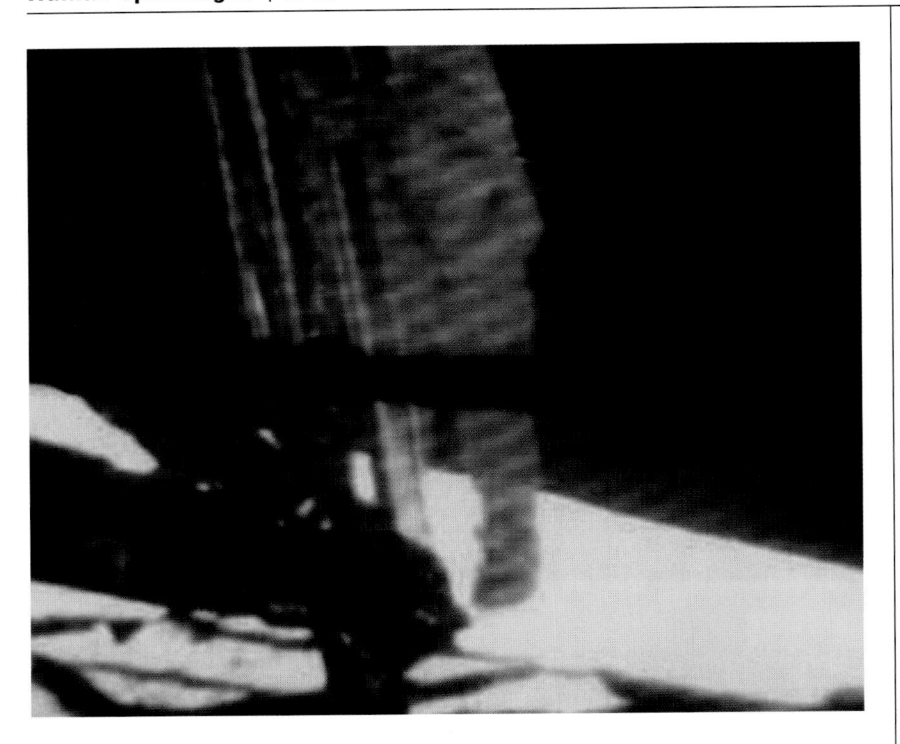

Neil Armstrong descends the ladder of the lunar module *Eagle* moments before becoming the first human to set foot on the moon. The event, broadcast to Earth by a television camera mounted on the side of the lander, was seen or heard by an audience estimated at 600 million people, one-fifth of the world's population.

The astronauts of Apollo 11 were all space veterans. Neil Armstrong, born August 5, 1930, had grown up in Wapakoneta, Ohio. A Navy fighter pilot in Korea, Armstrong went on to become a test pilot for the National Advisory Committee for Aeronautics, which later became **NASA.** Armstrong flew the X-15 rocket plane to the edge of space before being selected as an astronaut in 1962. As commander of the Gemini 8 mission in 1966, he piloted the first space docking, and narrowly averted disaster when a malfunctioning thruster sent his craft tumbling wildly through space. Armstrong was known for his coolness under pressure, his keen intellect, his dry wit, and his love of privacy.

Buzz Aldrin, from Montclair, New Jersey, was born on January 20, 1930. A graduate of West Point, he became an Air Force fighter pilot and also flew missions in Korea. Before being selected as an astronaut in 1963, he had earned a doctorate from MIT, doing his dissertation on piloting techniques for space rendezvous. During the Gemini program, Aldrin not only helped plan Gemini missions, but also copiloted the program's final flight, Gemini 12, in 1966, when he conducted three space walks. Known for a cerebral manner and an almost fanatical focus on technical areas of interest, Aldrin brought a wealth of experience to the Apollo 11 crew.

The third crewman, Mike Collins, would pilot the **command module** *Columbia.* Born on October 31, 1930, in Rome, Italy, Collins grew up in a military family and graduated from West Point. As an Air Force test pilot, Collins flew a variety of fighters at Edwards Air Force Base before his selection as an astronaut in 1963. As copilot on the Gemini 10 mission in 1966, Collins made two space walks and participated in two separate space rendezvous. Collins was known as much for his affable personality as for his skill as an astronaut.

On July 16, 1969, Armstrong, Aldrin, and Collins lifted off atop a **Saturn V** booster. Their trip to the moon was flawless, and on July 19 the men entered lunar orbit. The following day, Armstrong and Aldrin transferred to *Eagle*, leaving Collins alone in *Columbia*, and began their descent to the moon. On the way down, *Eagle*'s onboard computer became overloaded and threatened to abort the mission, but with help from Mission Control, Armstrong and Aldrin were able to keep going. Taking control of *Eagle*, Armstrong had to avoid a boulder-strewn crater the size of a football field before coming to rest on a level plain in the moon's Sea of Tranquillity at 4:17 p.m. eastern daylight time. At the moment of touchdown, there were just 20 seconds of fuel left before Armstrong and Aldrin would have been forced to abort the landing.

More than six hours later, Armstrong emerged from *Eagle* and descended the ladder on its front landing leg and, at 10:56 p.m., stepped onto the moon's powdery surface, saying, "That's one small step for man, one giant leap for mankind." Twenty minutes later Aldrin followed him to the surface, and the astronauts spent the next two hours exploring their landing site, taking pictures, and collecting samples of rock and dust for return to Earth. They also planted an American flag, and took a congratulatory phone call from President Richard Nixon. Meanwhile, in lunar orbit, Mike Collins kept a solo vigil, tending *Columbia*'s systems and waiting for his crewmates' safe return.

On July 21 Armstrong and Aldrin fired *Eagle*'s ascent **rocket** to lift off the moon and rejoin Collins. Later that day, inside *Columbia*, the men fired their **service module**'s main engine to leave lunar orbit and return to Earth. Splashdown came in the Pacific on July 24. Due to concerns that the men might harbor "moon germs"—concerns that proved groundless—Armstrong, Aldrin, and Collins were quarantined until August 10; they emerged as international heroes. For all three, Apollo 11 was their last space mission.

Apollo 13

Slated to be **NASA**'s third lunar landing attempt, Apollo 13 was launched on April 11, 1970, on a mission of scientific exploration. But that objective had to be abandoned after an oxygen tank aboard the Apollo 13 service module exploded almost 56 hours into the mission, when Apollo 13 was about 330,000 kilometers from **Earth.** The explosion, which knocked out the **service module**'s power-producing **fuel cells,** crippled the **command module. Astronauts** Jim Lovell, Jack Swigert, and Fred Haise suddenly found themselves in a life-or-death crisis. Without the service module's powerful engine, they could not return directly to Earth. They would have to continue toward the **moon,** then loop around it and begin the trip home. Their attached lunar module would serve as a lifeboat, with rocket engine, oxygen, and electrical power.

During the next four days, the astronauts battled one crisis after another. When levels of carbon dioxide in the cabin atmosphere began to climb, the men had to fashion an air scrubber using components that were not originally designed for the job. Fortunately, in this and other emergencies, they had help thanks to the heroic efforts of experts in Mission Control and at Apollo contractors around the country. One thing Mission Control could not alleviate, however, was the frigid temperature inside the **spacecraft.** Most of the electronic components had been turned off to save power, and thus did not provide the usual amount of heat. By the time the astronauts returned to Earth, on April 17, they were dehydrated and chilled to the bone, and Haise was suffering from a urinary tract infection. The men recovered from their ordeal, and NASA was able to determine the causes of the explosion and prevent it from recurring on future missions.

Apollo Explorations (Apollo 12, 14, 15, 16, 17)

In the wake of **Apollo 11**'s lunar landing, **NASA** moved to build on this success with more ambitious missions. During Apollo 12, in November 1969, Pete Conrad and Alan Bean landed on the **moon**'s Ocean of Storms, making two moonwalks while Dick Gordon circled overhead. The highlight of the mission was a visit to the unmanned Surveyor 3 probe, which had been on the moon since 1967. Apollo 12 proved the capability of landing at a pre-chosen spot on the moon's surface.

In February 1971, during the Apollo 14 mission, **Alan Shepard** and Ed Mitchell explored the Fra Mauro highlands, collecting the oldest lunar samples yet found, while Stu Roosa, in lunar orbit, took photographs and made scientific observations.

Scientific exploration of the moon reached a new level with the Apollo 15 mission in July and August 1971. Dave Scott and Jim Irwin made three separate excursions, each lasting up to seven hours, and were the first **astronauts** to use the **lunar rover** in their explorations. Collecting samples from the Apennine Mountains and adjacent plains, they discovered rocks dating back 4.5 billion years, almost as old as the moon itself. In lunar orbit, Al Worden gathered data with a new, sophisticated array of high-powered cameras and scientific instruments developed for the final Apollo missions.

Apollo 16, the first mission to the moon's central highlands, featured three moonwalks by John Young and Charlie Duke and the second use of the lunar rover. While Young and Duke explored the surface, Ken Mattingly surveyed the moon from orbit.

In December 1972, the final moon landing, Apollo 17, began. While Ron Evans remained in orbit, Gene Cernan and scientist-astronaut Jack Schmitt—the first professional scientist to reach the moon—touched down in the Taurus-Littrow valley. Their three moonwalks also featured excursions in the lunar rover. Among the mission's most exciting moments was Schmitt's discovery of orange soil, which scientists later determined had been formed by fountains of molten lava that erupted 3.5 billion years ago. Apollo 17's splashdown on December 19, 1972, concluded Apollo's lunar explorations. These first moon landings, though brief, had enormous scientific payoff: 382 kilograms of samples, thousands of photographs, and other scientific data transformed our understanding of both the moon and the evolution of the **solar system.** And Apollo showed that human beings had developed the capability to become a spacefaring species.

Lunar Rover

A key asset on the final three Apollo **moon** landings, the battery-powered lunar rover gave **astronauts** the ability to roam for miles across the lunar surface. Riding on wire-mesh wheels, the rover featured its own systems for navigation and communications and offered storage space for geologic tools, lunar samples, and other gear.

Command/Service Module (CSM)

History's first manned moonship, the **Apollo** command and service modules, was created to take astronauts to and from lunar orbit. The cone-shaped command module contained the **astronauts'** crew cabin, including the instruments used during the journey. Compared to previous U.S. **spacecraft,** the module cabin offered a bit more room and included a lower

equipment bay with navigation equipment, a food pantry, and storage areas. The command module's base was covered by the **heat shield** for **reentry,** and in its nose were stored the parachutes for landing.

Attached to the base of the command module was the cylindrical service module, which contained the large **rocket** engine, and the supply of fuel and oxidizer used to get the spacecraft into and out of lunar orbit. The service module also contained the **fuel cells** that provided electrical power, a set of maneuvering thrusters, and the communications antenna used for radio transmissions to **Earth.** The service module was jettisoned by the astronauts shortly before they reentered the **Earth's atmosphere;** the command module was the only portion of the Apollo spacecraft to return to Earth. In addition to their roles in the **moon** program, the Apollo command and service modules took three teams of astronauts to and from the Skylab space station in 1973 and '74, and aided in the international spaceflight of the **Apollo-Soyuz Test Project** in 1975.

Lunar Module (LM)

The craft that made the first manned lunar landings, the Apollo lunar module, has been called the first true spaceship, because it flew only in the blackness of space. The lunar module, whose mission was to carry two **astronauts** to and from the **moon**'s surface, was designed to be as light as possible. It had no need for aerodynamic streamlining, and the LM (pronounced "lem") looked like a giant mechanical spider.

The lunar module was composed of two sections. A boxy, eight-sided descent stage housed a **rocket** engine for the trip down to the moon. It included landing legs, as well as equipment the astronauts would use during their lunar explorations. Atop the descent stage sat the ascent stage, containing the crew cabin. Inside, the two astronauts stood side by side before an instrument panel and two triangular windows; because the LM flew only in **weightlessness** or in the moon's one-sixth gravity, there was no need for seats—a savings of hundreds of pounds. The ascent stage also contained the rocket engine used to take the astronauts from the surface of the moon back into lunar orbit.

During the **Apollo 13** mission, the lunar module took on a new role—that of a lifeboat. After their **service module** was crippled by an explosion, Jim Lovell and his crew used their attached lunar module's rocket engine, oxygen supply, and equipment to safely return to **Earth.**

High on the slopes of a lunar mountain, Jack Schmitt inspects a house-size boulder during Apollo 17's final moonwalk on December 13, 1972. During their three excursions, which marked the last time humans walked on the moon during the 20th century, Schmitt and mission commander Gene Cernan spent more than 21 hours exploring the moon's Taurus-Littrow valley, aided by a battery-powered lunar rover, which is visible to the right of the boulder.

International Space Programs

Apollo-Soyuz Test Project

Born of détente between the United States and the Soviet Union, the Apollo-Soyuz Test Project was the first international space mission. On July 15, 1975, a **Soyuz** spacecraft carrying cosmonauts Alexei Leonov and Valery Kubasov lifted off from the Baikonur launch center in central Asia. Hours later **Apollo astronauts** Tom Stafford, Vance Brand, and Donald Slayton were launched from the **Kennedy Space Center.** On July 17 Stafford and his crew rendezvoused with their Soviet colleagues and, using a specially designed docking system, the two craft were joined in orbit. For the next two days the astronauts and cosmonauts visited each other's **spacecraft,** exchanged gifts, and made television broadcasts to **Earth.** Soyuz landed in what is now Kazakhstan on July 21, and Apollo splashed down in the Pacific on July 24. The success of the mission paved the way for future joint operations, which resumed some 20 years later with U.S. missions to the Russian **Mir space station.**

During the first international spaceflight, the Apollo-Soyuz Test Project, on July 17, 1975, U.S. astronaut Tom Stafford (left) looks out from the U.S.-built docking module and Soviet cosmonaut Alexei Leonov floats at the hatchway of the Soyuz spacecraft's orbital module. The crews learned each other's languages during training for the flight.

European Space Agency

Formed in 1975, the European Space Agency (ESA) today has 15 members: Austria, Belgium, Denmark, Finland, France, Germany, Ireland, Italy, the Netherlands, Norway, Portugal, Spain, Sweden, Switzerland, and the United Kingdom. ESA **astronauts** have participated in missions of the United States, the Soviet Union, and Russia. ESA's accomplishments include the development of the Ariane series of **launch vehicles,** which have become dominant in the world of commercial space activities. ESA also maintains a program of robotic space exploration and has participated in a number of **solar system** missions. In 2003 the ESA budget was 2.7 billion euros, the equivalent of 3.1 billion U.S. dollars.

Guest Cosmonauts & Astronauts

Until the mid-1970s, only two countries were represented among the world's space travelers, the United States and the Soviet Union. In 1976 the Soviets selected "guest cosmonauts" to visit its **Salyut space stations.** Under the so-called Intercosmos program, guest cosmonauts from Cuba, Vietnam, Romania, and other Socialist nations, as well as France, were included on visiting crews, who spent about a week in space. These missions delivered new supplies to the stations and relieved the isolation of the Salyut crews during their months-long space marathons. International participation continued during the Mir program.

Meanwhile, in the U.S. space program, the **European Space Agency** was allowed to include astronauts on **space shuttle** flights, beginning with the first Spacelab mission in 1983. On that flight West Germany's Ulf Merbold became the first non-American to fly aboard a U.S. spacecraft. During the 1980s representatives of Canada, France, Saudi Arabia, the Netherlands, and Mexico also flew aboard the shuttle. Today, astronauts from Canada, Europe, and Japan are part of **NASA**'s astronaut corps, taking part in shuttle missions and visits to the **International Space Station.**

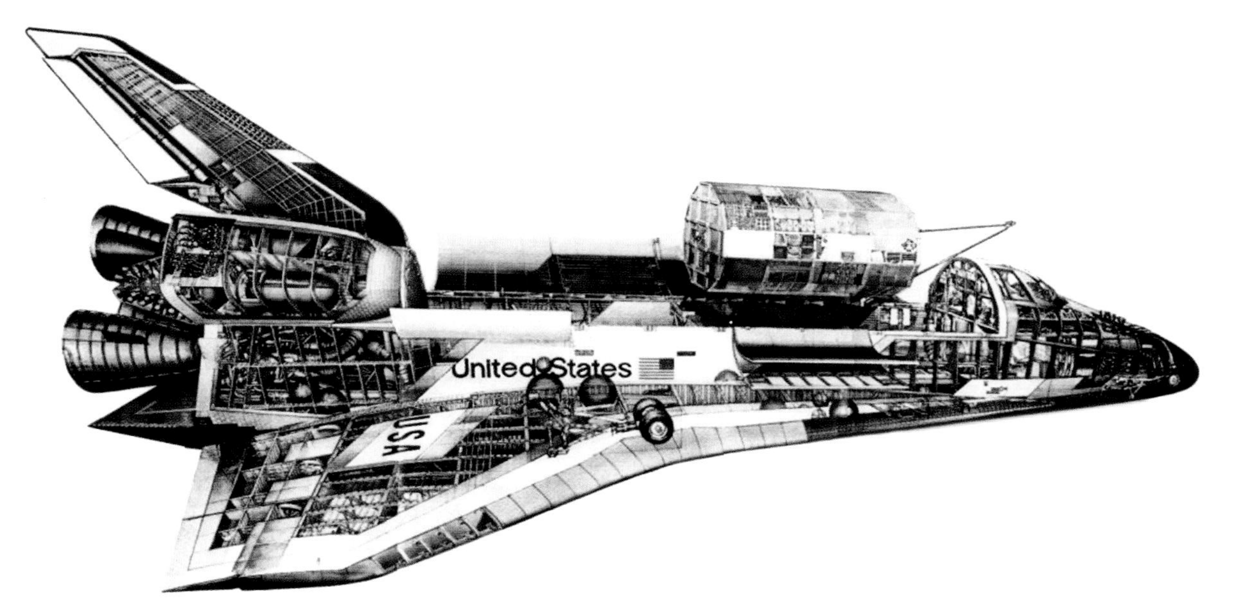

A cutaway view reveals the most complex flying machine ever created, the space shuttle orbiter. The first reusable spacecraft, the orbiter was designed to perform a wide variety of missions, including deploying and retrieving satellites, scientific and medical research, military reconnaissance, and space construction projects. Equipped with a robotic arm for manipulating large payloads, the orbiter can carry payloads of up to 25.4 metric tons inside its 18-meter-long cargo bay.

Space Shuttles

Space Shuttle

At **NASA** the dream of building a reusable spaceship began during the 1960s, even before the first **Apollo** moon landings. Apollo, like Mercury and Gemini before it, was a throwaway vehicle that ended up in museums after a single flight. Planners hoped that a new, reusable transport would make spaceflight economical and routine. Initially they envisioned the space shuttle as one element in a space infrastructure that would include a permanent **space station** in **Earth** orbit. But the Nixon administration approved only the shuttle, not the station. And by 1972, the year Congress gave the go-ahead to the shuttle program, budget restrictions had led NASA to compromise on the shuttle's design; in order to cut development costs, the craft became a mix of reusable and throwaway elements.

The centerpiece of the shuttle is the orbiter, a winged **spacecraft** as big as a mid-size airliner. About 37 meters long, with a wingspan of 24 meters, the orbiter includes the crew cabin, which can accommodate up to eight **astronauts;** an 18-meter-long cargo bay; and three liquid-fuel **rocket** engines used during launch, as well as a number of additional engines for maneuvering in space and returning to Earth. The secret of the orbiter's reusability is its thermal protection system, consisting of thousands of silica tiles, carbon panels, and other materials attached to the skin of the orbiter to protect the craft from the intense heat of **reentry.** During reentry and the final descent through the atmosphere, the orbiter has no propulsion system. It flies as an unpowered glider to a landing on a specially created runway at the **Kennedy Space Center.** There have been five orbiters: ***Columbia, Challenger,*** *Discovery, Endeavour,* and *Atlantis.*

For launch, the orbiter is mated to an external fuel tank some 47 meters long and some 8 meters in diameter. The tank supplies liquid hydrogen and liquid oxygen to the orbiter's main engines, which together generate more than 5,338,000 newtons of thrust at liftoff. In addition two solid rocket boosters, each providing 14,680,000 newtons of thrust, are attached to the external tank. They fire during the first two minutes of the flight, then separate and fall into the ocean, where they are recovered for reuse. Some eight minutes after **liftoff,** just before the shuttle reaches orbit, its main engines shut down and the external tank is jettisoned and destroyed during atmosphere reentry.

Since its first flight in April 1981, the space shuttle has performed a variety of missions. Shuttle crews have deployed **satellites** from the orbiter's cargo bay and have retrieved satellites for repair or return to Earth. The shuttle has been used as a scientific laboratory to study the effects of **weightlessness** on people and other living things and to investigate the possible commercial uses of space. It has also served as an observation platform for studying the Earth, the **sun,** and the universe. Shuttle flights have included secret missions for the Department of Defense, as well as visits to Russia's **Mir** space station. And the shuttle has been essential in the construction of the **International Space Station.**

Although the shuttle has not succeeded in reducing the cost of sending humans into space, it has proven itself to be an extraordinarily capable and versatile spacecraft. But it is also exceedingly complex and requires extensive maintenance between missions. The loss of two shuttles and their crews (*Challenger* in 1986, and *Columbia* in 2003), has underscored the risks of flying the shuttle. NASA plans to retire the shuttle fleet in 2010 following the targeted completion of the International Space Station.

In one of history's worst spaceflight disasters, the space shuttle *Challenger* explodes 73 seconds after launch on January 28, 1986; the shuttle's two solid rocket boosters careen away in twisted contrails. All seven crew members, including New Hampshire schoolteacher Christa McAuliffe, died. The cause of the disaster was traced to a faulty O-ring seal in one of the solid rocket boosters.

Sally K. Ride

Dr. Sally K. Ride, the first American woman in space, was born May 26, 1951, in Los Angeles, California. In college, science became her career choice, and by the time she earned her doctorate in physics from Stanford University in 1978, she wanted to be an **astronaut.** That year she was selected as one of 35 new astronauts for the **space shuttle** program. On June 18, 1983, Ride became the nation's first woman in space when she and her four crewmates were launched aboard the orbiter ***Challenger*** on the seventh shuttle flight. The mission included the first use of the shuttle's remote manipulator arm, which Ride operated and had helped design. Ride returned to space in October 1984, again aboard *Challenger,* for the eight-day mission STS-41-G. On that flight, Ride's crewmate Dr. Kathy Sullivan became the first American woman to walk in space.

Ride served on the presidential commission to investigate the 1986 *Challenger* disaster and was the first director of **NASA**'s Office of Exploration. Since 1989 she has been a physics professor at the University of California, San Diego, where she is director of the California Space Institute. She created and directed EarthKAM, an internet-based NASA project that lets middle-school students take pictures of the Earth from space and then download them over the Web.

Challenger

On January 28, 1986, the **space shuttle** *Challenger* was poised for **liftoff** with a seven-member crew that included, for the first time, a private citizen: teacher Christa McAuliffe. As *Challenger* lifted off and soared into a clear winter sky, all seemed well—until 73 seconds after liftoff, when the shuttle was engulfed by a tremendous explosion. The **spacecraft** was destroyed, and its crew perished.

During months of investigations, a presidential commission determined that the primary cause of the accident lay in one of the shuttle's twin solid rocket boosters, or SRBs. Each SRB is composed of several segments joined together, and the joints are sealed by large O-rings made of a rubberlike material. Because of unusually cold weather on **launch** day, one of the O-rings in the right-hand SRB lost its resiliency and failed to seal properly. Flame from the burning fuel inside the SRB leaked out and struck the side of the attached external fuel tank. By 73 seconds after launch, the tank structure had been so weakened that massive amounts of liquid hydrogen and liquid oxygen escaped, causing a huge explosion that destroyed the orbiter.

As **NASA** struggled to recover from the tragedy, it was mindful of the commission's finding that flaws in the agency's decision-making process, specifically with regard to flight safety, had helped cause the accident. The shuttle did not resume flying until September 1988.

Columbia

On February 1, 2003, the crew of the **space shuttle** *Columbia* prepared to end a 16-day mission. Aboard were mission commander Rick Husband; pilot William McCool; mission specialists Michael Anderson, Kalpana Chawla, David Brown, Laurel Clark; and **payload** specialist Ilan Ramone, Israel's first **astronaut.** As the time for *Columbia*'s **touchdown** passed, it became clear that something had gone terribly wrong. The orbiter had disintegrated over the southwestern United States during **reentry,** killing all seven astronauts.

In the months that followed, searchers recovered more than 84,000 pieces of debris. These, along with data radioed to **Earth** from the **spacecraft** before its breakup, became clues in a months-long investigation. The accident board determined that one of *Columbia*'s heat-protective panels had been knocked loose during **launch,** after being struck by a piece of insulating foam from the shuttle's external fuel tank. During *Columbia*'s fiery reentry into the **Earth's atmosphere,** hot gases penetrated the wing structure, leading to the orbiter's breakup and destruction. In the wake of the *Columbia* tragedy, NASA grounded the shuttle fleet, and worked to develop means of preventing future shuttle disasters. At this writing, the shuttle's return to flight was scheduled for the spring of 2005.

Hubble Repair Mission

When the Hubble Space Telescope was launched aboard the **space shuttle** *Discovery* in 1990, astronomers had great hopes for the orbiting observatory. High above the blurring effects of the **Earth's atmosphere,** the **telescope** was designed to see the universe ten times more clearly than any existing ground-based instrument. The first images sent down from Hubble were fuzzy; scientists realized there was a serious flaw—the shape of the telescope's 2.4-meter main mirror was slightly wrong.

Hubble scientists and engineers devised an alternate to returning the telescope to **Earth** to fix the flawed mirror. Spacewalking shuttle astronauts would place corrective optics—actually sets of specially shaped mirrors—inside Hubble to refocus light from the main mirror, much like a set of eyeglasses, before the light entered the instruments. (One of the instruments, the Wide-Field Planetary Camera, would be completely replaced by a new version that already contained the corrective optics.)

The repair mission, designated STS-61, took place in December 1993 as the shuttle *Endeavour* lifted off with a seven-member crew. They had trained for two years for the repair mission, which would include some of the most difficult space walks ever attempted. After rendezvousing with Hubble, the astronauts grappled it with the orbiter's remote manipulator arm and secured it in the cargo bay. During five space walks, the astronauts succeeded in repairing Hubble and restoring the telescope to full capability. Since then, three more missions to Hubble have updated its instruments and replaced faulty components. An additional servicing mission, originally scheduled for 2006, was canceled in the wake of the ***Columbia*** tragedy.

SPACE SHUTTLES

Space Shuttle	First Launch	Last Launch	Number of Missions Flown	Notable Highlights
Enterprise	15 Feb 1977	26 Oct 1977	16	Space shuttle test vehicle; not capable of space flight; tested capacity and responses of craft in flight; allowed for development of shuttles equipped for space orbit
Columbia	12 Apr 1981	16 Jan 2003	28*	1st space shuttle to orbit earth; crew and shuttle lost during reentry of last mission; pioneer in technology and development of space shuttles
Challenger	4 Apr 1983	28 Jan 1986	9	2nd orbiter shuttle; modifications to enable shuttle to carry Centaur†, but no Centaur flight was attempted; shuttle and crew lost at launch of 10th mission
Discovery	30 Aug 1984	2010 projected	30	3rd orbiter shuttle; upgraded to carry Centaur but no Centaur flight was attempted due to the loss of the *Challenger* shuttle
Atlantis	3 Oct 1985	2010 projected	26	4th orbiter shuttle; 165 modifications made to improve performance based on 3 previous shuttles and *Enterprise*
Endeavour	7 May 1992	2010 projected	19	5th orbiter shuttle; equipped with new hardware to improve/expand shuttle capabilities; rescued stranded communications satellite during first flight

**Lost on reentry of 28th mission, 2/1/2003; †a high-energy upper stage rocket.*

A Profound New Perspective

Senator Jake Garn

In November 1984 NASA Administrator James Beggs invited the Chairmen of congressional committees with jurisdiction over NASA authorization and appropriations to make a flight aboard the space shuttle. As chairman of the Senate Subcommittee on Appropriations (which had responsibility for NASA funding) as well as being a former Navy and Air Force pilot with more than 10,000 hours of pilot time, I gladly accepted the invitation and was assigned to fly aboard the shuttle *Challenger* on flight STS-51E as a payload specialist, with the mission responsibility for conducting a series of physiological and medical experiments. Due to technical problems with the TDRSS Satellite to have been launched on that mission, Flight STS-51E was canceled. I and most of the crew were reassigned to flight STS-51D aboard the space shuttle *Discovery,* which was launched on April 12, 1985, and landed on April 19, 1985.

I am often asked what I remember most about my flight in space or what is my fondest memory aboard the space shuttle *Discovery.* It is always a difficult question, because that experience so deeply affected everything about me. It created a life-long bond between me and my incredibly capable and talented crewmates and a kinship with all those who have traveled into space before and since; it gave me a sense of the true majesty of creation as I looked down at the indescribable beauty of the Earth, and outward at the dark vastness of space; it deepened my spiritual convictions in the power of the Intelligence that must have formed and organized it all; and it gave me an inkling of the truth of the phrase, "whatever the mind of man can conceive and believe, it can achieve."

The last point was especially true when I remembered sitting with my Dad, who had been a World War I pilot, as we watched Neil Armstrong set foot on the moon. Dad had seemed almost overcome by the emotion of the moment. When I asked, he said he was just remembering that, as a young boy of ten, he had sat with his father, who told him of the Wright brothers' first flight, which had taken place just the day before. Imagine! In my father's lifetime the belief in the possibility of flight resulted in both that short 120-foot first flight to men going to the moon and back and space shuttles circling the globe. And in my case, though my Dad was not alive to see it, to this little boy, who had learned about flight in my Dad's dirt-floored hanger in Richfield, Utah, getting my pilot's license before my driver's license, and then looking down on the incredibly beautiful blue planet, far above the air that had carried the Wright Flyer, the military aircraft that I had piloted for over 10,000 hours at that point, and the large airliners that had evolved since that first flight in 1903. And the historical circle was completed for me personally during 2003, when I had the unforgettable opportunity of flying a replica of the Wright Flyer for some of the celebrations commemorating a hundred years of flight.

A Spatial Perspective

One of the many enduring impressions from my spaceflight was the fact that, while looking down at the Earth, it was impossible to see the national boundaries that so often divide the inhabitants of our planet into warring camps, disputing territorial and national claims to property or preference. While I am extremely proud and grateful to be an American and believe strongly in our nation's ideals and freedoms, that feeling was overwhelmed when I was in orbit by the sense of simply being a part of the population of our planet—a planet that I could see so clearly was its own huge spaceship, traveling through the vastness of space and carrying all of us along as passengers on its journey around the sun.

When I looked down on parts of the Earth where I knew there was conflict, I could see no visible evidence of that, and it struck me as not only sad but also almost sacrilegious that anyone of us should do any-

Viewed from space, the Middle East is but a patchwork of land and water—its political borders and strife erased as distance reveals a shared environment.

thing to mar the beauty and peace that I could see before me. I wondered at the time—and I have often wondered since—how different life would be on our home planet if everyone could see the world from that vantage point in orbit. Would people be less ready to start a war against their neighbors? Would they have second thoughts about doing things that pollute our environment or destroy the abundant greenery that contrasted so beautifully with the crystal blue oceans and seas that we passed over in our 90-minute orbits? I often thought that if the leaders of warring factions or nations could somehow be brought to orbit for what would be the ultimate in a "summit meeting," that peace and harmony could be restored far more quickly.

A Place in Space

Looking at the ever moving horizon as we raced along in orbit, I was struck not only by the incredible beauty of the atmosphere that protects our planet and provides for our very life as a species, but also by how thin and fragile Earth seemed from that vantage point. The view gave me a strong feeling that we need to explore our solar system for other places in which human beings might be able to live, whether on distant planets we could one day visit or on places like Mars that might be turned into a habitable environment, possibly for a second time.

Finally, looking outward at the unbelievable vastness of space, with a view unobstructed by the atmosphere or the diffused lights of our planet, I was struck with the absolute certainty that, in all the many galaxies and solar systems out there, there must be other planets like Earth, populated with other people, making us not only members of a single human race on our own planet, but members of a galactic and universal family as well.

My spaceflight experience gave me a sense of truly being "right-size" in a way that I could have only experienced by being there. ■

Aboard the International Space Station, members of the Expedition 2 crew work in the Destiny science laboratory on April 5, 2001. American astronaut Susan Helms tests the effects of weightlessness; Russian cosmonaut Yuri Usachev, who served as commander of the ISS during the 165-day mission, records notes.

Surviving Space

The Body in Space

If you could travel into space, you would notice that your body changed in some subtle but important ways. You would feel a strange sensation of fullness in your head, as if you were catching a cold. If you looked in a mirror, you would notice that your face looked puffy, but that your legs were skinnier than they had been on **Earth.** And you would notice your posture had changed: when you relaxed your muscles, your shoulders would hunch, your knees would bend, and your arms would hang in front of you. All of these things are results of exposure to **microgravity,** otherwise known as weightlessness, and they are just the most visible signs of the effects of spaceflight on the human body.

At **NASA** it took many years of experience for researchers to identify these effects. The first U.S. space missions were brief, with the longest **Mercury** flight lasting just 34 hours. And on both Mercury and **Gemini, astronauts** had no room to move around within their cramped **spacecraft.** Not until the **Apollo** flights, when crewmen could float freely, did the problem of space motion sickness become recognized.

But it was the missions to the **Skylab space station** that gave NASA its first detailed look at physiological changes caused by flying in space. It became clear that weightlessness has a number of effects. Bodily fluids, no longer held down by gravity, migrate to the upper body, causing congested sinuses, puffy faces, and skinny legs. Skylab astronauts engaged in regular, strenuous exercise to counteract the deterioration of muscles and bones caused by prolonged exposure to microgravity.

Today, exercise is still among the most important means of dealing with the biomedical effects of spaceflight. These exercises include running on a microgravity version of a treadmill, pedaling a stationary bicycle, and using resistive exercise equipment to give bones and muscles simulated exposure to gravity. In fact, U.S. astronauts aboard the **International Space Station** spend more time on exercise—some two hours each day—than on any other activity except eating and sleeping. Russian cosmonauts follow a different fitness regime, with the major emphasis placed on exercise during the final weeks of a space mission in order to prepare for return to Earth.

One of the most fascinating changes that takes place in weightlessness has to do with perception. With no apparent gravitational force to dictate up and down, astronauts find that their sense of orientation can vary dramatically. The terms "floor," "walls," and "ceiling" are arbitrary and depend upon how an astronaut is oriented as he or she enters a room.

The space environment presents a number of deadly hazards, from the lack of atmosphere, to extremes in temperature, to solar and cosmic radiation. Each of these must be considered by engineers designing **spacecraft, space suits,** and other equipment used by astronauts.

Since Skylab, data gathered on Soviet and Russian space station missions, U.S. shuttle flights, and the International Space Station have increased the understanding of these and other effects of spaceflight. But there is much work that needs to be done before astronauts can safely venture to Mars or other worlds in our solar system, and biomedical studies continue to be a major focus of space research.

Life-support Systems

Living in space would be impossible without systems for life support and environmental control. **Astronauts** must not only have oxygen to breathe but also some means of removing the carbon dioxide they exhale from the cabin atmosphere. There must be drinking water, and that water must be free of bacteria and other contaminants. And whether the astronaut is wearing coveralls inside the **spacecraft** cabin, or a **space suit** in the vacuum of space, he or she must be surrounded by an atmosphere at a comfortable pressure, temperature, and humidity.

Some of the technologies for maintaining a spacecraft's atmosphere have been in use since the early days of spaceflight. Canisters of lithium hydroxide to remove carbon dioxide and activated charcoal to filter out other impurities were developed for the **Mercury** missions. Other systems, developed for Russian **space station** missions and now used aboard the **International Space Station,** include perchlorate "candles" that are burned to generate breathing oxygen. Oxygen is also generated by the breakdown of water by electrolysis, powered by the station's solar panels, with the resulting hydrogen vented overboard.

Water is a precious commodity aboard any spacecraft, and the life-support systems designed for the International Space Station have the ability to recycle drinking water from urine, water used for hygiene, and even moisture from the astronauts' breath. Although this process sounds unappetizing, it actually produces water that is cleaner than the tap water commonly available for drinking on **Earth.**

Ultimately, engineers hope to develop a so-called closed-loop life-support system that is virtually self-sufficient for use on interplanetary voyages. In such a system, plants grown onboard would provide oxygen, help purify water, and even process waste materials. For now, however, closed-loop life support is a goal yet to be realized.

Microgravity (Weightlessness)

Microgravity is the condition experienced by anyone in a state of free fall. **Astronauts** and their **spacecraft** orbiting the **Earth** are actually falling around the planet, and gravity is still acting on them, but the momentum of their free fall moves them as fast as gravity would, almost canceling the effect of gravitational pull on them. From their perspective, they are floating within a motionless spacecraft. The more common term for microgravity, "weightlessness," reflects the fact that even though objects in microgravity still have mass, their weight—defined as mass multiplied by acceleration due to gravity—appears to be zero. Weightlessness can also be experienced in an airplane flying a **ballistic trajectory;** this is one method **NASA** uses to test equipment and techniques for spacewalking astronauts.

In space, astronauts remain weightless unless they fire a **rocket** engine, which causes the spacecraft to accelerate. This is true whether they are orbiting the Earth, the **moon,** or another planet; it is also true when they are traveling from one celestial body to another. Microgravity can be a delightful experience. Astronauts find that they can move effortlessly within their spacecraft, and they can handle large objects with little difficulty. They can sleep, quite literally, "on air." However, working in weightless conditions takes some getting used to, especially for astronauts in pressurized space suits during a space walk. Without the effects of gravity everything behaves as if it were in a three-dimensional ice-skating rink, obeying **Newton**'s Third Law of Motion: "For every action there is an equal and opposite reaction." Astronauts must learn to work carefully, using special tools and equipment. As part of their training, they practice their tasks in a giant water tank to simulate weightlessness.

One of the most troubling aspects of microgravity is its negative effects on the body. More than half of all space travelers experience a condition called **Space Adaptation Syndrome** (SAS), whose effects resemble those of motion sickness on Earth. The symptoms vary from person to person and can range from headaches and nausea to more severe illness such as vomiting. Fortunately, SAS usually clears up during the first few days of a space mission.

Training for a mission to the International Space Station, U.S. astronaut Ed Lu and Russian cosmonaut Yuri Malenchenko float in a giant pool at the Star City cosmonaut training center outside Moscow in August 1999. Under the watchful eyes of safety divers, Lu and Malenchenko, spacesuited and precisely weighted to simulate the neutral buoyancy of microgravity, practice tasks they will perform during a 6.5-hour space walk in September 2000.

During spaceflights of a week or more, weightlessness has a more serious impact. Muscles atrophy, the cardiovascular system weakens, and bones lose calcium. To counteract these effects, astronauts undergo regular periods of exercise.

Eventually, for voyages to **Mars** and other destinations in the **solar system,** astronauts may be able to avoid the negative effects of microgravity by doing away with it altogether. Planners envision spacecraft that slowly rotate in flight, creating enough centrifugal force to provide some or all of the gravitational pull experienced on Earth.

Space Adaptation Syndrome

Space Adaptation Syndrome (SAS) is the term used to describe the body's initial reaction to prolonged **weightlessness.** The most troubling component of SAS is space motion sickness, whose symptoms are similar to motion sickness on **Earth.** Scientists believe space motion sickness is caused when the brain receives conflicting information from sensory organs. An **astronaut** floating inside a **spacecraft** may see the surroundings as "right side up," but the inner ear will register the sensations of free fall, while a glance out the window may show the Earth's horizon at some odd angle. The resulting "sensory conflict" can make the astronaut sick. About 60 percent of all space travelers have experienced space motion sickness; fortunately it clears up on its own within a few days after **launch.** Other symptoms of SAS, caused by shifting of fluids from the lower extremities to the upper body, include fullness in the head and a puffy face; these symptoms may last for the entire duration of a space mission.

Space Radiation

Space planners have been worrying about the hazards of radiation since before the first piloted spaceflights. Intense bursts of very high-energy photons—x-rays and gamma rays—can be deadly to humans. But an even greater threat to space travelers comes from subatomic particles speeding through space. **Solar flares** emit protons that travel at high speeds and can penetrate the walls of a **spacecraft.** If they strike the body, they can do considerable damage to human cells. A major solar flare could kill astronauts who were not properly protected. Even more deadly are **cosmic rays,** heavy atomic nuclei that are sprayed through the cosmos by exploding stars. Heavier and moving much faster than solar flare particles, cosmic ray particles can do even more damage to human tissue and can also harm electronic components aboard a spacecraft.

Scientists are still uncertain how to cope with space radiation, which poses the greatest threat to astronauts who venture beyond **Earth** orbit into deep space. Research has shown that materials rich in hydrogen, including water and some plastics, could help shield space voyagers from radiation. These materials may be incorporated into the design of spacecraft developed to carry humans to **Mars** and other destinations in the **solar system.** On the surfaces of the **moon** and Mars, astronauts could cover their habitats with a layer of soil to screen out deadly particles. However, they will still need to be protected when they work outside, and this is a challenge scientists and engineers designing **space suits** and other equipment for deep-space missions must face.

Space Suit

The development of space suit technology goes back to the 1930s, before the advent of pressurized cabins, when aviators like American Wiley Post created pressure suits to wear during high-altitude airplane flights. When **NASA** needed space suits for **Project Mercury,** it used a modified version of a pressure suit developed by the U.S. Navy for pilots of high-altitude jets. The suit consisted of an inner layer of Neoprene-coated nylon fabric to act as a pressure bladder, and an outer layer of aluminized nylon to keep the pressure bladder from ballooning in the vacuum of space. When pressurized, the suit was quite stiff, making it difficult for an **astronaut** to move his arms or legs. The Mercury space suit was intended only as a backup in case the **spacecraft** cabin lost pressure during flight; fortunately no such emergency occurred during the Mercury missions.

Similarly, Soviet cosmonauts who flew on the **Vostok** missions were never exposed to the vacuum of space and needed pressure suits only as a backup measure. That need changed on the Voskhod 2 flight, when cosmonaut Alexei Leonov became the first man to walk in space. His space suit kept him alive during his brief excursion, but it ballooned dangerously, almost preventing him from reentering the spacecraft cabin. Meanwhile, NASA developed a new space suit for the **Gemini** missions, which included the first U.S. space walks. Although the Gemini suit offered more mobility than its Mercury predecessor, it was still difficult to operate when pressurized, and several Gemini spacewalkers became tired or even exhausted in the course of their activities.

To create a space suit for **Apollo** moonwalkers, engineers used new techniques and materials. Bellows-like joints made of molded rubber were used for the suit's elbows, shoulders, hips, and knees, augmented by sets of cables. These modifications made the suit more flexible. Some 21 layers of materials, including Kapton, Mylar, and Teflon, protected against extremes of temperature and high-speed micrometeorites, and resisted tearing and abrasion. Underneath the suit, the astronaut wore a set of long underwear containing a network of plastic tubes that circulated cold water to keep him cool during the exertions of the moonwalk. A backpack supplied oxygen and cooling water and contained a radio transmitter for communications. An improved version of the Apollo suit, used on the last three lunar missions, allowed moon walks lasting more than seven hours.

For the **space shuttle** program a new modular space suit was developed. Instead of being individually tailored for each astronaut, the shuttle suit is assembled from components, each of which is manufactured in a number of different sizes, and the suit is reusable. Like the Apollo suit, the shuttle suit uses a liquid-cooled undergarment and a backpack for oxygen, water, and communications. Shuttle astronauts have made space walks lasting up to nine hours, repairing satellites, servicing the **Hubble Space Telescope,** and constructing the **International Space Station.** During long-duration missions to the International Space Station, spacewalkers often use the Russian "Orlan" (Eagle) space suit first developed in the late 1970s for the Soviet **Salyut space stations.** This one-piece suit features a hard-torso design, with a backpack that is hinged to allow the wearer to enter the suit from the rear.

Aviator Wiley Post models the first practical pressure suit, which he created in cooperation with the B.F. Goodrich Rubber Company for a high-altitude flight in 1934. A direct ancestor of the space suits used by astronauts, Post's pressure suit allowed him to survive in the thin air above 12,000 meters.

Today, engineers are working to develop advanced space suits that are lighter in weight, more comfortable to wear, and require less maintenance than current models. Such suits may be standard equipment by the time astronauts land on **Mars**—decades from now.

WALKING IN SPACE

Kathryn D. Sullivan

SPACE WALK. HOW DID IT EVER GET SUCH A MISLEADING NAME? EXCEPT FOR THOSE 12 incredibly lucky souls who actually did walk while wearing a space suit in an airless vacuum environment—on the surface of the moon—all the rest of us floated, maneuvering around the exterior of our spacecraft as if it were an odd jungle gym with an extraordinary view. Space walks are coveted and rather rare opportunities. As of mid-2004, only 148 of the 433 people who have flown in space (34 percent) have done one. I was overjoyed on the day in late 1983 when I got word that I was assigned not only to my first spaceflight but also to a space walk. Only three women had flown in space at that time (Valentina Tereshkova, Svetlana Savitskaya, and my classmate, Sally Ride), and my assignment marked the first time a woman was named publicly to an EVA. EVA, short for extravehicular activity, is NASA-speak for a space walk.

My friends and colleagues were thrilled that I would become the first woman in history to do a space walk, but I figured all along I would be second. Record book firsts were highly coveted by the Soviet space program, and I was quite sure that Savitskaya's arrival on the scene—some 20 years after Tereshkova and just as the six women in our class arrived—was no accident. My flight, STS-41G, was also 12 months off, giving the Soviets plenty of time to send Savitskaya on a second flight with a space walk. Sure enough, she flew again in July 1984 and did a 3 hour 35 minute space walk.

Training for a space walk is tough but fun. Through study and practice, you must master and then tightly interweave three strands of knowledge and skill: how the suit itself works, how to maneuver adeptly in microgravity, and how to use the tools and procedures for your tasks. This involves a good dose of schematics and technical manuals as well as vacuum chambers and lots of time underwater in one of the world's largest swimming pools.

In simple terms, the suit is a body-shaped spaceship. The garment, gloves, and helmet of the Extravehicular Mobility Unit (or EMU) seal you off from the vacuum of space. The backpack (or Portable Life Support System—PLSS) supplies oxygen, cooling and communications, and a package on your chest (the Display and Control Module, or DCM) holds the electronics and computer by which you run it all. Several weeks before flight, after the training manuals, engineering schematics, and system simulators have you up to speed on suit systems and operations, you don your flight suit for a vacuum test.

A vacuum chamber is basically a room whose walls are strong enough to allow all the air to be pumped out of it without buckling under the external pressure of the atmosphere. Outside the chamber, normal atmospheric pressure is 1,000 millibars (or 1.034 kg/cm^2). After pump-down the pressure inside is much less than one-millionth of this level.

Knowing that the room looks exactly the same with and without air in it, the technicians for each of my chamber flights brought in a simple prop to make the pressure change visible and impress upon me the importance of running my suit properly: a shallow pan of room-temperature water. As the air pressure in the chamber decreased, the water began to bubble and then boiled vigorously. Then, in one dramatic flash, the water making up the walls of the bubbles froze and dropped into the pan as a layer of slush. If a picture is worth a thousand words, this vivid demonstration of what would happen to all the fluids in your body if your suit depressurized in space was worth a million.

Neutral-buoyancy practice underwater is the main way of getting accustomed to maneuvering in microgravity and working out the choreography of your space walk. The pool currently used for this, Johnson Space Center's Neutral Buoyancy Laboratory, measures 61.5 by 31 by 12 meters and holds nearly 23.5 million liters of water. Real space suits are used, but ones that are dedicated to underwater training and will never fly in space. The normal flight PLSS and DCM are replaced by fake ones of the same size and shape that contain lead weights. These weights, plus others strapped onto your legs and arms, are needed to offset the buoyancy of the suit and its enclosed air. Scuba divers adjust the amount and placement of the weights, until you neither float, sink, nor tilt when they let you go in mid-water.

On the bottom of the tank are full-size mock-ups of whatever spacecraft you'll be working on, whether shuttle, space station, Hubble telescope, or something else. Where details are crucial to training, as in the location of major elements or features of things like connectors and fasteners, the mock-ups are very accurate; otherwise the mock-ups are tailored to immersion in water and simplified to save cost and weight. EVA astronauts spend a hundred hours or more refining and rehearsing every step of the entire space walk in the simulated weightlessness of the tank.

Most of our training focused on things that could go wrong, from fairly minor to quite severe. The intent was to prepare us to work around or overcome several problems simultaneously and to know our flight rules thoroughly. Many fairly minor things can trigger a flight rule and result in cancellation of a space walk. Indeed, the very first planned space walk from a shuttle, on STS-5 in 1982, was canceled because of a problem with the speed control of the suit's ventilation fan. There had been three successful space walks after that, and ours was next. With that backdrop, it was sometimes hard not to feel that we had a one-in-four chance of a cancellation.

In flight, the preparations for a space walk begin the day before, with adjustments to the cabin pressure to lower the risk of bends occurring and pretesting of all the equipment. EVA day itself is a special blend of mounting anticipation and solid discipline. Even with assistance from a third crew member, it takes about four hours to get fully suited up and ready to depressurize the airlock. My dad was watching all this from the viewing gallery at Mission Control in Houston. Recognizing him, our flight surgeon pointed out the trace of my heartbeat as I hooked up my tether and slipped out into the cargo bay. To this day he teases me about how at a moment of such great anticipation, drama, and risk, my pulse didn't even hit 80 beats per minute before settling back to something like 58.

Once out in the payload bay, Dave Leestma and I felt right at home and set about our tasks at a brisk pace. Our satellite refueling experiment went smoothly, as did the unplanned task of repairing a shuttle communications antenna and the filming of some scenes for the IMAX film *The Dream is Alive.* Having always schooled ourselves to meet or beat our timeline, we headed back into the airlock in what we felt was very good time. Only when I got home did I find out that the entire EVA team on Earth was hoping we would go slower, so that my spacewalk time would beat Svetlana Savitskaya's duration (I came in six minutes too early).

Dressed in a reusable space suit known as an EMU (Extravehicular Mobility Unit), Kathy Sullivan prepares to perform an EVA. On October 11, 1984, Sullivan became the first American woman to walk in space.

No matter how short in duration or distant in the past, my one EVA remains among my most vivid space memories. To this day, I'm sure I could don the space suit with my eyes closed. I was struck by how comfortably familiar I felt in the orbiter's payload bay. Indeed, until a brief pause let me pivot away from the spacecraft and take in the scene—the star-studded black of space, the blue and white of the Earth below (and not a scuba diver in sight!)—it could have been just another training run in the water tank. And to this day I can close my eyes and feel vividly as if I were hanging from the bar of an extraordinary jungle gym watching Venezuela slide by beneath my boots. ■

Space Stations

Space Stations

Decades before anyone ever went into space, science-fiction writers and theorists envisioned artificial outposts in **Earth** orbit. German space visionary Hermann Oberth first coined the term "space station" in the 1920s, describing "a sort of miniature moon" in which **astronauts** could study the Earth and the heavens, conduct military reconnaissance, and provide a communications relay for people around the globe.

By the mid-1950s **rocket** pioneer **Wernher von Braun,** a protégé of Oberth's, was thrilling American magazine and television audiences with his own version of the space station idea: A giant wheel-shaped facility, 76 meters in diameter that contained three separate levels where dozens of astronauts would live and work. By spinning slowly to create centrifugal force, the station would provide its occupants with artificial gravity. Although the Cold War environment led von Braun to stress the station's military uses, he and others envisioned using it as a way station for astronauts voyaging to the **moon** and the **planets.** In their minds, a space station was integral to human exploration of the **solar system.**

Making that vision a reality proved more difficult than von Braun and his colleagues anticipated. Although a space station was one of the options considered by President Kennedy in 1961, it was rejected in favor of putting men on the moon. By 1969, as the first **Apollo** lunar landings took place, a permanent space station was at the top of **NASA**'s wish list, along with a reusable **space shuttle** to ferry astronauts to and from the outpost. But faced with budgetary restrictions, NASA was compelled to put the station on hold until the shuttle could be completed. Still, in 1973 and 1974, the three missions to the **Skylab** space station, which was constructed from spare Apollo hardware, gave NASA a glimpse of what it would be like to live in space for months at a time.

In the Soviet Union, which had lost the "moon race" with the Americans, space stations became the focus of the human spaceflight program. Beginning in 1971 the Soviets launched a series of **Salyut** space stations on which cosmonauts pushed back the frontiers of long-duration spaceflight. These missions showed that humans could survive in space long enough to journey to the nearest planets, and produced new data on the medical effects of spaceflight. They also made the Soviets the world leaders in human spaceflight experience. In 1986 the Soviet space station program culminated with the first modular space station, **Mir.** During Mir's 15-year life, its crews set new space endurance records and met severe challenges.

In 1984 NASA received the go-ahead for its long-cherished space station, but the project was hampered by delays, budget overruns, and numerous redesigns. The **International Space Station** is now under con-

SPACE STATIONS

Space Station	Launch Date	Reentry Date	Average Orbital Height	Approx. Weight	Maximum Capacity	Notes
Salyut 1	14 Apr 1971	11 Oct 1971	197 km	18,500 kg	3	Launched by the Soviet Union, it was the first space station to be placed in orbit. Two crews were sent but only one was able to enter the craft. They lived in Salyut 1 for 23 days.
Skylab	14 May 1973	11 Jul 1979	436 km	91,000 kg	3	First American space station. Housed 3 Apollo crews over nine months for 28, 59, and 84 days.
Mir	20 Feb 1986	23 Mar 2001	367 km	122,500 kg	2 (long stay) 6 (short stay)	Launched by the Soviet Union. The station has hosted 104 humans (including 7 U.S. astronauts) who lived in Mir for up to six months at a time. The station survived a fire, collisions with spacecraft, and attack by glass/metal-eating microbes.
International Space Station	20 Nov 1998	still in use	375 km	470,000 kg (upon completion)	7 (upon completion)	This station is the cooperative effort of 16 nations and is in the process of becoming a permanent space station. The ISS is primarily an experiment with living in space; researchers are able to study human psychological and physical reactions to long stays in space.

struction. Although its completion date—and its ultimate impact—have yet to be determined, the space station concept has become an essential component of human activities in space and will likely remain so.

Salyut

In April 1971, ten years after **Yuri Gagarin**'s pioneering spaceflight, the Soviet Union launched the world's first **space station,** called Salyut—a name that means "salute," in tribute to Gagarin, who had died in 1968. Salyut consisted of a single module weighing 19 metric tons that offered 100 cubic meters of living space. In June 1971 three cosmonauts logged 23 days aboard the station, setting a new spaceflight endurance record. Their mission ended in tragedy when an air leak killed the men shortly before their **Soyuz spacecraft** made an automatic **reentry.**

Five more Salyuts were launched between 1974 and 1982. Two of these, Salyuts 3 and 5, were military space stations, and were equipped with high-resolution cameras to gather reconnaissance data from orbit. Aboard Salyuts 6 and 7, cosmonauts expanded the limits of long-duration spaceflight in an extraordinary series of space marathons. The longest of these, a mission to Salyut 7, lasted 237 days—nearly eight months in space. One cosmonaut, Valery Ryumin, logged a total of 360 days in orbit during two marathon stays, in 1979 and 1980, on Salyut 6.

In 1985 Salyut 7's electrical system failed, forcing a team of cosmonauts to stage a repair mission. In mid-1986, after two more crews had visited the station, Salyut 7 was abandoned for good. By that time, the Soviets were by far the world leaders in human spaceflight experience. With the newly launched **Mir** space station, they were poised to expand on this prodigious record of achievement.

Skylab

NASA's first **space station,** Skylab, was created from hardware developed for the **Apollo** moon program. The main compartment of the station, called the orbital workshop, was constructed inside the empty third stage of a **Saturn V** booster. Inside the workshop were sleeping, living, and working accommodations for three **astronauts,** along with a variety of scientific and medical experiments. Attached to the workshop were a docking compartment and a solar **telescope.** Altogether, Skylab had 210 cubic meters of habitable space, as much as a small house.

Aboard the Soviet space station Salyut 6, cosmonauts pose for a self-portrait in March 1978. From left: Czech cosmonaut Vladimir Remek, Soviet cosmonauts Alexei Gubarev, Georgi Grechko, and Yuri Romanenko. Remek was the first of a series of "guest cosmonauts" selected from Socialist nations as part of the Soviets' Intercosmos program.

Skylab's mission, however, almost ended before it could begin. During its launch on May 14, 1973, a shield designed to protect the station from micrometeorite impacts was torn off by aerodynamic stresses, taking one of Skylab's two solar-power arrays with it. The remaining array was pinned to the side of the workshop by a piece of **space debris.** Starved for electrical power and overheating, Skylab needed help. It was up to Skylab's first crew, astronauts Pete Conrad, Joe Kerwin, and Paul Weitz to save the mission. After arriving at the station on May 25, the astronauts installed a sunshade on the exterior, and during a daring space walk managed to free the stuck solar array. Their 28-day mission set a U.S. space endurance record.

Two more crews visited Skylab, living aboard the station for 59 and 84 days. The Skylab missions gave NASA its first understanding of the medical effects of long-duration spaceflight; they also produced new data on the **sun,** the **Earth,** and on the behavior of materials in **weightlessness.** A second Skylab, built but never launched, is on display at the National Air and Space Museum in Washington, D.C.

Mir

In February 1986 the core of a new type of orbiting outpost—the first modular **space station,** named Mir ("peace")—was launched by the Soviet Union. The Mir core module, weighing 20 tons, had six docking ports to accommodate additional modules, as well as **Soyuz** ferry craft and Progress supply ships. Over the next decade, Mir grew into an orbital outpost as new modules were added. The first of these, called Kvant, contained **telescopes** for astronomical observations and reached the station in March 1987. Priroda, the last module, dedicated mostly to remote sensing, was added in April 1996, bringing Mir's total habitable volume to about 380 cubic meters and its total weight to 122,500 kilograms.

Aboard Mir, which offered more creature comforts than **Salyut,** including individual sleeping compartments, cosmonauts set new records for long-term spaceflight. In 1987 and 1988 Vladimir Titov and Musa Manarov spent a year aboard Mir. And in 1995 physician-cosmonaut Valery Polyakov completed 14 months aboard the station, a single-spaceflight record that still stands.

In a joint program between **NASA** and the Russian space agency, U.S. **astronauts** lived aboard Mir between 1995 and 1998. These missions gave the United States its first experiences with long-duration spaceflight since **Skylab.** Astronaut **Shannon Lucid** spent six months aboard Mir in 1996, more than double the length of the longest Skylab mission.

Mir was the scene of some of the most harrowing moments in spaceflight history, which occurred during a series of long-duration stays on Mir by U.S. astronauts. In February 1997 an oxygen generator in one module malfunctioned, spewing flame and dense smoke for long minutes. The fire nearly forced U.S. astronaut Jerry Linenger and his Russian crewmates to evacuate. Even after the smoke had cleared, more difficulties arose aboard Mir; the station's plumbing leaked toxic coolant, and the cosmonauts were besieged by a relentless workload to save themselves.

Then, in June, came the most serious crisis of all, when an automated Progress supply spacecraft struck the station during a docking exercise. One module was punctured by the collision, causing the station to lose air. The leak was stopped by the quick action of astronaut Mike Foale and his crewmates, who sealed off the damaged compartment. The crash also damaged one of Mir's solar panels, shutting off much of its electrical power system. The station drifted, starved for power, until the crew managed to bring it back to life.

In many ways, the crises faced by the Mir crews and by mission controllers on **Earth** anticipated the challenges that will someday confront the first humans who journey to **Mars.** And these difficulties were outweighed by the accomplishments of the Mir program. In all, 104 cosmonauts and astronauts from 12 countries logged a total of 4,591 days—more than 12 years—on Mir. The station's crews made 78 space walks totaling 352 hours and performed 23,000 scientific experiments. From September 1989 to August 1999 the station was continuously manned. Mir survived a total of 15 years in space—three times longer than its designers had envisioned—before reentering the atmosphere on March 23, 2001.

Shannon Lucid

One of the champions of long-duration spaceflight, Dr. Shannon Lucid was born on January 14, 1943, in Shanghai, China, and grew up in Oklahoma. Trained as a biochemist, Lucid was selected as one of **NASA**'s first women **astronauts** in 1978. Between 1985 and 1993 she flew on four **space shuttle** missions, including a highly successful 14-day flight devoted to medical experiments. In March 1996, after her fifth shuttle launch, Lucid arrived at the **Mir space station** to begin a marathon mission. Aboard Mir she conducted a number of scientific experiments in life sciences and physical sciences. She made her return to **Earth** aboard the shuttle *Atlantis* in September, after 188 days in space—a U.S. spaceflight endurance record.

International Space Station

In 1984 President Ronald Regan directed **NASA** to construct a permanently manned **space station** in **Earth** orbit. NASA had been longing for such a mandate ever since the **Apollo** moon program, but the project was soon mired in delays, cost overruns, and bureaucratic complexity. By the early 1990s the station effort—now a partnership between NASA, Canada, the **European Space Agency,** and Japan—was in trouble. Most of the money allocated to the project had been spent on designing rather than building hardware. Several times the station had been redesigned, and still criticism and uncertainty about its purpose remained. Named Freedom, the space station faced cancellation by an ambivalent Congress. In 1993, however, the project was recast by the Clinton Administration as the International Space Station (ISS), now including Russia, which was slated to provide key elements of the station hardware.

Physician-cosmonaut Valery Polyakov peers through one of Mir's windows on February 6, 1995, as the space shuttle *Discovery* makes a close approach to the station. Polyakov became the world's space endurance champion when he logged 14 months aboard the Mir space station beginning in January 1994.

At the same time, plans were made for a series of U.S. missions to the **Mir** space station to pave the way for future joint space operations by the two nations.

As envisioned by NASA, the ISS would be a world-class, international scientific laboratory in space. When completed, with more than a hundred components, it would be the largest structure ever assembled in space, weighing 453 metric tons and covering an area the size of a football field. Its habitable volume—including laboratory modules from the U.S., Europe, and Japan—would be equivalent to the passenger cabin of a 747 jumbo jet.

In 1998 a Russian **Proton rocket** launched the first ISS element, the Russian Zarya ("sunrise") module, containing systems for navigation and control of the station. It was followed by the U.S. Unity module, designed as a connector between Zarya and other modules, carried into orbit by the **space shuttle.** The service module Zvezda (Star), which also served as living quarters for the crew, arrived in July 2000. In November 2000 the station received its first long-duration crew, officially designated Expedition One, when **astronaut** Bill Shepherd and cosmonauts Yuri Gidzenko and Sergei Krikalev arrived in a **Soyuz** ferry craft for a four-month stay. By the end of 2001 the station had grown to six modules, including the U.S. Destiny laboratory, and ISS crews divided their time between maintenance of the station and scientific research. But even as the ISS grew, new problems surfaced, as NASA revealed cost overruns, forcing some key elements of the station to be delayed or canceled. One of these deferred elements, the Crew Return Vehicle, was required before the size of station crews could be increased from three people to six or seven, the number deemed necessary to conduct the program of scientific research envisioned.

Still, construction of the ISS continued—until the shuttle ***Columbia*** was destroyed during **reentry** on February 1, 2003. In the aftermath of the *Columbia* tragedy, ISS crews were reduced to two people, launched aboard Soyuz **spacecraft** for six-month tours. NASA was still working toward returning the shuttle to flight and resuming construction of the ISS when the mission of the station was redefined once again. In January 2004 President George W. Bush announced a new space exploration program for NASA, including human missions to the **moon** and **Mars.** As part of this new initiative, research aboard the ISS would be focused on solving the biomedical problems of interplanetary human voyages. As yet the fate of the Bush space initiative, like so much about the International Space Station, remains uncertain.

The waters of the Pacific Ocean serve as backdrop for the International Space Station as seen from the space shuttle *Endeavour* in December 2000. The large solar array at top, which provides electrical power, measures 73.15 meters from tip to tip. Construction of the station, halted following the loss of the shuttle *Columbia,* is slated to resume in 2005 and be completed by 2010.

Today & Tomorrow

Space Tourism

In the spring of 2001 a new kind of space traveler was launched aboard a Russian **Soyuz spacecraft:** Dennis Tito, a California businessman, paid a reported 20 million dollars for an eight-day orbital "vacation" that included a stay aboard the **International Space Station.** Tito's flight, arranged by a Virginia-based company called Space Adventures, aroused controversy,

California businessman Dennis Tito relays his readiness to be the first space tourist on April 28, 2001. Launched aboard a Soyuz ferry craft, Tito spent six days aboard the International Space Station. The price for Tito's orbital "vacation" was reported to be about 20 million dollars.

especially inside **NASA,** where managers worried that his visit would delay critical work scheduled aboard the ISS. But those fears proved ungrounded, and the flight marked the beginning of space tourism. Tito's excursion was followed in the spring of 2002 by a second space tourist, South African technology entrepreneur Mark Shuttleworth, and more tourist visits to the **space station** are planned.

Today, orbital voyages are still the rarest of adventures, available only to a tiny handful of wealthy individuals. With the development of new space vehicles, that could change, so that a trip to **low Earth orbit** would be comparable to today's "extreme vacations," for example, climbing in the Himalaya or journeying to Antarctica. Meanwhile, within the next few years Space Adventures hopes to offer suborbital flights at a relatively affordable cost (about U.S. $100,000) once the necessary vehicles are developed. Designs for such vehicles are in the works at several private companies, and it is possible that these could yield a practical suborbital passenger vehicle soon.

As for space tourism beyond Earth orbit—vacations on the **moon,** for example—that will probably remain just a dream for many decades to come.

Space Plane

The holy grail of **spacecraft** designers is the space plane—a single-stage, reusable orbital vehicle that could take off and land like an airplane. Such a craft, many have maintained, would make human spaceflight safer, more affordable, and routine. But developing such a vehicle has proved beyond the reach of engineers—so far.

The most daunting problem is one of weight. Using today's liquid-fuel **rockets,** a single-stage orbital vehicle must reserve about 89 percent of its takeoff weight for fuel. The remaining 11 percent would have to include the vehicle itself, along with its cargo and crew. With this restriction in mind, it seems clear that new materials, as well as advances in propulsion, will be required before the dream of a space plane can become reality.

Chinese Space Program

On October 15, 2003 (Beijing time), some 33 years after launching its first **satellite,** China became the third nation in history to **launch** a human into space. The 21-hour Earth-orbit flight of Lt. Col. Yang Liwei represented the culmination of years of effort, and seems to mark the beginning of an ambitious human spaceflight program. According to public statements by Chinese space officials, plans call for a second manned orbital flight in the fall of 2005, which will last five to seven days, to be followed by a third flight before 2010. In the longer-term, China plans to build a **space station** in **Earth** orbit. According to several reports, there is a **moon** program under way, with plans for a lunar satellite by 2006 and robotic exploration by

2010. Such probes could pave the way for human missions in perhaps two decades.

Shenzhou

China's manned **spacecraft,** called Shenzhou (Divine Vessel), is similar to Russia's **Soyuz,** and, indeed, was developed with assistance from Russian space experts. Like Soyuz, Shenzhou consists of a descent module containing the main crew cabin, a service module containing the craft's **retro-rockets,** and an orbital module for conducting scientific experiments. Only the descent module returns to **Earth.** Unlike Soyuz, however, the orbital module contains its own propulsion system and can remain in orbit, unpiloted, to conduct research. Four unmanned test flights of Shenzhou were conducted from 1999 to 2003. In October 2003 Yang Liwei piloted Shenzhou 5 on the first manned Chinese spaceflight. More Shenzhou missions are planned, likely leading to the construction of an Earth-orbit **space station.**

Moon, Mars, & Beyond

On January 14, 2004, President George W. Bush called on **NASA** to revamp its space activities for the coming decades to focus on human exploration beyond **Earth** orbit. Bush's new space initiative includes sending humans back to the moon by 2020, and then to other destinations in the **solar system,** including the surface of Mars. Unlike the **Apollo program,** which focused on a specific destination—the moon—and a deadline for getting there—1969—the new initiative is based on a pay-as-you-go approach, in which new missions are developed as funding and technology allow. The goal of the initiative is not a specific destination but a broad program of human and robotic exploration that can expand indefinitely.

In his speech, which came almost a year after the ***Columbia*** tragedy, Bush called for retiring the **space shuttle** by 2010, following the completion of the **International Space Station.** Funding freed up by the shuttle and station programs would be applied to the new exploration effort. To replace the shuttle as a means of carrying astronauts to and from **low Earth orbit,** NASA would create a **Crew Exploration Vehicle,** which would also become the basis for elements of piloted lunar and interplanetary craft. As of now, the fate of the Bush space initiative is uncertain, pending funding decisions in Congress. But even if the plan is not fully implemented, it appears to mark the end of one era in NASA's human space activities—the shuttle/station era—and the beginning of another, which could at last see humans explore Mars and other worlds.

Crew Exploration Vehicle

The Crew Exploration Vehicle, or CEV, will be the first new piloted **spacecraft** to be developed under the space initiative outlined by President Bush in January 2004. The CEV is one component of a new generation of vehicles, under the name Project Constellation, that would allow humans to travel beyond **Earth** orbit for the first time since the **Apollo** missions. Plans call for the CEV to be produced in several versions. One would be used for ferrying crews to and from the **International Space Station,** another would be designed for missions to the lunar surface, and others would be created for interplanetary voyages. For missions beyond Earth orbit, the CEV would be mated in space to other components, such as modules for propulsion and supplies, to form a fully capable lunar or planetary craft. Plans call for the first unpiloted test flights to take place by 2011, followed by the CEV's manned debut in 2014.

China's first man in space, 38-year-old Yang Liwei, stands before his Shenzhou 5 spacecraft after landing on October 16, 2003. With the successful Earth-orbit mission, which lasted 21 hours, China became the third nation in history to conduct a manned spaceflight.

AN ADVENTURE WITHOUT END

Sean O'Keefe

THE DREAM OF INTERPLANETARY EXPLORATION IS ALIVE AND WELL. AS THE second century of flight unfolds, human pioneers working with robotic explorers will extend the reach of civilization throughout the solar system and beyond, and search for life on other worlds. I envy those who will have the opportunity to participate in these grand quests. Despite all the challenges we face on our home planet, we live in an age of heroic potential.

Inspired by the achievements of those in the past century who explored both poles, climbed Mount Everest, descended to the depths of the oceans and walked on the moon, our next generation of explorers will carry the torch of exploration to heights unimagined and into frontiers unknown.

Our 21st-century space explorers will dramatically expand scientific knowledge and promote the development of revolutionary technologies and capabilities to benefit all of us here on Earth. Indeed, I'm optimistic the continued exploration of space will boost our opportunities to become a smarter, safer, healthier world on a scale never seen before in the planet's history, and at a pace hardly thought possible.

As has always been the case with space exploration, the men and women of NASA are honored to play a leadership role in bringing new worlds into our grasp. The National Geographic can also be proud of its work to faithfully chronicle humanity's space achievements and for its longstanding efforts to make certain that exploration adventures do not end at land's end. In fact, the first great voyage into near space was the National Geographic Society–U.S. Army Air Corps sponsored 1934 balloon flight into the stratosphere.

Today, our human spaceflight program, for all its accomplishments, is still in its infancy. In technological terms, space exploration is in its age of sail. Promisingly, our age of steam is just around the corner. Using a variety of new technologies (for example, advanced power and propulsion systems, laser communications, and unique micro-devices) and a stepping-stone approach aimed at achieving sustainable, affordable exploration progress, we are charting a course that will take inquisitive, brave people back to the moon, and then on to Mars, and beyond.

This time around, the humans who set foot on other celestial bodies will explore well beyond their landing zones for weeks to months at a time, perhaps digging deeper into the mysteries of the subsurface as well. The next great human migration will be a vertical one. These space pioneers will be privileged to explore answers to fundamental questions of importance to science and society. Questions such as: How did we get here? Are we alone in the Universe? Where are we going?

When we return to the moon, our explorers will experiment with resources found within the lunar surface in order to learn to live off the land, as well as to seek answers to when life may have arisen on Earth. The information gained from our new lunar experiences will give us the confidence to extend our human exploration horizons to Mars.

The red planet is filled with fascinating sites for human exploration. In 2004, the world was enthralled by the discovery of an ancient salty sea where our Mars Exploration Rover Opportunity landed. At about the same time NASA's orbiting Mars Global Surveyor took images that clearly suggest the presence of a "frozen in stone" river delta 25 degrees south of the Martian equator, where a long-standing body of liquid water must have existed in the distant Martian past. These spots, as well as countless others, will be prime locations for human explorers to search for evidence of past life on Mars and to learn how a planet so seemingly similar to our own evolved in starkly different ways, perhaps foretelling our own planet's destiny. But before our courageous human explorers set foot on Martian soil, an incredible robot will journey to the most promising bedrock of Mars, collect priceless samples, and return

Envisioning the future of man in space, an artist's rendering depicts an astronaut working in concert with a robot to build a telescope in space. Once thought impossible, such a scenario is now probable given the technological advances and humankind's continued thirst for exploration.

them to Earth, where scientists will probe the real Mars for clues to its life-bearing capacity, and to ensure its safety to future human explorers.

Other Mars explorers will have the opportunity to climb the Olympus Mons volcano, which at more than 26,000 meters tall is more than three times higher than Mount Everest, and to traverse the 4,000-kilometer-long expanse of the great chasm, Valles Marineris, which is five times longer and four times deeper than the Grand Canyon of Arizona.

More mysteries await us beyond Mars, both in the vicinity of Jupiter, and elsewhere. In *2010,* his sequel to the science fiction classic *2001,* author Arthur C. Clarke placed his fictional monolith in the vicinity of Jupiter's moon Europa for good reason. Beneath Europa's icy, fractured crust a potentially life-hospitable slushy or liquid-water ocean might exist. Europa may well become the venue for the first exploration submarine to function outside of Earth's oceans, seas, and deep-water lakes. Likewise, Saturn's giant moon Titan may hold the secrets to Earth's early atmosphere, from which life itself sprung forth billions of years ago.

"Mystery creates wonder, and wonder is the basis of man's desire to understand," said Neil Armstrong, the first explorer to set foot on an extraterrestrial body. Consider these other wonder-inducing exploration targets that for the foreseeable future will be scouted by robotic explorers: the Great Red Spot on Jupiter (before it goes away); Jupiter's other icy moons, Ganymede and Callisto, and the active volcanoes on their sister moon Io; the atmosphere of Saturn's moon Titan, composed of many of the same chemicals as Earth's early atmosphere, and believed to potentially contain complex, prebiotic chemistry; and the large main-belt asteroids, perhaps remnants of failed planets as relicts from the earliest days of our solar system, some large enough to have harbored liquid water.

Our exploration vision extends to the stars. Ten years ago we didn't even know for certain if there were planets beyond our own solar system. Working with new scientific observation capabilities, astronomers have now found more planets orbiting other stars—and the number continues to climb with new discoveries and innovative techniques. Our quest for knowledge about these planets has barely begun. Now on the drawing board is the Terrestrial Planet Finder, a large space telescope capable of finding Earth-like planets and detecting the chemicals in their atmospheres that may indicate the presence of life. Future telescopes will be able to map continents on these distant bodies.

Five centuries ago, when Columbus voyaged across the Atlantic, his ships carried the inscription, "Following the light of the Sun, we left the Old World." I look forward to the adventures ahead as we follow the light of the planets and stars into the new worlds of the 21st century, and learn finally that we are not alone. ■

5 | Earth Science & Commerce from Space

The European Retrievable Carrier (EURECA) spacecraft is silhouetted against the Atlantic Ocean just offshore Cape Canaveral and the NASA Kennedy Space Center in Florida. The satellite had just been deployed from the cargo bay of the space shuttle *Atlantis* on a mission in July 1992. Shuttle astronauts took the photograph using a handheld camera; it reveals the unique vantage point space affords of Earth's features.

THE COLLECTION OF SCIENTIFIC KNOWLEDGE AND THE EXPANSION OF commerce are not new to mankind. Thousands of years ago our ancestors became tribes of hunter-gatherers and then settled into communities. This breakthrough—the reliance on the strength and diversity of the group for the welfare of the community—brought about the specialization of individuals into farmers, hunters, soldiers, or merchants. Once the value of such specialization was demonstrated, what began as natural human curiosity and the challenge of survival evolved into scientific research and exploration, and what began as simple trading for the necessities of life became commerce for profit and wealth.

An early example of civilization, during the first millennium B.C., is the Phoenician ships that plied the Mediterranean Sea, exploring its secrets and carrying goods through the ancient world. Each ship carried lookouts high in its masts, and their sharp eyes would be alert for shoals, approaching squalls, schools of fish, or the sails of enemy ships—the higher the mast and the sharper the eyes of the lookout, the safer the voyage. Mastheads carried fluttering signal flags needed to communicate with distant ships and to distinguish friend from foe.

From those ancient times to today, we have placed our lookouts and signals higher and higher, first on bigger ships, then in aircraft, and then on satellites flying in space. We have replaced eyes with cameras and flags with radios, but the concept remains the same. Exploitation of the ultimate heights of space for discovery, communications, and navigation has been the natural progression of human ingenuity and common sense.

Space systems are also used for military applications, and these have their own profound impact on commerce, but this chapter will concentrate principally on the scientific and civil applications of space systems and define exactly what we mean by "Earth Science and Commerce from Space."

Three important avenues are explored in the chapter: Earth science and remote sensing—how satellites and the instruments they carry "remotely sense" parameters that allow us to monitor the Earth's atmosphere, oceans, and land; satellite communications—how satellites are used to convey information by radio or laser from one place to another, allowing us to communicate from anywhere at any time we choose; and satellite navigation—how satellites are used to determine the precise location of any person or object on the Earth, in the air, or even in space.

This Geostationary Operational Environmental Satellite (GOES) image of Hurricane Fran was taken in early September 1996. Forecasters use GOES images to track storms and provide early warnings. Fran slammed into North Carolina's coast with gusts up to 201 kilometers an hour. Satellite imagery saves lives and helps reduce property loss.

Earth Remote-sensing Basics

Earth Remote Sensing

The term "remote sensing" has been synonymous with observations of the **Earth** from aircraft and **satellites** for decades. It has generally been used to make the distinction between the observations of a human observer actually looking down from a certain vantage point and the observations made by some type of camera or other instrument placed at high **altitude,** remote from the person who will later look at the picture.

Even our own eyes and ears can be considered remote-sensing devices. When we look at an object, such as another person's face just a few feet from us, or at a faint light in the night sky that may, in fact, be a **planet,** a star, or even a **galaxy** billions upon billions of kilometers away, our eyes are remotely sensing the face or the galaxy. And the more closely we observe those things, the more we learn about them.

Remotely sensed images vary in resolution—spatial, spectral, or temporal. In the top row, the Pentagon is more distinguishable the higher the spatial resolution (the smaller the individual picture element). Spectral resolution reveals different features of Cape Canaveral, Florida, in visible light, near infrared, and thermal infrared wavelengths. Changes in the ice pack off Alaska show the effect of temporal spacing of sequential images.

SPATIAL (The Pentagon, Arlington, Virginia)

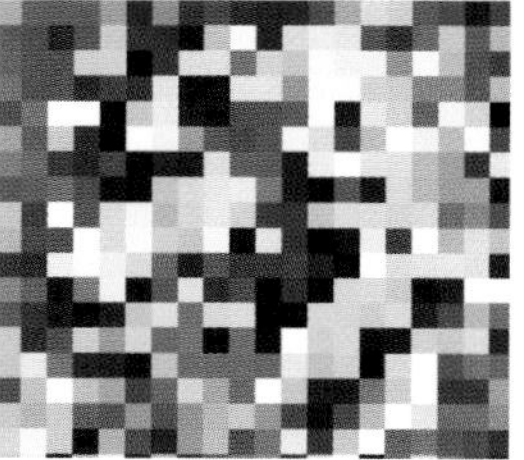

30 meter

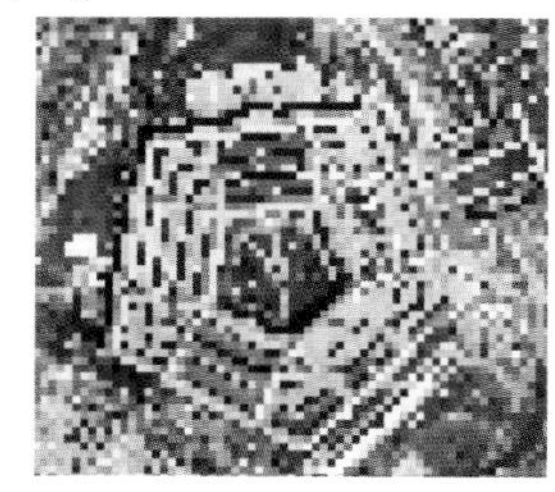

10 meter

1 meter

SPECTRAL (Cape Canaveral, Florida)

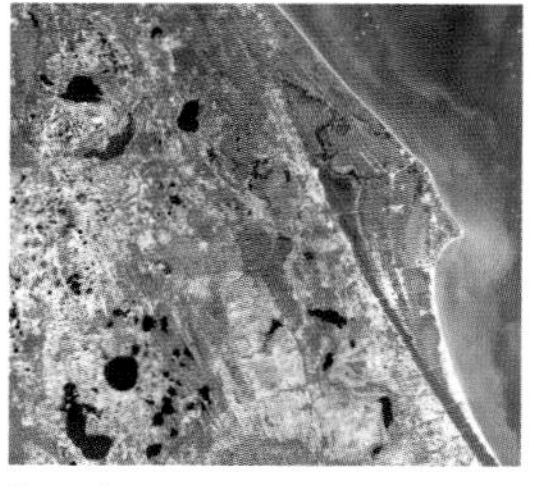

Natural

Near Infrared

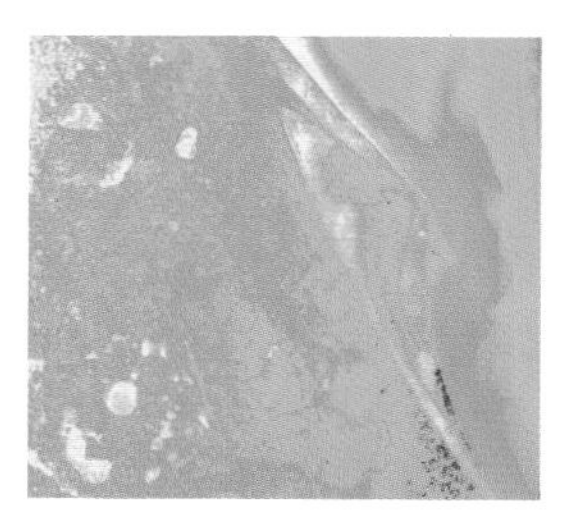

Thermal Infrared

TEMPORAL (Point Barrow, Alaska)

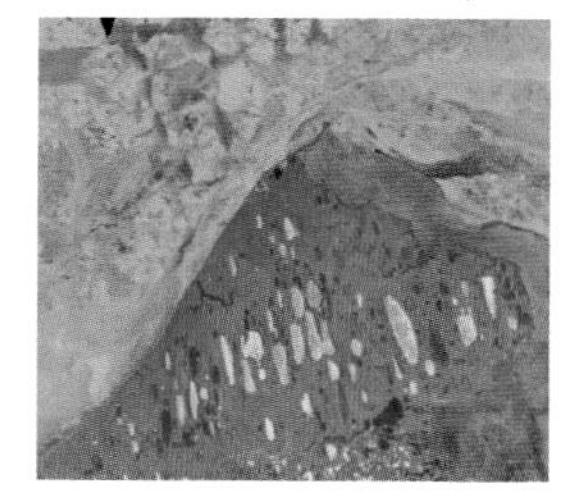

February 4, 1992

February 7, 1992

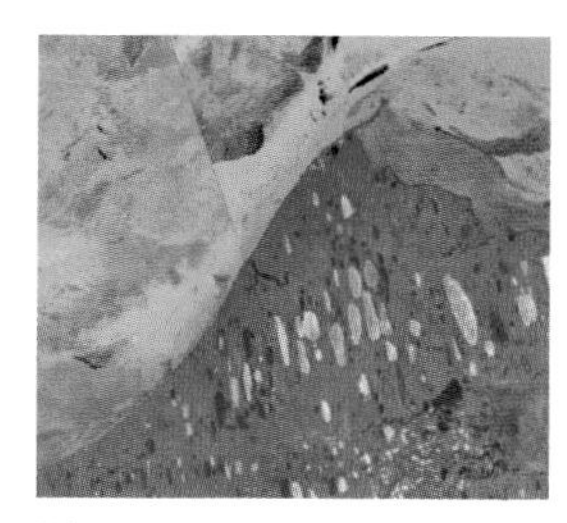

February 10, 1992

The view of the Earth from space is fantastically beautiful, our blue planet being the aesthetic superstar of our **solar system.** But, in addition to that beauty, the images taken on all American human spaceflight missions since **John Glenn**'s historic orbital flight also demonstrate the power of space-based observations in revealing the physical nature of the Earth's sky, seas, and landmasses. Cloud formations, discolorations of the oceans, sea ice and the ice sheets covering high latitude landmasses, and complex geology of the continents are all captured in the **astronauts**' photography.

Remote sensing, at its most basic level, is what all humans do so that we can collect information through a variety of sensors and process that data into mental images or messages of comfort or danger. Remote sensing from space is simply an extension of the same process. Astronauts use their eyes and handheld cameras, but a whole range of satellites now employ remotely operated cameras, which we call imaging instruments, or other sensors that collect information and relay it to people on the ground for study. This is satellite remote sensing.

Satellites monitor atmospheric conditions (clouds, wind, moisture content, precipitation, and chemistry), oceans and seas (currents, waves, biological content, bathymetry, salinity, color, and clarity), land cover and land use, and cryology (sea and land ice cover, thickness, and motion).

Research Satellites

Research satellites, which are designed to collect scientific data on previously unmonitored or poorly understood environmental phenomena, are often one-of-a-kind and experimental in nature. They may be used for experimental flights of new instrument technology, **remote sensing** from a different perspective (for example, a new angle or wavelength) or looking with an improved capability (for example, higher spatial or spectral resolution, or more frequent sampling).

Accordingly, research satellites are funded by a research agency such as **NASA,** and often their missions are carried out at some risk of failure. Examples include NASA's **Earth Observing System** (EOS) satellites, each very different from the other and carrying multiple sensors. They are part of a broad **Earth** science program, providing huge amounts of data on dozens of different environmental parameters to researchers so that the complex interactions of the Earth's systems can be better understood.

Operational Satellites

Some space systems are mature in their design, having been built with the benefit of previous experience in the use of their sensors. These operational satellites provide observations of the **Earth** that are used by either government or commercial organizations in allowing regular, methodical collection of data in support of such activities as weather forecasting, monitoring ocean pollution, oil exploration, agriculture, mapping, and forest fire location and suppression.

These satellites are generally funded by government agencies such as the National Oceanic and Atmospheric Administration (NOAA) or by commercial companies. Their designs are usually based on previously demonstrated technologies; by the time they are launched, most elements of risk in the programs to build and launch them have been reduced to a minimum. These satellites are designed to be placed in orbit and put to work, with replacements for them either waiting to be quickly launched upon the failure of a satellite or one of its mission-critical sensors, or with a replacement launched and stored in orbit in semi-sleep mode until it is needed. Data continuity from operational satellites is a critical element of their overall system design.

Active & Passive Remote Sensing

Space-based instruments for **Earth** observation are designed for either passive or active remote sensing. A passive sensor, by far the most commonly used, collects data by receiving either reflected or radiated energy from the object being observed. An active sensor actually transmits energy from the instrument to the object and then observes the characteristics of the "echo" coming back.

A common example of passive sensing in everyday life is a photograph taken by a camera without use of flash: It simply collects the light coming from the imaged scene. Passive remote sensing mimics human eyesight in that, like eyes, the instruments collect the light that is reflected from what they are observing. Many sensors operate in the optical bands and cannot see in the dark, but operating in other frequency bands allows night vision.

An example of active sensing is a flash camera, which sends its own light from the flash to an object and then records the reflected light. Active remote sensing is what bats do when they fly in the dark. They transmit high frequency sound and then navigate and capture their sources of food—bugs—by processing the echoes their very sensitive ears receive. Active remote-sensing devices work in the same way, except that they generally use radar waves or high-intensity light to probe the Earth from above.

Electromagnetic Spectrum Used in Remote Sensing

Electromagnetic energy is often referred to as "light," but this is a misleading term because it confuses what we know as visible light with the whole **spectrum** of energy transmitted from one place to another in the form of photons, traveling at the speed of light. It is more appropriate to refer to electromagnetic energy in terms of its spectrum: the classification of energy by the frequency or wavelength in which it is found. The electromagnetic spectrum defines the entire range of wavelengths—from the very shortest, gamma rays, and x-rays, into the ultraviolet (UV) and to the longer wavelengths of infrared (IR) and microwave, increasing to TV and radio waves. Spectrum charts almost always place visible light in their center because it is from this most well-known characteristic of light that we compare wavelengths, and to which we as humans can relate.

The **Earth** is bombarded at all these wavelengths by emissions from the **sun** and distant stars, **quasars,** and other objects in space. To gather as much information as possible, the remote-sensing community has developed techniques for observing the Earth in all these bands, using both **active** and **passive remote-sensing** systems. However, the UV, visible, IR, and microwave bands are most useful. UV radiation from the sun is absorbed by chemicals in **Earth's atmosphere,** not the least of which is ozone; hence UV instruments are used to measure its properties. The discovery and characterization of the ozone hole over the South Pole and general thinning of stratospheric ozone was made by passive instruments on NOAA and **NASA** satellites.

Visible light reflected from clouds and the Earth's surface is used to observe weather, vegetation, biological content of the ocean, and many other useful parameters. IR radiation is monitored to assess the temperature of cloud tops—an indication of their height in the atmosphere—the sea-surface temperature of the ocean, and land properties. **Microwave sensors** are used to passively monitor sea ice, precipitation, water content in the atmosphere, and ocean properties, and active microwave sensors (**radar**) are used to observe ice, ocean currents, land topography, and stresses in the Earth's faults.

Orbits Employed for Remote Sensing

Remote-sensing satellites fly in **low Earth orbit** (LEO), **geostationary Earth orbit** (GEO), **highly elliptical orbit** (HEO), **medium Earth orbit** (MEO), or in an Earth-sun libration point. The higher the orbit, the greater the area on the Earth visible, but the less clear the features become. The **International Space Station** flies in LEO at an altitude of about 375 kilometers, in a west-to-east **trajectory** covering a **ground track** between 51.65° north and south latitude. The station may be able to observe almost half of the Mediterranean Sea, but it cannot observe an entire continent or ocean basin and cannot observe either pole. Therefore, its orbit is not optimized for Earth remote sensing.

Polar-orbiting satellites, also at LEO, fly essentially north and south and orbit the Earth about once every hour and a half. They generally fly a little more than twice as high as human **spacecraft:** between 700 and 900 kilometers. This higher orbit not only allows a larger viewing area, but also provides the same sun illumination for each band of latitude if the satellites are placed in a **sun-synchronous orbit.** This specific orbit crosses the Equator at exactly the same local time on each and every orbit.

Satellites in GEO are placed at 35,786 kilometers above the surface of the Earth at the Equator and orbit at exactly the same angular rate at which the Earth turns. From GEO, a satellite sees the whole disk of one side of the **planet;** however, both the view of the poles and around the Earth's limb (horizon) are at very shallow angles, so a GEO satellite cannot really observe "half" of the planet in a useful way. It takes several such satellites arranged around the Equator to observe the whole Earth, except for the area at the poles.

To get the same kind of wide field, but with a view of the polar regions, satellites can be placed into HEO. An orbit of this type passes close to Earth near one pole, generally the South Pole, at its **perigee** and then flies out to an **apogee** of several thousand kilometers.

Satellites located between 1,000 kilometers and 15,000 kilometers are said to be operating in MEO. Remote-sensing satellites do not generally use this orbit, which is more optimized for other applications, such as satellite navigation.

A final and interesting orbit used for remote sensing is the Earth-sun libration point, often referred to as **L1.** This is an orbit around the sun, closer to the sun than the Earth's orbit, at which a satellite is always looking at the sun without being blocked by Earth and is always looking backward at the illuminated side of Earth. L1 is currently used only for solar-monitoring satellites, but Earth-viewing satellites have been proposed for this unique orbit.

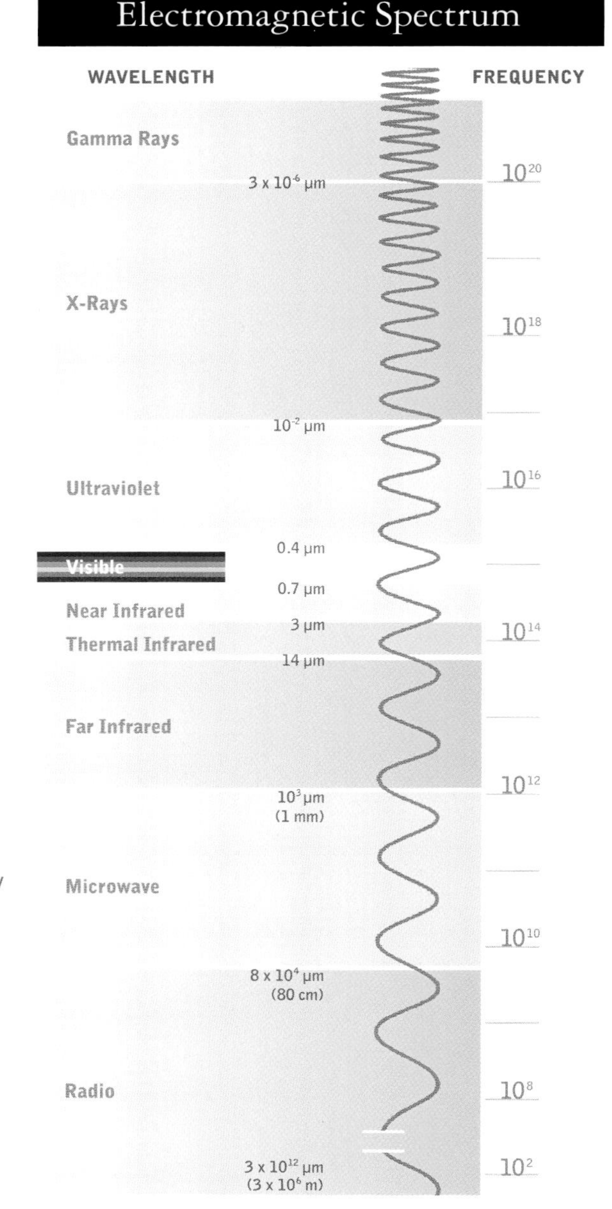

Electromagnetic energy travels in waves that can be described either by wavelength (the distance from peak to peak, usually measured in microns to meters) or by frequency (measured in cycles per second). Earth remote-sensing systems, which take observations of the ocean, land, and atmosphere, use wavelengths from ultraviolet through microwave. Light and radio wave spectra play roles for communications.

Field of Regard

The geographic area of **Earth** that a sensor can observe over its normal span of motion is called the field of regard. This does not include platform movement for a three-axis-stable platform normally pointed at the Earth. The field of regard refers to the instrument, or collectively to the suite of instruments on a remote-sensing satellite. The field of regard can be expanded by rotating the satellite to change the direction in which its instruments are pointing. It may be referred to in terms of a number of degrees of angle, or by an area on the Earth's surface, expressed in square kilometers.

Field of View

From any orbit, the field of view (FOV) is the area on the **Earth** that the **remote-sensing** instrument is physically capable of collecting data from at any given moment, without moving the sensor. Like **field of regard,** the FOV may be referred to in terms of a number of degrees of angle, or by an area on the Earth's surface, expressed in square kilometers.

Footprint

The footprint of a **remote-sensing** instrument is the geographic area on the Earth's surface within the **field of view** that the instrument collects data from as it flies along its orbital track. It defines the geographic boundaries of what the remote-sensing instrument will either take a picture of, or collect data from.

Slant Range

Although the shortest distance between the surface of the **Earth** and the **satellite** is its **altitude,** a **remote-sensing** instrument is often designed to look to one or both sides, ahead, or behind the spot immediately beneath the satellite, out to the edge of the **field of regard.** The distance between the instrument and any specific point on the Earth being observed is referred to as the slant range. The farther away from the point directly beneath the satellite, called the **nadir,** that an instrument observes, the greater the slant range. Observations at different slant ranges will differ because both the distance to the object and the angle at which the object is viewed are different. Instruments can be designed, however, to correct for this in order to allow seamless data collection within an entire **field of view.**

Swath

As a **satellite** flies along, its instruments generally make a continuous series of observations. The geographic area which is swept by the **footprints** of the satellite's instruments is referred to as the swath of the sensors.

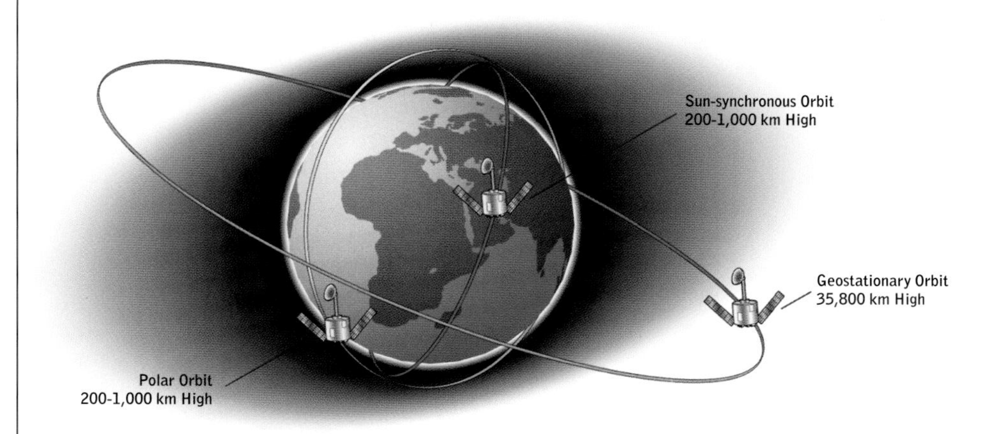

Geostationary and polar orbits (including sun-synchronous)—as well as other orbits not pictured—all offer different perspectives and applications in remote sensing, communications, and satellite navigation.

Picture Element (Pixel)

A picture element, or pixel, is the smallest spatial unit of measurement in a digital image, the most basic element of an image containing information.

It is analogous to a film camera, which creates a picture by collecting light through its lens and exposing a rectangle of film to that light. The photons of light cause a reaction in the chemical particles on the film. When the negatives are developed and then light is projected through them onto photographic paper that has its own light-sensitive chemicals, the picture becomes visible. The printed picture is an arrangement of tiny dots of ink, with each dot having a different darkness and/or a different color from surrounding dots. The more closely packed the dots, the sharper the image and the more information it reveals about the object that was photographed. The dots of ink from chemical film are the equivalent of the pixels in a digital image.

Images collected by digital cameras on **satellites** observe clouds, land, or sea in thousands or even millions of individual pixels by collecting photons reflected from those spots and measuring their intensity (the relative number of photons received) or their wavelengths (the specific characteristics of the photons depending upon the wavelength of the light collected).

These pixels can then be electronically reassembled into a digital image. The more pixels observed in a geographic area, the greater the spatial resolution achieved. The greater the spatial resolution, the smaller the individual feature on the Earth that can be identified in the picture.

Remote-sensing Sensors & Techniques

Optical-imaging Sensor

An optical-imaging sensor records a picture using the light coming from the object or scene being imaged. The earliest **remote-sensing** instruments were film cameras, with film returned to Earth in capsules and retrieved in order to process the film. Later, electronic cameras were used, with images transmitted electronically to the ground by radio. The most commonly used sensors for weather, oceanography, and Earth resources sense reflected light in the optical band. These sensors require daylight to operate, with the exception of low-light-level cameras that can record features such as city lights on the dark side of the Earth.

A compilation of several data sets of the type produced by NASA's Earth Observing System satellites includes cloud cover, vegetation, and the temperature at the sea surface. The composite image reveals the complex system of interdependent natural phenomena that affect Earth. El Niño, a huge area of abnormally warm water, appears red.

Electro-optical Scanner

An electro-optical scanner is an electronic optical system that senses **electromagnetic radiation** at optical (UV, visible, or IR) wavelengths. Its major elements include an optical system consisting of lenses, mirrors, apertures, modulators, and dispersion devices; detectors; and a signal processor, a computer that turns the electrical signals from the **detectors** into the desired output data.

Detector

A detector provides an electrical signal proportional to the intensity of radiation on its surface, generally some kind of semiconductor. Any device sensitive to electronic energy that processes information about the energy reaching it is a detector. For optical **remote sensing,** a detector would be an electronic device such as a **charge-coupled device** (CCD). In optical remote sensing, a **telescope**'s mirrors or **lenses** direct light to the detector where the photons are sensed. In microwave remote sensing, the antenna is the equivalent of a telescope, directing electromagnetic waves to the "feed," or detector, where the received energy is then sensed.

Charge-coupled Device (CCD)

A CCD **detector** is an array of closely spaced and microscopically small light-sensitive solid-state devices on the surface of a semiconductor, such as silicon carbon. The devices are sensitive to photons of light and emit an electrical current proportional to the intensity of the photon barrage being received. The resultant currents are sorted by a microprocessor, transmitted as a radio signal, and re-created into an image.

Imaging Techniques (Staring and Scanning)

Optical sensors create images by collecting light using **detectors,** or optical elements. This can be done by staring or scanning. Staring—pointing at one spot on the **Earth** and taking what is essentially a still picture—produces the highest resolution imagery. This technique requires a detector array with a two-dimensional arrangement of elements. Moving either the entire camera or just the optics (lenses or mirrors) from side to side or in a circular motion and taking a continuous series of images in order to create a continuous mosaic of images is called scanning. If the device has only a single line of detectors, a two-dimensional image can still be created through scanning motions of the camera's optics, or simply by holding the detector array arranged at right angles to the direction of **satellite** motion and collecting the image as the **spacecraft** flies along. This is called "push-broom" scanning.

Panchromatic Imaging

Panchromatic imaging is a **remote-sensing** technique that collects all visible light, regardless of color, in wavelengths from 0.4 to about 0.7 microns. Its common analogy is black-and-white photography, in which each **pixel** of the photograph is either black, white, or a shade of gray. Since this device collects all photons received, without rejecting those of various wavelengths, it is more sensitive than cameras designed to record only one light band, and creates the sharpest, highest resolution pictures.

Multispectral Imaging

Multispectral imaging is a technique in which a **remote-sensing** instrument simultaneously collects bands of light in multiple wavelengths. While black-and-white imagery may be the sharpest and easiest to interpret, it cannot offer information revealed in the colors of light being brought through the lens. For instance, green seawater is an indication of high biological content, such as algae; blue water is relatively low in such content. A camera that can collect both blue and green light separately and then create an image in which each ocean **pixel** is evaluated for its shade of blue or green allows oceanographers to study such things as coastal pollution, and lets the fishing industry find areas likely to be rich in fish.

Multispectral imaging means that the instrument distinguishes between several fairly broad spectral bands. It is a powerful tool in a host of applications, including studying vegetation and mineralogy.

Hyperspectral Imaging

Hyperspectral sensors collect and record data in hundreds of spectral bands. This much larger number of individual bands provides an opportunity to detect very subtle spectral differences in color that would be missed in **multispectral imagery.**

Computer algorithms can be used to automatically detect certain phenomena in multispectral imagery, and an enhanced multispectral image can be visually interpreted by a trained analyst. Computer analysis is virtually always required for hyperspectral imagery, but a great deal more information can be obtained from a single image.

Types of Spectral Imaging

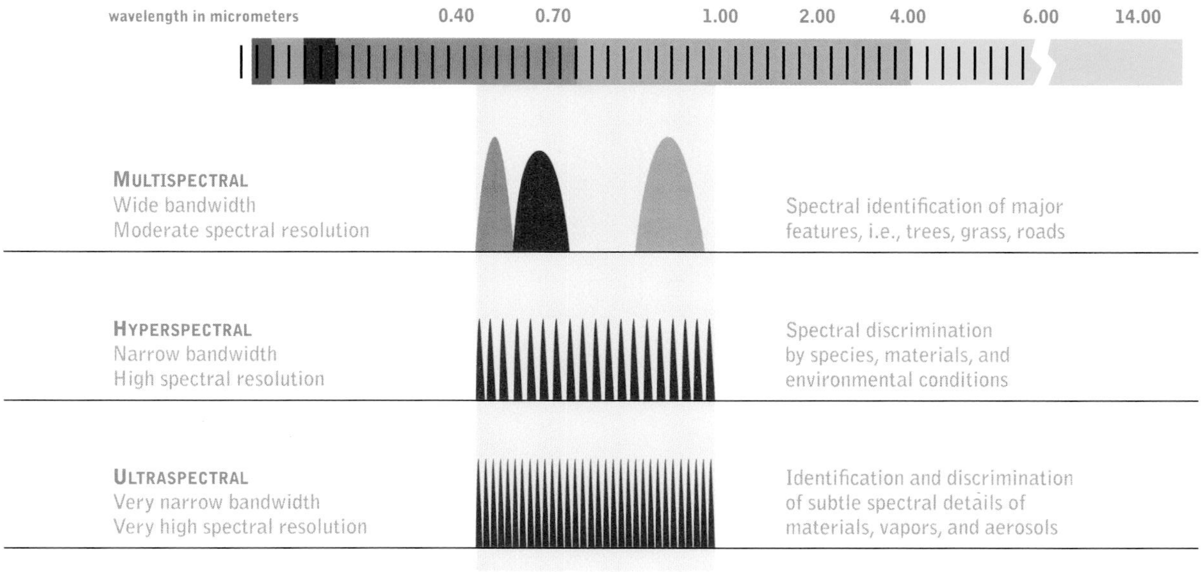

Modern remote-sensing instruments collect light at discrete frequencies and save multiple images, one for each spectral band, then display them individually or digitally combine them to discern different pieces of information. Depending upon how many different spectral bands are collected, an imaging system is classified as multispectral (a few bands), hyperspectral (hundreds of bands), or ultraspectral (potentially thousands of bands).

Ultraspectral Imaging

An emerging field of research still in the conceptual stage, ultraspectral imaging will use thousands of very narrow **bandwidths** for very precise **remote sensing.** When this technology is incorporated into remote-sensing instruments, it will allow analysis of thousands of extremely narrow bandwidths, which will allow the identification of atmospheric aerosols (such as smoke) and gas plumes (such as effluent gases from factories).

Infrared Sensor

An infrared (IR) sensor is one designed to collect radiated energy from the **Earth** in the infrared band, which indicates heat. Some phenomena, such as forest fires, can generally be seen both in the visual bands because of the sunlight reflected from smoke, and in the IR bands due to heat radiated by the fire. Another common application of infrared sensors is the mapping of sea-surface temperature from the world's oceans.

Some infrared instruments require cooling of their optics and **detector** elements so that the longer wavelength infrared photons can be detected. In the near infrared band (wavelengths from 0.7 to 1.1 microns), radiators are used to passively cool the instrument, but for the thermal infrared bands (wavelengths from 1.1 to 14 microns) the instrument must be made very cold.

This process is referred to as cryogenic cooling, which can be done either through the use of dewars (essentially large thermos bottles containing liquid inert gases, such as nitrogen) or through cooling devices that pump heat away from the optics. These cooling devices are high-technology versions of the mechanisms used in refrigerators.

Ultraviolet Sensor

Ultraviolet systems sense in the ultraviolet (UV) light band—wavelengths shorter than 0.4 microns.

Various constituents of the atmosphere, such as gaseous chemicals like ozone, absorb UV rays from the **sun,** providing a natural protective shield to living things on the **planet.**

Different chemical constituents of the atmosphere, and variability of the concentration of chemicals in either geographic coverage or varying altitudes, cause variability in the absorption of ultraviolet radiation. UV sensors provide powerful tools to observe the chemical properties of **Earth's atmosphere.**

Radiometer

A radiometer is an instrument that measures intensity of **electromagnetic radiation** in some part of the **spectrum.** When the radiation being measured is light from the narrow visible-light spectral band, the term photometer can be substituted, since the device is essentially measuring the intensity of the incoming photons of light.

Spectrometer

A spectrometer is a **radiometer** that includes a component such as a prism or a grating to break the incoming radiation into discrete wavelengths, separate them, and send them to detectors to measure the radiation at different wavelengths. A spectrometer can be used to determine various properties, such as chemical composition, of the object being viewed. One type of spectrometer, called a spectroradiometer, passes multiwavelength radiation through a slit and then disperses the radiation in narrow wavelength bands that can be separately measured. Most space-based optical **remote-sensing** systems are spectroradiometers.

Sounding Sensor

A sounding sensor, or sounder, is an infrared **radiometer** or a **spectrometer** (or a hybrid instrument called an imaging spectrometer) that is used to measure the temperature of the atmosphere and/or its moisture content (humidity) at various altitudes. This technique is a high-technology equivalent of sending aloft thousands of radiosonde sounding balloons from spots all over the world. Balloon soundings—measurements of temperature and humidity made by small sensor packages carried up though the atmosphere by balloons—are highly accurate because they have very sensitive instruments that are actually in the air mass being measured. They are, however, expensive and logistically impossible to launch simultaneously all over the **planet,** especially from the sea. **Satellite** sounding is a **remote-sensing** alternative that measures with lower accuracy and at a limited number of levels, but creates a worldwide sampling of atmospheric data critical to modern numerical weather-prediction models.

Laser

A laser, which stands for Light Amplification by Stimulated Emission of Radiation, is a device that emits a highly focused and directional beam of light at one very discrete wavelength. The beam is coherent, meaning it is "organized" with each photon moving in coordination with other photons.

The ubiquitous use of such simple devices as laser pointers belies how sophisticated such devices actually are. They work by "pumping" a lasing medium (generally a homogeneous gas or crystal of one specific element or chemical) by exciting the atoms with either a flash of light or an electrical jolt. When the atoms are sufficiently energized, they emit photons at a specific wavelength and phase, which are then focused by a pair of mirrors on either end of a tube. One of those two mirrors is only half-silvered, so that it does not reflect light. The photons move back and forth between the pair of mirrors, escaping the half-silvered end in the form of an intense beam of monochromatic (one wavelength) light.

Lidar

Light Detection and Ranging (lidar) is an **active remote-sensing** instrument that uses one or more lasers to generate a beam or beams of monochromatic light at hundreds or thousands of pulses each second down from a **spacecraft** through the atmosphere to the **Earth**'s surface; some of the light is reflected back from the surface to the spacecraft. A radiometer is then used to collect the reflected light, carefully measuring the time it took for the pulse to travel round-trip.

Lidars are used to measure such features as the depth of the ocean in shallow water by comparing the arrival times of the light reflected from the sea surface versus the light that penetrated the water, reflected off the seafloor, and returned to the instrument. They can also be used to measure the properties of materials carried in the atmosphere, such as clouds, smoke, and other pollutants.

Laser Altimeter

An application of **lidar** is a laser altimeter, an instrument in which a **laser** beam is directed down to the **nadir** location directly beneath a **spacecraft,** and precisely measures the time required for the light to arrive back at the spacecraft.

Such a device is used in developing precise topographic maps or three-dimensional models of city buildings, as well as for monitoring long-term changes in the topographic shape (and thickness) of the polar and Greenland ice sheets, an indication of climate change.

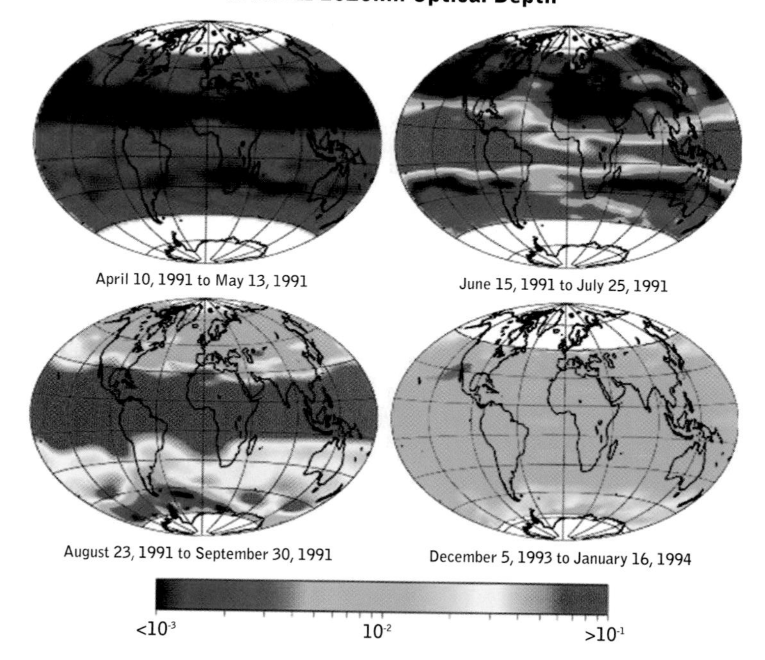

In June 1991, the Philippines' Mount Pinatubo spewed dust particles into the atmosphere. The NASA Stratospheric Atmosphere & Gas Experiment (SAGE II) sensor tracked the concentration of aerosols over a three-year period, proving that they caused a global temperature decrease and influenced ozone levels in the upper atmosphere.

Microwave Sensor

Microwave sensors measure the radiance from the **Earth** (land and sea) and atmosphere in the microwave band, generally at frequencies of 10.7 to 85 gigahertz. Just as various phenomena on the Earth, such as fires, ocean currents, and vegetation, radiate infrared energy that can be observed with IR sensors, features such as atmospheric moisture, sea ice, ocean water, and land radiate very low power energy—referred to as microwave energy—at wavelengths much longer than IR. Passive microwave sensors are essentially radio **antennas** that scan the Earth's surface, recording the

OBSERVING EARTH IN A NEW WAY

Diane L. Evans

RADAR IS A LOWERCASE ACRONYM FOR RADIO DETECTION AND RANGING. SYNthetic aperture radar (SAR) is a mapping technique that makes use of the Doppler shift resulting from the motion of a platform to synthesize the effect of a large antenna aperture. SAR transmits pulses of microwave energy toward Earth and measures the strength and time delay of the energy that is scattered back to the antenna. One of the most useful features of SAR is its ability to collect data over virtually any region at any time, regardless of weather or sunlight conditions. The radar waves can penetrate clouds, and under certain conditions can also see through vegetation, ice, and extremely dry sand. In many cases, using radar provides the only way scientists can explore inaccessible regions of Earth's surface.

Conditions on the Earth's surface influence how much radar energy is scattered back to the antenna. An area with a variety of surface types, such as hills, trees, and large rocks, scatters more energy back to the radar than a smooth surface. Using different wavelengths of radar makes it possible to differentiate different scales of roughness and to penetrate different amounts into surfaces.

The longer L band and P band radar wavelengths, of approximately 25 centimeters and 70 centimeters, respectively, are particularly useful for looking beneath surfaces and penetrating vegetation canopies, while X band and C band, with wavelengths of about 3 centimeters and 5 centimeters, respectively, are sensitive to finer scales of surface roughness and the tops of vegetation canopies.

It is also possible to transmit and receive horizontally and vertically polarized radar waves. For example, when data are acquired with HH polarization, the wave is transmitted from the antenna in the horizontal plane, and the antenna receives the backscattered radiation in the horizontal plane. With HV polarization, the wave is transmitted horizontally and received by the antenna in the vertical plane. It is the interaction between the transmitted waves and the Earth's surface that determines the polarization of the waves received by the antenna. Multipolarization data contain more specific information about surface conditions than single polarization data. Multipolarization data are particularly useful to scientists studying vegetation because the data allow them to discriminate different types of crops and to measure the volume of trees contained under the canopy of a forest.

Synthetic Aperture Radar

The first civilian spaceborne SAR was launched aboard the Seasat satellite on June 26, 1978. Seasat operated at L-band with HH polarization and a fixed incidence angle of approximately 23 degrees. Although the mission ended prematurely on October 10, 1978, due to a failure of the vehicle's electric power system, it laid the groundwork for future SAR missions. Based on the results of Seasat for both ocean- and land-surface mapping, SARs were selected to be flown on the space shuttle, including NASA's Shuttle Imaging Radar-A (SIR-A) in 1981, and SIR-B in 1984, which had the capability to view Earth's surface at variable angles.

The most advanced civilian SAR ever built, the Spaceborne Imaging Radar-C and X-Band Synthetic Aperture Radar (SIR-C/X-SAR), flew aboard two space shuttle flights in April and October 1994. Data acquired during these flights provided scientists with a wealth of information about Earth's changing environment and opened up new application areas such as archaeology.

SIR-C was a two-frequency fully polarimetric radar system that operated at L band and C band. It was the first spaceborne radar with the ability to transmit and receive horizontally and vertically polarized waves at both frequencies, enabling SIR-C to acquire simultaneous images with HH, VV, HV, and VH polarizations. The SIR-C antenna was an active phased array, with an electronically steerable beam that allowed images to be acquired from 15- to 55-degree angle of incidence. SCANSAR was demonstrated with SIR-C by steering the antenna to four different elevation angles during each synthetic aperture interval, resulting in a swath width of 225 kilometers.

X-SAR, which was provided by the Deutsche Forschungsanstalt für Luft- und Raumfahrt e.V. (DLR), and the Agenzia Spaziale Italiana (ASI), operated at X band with VV polarization. The X-SAR antenna was mounted on a supporting structure that was tilted mechanically to align the X-band beam with the L-band and C-band beams. SIR-C and X-SAR could be operated as stand-alone radars or in conjunction with each other, resulting in a three-frequency capability for the total SIR-C/X-SAR system. The width of the ground swath varied from 15 to 90 kilometers, depending on the orientation of the antenna beams; the resolution of the radars varied from 10 to 200 meters.

SIR-C/X-SAR data have been used to produce maps of vegetation type and biomass; snow, soil and vegetation wetness; and the distribution of wetlands. In addition, SIR-C/X-SAR science team members experimented with SAR interferometry (InSAR) to produce three-dimensional images of the Earth's surface. During the October flight, the radars were flown twice over nearly identical orbit passes to generate interferometric data at dozens of sites around the world, and digital elevation models were generated at all three radar frequencies simultaneously.

Based on the success of SIR-C/X-SAR, NASA, DLR, and ASI continued their collaboration—the Shuttle Radar Topography Mission (SRTM)—with the National Geospatial-Intelligence Agency (NGA) as an additional partner. SRTM collected topographic data over all of Earth's land surface that lies between 60° north and 56° south latitude, creating the first-ever data set of elevations for nearly 80 percent of Earth's land surface. To obtain two radar images taken from different locations, SRTM used single-pass interferometry, with the SIR-C C-band and X-SAR antennas in the shuttle payload bay and two additional radar antennas attached to the end of a mast that extended 60 meters from the shuttle.

The next step in the NASA SAR Program is a dedicated InSAR mission to provide repeat pass interferometry for seismic and volcanic deformation mapping, ice sheet and glacier velocity measurements, and hazard monitoring and assessment.

DLR, ASI, Russia, Japan, Canada, and the European Space Agency have flown and/or plan to fly spaceborne SARs in the future, leading to the potential of a multi-platform constellation. Plans are also being formulated for radars for exploration of the moon, Mars, and Jupiter's moon Europa. ■

The space shuttle *Endeavour* uses SRTM radar to map Earth's topography from 233 kilometers in a composition that draws on art, radar imagery, and Earth-orbital imagery. SRTM uses radar so it can penetrate clouds, and operate without daylight.

variability of this radiated energy. They are generally used to measure rain, sea ice coverage and temperature (an indication of ice age and thickness), and ocean properties such as sea-surface temperature.

Space-based Radar

Radio detection and ranging (radar), invented just before World War II, is a device that transmits high-power microwave energy that moves away from the transmitter, reflects off a target (such as a ship, an airplane, land, or a rain cloud) and is received by an antenna. Radars generally transmit microwave energy with wavelengths in the range of one centimeter to one meter, which corresponds to a frequency range of 300 megahertz to 30 gigahertz. Active microwave **remote sensing** is conducted by space-based radar that uses the same process as ship or airborne radar. Space-based radar transmissions are directed downward in order to bounce the radio waves off the phenomenon being observed, such as the Earth's surface land, sea, or ice; water droplets in clouds; or even water vapor being carried by wind.

Synthetic Aperture Radar (SAR)

Radar beam width is inversely proportional to the size of the antenna used. A small antenna, therefore, results in a large **footprint** and a low-resolution image. A synthetic aperture radar, or SAR, is a small imaging radar that uses a spacecraft's motion to "synthesize" a very long antenna by combining reflected signals along its line of flight into a combined image. This process can create very high resolution images of the **Earth** from a relatively small antenna.

About 1,500 high-power pulses a second are transmitted toward the target or imaging area, with each pulse lasting 10 to 50 microseconds. When a pulse hits its target, the energy is scattered in all directions, with some reflected back toward the radar's antenna. These echoes are processed to determine their strength and possible changes in polarization (horizontal or vertical) from the transmitted pulse, and these pieces of information are converted to digital data for processing into an image. SAR imagery has been helpful in dozens of applications, including **oceanography, mapping and charting,** sea ice monitoring, and **agriculture.**

Interferometric SAR

Interferometric SAR is a technique that processes the digital image data from two **synthetic aperture radar** (SAR) images of the same place on **Earth** to compare both the amplitude and phase information in the **radar** echo. This will yield more information than is possible

by viewing one or the other image alone. Combining the "phase measurements" from two images can be used to infer such information as the topography, or precise elevation and shape, of the Earth's surface.

If there has been some passage of time between the two SAR images, interferometry comparison canbe used to determine slight changes in the shape of the surface, an indication of distortion of the Earth, such as can be found near fault lines or earthquake areas. These techniques are enormously valuable in **mapping and charting** and in warnings of potential natural disasters.

Scatterometer

A scatterometer is a specific application of radar technology to measure the speed and direction of wind near the ocean surface. It does this by transmitting pulses of microwave energy at the ocean surface in multiple beams, typically with frequency of 14 gigahertz or another suitable frequency according to the instrument's design. These beams strike the ocean surface and are scattered by surface waves—waves that are driven by wind. The resultant echoes are received by multiple antennas, and compared for Doppler shifts (frequency changes caused by motion of the waves).

Processing of these multiple echoes yields information about the direction and height of the waves, which can then be used to infer the speed and direction of the wind at the surface of the ocean. These observations are the functional equivalent of having wind measuring devices, such as anemometers, at thousands or millions of locations all over the Earth's oceans—providing enormous benefit to **meteorology** and **oceanography from space.**

Radar Altimeter

A radar altimeter directs its transmission of microwave energy directly downward and senses the return in order to measure the distance between the surface of the **Earth** and the **satellite.** The measurement of the time between the transmitted pulse and the received echo of reflected energy by the antenna is used to compute the precise **altitude** of the satellite. This procedure may be used to find the **spacecraft**'s altitude in order to determine the orbital characteristics of its flight, but can also be used to measure the changing shape of the Earth below the satellite's orbital path.

This continuous profile of the shape of the Earth's surface can then be used to develop topographic maps of the land, identify the locations of sea ice, or measure the shape of the ocean surface, which at varying scales can be an indication of bottom topography (bathymetry), ocean currents, or surface waves (wind-driven seas and swells).

Two digital data sets collected by satellites flying directly overhead at different times—a digital elevation model database created by the Shuttle Radar Topographic Mapping (SRTM) system aboard the space shuttle and a multispectral image taken by Landsat—were combined to create this false 3-D image of Mount Ararat in the plains of eastern Turkey.

Remote-sensing Applications

METEOROLOGY

Meteorology, the science of understanding and predicting the **Earth**'s weather, has been profoundly advanced by satellite remote sensing. As recently as two centuries ago, people had very little information to help them in forecasting: a view of the clouds overhead, barometric pressure (whether it was rising or falling), temperature, and local winds. Sailors had only these clues, plus the arriving waves generated by distant storms. Weather balloons were introduced to observe conditions aloft, but these were still only local observations. With the advent of the telegraph and then radio, weather observations from distant reporting stations could be analyzed, and this accomplishment helped weather forecasting immensely.

On April 1, 1960, meteorology was changed forever with the very first television picture from space taken by the world's first meteorological **satellite,** called **Television Infrared Observation Satellite** (TIROS). TIROS was a **low Earth orbit** remote-sensing system. On December 7, 1966, the first meteorological satellite in **geostationary Earth orbit,** the **Applications Technology Satellite** (ATS), was launched. The innovation of observing clouds from orbit allowed forecasters to see for the first time the big picture of complex weather patterns.

Later, the invention of satellite data-collection systems, such as temperature and moisture sounders, ozone mappers, and wind-observing **scatterometers,** enabled satellites to do more than just collect images of the clouds.

Sensors on meteorological satellites collect observations of many different parameters that are transmitted to supercomputers. These computers compose three-dimensional numerical models of the Earth's entire atmosphere. The computers combine satellite observations with surface and aircraft observations to determine the current state of the atmosphere. Then, using fluid dynamics equations, they "move" the atmosphere into what it "should" look like hours, days, or even weeks in the future.

Meteorologists use these predictive presentations of the atmosphere, provided in the form of computer-drawn charts, to produce their forecasts.

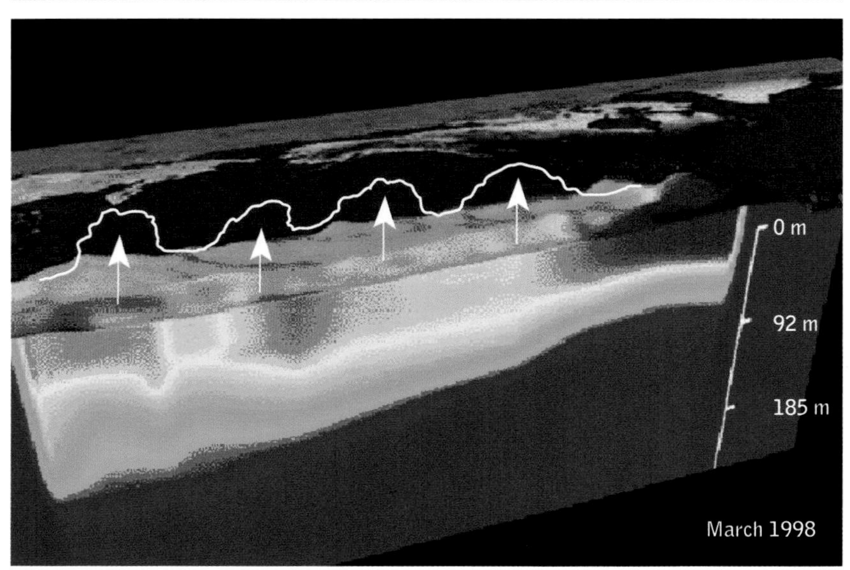

Satellites collect observations of clouds and ocean temperatures to create models of their connections. A two-dimensional east-west "slice" of the Pacific Ocean (top) shows a normal formation of rising clouds over warm water (red) and clear air subsidence over cooler water (blue). In El Niño (bottom), low cloud cells extend across the area.

CLIMATE

Climate-change research is multifaceted and attempts to answer questions such as: What are the trends and patterns in the **Earth**'s climate system, including the atmosphere, oceans, glaciers, sea ice, and the **biosphere?** What are the processes that affect the dynamics of climate, including internal factors such as water vapor, clouds, and heat transfer by the atmosphere and oceans, and external factors such as solar variability and volcanic activity?

Large-scale processes, such as the much publicized El Niño features that occur in the Pacific basin, the ozone hole over the South Pole, and changes in the polar ice caps, are all indicative of climate changes that are affecting human lives and economies around the globe. All are still poorly understood due to the complex interrelationship of natural forces and human-produced effects. Given that satellite **remote sensing** has developed systems and techniques for monitoring virtually all the most important atmospheric, oceanographic, and terrestrial phenomena that are indicative of climate change, research using space systems has become a critical element in the international climate-change research program.

The two keys to effective research of climate change are calibration of remote-sensing systems so that observations by the same or similar instruments taken over time can be compared for changes, and the continuity of observations over many years so that these changes can be observed and studied. NASA's **Earth Observing System** is an example of just such a research capability.

Oceanography

Oceanography, the study of the world's ocean, encompasses many different but closely related scientific disciplines, among them fluid mechanics and thermodynamics (physical oceanography), biology, chemistry, and geology. Knowledge in all these fields has been greatly advanced through the advent of **satellite remote sensing.** Given that more than 70 percent of the **Earth** is covered by its oceans, the interactions between the atmosphere and the ocean affect both land and sea. When the field of **meteorology** evolved with the coming of the space age, oceanography from space was born almost at the same time. The **Nimbus** satellites, first launched in 1964, were **research satellites** flown by NASA to increase our understanding of the atmosphere and oceans and to test new instrument technology that could be used in future **operational satellites.**

The **Coastal Zone Color Scanner** (CZCS) flown on this series of satellites was the world's first space-based ocean remote-sensing system. In 1978 NASA's evolution of ocean remote sensing resulted in the **Seasat** mission, which carried multiple sensors—all focused on ocean sensing. Among the many techniques for ocean remote sensing developed through the years are infrared sensing (sea-surface temperature), multispectral optical imaging for ocean color (coastal dynamics, biology, pollution monitoring, sea ice, and bathymetry), passive microwave sensing (sea ice, wind speed, and ocean temperature), **lidar** (bathymetry, water clarity, and sea ice), **synthetic aperture radar** (sea ice, waves, coastal dynamics, and ship wakes), radar altimetry (currents, water mass fronts and eddies, wind speed, and sea ice), and scatterometry (near-surface wind direction and speed). Satellites are also used to relay data collected from ocean buoys and other surface-based sensors.

Agriculture

Agriculture from space is the observation, using **satellite remote-sensing** systems, of natural and man-made conditions that affect farming. Farmers have many critical decisions to make, such as when and where to plant which crops, when and where to apply fertilizer or pesticides, and when and where to irrigate. Timber companies, ranchers, and dairy farmers share many of the same interests. All these decisions can have profound consequences in the financial success or failure of a farmer's business, given the profound sensitivity of farming to the environment and the extremely competitive nature of the market for his crops or products.

The national and global economic issues involved in agriculture also lead governments, economists, and commodity traders to carefully monitor environmental conditions that affect agriculture. Space systems that are used by farmers and others include **multispectral optical imaging** (vegetation health and maturity, mapping, forest fires, soil analysis, and terrain), **meteorology** from space systems (precipitation, flooding, and storms), climate from space systems (draught, seasonal to interannual climate change), and imaging radar and microwave sensing (soil moisture, terrain, and flooding). The **Landsat** program and its international counterparts, **SPOT** and **INSAT,** have all become common tools in agricultural applications.

Forestry

A specific application of **agriculture** from space is forestry. Satellite **remote-sensing** systems are commonly used by governments and the timber and paper industries in monitoring the health and mapping the extent of forest areas, including the variety of trees and other vegetation, and in detecting and monitoring forest fires and supporting their control.

Optical imaging instruments, particularly multispectral and infrared sensing systems, are used for these purposes. **Synthetic aperture radar** has been shown to be particularly useful in assessing tree density, health, and type. **Lidar** can be used for assessing forested areas for tree canopy height, vegetation type, and terrain.

Conservation

Satellite **remote-sensing** instruments and handheld photographs taken by astronauts of ecosystems, natural resources, and human-developed infrastructure and activities that have an impact on the natural environment are very powerful tools for conservation.

Digital remote sensing using **optical imaging** systems, particularly **multispectral imaging** (such as **Landsat**) and **synthetic aperture radar** (SAR), have been demonstrated as capable tools in monitoring important resources such as wetlands, rivers, streams, lakes, forests, and mineral deposits.

These systems, plus others capable of **mapping and charting** from space, are used to monitor and plan for population expansion, construction, water management projects, and mining. The systems also support governmental and private-sector programs seeking to regulate this kind of activity so that sustainable economic development can proceed without depleting natural resources.

Conservation of these resources, the key to sustainable development, often begins with an assessment of current conditions and monitoring of changes. Digital images of a geographic location taken at varying intervals of time can be compared, **pixel** by pixel, and changes in such characteristics as vegetation and surface soil exposure and the building of roads or structures can be easily identified and quantified for analysis. Before-and-after images can be striking when placed side by side, but even slight changes in land use or land cover can be detected using these digital techniques for comparing remote-sensing data.

Hydrology

Hydrology from space is the use of satellite-borne sensors to monitor environmental or geophysical parameters that are used in the study of all aspects of water, including precipitation, the condition of wetlands, streams, rivers, and lakes, and flooding.

Also included in this field is water management, such as planning for and building dams and reservoirs, and draining wetlands for construction or agriculture. Satellite **remote-sensing** systems are used by hydrologists to monitor weather from space, to map terrain drainage features through **mapping and charting** from space, and to directly measure lake and river levels through **optical imaging** and **synthetic aperture radar** observations.

Passive microwave sensing has also been shown to be useful in measuring soil moisture.

Surveying several hundred square kilometers, the Enhanced Thematic Mapper instrument aboard Landsat 7 captured the sands and seaweed visible through crystal clear water surrounding the Bahamas. Tides and currents created the beautifully sculpted patterns in much the same way that winds shape the vast sand dunes of the Arabian Desert.

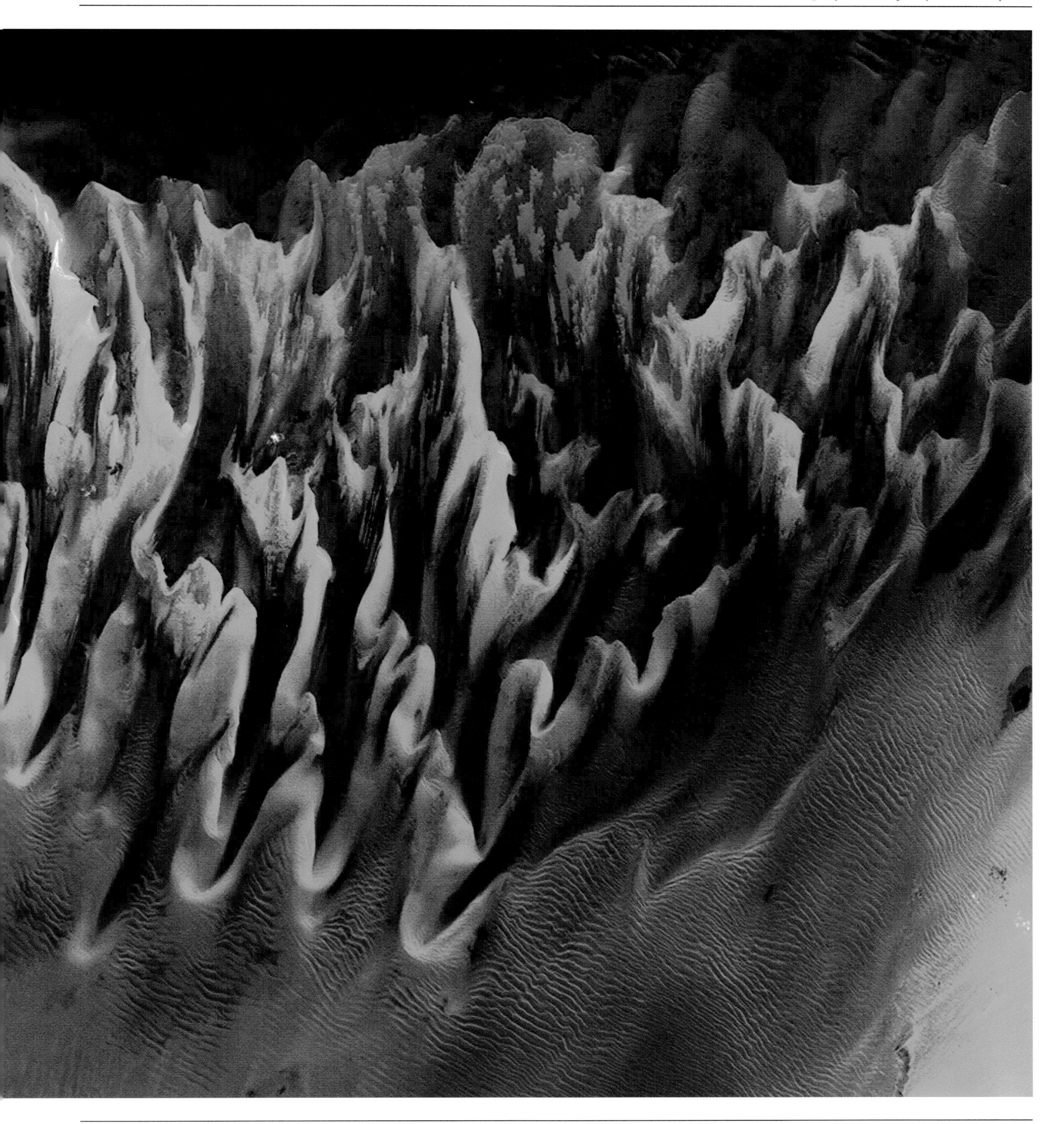

Geology

Analysis of the **Earth**'s surface by **remote sensing** is geology from space. Optical imaging has proven to be extremely effective in determining the mineral content of rock formations and soil because various minerals reflect the **sun**'s light at different wavelengths, providing a signature color for each. Color photographs taken by **astronauts** are very effective in describing geological structure—particularly such features as faults and fracture zones.

The use of **multispectral** and **hyperspectral** imaging sensors to compare the narrow bands of backscattered light allows a much more precise mapping of specific minerals, limited only by the spatial and spectral resolution of the sensor.

The computer-enhancement of multispectral images allows the assignment of strikingly different colors to areas that may be reflecting only slightly different wavelengths of light. These slight differences are very significant because they indicate the presence of different minerals. This artificially created contrast in the displayed multispectral image of a particular geographic area makes the variety of minerals clearly visible to the geologist studying the image.

In hyperspectral sensing, analysts can use a computer to create a detailed geological and vegetation map based on the extremely narrow spectral band combinations sensed.

In addition to optical remote sensing, **synthetic aperture radar** has proven effective in understanding geological features not visible in the optical band. Synthetic aperture radar transmits beams of microwave energy that penetrate very dry soil or sand, reflect off buried solid rock formations, and are received by the radar's antenna. Using this procedure, scientists have observed features such as bedrock or ancient riverbeds under desert sands.

Natural Disaster Warnings

The danger to lives and the economic impact of natural hazards—earthquakes, volcanic eruptions, tsunamis, forest fires, tornados, hurricanes, and floods—can be mitigated to varying degrees based on disaster warnings received from space. Many such hazards are simply naturally occurring worst-case examples of geophysical phenomena already being studied by **remote-sensing** systems.

Operational meteorological satellites—particularly the **Geostationary Operational Environmental Satellite** (GOES)—are specifically designed to immediately detect conditions favorable to tornado development in order to issue warnings to the public. These same satellites are also critical in tracking hurricanes so that coastal areas can be evacuated and lives saved.

Other infrared imaging satellites, including some designed and operated primarily for military purposes, are commonly used to first detect the heat signatures of volcanic eruptions and forest fires. Visible band meteorological sensors can often detect smoke or dust plumes from these same features. Ultraviolet imaging sensors are capable of determining the presence of volcanic dust clouds or smoke, both of which can be difficult to detect because of the similarity of appearance to naturally occurring moisture clouds.

Interferometric synthetic aperture radar (SAR) can detect the deformation of the **Earth**'s crust, such as the bulging shape of a volcano before it erupts or a shifting fault zone, a common precursor to an earthquake. No satellite remote-sensing system has yet to observe a tsunami, or tidal wave, but such an event would be clearly visible in SAR imagery if a person happened to to be observing in the right place at the right time.

The U.S. Geological Survey under the Department of the Interior and the National Interagency Fire Center in Boise, Idaho, use satellite remote-sensing data as an important tool in issuing earthquake and forest fire warnings and to support federal, state, and local government disaster relief efforts.

Mapping & Charting

The use of digital images taken by optical sensor or **synthetic aperture radar** (SAR) to augment or replace surface and airborne surveys has become an accepted method of mapping and charting from space. Maps and charts are used for navigation on land and water, planning construction projects, monitoring coastal ecosystems, and innumerable other applications.

Historically, cartographers (mapmakers) had to rely on surveys made by observers on the ground and lead-line soundings taken from boats and ships. With the advent of aerial photography, surveyors were afforded a powerful tool in mapping, particularly with the help of stereo photographs to depict terrain elevation features. Space systems have simply taken this technique to much higher altitudes, allowing broad-area mapping on a global basis.

All photographs, however, contain distortion caused by terrain undulations, camera characteristics, and viewing angle (which varies for each **pixel** across an image). Orthorectification is the process that

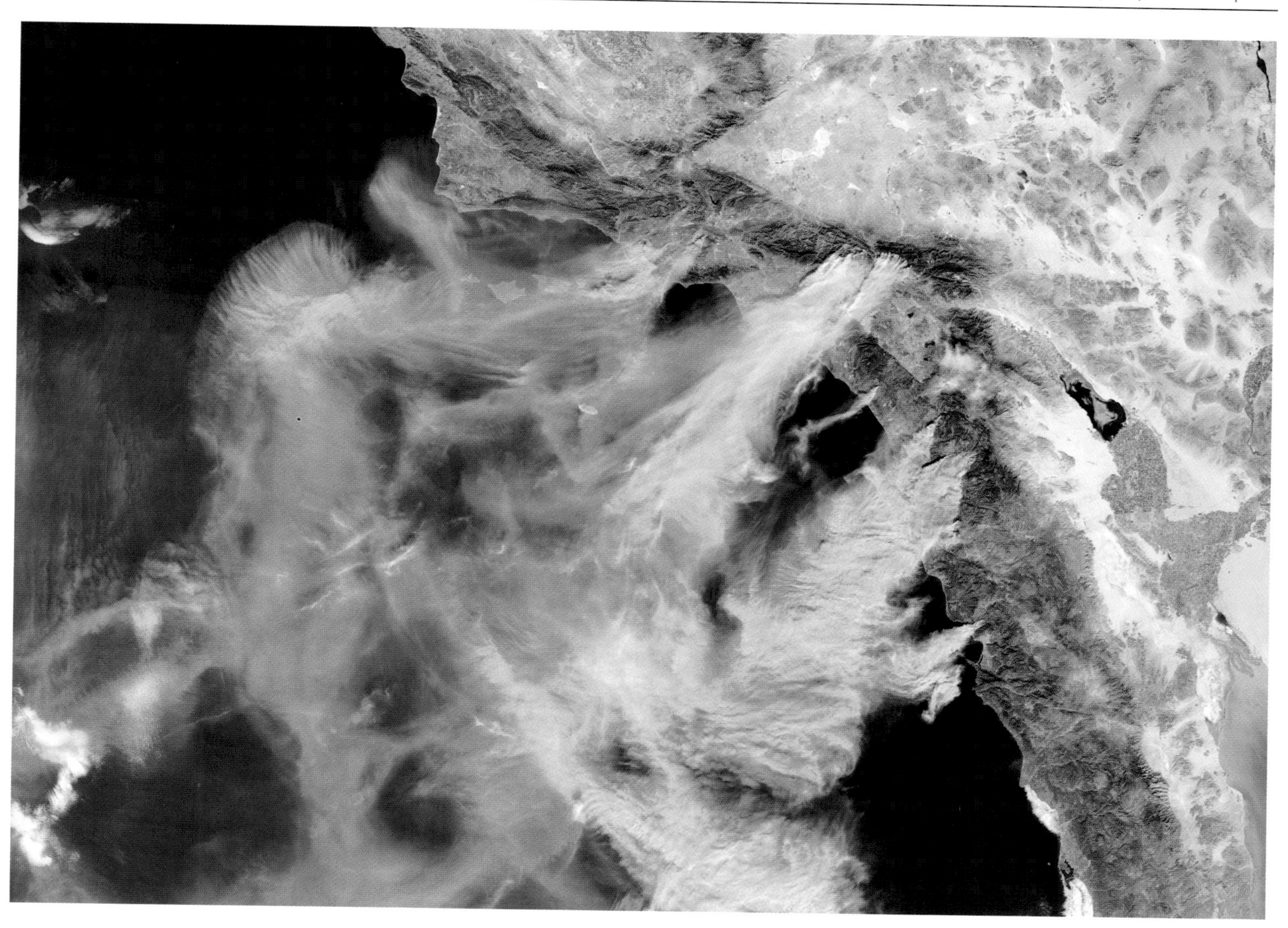

integrates the **satellite** orbital and sensor information, ground control points, and digital elevation data so that an image is produced that can be overlaid on a map—or can be used to create a map. Orthorectification ensures that the image has a uniform scale, a known accuracy, and is geo-referenced (meaning that each pixel's location is precisely known with respect to the **Earth**'s surface).

With varying geographic coverage, spatial resolution, and spectral resolution, a number of satellite systems are used to collect orthorectifiable images for mapping and charting. **Landsat** and its international counterparts **SPOT** and **INSAT** are used for low- and medium-resolution mapping (as compared to airborne observations), and commercial remote-sensing systems—including **IKONOS, QuickBird,** and **OrbView**—are used for high-resolution mapping applications.

Stereo images taken by these optical systems can be used to create three-dimensional images for terrain analysis. SAR can also be used for mapping, with the added benefit of being able to create maps of areas that are almost always shrouded by clouds. Clouds prevent optical imaging, but they are invisible to radar except when they contain heavy amounts of precipitation. **Interferometric SAR** can be used to create terrain maps similar to stereo-optical images. A near-global terrain map of the Earth is currently being created using data collected by the Shuttle Radar Topography Mission (SRTM) flown in 2000.

The Moderate-resolution Imaging Spectro-radiometer (MODIS) aboard the NASA Earth Observing System satellite Terra captured this image of huge wildfires that raged across southern California on October 26, 2003. The visible spectral band shows the smoke; the infrared band the heat from the fires. Hundreds of homes were destroyed.

Gravity & Magnetic Field Measurements

Measurements of the **Earth**'s gravity field and magnetic field are made from space. In fact, only by using space

EXPLORING THE OCEAN FROM SPACE

Sylvia A. Earle

OWING LARGELY TO THE USE OF SPACE TECHNOLOGIES, MORE HAS BEEN LEARNED about the waters of the world in the past few decades than during all preceding history. Through the eyes of astronauts, cosmonauts, and instruments lofted into space, the view of Earth as a "blue planet," dominated by water, fundamentally transformed the way humankind regards the ocean—a vast and vital liquid realm that governs the way the world works.

Looking at ourselves from afar we came to see the planet as a whole, with all landmasses embraced by one enormous, flowing body of water, laced to the land by ribbons of fresh water, capped by frozen water at both poles, and linked to the skies above by wreaths of water as vapor. Use of manned and unmanned spacecraft greatly accelerated acquisition of information and insights that in turn provided new understanding of the enormous influence the ocean has on temperature, weather, climate, oxygen production, and other grand planetary processes. Scientists previously had been limited to looking at small aspects of the ocean.

More than two centuries ago, Benjamin Franklin used the ships' logs from sea captains to piece together a pattern that led to the first known map of the Gulf Stream, a prominent but previously undefined ocean feature. When the H.M.S. *Challenger* set out on the first global expedition to explore the ocean in 1872, information was gathered by the scientists aboard, one data point at a time, often from the deck of their rolling ship. For many decades thereafter, simple things such as measuring the temperature of surface waters in the open sea were accomplished by scooping a sample of water, reading a calibrated thermometer, logging the information, and noting the location of the ship based on use of a compass, a clock, and a handheld sextant. Prior to the advent of satellites that provide accurate positioning, just getting back to the same precise place in the ocean required a fair measure of luck as well as significant navigational skill.

Most ships today are equipped with global positioning systems linked to orbiting satellites that provide unprecedented accuracy about their location. Charts defining land, sea, and undersea terrain now combine acoustic survey data from ships with information gathered by instruments aboard spacecraft. In recent years, instruments orbiting 800 kilometers above the Earth have even been used to chart an astonishing number of previously unknown features that are far below the ocean surface. Satellites measure the subtle differences in sea level that mirror the shape of seafloor features and gravity anomalies that indicate underwater masses of rock.

Comprehensive knowledge of surface ocean temperature is now known from decades of data acquired globally by satellites. Energy in the microwave band of the electromagnetic spectrum is able to penetrate clouds—unlike visible light and infrared energy—thus providing a clear view of the ocean below. Ocean-sensing radars aboard satellites using the microwave band actively probe the sea surface and measure ripples that in turn can be used to estimate wind speed and direction—from hundreds of kilometers in the sky.

Views from the skies above also give new insights concerning surface currents, wave patterns, salinity, biological productivity, and even the migration of certain large sea animals. Bluefin tuna, sea turtles, sea lions, elephant seals, and whales have been outfitted with devices that monitor the animal's temperature, location and depth when diving, and water temperature, with data transmitted to satellites that relay the information to researchers. Transmitters aboard ships are also being used in some places to monitor fishing activity and other ship traffic.

To gauge the nature and extent of biological productivity near the sea surface and locate and assess polluted waters, satellites sense the presence of pigments, especially chlorophyll in microscopic plankton, as well as measure shifts in temperature. Different types of plankton as well as intertidal sea grass beds and kelp forests show up at different frequencies in the electromagnetic spectrum.

Since electromagnetic radiation does not penetrate far into the sea, other methods determine what is going on within the ocean itself. In the 1990s, special floating buoys named Argo were developed to collect data below the surface to complement observations from the orbiting Jason 1 satellite. Argo floats are deployed from ships or aircraft and sink to a preset depth, usually 1,000 to 2,000 meters, where they drift with deep currents. Every few days a float ascends, measuring temperature and salinity on the way, then transmits the data and its position before sinking back down to resume its journey. The combined system—the Integrated Global Observing Strategy (IGOS)—is part of an international effort to develop a network of monitoring stations. An international initiative known as "GOOS"—the Global Ocean Observing System, involving satellites in the sky and instruments in the sea—will eventually monitor the ocean globally, also assess biological activity, water chemistry, and, over time, provide data to evaluate climate change and other basic Earth processes.

Already, oceanographic data acquired from spacecraft, coupled with long-term observations on the land and at sea, have made it possible for scientists to unravel mysteries concerning the causes of El Niño and La Niña, hurricane cycles, patterns of drought, flooding, and other events that demonstrate the inextricable connections among land, air, ocean. We now understand that the sea is a dynamic, living system that makes life as we know it possible.

We also now know, based on the unprecedented global overview of ocean processes, that humankind has the power to alter the nature of the sea. Compelling documentation exists of recent human impacts on the ocean in terms of shoreline modification, coastal and open ocean pollution, decline of coral reefs, and massive reduction of marine wildlife. New computerized mapping programs that accurately connect information stored in a database to the correct points on a chart are yielding new insights into natural and human-induced changes in the sea as well as on land.

The heightened awareness about the significance of Earth's ocean has helped drive the search for water, and thus life, elsewhere in the universe. Knowing that the red planet, Mars, was once blue with an ocean raises questions about the possibility that life was once abundant there, and may still be present. The existence of liquid water under a thick layer of ice on one of Jupiter's moons, Europa, has aroused considerable enthusiasm for sending probes to explore that distant world.

The search for water and for life elsewhere in the universe is creating profound appreciation for services Earth's ocean provides. Without it Earth would resemble bleak and barren Mars. Far more valuable than a place to play or extract food, oil, or minerals or put wastes, the ocean's role as the backbone of our life support system is now being acknowledged. While more has been discovered about the blue part of Earth since the middle of the 20th century than during all previous time, the greatest era of exploration has just begun. ■

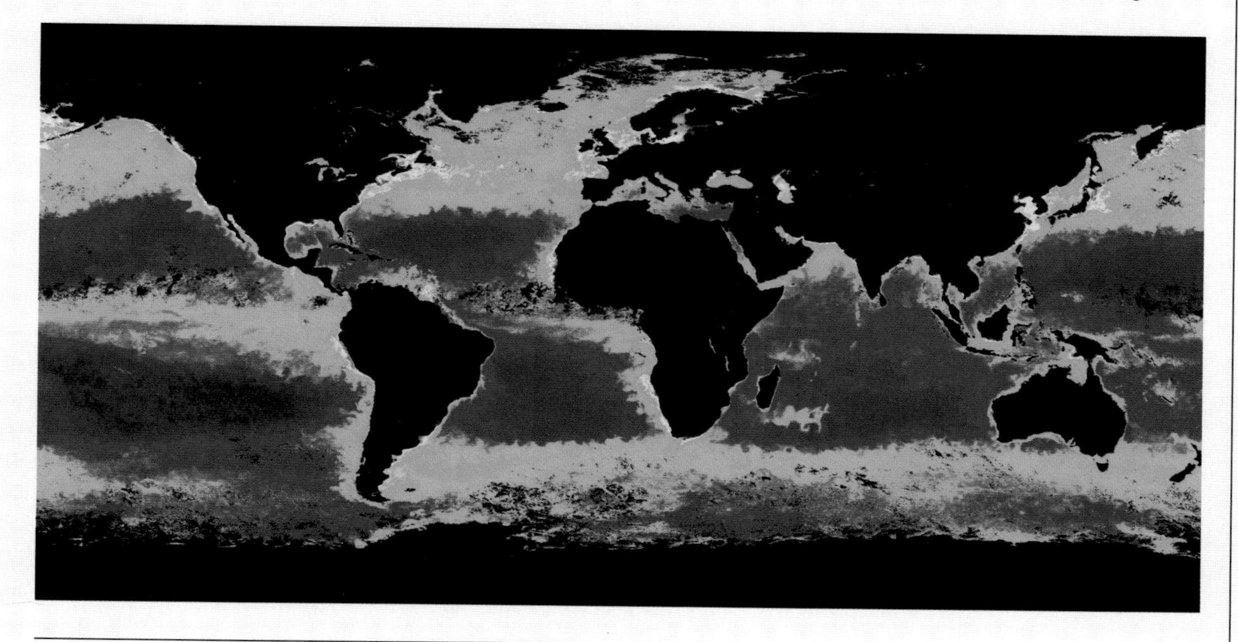

SeaWiFS data was instrumental in creating this false-color map of the ocean's phytoplankton chlorophyll concentration for September 1997 through August 1998. Red indicates areas of highest concentration; dark blue, the lowest.

systems can these measurements be made for the entire **planet.** Gravity is, of course, the invisible force that pulls two masses together.

Sir Isaac Newton discovered that gravitational attraction increases with an increase in an object's mass, hence a container of soft drink is heavier, that is,

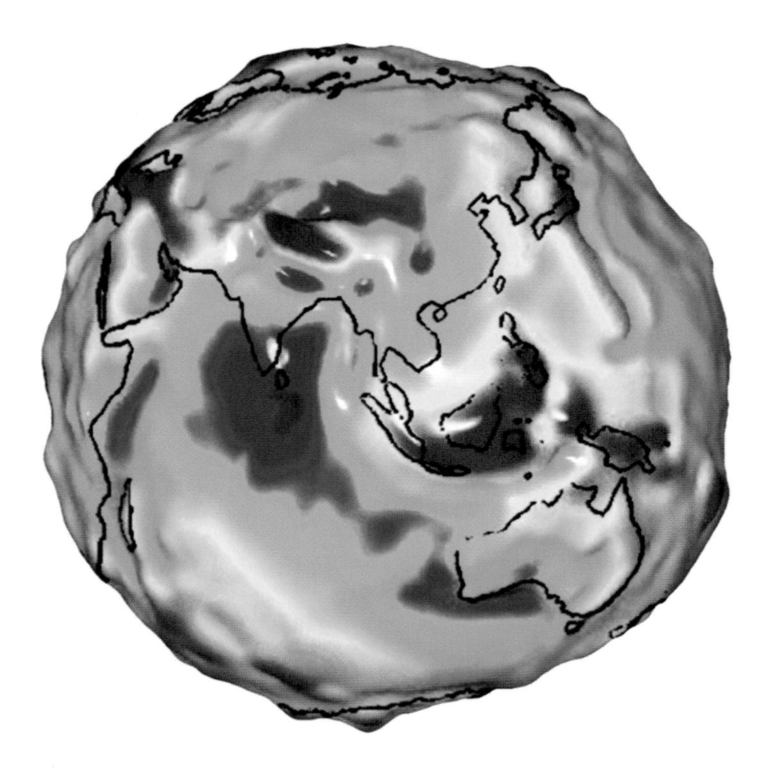

The Gravity Recovery and Climate Experiment (GRACE)—sponsored by NASA, Deutsche Forschungsanstalt für Luft- und Raumfahrt e.V. (DLR, the German space agency), and other agencies—uses two small satellites flying in formation to measure the distortions of Earth's gravity field. The satellites also map the geographic variations in the entire planet's gravity field, shown here in color and in 3-D.

attracted toward the **Earth** with greater force, than the same size container of popcorn. Variations in mass and density cause variations in the pull of gravity.

We may think that this gravitational pull is the same, no matter where we are on the planet, but the Earth's topography is highly variable. Its landmasses vary, with mountains, valleys, and plains. Beneath the ocean the bathymetry also varies, with trenches, abyssal plains, and seamounts. As a consequence of this variability, the Earth's mass, and its consequent gravitational pull, vary depending on the exact position on the planet.

The variability of the Earth's gravitational field has been assessed for its ocean areas by precise measurements from **radar altimeters.** "Sea level" actually is not level, but varies considerably from ocean currents, the **coriolis effect,** and the local gravity field. By measuring the length and shape of the ocean surface—which is affected by the gravitational pull at that location—radio altimeters were able to measure the marine geoid (mean sea level).

A direct measurement of the variability in Earth's gravity is being made by a pair of **research satellites** that comprise the Gravity Recovery and Climate Experiment (GRACE), launched in 2002. By carefully measuring the changes in relative position (the distance between them) caused by differing gravitational pull on each of the satellites, researchers have been able to map the spatial variability of gravity over the entire planet.

The Earth's magnetic field is another geophysical property that we take for granted, except for the usefulness of magnetic compasses. But our very survival as living creatures is due to the **magnetosphere**'s effect in bending the incoming **solar wind**—radiation from the **sun**—around us, shielding us from radiation danger.

The Earth's magnetosphere is formed from two essential ingredients. The first is the Earth's magnetic field, generated by currents flowing in the Earth's core, and above the Earth's surface taking the form of a bar magnet, a dipole field, aligned approximately with the Earth's north-south spin **axis.** The second ingredient is the solar wind, an ionized hydrogen/helium plasma that streams continuously outward from the sun into the **solar system** at speeds of about 300 to 800 kilometers a second.

Satellite-borne magnetometers are instruments designed to measure the magnetosphere, providing forecasters with information on the field's variability. This can be combined with solar observations to predict "space weather" radiation effects that can be dangerous to humans and may disrupt satellite operations and electrical power distribution systems on Earth.

Resource Exploration

Multispectral and **hyperspectral optical imaging** systems can be used for **geology** from space that supports exploration for natural resources.

Analysis of the backscattered spectral signatures of imaged areas is used to determine the presence of minerals that are sought or that are indicative of the likely presence of a desired material, such as oil or natural gas. Stereo-optical images and **interferometric synthetic aperture radar** can also be used to create topographic maps that are needed for surveys or earth-moving mining operations.

Precise **geolocation** of features observed in **satellite** imagery is often accomplished through the use of in-situ observers who are equipped with satellite navigation devices such as **Global Positioning System** (GPS) receivers.

Photography Satellites

Landsat

Since 1972 Landsat Earth-observing optical imaging satellites in **low Earth orbit** have continuously supplied information to support environmental research, scientific research, commercial enterprises, education, and national defense.

Multispectral imagery in the optical and thermal infrared bands is used for many applications. Among them are geology, petroleum and mineral exploration, land use studies, agricultural monitoring, coastal ecosystem studies, coastal bathymetry, and water-pollution research.

Millions of images have been collected from the six Landsat satellites, which have provided the longest, relatively high-resolution multispectral record of the **Earth**'s continental surfaces seen from space.

These images are archived at the Earth Resources Observation Systems (EROS) Data Center in Sioux Falls, South Dakota, operated by the U.S. Geological Survey of the Department of the Interior, and at Landsat receiving stations around the world. They are a uniquely valuable resource for global climate change research and other applications.

The primary instrument on the current Landsat (Landsat 7) is the Enhanced Thematic Mapper Plus (ETM+), a seven-channel imaging system with a 185-kilometer swath width that creates images with a spatial resolution of 15 meters in panchromatic, 30 meters in its seven visible and near-infrared bands, and 60 meters in the thermal infrared. Worldwide coverage of the Earth is provided every 16 days between 81° north latitude and 81° south latitude. Landsat 7 flies at an **altitude** of 705 kilometers in a **sun-synchronous polar orbit.**

Landsat began as a government research program and then transitioned into an experimental quasi-commercial endeavor. The U.S. government has paid for the construction and launch of every Landsat satellite. Some of the satellites were then operated by private companies, which then provided data to the U.S. government as well as sold it for commercial or foreign government use.

The most recently launched satellite at the time of this writing is Landsat 7. It is fully operated by the U.S. government and data are sold for a nominal fee, compared to the pricing of the earlier commercial operation.

Earth Resources Technology Satellite (ERTS)

The Earth Resources Technology Satellite was a **NASA research satellite** program that evolved into a series of multispectral remote-sensing systems later renamed **Landsat.** The first satellite (ERTS 1) was launched July 23, 1972. It was modeled after the **Nimbus** meteorological satellite.

IKONOS imagery reveals a 17th-century fortress in the Netherlands. The entire scene (top) was collected by the sensor, which is capable of zooming in (bottom) with a very high resolution.

IKONOS

While the **Landsat** program is currently operated by the U.S. government, private companies—in some cases as subsidiaries of satellite-building aerospace companies—have been building, launching, and operating imaging **satellites** for purely commercial data sales.

The IKONOS satellite, built by Lockheed Martin and launched on September 24, 1999, was at that time the most sophisticated remote-sensing **spacecraft** ever operated for the commercial market.

This satellite is owned and operated by Space Imaging Corporation, and it is capable of photographing objects on the ground as small as one meter in the panchromatic band and three meters in **multispectral imagery**. Panchromatic and multispectral images can be digitally combined to produce images that offer the advantages of both techniques—color pictures with very sharp edges allowing more accurate interpretation of objects.

IKONOS imagery can be orthorectified for mapping, and stereo imagery is available for topographic mapping and three-dimensional views.

QuickBird

QuickBird is a fully commercial **satellite,** producing panchromatic imagery with a spatial resolution of 0.61 meter at nadir and 2.44 meters in multispectral. Built by Ball Aerospace & Technologies Corp. and launched on October 18, 2001, the QuickBird imaging satellite is operated by the DigitalGlobe Corporation (founded in the mid-1990s as the Worldview Imaging Corporation).

The imagery is sold by DigitalGlobe in digital form, primarily over the Internet. Because of the potential for military or intelligence use of imagery that has this very high resolution, commercial satellites such as QuickBird and IKONOS are operated under license from the U.S. government, which regulates how the satellites are operated, how and to whom data can be sold, and how data are to be archived.

SPOT

The SPOT (Système pour l'Observation de la Terre) program is similar to the **Landsat** program in design and applications. The first SPOT **satellite** was launched in 1986 by France's space agency, Centre National d'Études Spatiales (CNES), with support from Sweden and Belgium.

SPOT satellites are in **sun-synchronous,** near **polar orbits** at altitudes around 830 kilometers above the Earth, which results in orbit repeat every 26 days. SPOT satellites were the first to use "push-broom" along-track sensor scanning.

Unlike Landsat, SPOT satellites are capable of tilting their orbital position to point at different slant angles, allowing the collection of pictures on successive orbits for stereo imagery. This capability allows topographic mapping. The SPOT-5 satellite has the capability of collecting images with a spatial resolution of five meters in the panchromatic band.

SPOT is a quasi-commercial program, with the French government funding the satellite and its launch and a private company, SPOT Image, operating the satellite and selling its data commercially.

OrbView

Orbimage, Inc. operates two commercial imaging **satellites,** OrbView-2 and OrbView-3. The OrbView-2 satellite, launched in August 1997, was originally called the SeaStar satellite by its builder, Orbital Sciences Corporation. It carries the **Sea-viewing Wide Field-of-view Sensor** (SeaWiFS), a multispectral instrument designed primarily to observe the color of the ocean to determine its biological content for NASA research.

NASA funded the development of SeaStar as a research satellite, but it was launched and then operated commercially in a unique program that provided NASA with access to data for research purposes and allowed its operator to also sell data for commercial purposes on the domestic and international market.

By detecting subtle color changes on the **Earth**'s surface, OrbView-2's imagery is valuable for monitoring plankton and sedimentation levels in the oceans and assessing the health of land-based vegetation on a global basis. It can be used by fishermen for monitoring the locations of plankton populations to pinpoint productive areas for commercial fishing.

The multispectral SeaWiFS instrument has eight channels, six in the visible and two in the near-infrared **spectrum,** and has a spatial resolution of 1.1 kilometers. It provides daily coverage of the entire Earth with a swath width of 2,800 kilometers in its near **polar, sun-synchronous orbit.**

OrbView-3, launched in 2003, is a high-resolution commercial imaging satellite, also operated by Orbimage. The satellite takes panchromatic images with a spatial resolution of one meter, and at four meters in its four-channel multispectral mode.

Believed to have been built about 2650 B.C., the Great Pyramid at Giza was imaged by QuickBird. The multispectral color image was sharpened using the satellite's 61-centimeter resolution panchromatic band to create a picture so detailed that it shows individual stones.

U.S. Environmental Satellites

Geostationary Operational Environmental Satellite (GOES)

GOES is an **operational satellite** program operated by the National Oceanic and Atmospheric Administration (NOAA), an agency of the U.S. Department of Commerce. The normal constellation consists of two functionally identical **spacecraft** in **geostationary Earth orbit.** GOES-East is positioned at 75° west longitude and GOES-West at 135° west longitude. From these two **satellites,** environmental conditions for the entire United States, including Alaska and Hawaii, can be monitored. These orbital positions are particularly important as they allow forecasters to monitor weather systems approaching the western U.S. from the Pacific Ocean and the development and motion of tropical storms in the tropical Atlantic and Pacific Oceans.

Each satellite is equipped with a **multispectral imaging** sensor, a **sounding sensor,** and space environmental monitoring sensors, along with a data collection system and search and rescue receivers. At the time of this writing, the most recently launched GOES satellite (GOES-M, renamed GOES-12 after being launched successfully) also carries a solar x-ray imager (SXI) sensor to monitor **solar flare** events.

The primary value of GOES is its ability to observe the entire Western Hemisphere, either as a full disk or in selected subsets of geographic coverage, for the precursor signs of dangerous weather development such as hurricanes or tornados. While the American public sees GOES imagery in virtually every television weather forecast, they may be unaware that NOAA's National Weather Service forecasters at regional forecast offices, the National Hurricane Center/Tropical Prediction Center in Miami, Florida, and the National Severe Storms Laboratory in Norman, Oklahoma, are using GOES imagery around the clock as a primary tool for developing forecasts and severe weather warnings that save lives.

NOAA is currently entering into the procurement process for the next generation of GOES satellites. These are referred to as GOES-R, and they will have far greater capabilities than their predecessors, including higher spatial and temporal resolution imagery, with expanded multispectral capabilities for oceanography from space.

Applications Technology Satellite (ATS)

The **National Aeronautics and Space Administration** (NASA) established the Applications Technology Satellite (ATS) program to develop and conduct test flights of experimental **payloads** and to study the effect of the space environment on space systems flying in **geostationary Earth orbit.** Five ATS satellites in three configurations were built for NASA by Hughes Aircraft Corporation (which has now become part of the Boeing Company) from 1966 to 1969.

ATS-1 was launched on December 7, 1966, and stationed above the Pacific Ocean. The satellite successfully collected photographs of the **Earth** with a spin-scan cloud camera and became the first geosynchronous meteorological satellite, a precursor to today's **Geostationary Operational Environmental Satellite** (GOES) program.

Synchronous Meteorological Satellite (SMS)

After the successful demonstration of collecting weather imagery from **geostationary Earth orbit** by **NASA's Applications Technology Satellites** ATS-1 and ATS-3, NASA initiated the Synchronous Meteorological Satellite (SMS) program, launching SMS-1 on May 17, 1974.

After two **geosynchronous** Earth-orbit satellites were launched, the National Oceanic and Atmospheric Administration (NOAA) took responsibility for the operation of this satellite program, renaming it the **Geostationary Operational Environmental Satellite** (GOES). NASA launched the GOES-1 satellite on October 16, 1975.

Television Infrared Observation Satellite (TIROS)

NASA launched the first Television Infrared Observation Satellite (TIROS) on April 1, 1960, to demonstrate the feasibility of observing the entire **planet**'s cloud cover and weather patterns from space. While satellites in **geostationary Earth orbit** can see the entire full disk of the Earth from their orbits above the Equator, they cannot observe the whole planet.

Therefore, a series of ten experimental TIROS satellites were flown in **low Earth orbit** between 1960 and 1965. Success of this program led to the TIROS

Operational Satellite (TOS) program, operated by the Environmental Sciences Services Administration (ESSA), which later became part of the National Oceanic and Atmospheric Administration. The first TOS satellite, designated ESSA-1, was launched on February 3, 1966.

Polar Operational Environmental Satellite (POES)

The Polar Operational Environmental Satellite (POES) program has been operated by the National Oceanic and Atmospheric Administration (NOAA) since 1978 as a constellation of two **satellites** in **circular,** near **polar, sun-synchronous, low Earth orbit** (LEO). This program evolved from the TIROS program, beginning with the launch of the Improved TIROS Operational Satellite (ITOS-1) on January 23, 1970, and followed later that year, on December 11, by the newly redesignated NOAA-1 satellite.

The POES program continues today, with the two **operational satellites** designated NOAA-17 and NOAA-18. These **spacecraft** are of the Advanced TIROS (TIROS-N) design. Each satellite carries an instrument suite for imaging and measuring the **Earth's atmosphere** and cloud cover in addition to its surface, particularly vegetation coverage and sea-surface temperature measurements.

Primary instruments are the Advanced Very High Resolution Radiometer (AVHRR), the TIROS Operational Vertical Sounder (TOVS), and the Solar Backscatter Ultraviolet (SBUV) instrument. The SBUV **ultraviolet sensor** is used to monitor atmospheric ozone, including the ozone hole observed over the South Pole.

Data collected by POES spacecraft are fed directly into short-, medium-, and long-range forecast models as well as climate models. These satellites also have instruments for various other secondary missions, including the **Search and Rescue Satellite Aided Tracking (SARSAT) System.**

POES satellites collect and record global data and downlink it to NOAA receiving stations in the United States for use by computers at the National Centers for Environmental Prediction at Camp Springs, Maryland, and by numerical prediction centers operated by the Navy and Air Force at Monterey, California; Bay St. Louis, Mississippi; and Offutt, Nebraska.

The satellites also broadcast observations in real time as the measurements are sensed. These data are unencrypted for use by anyone with an inexpensive satellite data receiving system, a computer, and commercially available software.

Electrical energy shimmers on a digitally created mosaic of hundreds of separate nighttime images of cloud-free areas around the entire globe. The low-light imaging capability of the U.S. Defense Meteorological Satellite Program (DMSP) was designed to allow weather forecasters to observe cloud formations even on the dark side of the planet to support military operations.

Defense Meteorological Satellite Program (DMSP)

The Defense Meteorological Satellite Program (DMSP), operated by the U.S. Department of Defense, is a dual-satellite constellation very similar to the National Oceanic and Atmospheric Administration (NOAA) **Polar Operational Environmental Satellite** (POES) program. DMSP consists of two satellites in **circular,** near **polar, sun-synchronous, low Earth orbit.**

In fact, the **spacecraft** for both programs are nearly identical, manufactured by Lockheed Martin Corporation. Their instrument suites are also similar, with their primary sensors being visible and infrared imaging **radiometers** with similar characteristics. The DMSP satellites have slightly better spatial resolution. They are able to resolve cloud elements as small as about 520 meters with the Optical Linescan Sensor (OLS) instrument, as opposed to the POES resolution of 1,100 meters.

The DMSP systems have much less capability in measuring sea-surface temperature than POES, though, due to a different selection of infrared spectral bands. DMSP satellites carry the Special Sensor Microwave Imager (SSM/I) passive microwave radiometer, which measures atmospheric moisture, the heaviness of precipitation, and the coverage of sea ice in polar regions.

Data **downlinked** from DMSP satellites can be encrypted for use only by authorized military and other government users in time of war. Direct broadcast data can be received by mobile ground stations on land and on ships for immediate use in weather forecasts by meteorological personnel in the field or at sea.

National Polar-orbiting Operational Environmental Satellite System

On May 5, 1994, U.S. President Bill Clinton made the decision to merge the United States' military and civilian operational meteorological **satellite** systems, the **Polar Operational Environmental Satellite** (POES) and the **Defense Meteorological Satellite Program** (DMSP).

An Integrated Program Office was established, with experts in space-system acquisition and environmental **remote sensing** from the National Oceanic and Atmospheric Administration, the Department of Defense, and **NASA.** The consolidated satellite program is called the National Polar-orbiting Operational Environmental Satellite System (NPOESS). NPOESS will have an all-new satellite design that is being built by Northrop Grumman Corporation in collaboration with the Raytheon Corporation. The instrument suite of NPOESS will include all new sensors. They will be similar to their predecessors, but will have much greater capability. The first NPOESS satellite is scheduled to be launched in 2012, and it will serve both military and civilian government users.

Aeros

The Aeros **satellites** were examples of early collaboration between **NASA** and foreign partners, in this case the German Ministry of Science and Technology. Aeros-1 and Aeros-2 were launched into **elliptical, polar, nearly sun-synchronous** Earth orbits on December 16, 1972, and July 16, 1974, on Scout **rockets** from Vandenberg Air Force Base, in California.

The purpose of these missions was to study the upper atmosphere and **ionosphere,** especially with regard to the influence of solar ultraviolet radiation. Five experiments provided data on the temperature and density of electrons, ions, and neutral particles, the composition of ions and neutral particles, and solar ultraviolet flux.

Nimbus

Seven Nimbus **research satellites** were launched by NASA between 1964 and 1978. Their purpose was to test new sensing devices. These instruments were later evolved and incorporated into **operational satellites,** including the **Polar Operational Environmental Satellite** (POES), **Defense Meteorological Satellite Program** (DMSP), and **Landsat.** Nimbus satellites were butterfly-shaped, 3 meters long and 3.4 meters across their wing-like solar panels. Each had a circular platform 1.5 meters wide that carried sensing instruments designed to collect observations at various wavelengths from the infrared to the ultraviolet.

These satellites were used to take images of the **Earth**'s clouds, ice caps, and land cover. The Nimbus satellites also made measurements of temperature and humidity at various levels of the atmosphere with their **sounding sensors.**

Coastal Zone Color Scanner

One of the many successful instruments demonstrated on the **Nimbus** satellites was the Coastal Zone Color Scanner (CZCS). It flew on Nimbus-7, which was launched on October 24, 1978. This instrument was a six-channel scanning **radiometer** designed to observe color, principally relative blue and green shading, in coastal areas and very large lakes.

Ocean color from space measurements are used to determine chlorophyll concentration, sediment content, and such important phenomena as red tide, which can be disastrous to marine mammal populations and fisheries. The success of CZCS led to the later development of the **Sea-viewing Wide Field-of-view Sensor** (SeaWiFS) and other instruments.

Seasat

Seasat, launched on June 28, 1978, was the first **satellite** designed specifically to observe the Earth's oceans. It carried the first civilian space-based **synthetic aperture radar** (SAR). This **research satellite,** developed for **NASA** by the Jet Propulsion Laboratory of the California Institute of Technology, flew in a nearly circular orbit at an altitude of 800 kilometers, inclined at 108 degrees. The satellite operated for only 105 days, until October 10, 1978, when a massive short circuit in the satellite electrical system ended the mission. All five of the instrument's sensors, however, performed entirely as intended and led the way for later flights of sensors that evolved from them on follow-on research and operational satellites.

Seasat's **radar altimeter** precisely measured the **spacecraft**'s height above the ocean surface, leading to the development of the U.S. Navy's **Geodesy Satellite** (Geosat-A). Similar versions of its SAR flew on four space shuttle missions as the Shuttle Imaging Radar (SIR-A/B/C).

The scanning multichannel microwave radiometer evolved into an operational system on the **Defense Meteorological Satellite Program** (DMSP). The Seasat microwave **scatterometer** observed wind speed and direction, measurements being made today by NASA's **QuikSCAT** mission. Finally, Seasat carried a visible and infrared **radiometer** to identify cloud, land, and water features.

Earth Observing System (EOS)

The Earth Observing System (EOS) program is, as its name implies, a long-term, multifaceted effort to observe the Earth using space systems. Its goal is to manage the massive amounts of data being generated so as to help scientists across many disciplines understand the complex processes that define our environment. The program is managed by the **NASA** Space Science Enterprise, in close cooperation with the international community, particularly the European Community and Japan.

Although the program was formally established in 1991 as a presidential initiative and continues today, it can trace its heritage to the experimental Earth-viewing **satellites** of the 1960s and 1970s. In 1979 the World Climate Research Program was created as an international scientific effort to understand the fundamental processes of climate, an endeavor born of a realization that society was experiencing the impact of a seemingly changing global climate.

Then, starting in 1985, the U.S. National Research Council of the National Academy of Sciences produced the first of what became regularly updated reports on global climate change. The council called for a research program that would require simultaneous, calibrated,

EARTH OBSERVING SYSTEM MEASUREMENTS

Discipline	Parameters Measured
Atmosphere	**Cloud Properties** (amount, optical properties, height)
	Radiative Energy Fluxes (top of atmosphere, surface)
	Precipitation
	Tropospheric Chemistry (ozone, precursor gases)
	Stratospheric Chemistry (ozone, ClO, BrO, OH, trace gases)
	Aerosol Properties (stratospheric, tropospheric)
	Atmospheric Temperature
	Atmospheric Humidity
	Lightning (events, area, flash structure)
Solar Radiation	**Total Solar Irradiance**
	Ultraviolet Spectral Irradiance
Land	**Land Cover and Land Use Change**
	Vegetation Dynamics
	Surface Temperature
	Fire Occurrence (extent, thermal anomalies)
	Volcanic Effects (frequency of occurrence, thermal anomalies, impact)
	Surface Wetness
Ocean	**Surface Temperature**
	Phytoplankton and Dissolved Organic Matter
	Surface Wind Fields
	Ocean Surface Topography (height, waves, sea level)
Cryosphere	**Land Ice** (ice sheet topography, ice sheet volume change, glacier change)
	Sea Ice (extent, concentration, motion, temperature)
	Snow Cover (extent, water equivalent)

and continuous measurements of key physical parameters of the Earth's environment.

The EOS program became a critical element in what became the U.S. Global Change Research Program, and more recently, the Climate Change Research Program. Science panels were convened to define the requirements for this program, and a list of 24 measurements was compiled, grouped into the following categories: atmosphere, solar radiation, land, ocean, and cryosphere.

Data from what has become a program of more than 25 satellites with multiple **remote-sensing** instruments are being received, processed, distributed, and archived by the EOS Data Information System. Three flagship EOS spacecraft were developed, named **Terra, Aqua,** and **Aura** (due to their respective targeting of land, sea, and atmospheric phenomena), along with Earth Probe missions, such as the **Total Ozone Mapping Spectrometer,** and international collaboration missions, such as the U.S.-France **TOPEX/Poseidon** ocean topography mission.

Terra

Terra, the word for "land" in Latin, is a school bus–size spacecraft built by Lockheed Martin Corporation for **NASA.** Terra was launched on December 18, 1999. It is known as the flagship satellite of the **Earth Observing System.** It carries five sensors: Advanced Spaceborne Thermal Emission and Reflection Radiometer (ASTER), Clouds and the Earth's Radiant Energy System (CERES), Multi-angle Imaging Spectro-radiometer (MISR), **Moderate-resolution Imaging Spectroradiometer** (MODIS), and Measurements of Pollution in the Troposphere (MOPITT).

Together, these **passive remote-sensing** instruments represent the first major step in a multidisciplinary and multinational effort to collect a 15-year global data set of measurements of the Earth's physical characteristics in order to understand our climate and climate changes.

Aqua

The second major multi-instrument mission of the **Earth Observing System** (EOS) is Aqua, the Latin word for "water."

The Aqua mission, launched on May 4, 2002, is a **research satellite** designed to collect a huge amount of data to study the **Earth**'s water cycle, including evaporation from the oceans, water vapor in the atmosphere, clouds, precipitation, soil moisture, sea ice, land ice, and snow cover on the land and ice.

The Japanese-built Advanced Space-borne Thermal Emission and Reflection Radiometer (ASTER) instrument on NASA's Terra satellite captured this image of sand dunes in Namib-Naukluft National Park in Namibia on October 14, 2002. The 3-D perspective image was created in computer image processing by "draping" a color image taken from directly overhead over a digital elevation model—a terrain database.

In addition, Aqua is measuring radiative energy fluxes, aerosols in the atmosphere, vegetation cover on land, phytoplankton and dissolved organic matter in the oceans, and the temperature of the land, air, and sea-surface. This multifaceted observation capability is provided by a suite of six sensors: the Atmospheric Infrared Sounder (AIRS), the Advanced Microwave Sounding Unit (AMSU-A), the Humidity Sounder for Brazil (HSB), the Advanced Microwave Scanning Radiometer for EOS (AMSR-E), the **Moderate-resolution Imaging Spectroradiometer** (MODIS), and Clouds and the Earth's Radiant Energy System (CERES). The very large **spacecraft** was built by TRW, now part of Northrop Grumman Corporation, for **NASA.** Aqua flies in **sun-synchronous orbit** similar to that of its sister mission **Terra,** except that the orbit of Aqua crosses the Equator on each ascending orbit in the afternoon, while Terra crosses in the morning. Hence their original mission names of AM-1 and PM-1 for Terra and Aqua, respectively.

Aura

Aura, originally named CHEM-1, is an **Earth Observing System** (EOS) mission launched on July 15, 2004, to study the **Earth**'s ozone, air quality, and climate. The third of three multisensor EOS missions, Aura carries four instruments: the High Resolution Dynamics Limb Sounder (HIRDLS), the Microwave Limb Sounder (MLS), the Ozone Monitoring Instrument (OMI), and the Tropospheric Emission Spectrometer (TES). The OMI was developed in the Netherlands as a European contribution to the mission. This suite of sensors will make multiple measurements of the composition, chemistry, and dynamics of the Earth's upper and lower atmosphere.

Aura will build upon research begun with NASA's Upper Atmospheric Research Satellite (UARS), launched by the **space shuttle** *Discovery* in 1991. Aura will continue the collection of atmospheric ozone observations begun by the **Total Ozone Mapping Spectrometer** (TOMS) missions.

Moderate-resolution Imaging Spectroradiometer (MODIS)

MODIS is a passive optical and infrared **multispectral imaging** instrument carried by the two major **Earth Observing System** missions **Terra** and **Aqua.** The MODIS instruments on these two missions in near **polar, sun-synchronous, low Earth orbit** are able to observe the entire Earth in 2,300-kilometer-wide **swaths** every one to two days in 36 discrete spectral bands. MODIS's observations of multiple parameters, together with measurements of other sensors on the same **spacecraft,** are intended to determine the impact of clouds and aerosols on the Earth's energy budget, and in so doing support climate change research. MODIS is capable of measuring the photosynthesis activity of both land vegetation and phytoplankton in the ocean and therefore indirectly monitoring large-scale changes in the Earth's **biosphere** and carbon cycle. The instrument's ocean measurements can provide insights into such important phenomena as El Niño and La Niña, climatic features that affect the Pacific Basin and North, Central, and South America as well as the global circulation patterns of both the atmosphere and the world's oceans.

Sea-viewing Wide Field-of-view Sensor (SeaWiFS)

SeaWiFS is an ocean-color imaging instrument designed for **oceanography** from space. It is similar in design and function to the earlier successful flight of the **Coastal Zone Color Scanner** (CZCS) on the NASA Nimbus-7 spacecraft. SeaWiFS, a NASA Earth Probe mission, provided the world's first daily color imagery of the Earth. It was developed in a unique contract between NASA's Goddard Space Flight Center and Orbital Sciences Corporation (Orbital). Under this arrangement, Orbital procured the SeaWiFS instrument from its manufacturer, Raytheon Santa Barbara Remote Sensing.

Orbital launched SeaWiFS on August 1, 1997. The **satellite** is operated by the Orbimage, Inc., a commercial remote-sensing company, as the **OrbView-2** mission. NASA provided the instrument's required specifications and, on behalf of itself and other government agencies, has bought access to regional real-time direct **downlink** and global stored data for research purposes through a data-buy arrangement. Meanwhile, Orbimage is able to sell imagery data and derived information, such as maps of biologically productive ocean sites for commercial fishing, for operational use.

The SeaWiFS' **multispectral imaging** instrument has eight channels, six in the visible and two in the near-infrared **spectrum,** with a spatial resolution of 1.1 kilometers. This capability allows observation of subtle changes in ocean color (blue-green shades), which indicate the types and amounts of microscopic marine plants called phytoplankton. Understanding of such biological activity has applications in scientific research as well as practical use, and it is this broad value that has made the government-private industry cost-sharing arrangement both unique and successful.

Tropical Rainfall Measuring Mission (TRMM)

A joint mission between **NASA** and the Japan Aerospace Exploration Agency (JAXA), the Tropical Rainfall Measuring Mission (TRMM) is a research satellite designed to help scientists understand the water cycle of the **Earth**'s climate system. Using both **active** and **passive remote sensing,** the mission carries five instruments: the Precipitation Radar, TRMM Microwave Imager, Visible Infrared Radiometer, Clouds and the Earth's Radiant Energy System (CERES), and the Lightning Imaging Sensor.

Together these instruments provide rain radar and microwave radiometric data that measures the distribution of precipitation over the Earth's tropics between 35° north latitude and 35° south latitude. These measurements are allowing researchers to make computer models of the interactions between the sea, air, and landmasses that produce changes in global rainfall and climate. The TRMM **satellite** was successfully launched from the Tanegashima Space Center in Japan in November 1997. By July 2004, the satellite had greatly exceeded its life expectancy, and NASA and JAXA were discussing ending the mission and **deorbiting** the satellite in 2005.

The Ocean Topography Experiment (TOPEX)/Poseidon

TOPEX/Poseidon is a **radar altimeter** mission to conduct **oceanography** from space. It was jointly developed by **NASA** and the Centre National d'Études Spatiales (CNES, the French space agency). The mission is managed in the U.S. by the Jet Propulsion Laboratory of the California Institute of Technology.

Launched in August 1992 from the **European Space Agency**'s space center in Kourou, French Guiana, TOPEX/Poseidon flies in an orbit 1,336 kilometers above the **Earth** and precisely measures the sea level directly beneath its orbital path every ten days using a dual-frequency radar altimeter developed by NASA and a single-frequency radar altimeter developed by CNES. This technique of using space-based altimeters to measure ocean topography was first demonstrated by NASA's **Seasat** mission in 1978 and the U.S. Navy's **Geosat-A,** which operated from 1985 to 1989. Although

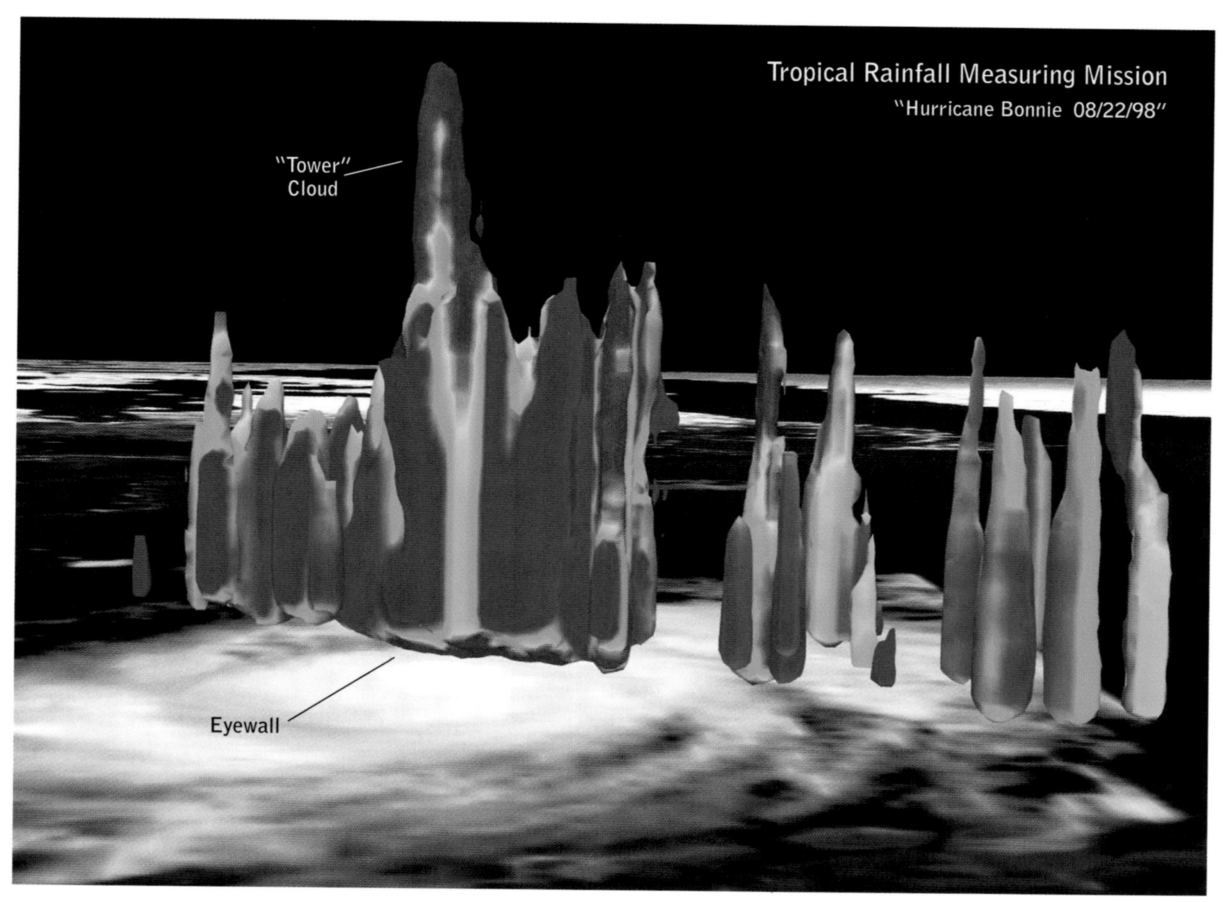

The TRMM spacecraft overflew Hurricane Bonnie as it bore down on the U.S. East Coast on August 22, 1998. The satellite's first-of-a-kind precipitation radar created this 3-D image of the eyewall structure. Exaggerated in its vertical dimensions for clarity in research applications, the cloud structure is colored to distinguish precipitation amounts from light (blue) to heavy (red).

The View from Space as a Window into the Earth System

Ghassem R. Asrar

The dawn of the space age coincided with the first International Geophysical Year (IGY) in 1957-58. While precipitated largely independently, the two occasions inaugurated intertwining futures, one enabling and strengthening the other. The IGY promoted the study of the remote regions of the Earth in the growing recognition that changes there were bound in complex ways to changes in the inhabited regions. In the era preceding IGY, Earth science proceeded along discipline-specific paths—oceanography, geology, atmospheric chemistry, ecology—with only occasional collaboration among them. Analogously, in the era preceding the space age, tools for measuring global-scale change were limited to aggregation of data from sparse in situ networks. The World Weather Watch provided a well-organized network for weather-forecasting data in populated regions, but not for remote locations, nor for research-caliber observation. The Keeling observations were taken from Mauna Loa as a representative proxy for global distribution of carbon dioxide.

The space age opened a paradigm-shaping view of the Earth that brought the means to birth and carry out the scientific vision nascent at IGY. In 1960 the first weather satellite began providing global views of atmospheric dynamics leading to the advance from two-day to five-day weather forecasting. In 1972 the first civilian land-imaging satellite demonstrated the capability to track changes in land cover globally, allowing the scientific assessment of changes in the world's great deserts and rain forests. In 1978 the Seasat satellite demonstrated the ability to do oceanographic research on a global scale. Also in the late 1970s, scientists began monitoring global ozone concentrations, revealing the vulnerability of our natural shield against ultraviolet radiation and prompting international action. The Montreal Protocol on Substances Depleting the Ozone Layer, with its subsequent amendment, endures as the prototype of global observation augmenting scientific understanding that led to effective international policy decisions and action for the protection of planet Earth.

What began as a series of space-technology demonstrations led the broader scientific community to think in new ways about Earth science. Just as satellites view the Earth without regard to national boundaries, they see the Earth without regard to the traditional boundaries of scientific disciplines. From space, the interactions among continents, oceans, atmosphere, ice caps, and life itself are most striking. The transitions from circulating ocean to sea ice to ice caps across seasons and years; the global respiration of terrestrial and marine vegetation; the transport of Saharan dust across the Atlantic; the regional differentiation as global average temperatures rise; the human influences on global change and their consequences—these emerge as questions of interest that no one Earth science discipline can adequately address. In the 1980s the science community, under NASA sponsorship, developed the interdisciplinary concept of Earth System Science as the framework within which to pursue research on questions of global change and regional changes in their global context.

The challenge and the promise of Earth System Science in turn demanded the progression from a series of remote-sensing technology demonstrations to an Earth Observing System (EOS) designed specifically to measure parameters key to the interactions among the major components of the Earth system—atmosphere, oceans, continents, ice, and life. Conceived in the late 1980s and adopted as a presidential initiative for fiscal year 1991, NASA's Earth Observing System, with the directly contributing and compatible programs of other nations, is the answer to this challenge. With the launch of the Aura satellite in July 2004, the first series of

EOS is complete—the world has its first capability to systematically measure and monitor Earth's vital signs.

EOS and its precursors have already achieved remarkable breakthroughs. We now understand, and can quantify uncertainties in, Earth's energy balance—that is, how the Earth responds to incoming solar radiation and its variability that powers the climate system. We understand the mechanics behind the El Niño–La Niña phenomenon, though what triggers it is still under investigation. We are following the cycling of carbon and of water in the Earth system, which is what makes Earth hospitable for life. New satellites planned to measure global atmospheric carbon, sea-surface salinity, and global soil moisture will help to close the remaining gaps in our understanding of these cycles. We have produced the first globally consistent topographic maps of the inhabited regions of the Earth, and we have generated a multidecade record of land-cover change. We routinely measure global ocean temperature and topography, circulation, and wind speeds and directions at the ocean surface; and we have a decade-long record of sea level change. We have mapped the Greenland and Antarctic ice sheets and are making measurements to establish rates and patterns of change (Greenland ice, for example, is thinning along the eastern coast and gaining in the interior). We have captured from space the variations in sea ice extent for two decades to understand its influence on sea level rise.

Beyond their scientific utility, data from Earth-observing satellites are increasingly employed in routine decision making of organizations providing essential services to society. These include weather forecasters, agriculture and natural resource managers, emergency-response officials, air traffic controllers, and energy suppliers. Space research and environmental agencies are working together to ensure transition of these unique remote-sensing capabilities to operational systems of the future to combat forest fires, provide warning of volcanic eruptions and earthquakes, and enhance forecasting of severe weather events.

The recognition of such societal benefits prompted new collaborative international endeavors. The G-8 nations have made Earth observation one of their three top science and technology priorities for the next decade, along with hydrogen energy and biotechnology for agriculture. More than 40 nations and more than 20 international organizations have together formed an ad hoc international Group on Earth Observations that is crafting a ten-year plan to produce a comprehensive, coordinated, and sustained system of plans for Earth observations.

We have made enormous progress in Earth observation from space, from qualitative to quantitative understanding; from discipline-oriented to interdisciplinary science; from passive to active remote sensing that enables imaging in four dimensions; from process simulation models to coupled, data-driven Earth system models capable of simulating and predicting Earth system change; from purely scientific inquiry to science in service to society. Earth system science was enabled by the space age. Now Earth system science is poised to enable the next great steps in space exploration by providing our planetary-scale observing techniques, scientific understanding of Earth's planetary processes, and large-volume data management and modeling approaches to those embarking on the detailed study of other planets in our solar system and beyond. ■

A wide variety of remote-sensing satellites observe the Earth. The data they provide are used in meteorology, oceanography, geologic studies, and more, rapidly increasing our understanding of how Earth's dynamics interrelate.

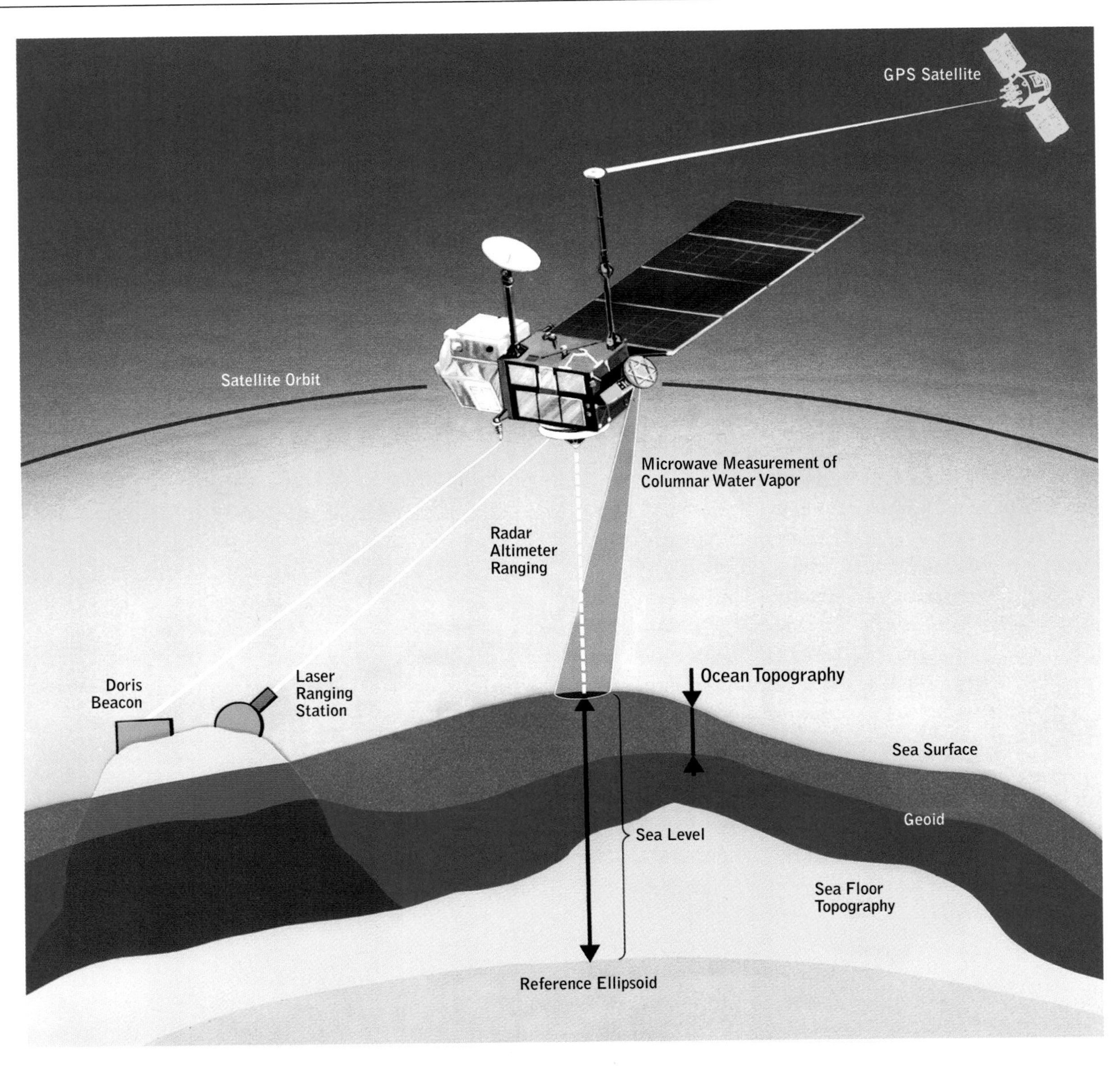

The U.S.-French TOPEX/Poseidon radar altimeter mission is designed to measure the ocean's shape, or dynamic topography. The satellite's height is measured using its downward-looking radar, corrected for the effects of moisture and the ionosphere. Once the satellite's position is accurately determined using GPS, laser tracking, and DORIS radio Doppler tracking, the surface-to-satellite range becomes a measurement of sea-surface height. Radar altimetry has proven a powerful tool in observing seasonal and long-term climate change features.

the **satellite** continues to operate, it has exceeded its five-year design life and has been succeeded by another NASA-CNES joint radar altimeter mission, called Jason-1, launched in late 2001.

SeaWinds on QuikSCAT

NASA proved the ability of an active microwave **scatterometer** instrument to measure wind speed and direction from space on the 1978 **Seasat** mission, which operated briefly before failing due to an electrical short circuit. Then, after several attempts to fly another scatterometer instrument on the National Oceanic Satellite System (NOSS) and the Navy Remote Ocean Sensing System (NROSS) missions, both of which were cancelled by the U.S. government before they were built, NASA succeeded in flying the NASA Scatterometer (NSCAT) instrument on the Japanese **Advanced Earth Observing Satellite** (ADEOS-I), launched on August 17, 1996. Again, this satellite's premature failure left NASA without a wind-observing instrument. The SeaWinds instrument was then built at the California Institute of Technology's Jet Propulsion Laboratory and launched on a single-instrument spacecraft, built by Ball Aerospace & Technologies Corporation and named Quick Scatterometer (QuikSCAT). This fully successful mission was launched on June 19, 1999. Unlike earlier "stick scatterometers" which had multiple fixed-position **antennas,** SeaWinds uses a rotating dish antenna that sweeps in a circular pattern, making nearly 400,000 measurements of maritime wind direction and speed over 90 percent of the **Earth**'s oceans every day. Data observations from QuikSCAT are used for both research and operational weather forecasting.

Total Ozone Mapping Spectrometer (TOMS)

For more than 30 years, **NASA** has been collecting observations of atmospheric ozone from space, a global data set that began in 1978 with the flight of a TOMS instrument on the **research satellite** Nimbus-7. TOMS is a six-channel ultraviolet (UV) **remote-sensing** instrument that measures total ozone amounts from space by measuring backscattered UV radiation (UV light reflected from gases in the atmosphere). TOMS measures important climate features such as the Antarctic ozone hole and the sulfur dioxide that is released into the atmosphere from volcanic eruptions.

The **Nimbus-7** mission provided good data until May 1993, overlapping a second TOMS that was carried by a Russian **Meteor-3** weather satellite launched in August 1991. After an 18-month gap in data continuity, another TOMS was launched on the Japanese **ADEOS** mission, which provided data from its launch in August 1996 until its failure in June 1997. This period overlapped a NASA Earth Probe (EP) TOMS mission launched on July 2, 1996. A fourth TOMS instrument was lost when the NASA QuikTOMS satellite was destroyed in a **launch vehicle** failure on September 21, 2001. Although well beyond its planned mission life, EP-TOMS continues to provide data continuity. Instruments on **Aura,** an **Earth Observing System** (EOS) mission, and the **National Polar-orbiting Operational Environmental Satellite System** (NPOESS) will continue to make important ozone observations in the future.

Geodesy Satellite (Geosat-A) & Geosat Follow-on (GFO)

The **Skylab space station** launched by the United States in 1973 was the first successful flight of a space-based **radar altimeter** for geophysical research. Based on that experience, the **Geodynamics Experimental Ocean Satellite** (GEOS-3) was built by the Applied Physics Laboratory (APL) of the Johns Hopkins University, and was launched in April 1975. It provided measurements of the marine geoid (distortion of sea level due to regional variations in the **Earth**'s gravity caused by such features as sea-bottom ridges and seamounts). The U.S. Navy contracted with APL for another **satellite,** officially designated Geosat-A (referred to as Geosat), which was launched on March 12, 1985. During its 18-month-long Geodesy Mission, Geosat collected ocean topography measurements used by the Naval Oceanographic Office to create classified geodetic data sets for the Navy's submarine force. This data is used to correct launch parameters of strategic **ballistic missiles** to account for the deflection of local vertical caused by gravity. The entire Geosat data set was declassified in the 1990s, allowing development of the world's most accurate maps of ocean bathymetry from 72° north latitude to 72° south latitude.

Geosat's initial mission was completed in September 1986. In November, for the Exact Repeat Mission (ERM), its orbit was changed to a 17-day exactly repeating orbit. It flew precisely over the ground track of **NASA**'s 1978 **Seasat** mission in order to conduct **oceanography** from space measurements. These measurements, useful for operational support to the Navy and for research applications, include the location of major currents, water mass fronts and eddies, wave height, wind speed, and sea ice location.

The success of this ERM led the Navy to fund development of the Geosat Follow-on (GFO) satellite, built by Ball Aerospace & Technologies Corporation, and launched in February 1998 to continue these measurements as an operational Navy mission. GFO and the joint U.S.-French **TOPEX/Poseidon** and Jason missions are similar in design and function, although GFO is capable of transmitting its data in an encrypted format for military-only use, if needed.

Geodynamics Experimental Ocean Satellite (GEOS-3)

Launched on April 9, 1975, into an 843-kilometer **circular orbit,** GEOS-3 was the third in a series of geodetic **research satellites** sponsored by **NASA.** This satellite was designed to determine the feasibility of using a **radar altimeter** to map the topography of the ocean surface with an absolute accuracy of plus or minus five meters and a relative accuracy of one to two meters.

Such precise measurements were successfully demonstrated, proving that it was possible to measure the deflection of local vertical due to gravity and the height of wind-driven waves (sea and swell). The mission led to the NASA **Seasat** altimeter, the Navy **Geosat** satellite, and the joint U.S.-French **TOPEX/Poseidon** mission. Radar altimeters have also been successfully flown on the **European Space Agency**'s **ERS-1** and **ERS-2** missions.

International Environmental Satellite Programs

European Space Agency (ESA) Satellite Programs

The success of the European quasi-commercial imagery **satellite SPOT** (Système Pour l'Observation de la Terre), launched in 1986 by France's space agency, the Centre National d'Études Spatiales (CNES), has led the way for a robust **Earth remote-sensing** program in Europe, not only supported by France, but also by the multinational ESA. European nations continue to work closely with the United States in collaborative research efforts. These include the U.S.-French **TOPEX/Poseidon** and Jason **radar altimeter** missions and the soon-to-be-launched Cloud-Aerosol Lidar and Infrared Pathfinder Satellite Observations (CALIPSO) satellite, being developed by **NASA** and CNES to help scientists answer significant questions and provide new information about the effects of clouds and aerosols (airborne particles) on changes in the Earth's climate.

ERS-1 & ERS-2

Building upon the basic large **spacecraft** design used for **SPOT,** the **European Space Agency** launched the ERS-1 and ERS-2 satellites for **Earth** observation in 1991 and 1995, respectively, on Ariane IV launch vehicles. These highly successful **satellites** were very large, weighing about 2,400 kilograms and measuring 12 meters by 12 meters by 2.5 meters, making them the largest and most sophisticated satellites ever built in Europe at that time.

Launched into near **polar orbits** with a mean altitude of 780 kilometers, ERS-1 and ERS-2 were similar to the NASA **Earth Observing System** (EOS) satellites **Terra** and **Aqua.** The instruments, all built in Europe, were the Active Microwave Instrument (AMI)—combining the functions of a **synthetic aperture radar** (SAR) and a **scatterometer**—a **radar altimeter,** an Along Track Scanning Radiometer (ATSR), a Precise Range and Range-rate Equipment (PRARE), and Laser Retro-reflectors (LRR). The satellites also carried an ozone-mapping instrument called the GOME, and a microwave radiometer (MWR) to provide atmospheric moisture corrections for the radar altimeter. Both missions exceeded their planned mission life; ERS-1 was decommissioned in 2000 and ERS-2 still continues to operate.

Envisat

In March 2002 ESA took its next step in Earth observation: the advanced polar-orbiting, multi-instrument Envisat mission. Envisat continues the mission begun by **ERS-1 and ERS-2,** taking measurements of the atmosphere, ocean, land, and ice. An even larger **spacecraft**

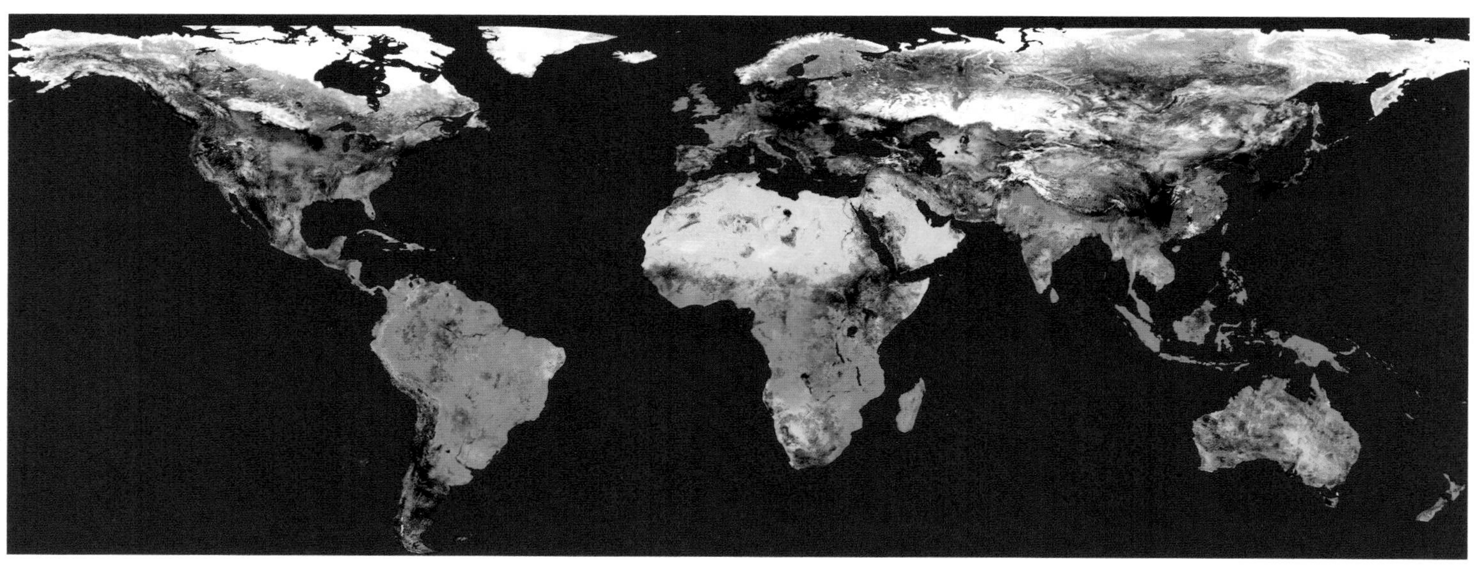

Piecing together hundreds of daytime cloud-free images taken during a Northern Hemisphere winter, this SPOT image reveals in natural green, brown, and white Earth's forests and fields, deserts and tundras, and fields of snow and ice. Such images help widely diverse users, from farmers and commodities brokers to intelligence agencies.

than its two predecessors, with overall dimensions of 26 meters by 10 meters by 5 meters and weighing 8,210 kilograms, Envisat carries nine instruments: the Michelson Interferometer for Passive Atmospheric Sounding (MIPAS), the Global Ozone Monitoring by Occultation of Stars (GOMOS), the Scanning Imaging Absorption Spectrometer for Atmospheric Chartography (SCIAMACHY), the Medium Resolution Imaging Spectrometer (MERIS), the Advanced Along Track Scanning Radiometer (AATSR), the Advanced Synthetic Aperture Radar (ASAR), a **radar altimeter** (RA-2), a microwave radiometer (MWR), and the Doppler Orbitography and Radiopositioning Integrated by Satellite (DORIS) system for very precise orbit determination.

Meteosat

In the United States, **NASA,** the developer of research satellites, has a counterpart (NOAA) in the managing of **operational satellites** for environmental monitoring and prediction. Similarly, in Europe, the **European Space Agency**'s counterpart is the European Organization for the Exploitation of Meteorological Satellites (EUMETSAT), created in June 1986 to take over the ESA Meteosat geostationary operational weather satellite program. Meteosat-1 was launched on a Delta rocket from Cape Canaveral in November 1977. Today, EUMETSAT consists of 18 member states with seven cooperating states, and has its headquarters in Darmstadt, Germany.

In the middle of this decade, the Meteosat geostationary satellites will be augmented by the first of three planned polar-orbiting weather satellites: Metop-1 through Metop-3. These satellites, now being jointly developed by ESA and EUMETSAT, will be integrated into the constellation of **National Polar-orbiting Operational Environmental Satellite System** (NPOESS) satellites currently under development in the U.S.; data from both of these systems will be shared freely with users worldwide.

Russian Satellite Programs

Russia has had an active **Earth remote-sensing** program for decades; their primary weather **satellite** system, called Meteor-3, is a series of near **polar circular orbit spacecraft** flying at an altitude of 1,200 kilometers, inclined to 82.5 degrees. In addition to their normal complement of cloud-imaging sensors, these spacecraft often carry experimental and research instruments. The Meteor-3 satellite number five, launched on August 15, 1991, carries a **Total Ozone Mapping Spectrometer** (TOMS), and satellite number eight carries a Stratospheric Aerosol & Gas Experiment (SAGE III). Both of these instruments were provided by NASA. A Russian Spectrophotometer (SFM-2) provides observations of vertical ozone profiles between the altitudes of 5 kilometers and 80 kilometers.

Geostationary Operational Meteorological Satellite (GOMS)

On October 31, 1994, Russia expanded its meteorological **satellite** program to include the Geostationary Operational Meteorological Satellite (GOMS), which

On March 18, 2002, ESA's Envisat satellite captured this image of the Larsen ice shelf on the Antarctic Peninsula. The satellite's synthetic aperture radar observed the collapse of the ice shelf and its breakup. The event, which might have gone unobserved if not for Envisat's radar, may be another sign of global warming.

is located over the Equator at 76° 50' east longitude. GOMS carries instruments for real-time visible and infrared images of cloud cover, as well as space environment instruments to measure solar and galactic particles and variations in the **Earth**'s magnetic field.

OKEAN-01

Operational since 1983, the OKEAN-01 series of ocean **remote-sensing** satellites is equipped with **radar,** microwave, and optical instruments for **oceanography** from space. A main task of the OKEAN satellites is to map the coverage of sea ice in the Arctic and high latitude seas to support shipping operations around Russia's far northern ports.

RESURS-01

Russia also operates the RESURS-01 series of Earth Remote Sensing System multispectral imaging satellites, comparable to the U.S. **Landsat,** French **SPOT,** and Indian **IRS** satellites. RESURS-01 was launched on November 4, 1994, from the Baikonur Cosmodrome aboard a Zenit vehicle. Data from these Russian **remote-sensing** satellites are managed by the Space Monitoring Information Support laboratory (SMIS) of the Space Research Institute (IKI RAN), in Moscow.

Indian Satellite Programs

The government of India established a Space Commission and Department of Space (DOS) in June 1972 and then established the Indian Space Research Organization (ISRO), which operates a wide-ranging space program that began with the launch of the Aryabhata technology **satellite** on April 19, 1975. India's first **Earth remote-sensing** satellites, Bhaskara-I and Bhaskara-II, were launched by Russian vehicles in July 1979 and November 1981, respectively. Since then, ISRO has developed and launched dozens of **spacecraft** for satellite communications, meteorological observations, and **Earth** resources imagery.

INSAT-1A

INSAT-1A, a geostationary **satellite** launched by a U.S. Delta vehicle on April 10, 1982, was India's first multipurpose **operational satellite** used for cloud imagery. It has been followed by additional satellites in this series.

Indian Remote Sensing (IRS) Satellites

In 1988 a series of near **polar orbiting** Indian Remote Sensing (IRS) satellites flying at an altitude of 905 kilometers began operation, with IRS-P2 launched for the first time by an indigenous Indian **launch vehicle.** Primary instruments of the IRS satellites have been the LISS-3 multispectral sensor that provides a 23.5-meter spatial resolution, and a two-channel wide-field sensor that provides 190-meter resolution.

Beginning with IRS-P4, India began to expand the capabilities of its **Earth** resources satellites. IRS-P4, launched in May 1999, was the first Indian satellite built primarily for ocean applications, earning it the designation OceanSat-1. This 1,036-kilogram satellite was placed into an orbit of 720 kilometers with a multispectral Ocean Color Monitor (OCM) and a Multifrequency Scanning Microwave Radiometer (MSMR). The IRS-P6 satellite, launched on October 17, 2003, is designated as ResourceSat-1. It carries the advanced LISS-4 high-resolution camera that can be operated in either a panchromatic or multispectral role, providing support to **agriculture,** disaster management, and land and water resources monitoring.

On January 27, 2004, Antrix Corporation, the commercial arm of ISRO, signed a contract with the U.S. commercial remote-sensing company Space Imaging Corporation, through which Space Imaging will market imagery data from the IRS satellites, ResourceSat-1, and (when launched) the follow-on IRS-P5 CartoSat-1 satellite. CartoSat-1 will have two panchromatic cameras for in-flight stereo viewing with a spatial resolution of 2.5 meters.

Japanese Satellite Programs

The Japan Aerospace Exploration Agency (JAXA)—known until October 1, 2003, as the National Space Development Agency of Japan (NASDA)—manages scientific and technical space activities for that nation. In addition to its own meteorological and **Earth remote-sensing** programs, JAXA has collaborated with **NASA** in the United States to develop the **Tropical Rainfall Measuring Mission** (TRMM). It was launched in November 1997 to monitor the Earth's water cycle using a combination of established and experimental sensing instruments. The close collaboration between JAXA and NASA continued with the flight of the Advanced Microwave Scanning Radiometer for EOS (AMSR-E), a passive microwave scanner with a 1.6-meter-diameter antenna, carried aboard NASA's **EOS research satellite Aqua.** The AMSR-E is helping obtain data to understand Earth's global-scale water and energy cycles.

JAXA is currently planning its future missions, which include the Advanced Land Observing Satellite (ALOS), designed for precise land-coverage observations with

The Japan Aerospace Exploration Agency (JAXA) launched Midori-II, an Earth observation satellite, in December 2002 and collected observations from its multiple sensors until it ceased operating in October 2003. Japan has maintained a close working relationship with NASA for many years and Midori-II is an example of this close partnership.

optical imaging sensors and a **synthetic aperture radar.** A mission planned for 2007, called the Greenhouse Gases Observing Satellite (GOSAT), will help monitor the density of carbon dioxide and methane, both greenhouse gases that affect climate.

Geostationary Meteorological Satellite (GMS)

As a participant in the World Meteorological Organization (WMO), an agency of the United Nations, Japan has operated a geostationary weather **satellite** since 1977, and has made its data available for international use through the WMO's World Weather Watch program. Four U.S.-built GMSs had been launched by NASDA (now JAXA) through 1995 when GMS-5, the last of this series, was placed into operation.

The first of the next generation of Japanese weather satellites, the Multifunctional Transportation Satellite (MTSAT-1), was destroyed in a launch failure in 1999, but its replacement, MTSAT-1R, has been delivered by its manufacturer, the U.S. company Space Systems Loral, and is expected to be launched by the end of 2004.

Japanese Earth Resources Satellite (JERS-1)

Japan's first domestically built **Earth remote-sensing** satellite, the Japanese Earth Resources Satellite (JERS-1), also known as Fuyo-1, was launched on February 11, 1992. This highly successful mission carried a **synthetic aperture radar** (SAR) and an optical sensor (OPS), operating in visible and near infrared bands.

Advanced Earth Observing Satellite (ADEOS)/Midori

On August 17, 1996, Japan launched the Advanced Earth Observing Satellite (ADEOS) on an H-II vehicle from the Tanegashima Space Center and renamed it Midori, the Japanese word for "green." This large satellite, weighing 1,200 kilograms, was placed into a **sun-synchronous near polar orbit** at 800 kilometers. It carried eight sensors, two of which, the **TOMS** and NASA Scatterometer (NSCAT) were provided by **NASA.** The Japanese-built sensors were the Ocean Color and Temperature Scanner (OCTS), the Advanced Visible and Near-Infrared Radiometer (AVNIR), the Polarization and Directionality of the Earth's Reflectances (POLDER), the Improved Limb

Atmospheric Spectrometer (ILAS), the Retroreflector in Space (RIS), and the Interferometric Monitor for Greenhouse Gases (IMG). Midori operated for only about ten months when it went out of control in June 1997.

Japan launched a second Earth-observing satellite—ADEOS-II, also known as Midori-II—in December 2002 aboard an H-II vehicle. This satellite carried an Advanced Microwave Scanning Radiometer (AMSR), a Global Imager (GLI), an Improved Limb Atmospheric Spectrometer-II (ILAS-II), the NASA-provided **SeaWinds** scatterometer, POLDER, and the Data Collection System (DCS). The satellite malfunctioned in October 2003.

NASA's Terra satellite has allowed global monitoring of carbon dioxide. In April and October 2000 the Canadian-built instrument known as the Measurements of Pollution in the Troposphere (MOPITT) captured industrial pollution from Asia (top) and immense plumes of carbon dioxide emitted from the burning of forests and grasslands in South America and southern Africa (bottom).

Canadian Remote Sensing

The Canadian space program, managed by the Canadian Space Agency (CSA), has long had close ties with its counterpart in the United States, supplying robotic arms for the U.S. Space Shuttle Program and the **International Space Station** (ISS). However, Canada also has a very active program in **Earth remote sensing** and scientific research. The Canadian instrument, called the Measurements of Pollution in the Troposphere (MOPITT), is an eight-channel infrared spectrometer that was launched on the NASA **Earth Observing System** (EOS) **Terra** spacecraft on December 18, 1999. MOPITT provided the world's first long-term global measurements of carbon monoxide and methane gas in lower levels of the atmosphere—both important measurements in climate research.

As a member of the **European Space Agency** (ESA), Canada has close collaborative ties to Europe as well. The Canadian-built Optical Spectrograph and InfraRed Imager System (OSIRIS), an instrument designed to study ozone depletion in the atmosphere, was launched on the Swedish Space Corporation's Odin **satellite.** This marked the third mission in which Canada and Sweden collaborated on space programs. On the previous two joint missions, Viking in 1986 and Freja in 1992, CSA provided ultraviolet imagers that took pictures of the aurora at high northern latitudes.

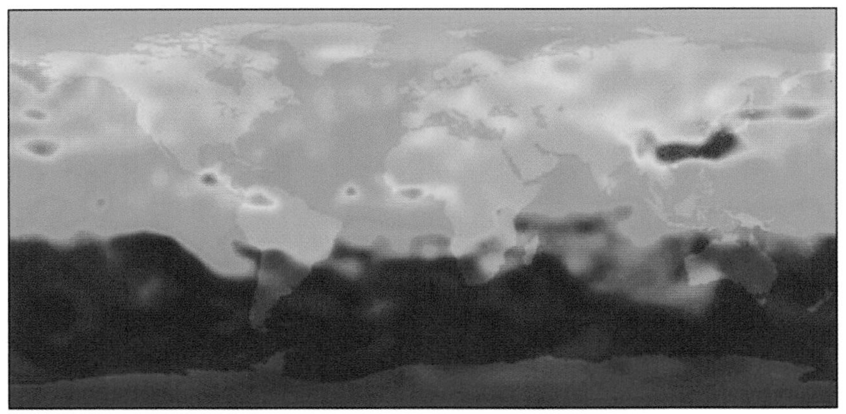

April 30, 2000

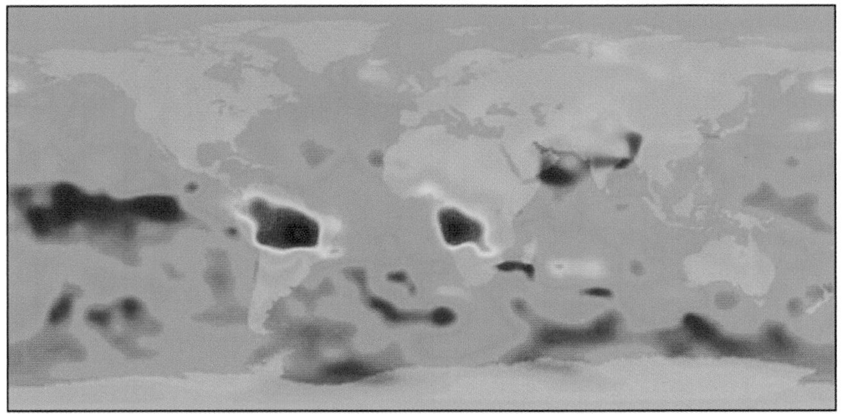

October 30, 2000

Carbon Monoxide Concentration (parts per billion)

50 220 390

RADARSAT

Due to the practical and economic impact of sea ice on Canadian commerce, not only for access to its coastal ocean ports, but also in the Great Lakes, the ability to monitor ice coverage and motion is a central focus of Canada's remote-sensing program. The centerpiece of this effort since its launch in November 1995 has been the RADARSAT-1 program. The satellite "bus" for this program was procured from Ball Aerospace & Technologies Corporation in the United States, but the satellite's **synthetic aperture radar** and overall system design and integration were done in Canada, with SPAR Aerospace of Montreal as the primary contractor.

RADARSAT images are used primarily for ice monitoring, but imaging radar observations have many applications in **oceanography,** coastal ecosystem monitoring, **mapping and charting, agriculture** and **forestry,** hydrology, and disaster management. A private company, RADARSAT International, was created to market radar-imagery data to commercial users and foreign governments around the world, the revenue from which is intended to offset the cost to the government of Canada for development and operation of the **satellite.**

In order to maintain data continuity for this important program, CSA has contracted for its follow-on satellite, RADARSAT-II, now being built. This new system is being designed for improved image quality and increased observations.

Space-communications Basics

Space Communications

The need to develop satellites for communication arose from three factors: the **Earth's atmosphere** and **ionosphere,** the radio frequency of the broadcast, and the economic advantage of satellites over wires.

Electromagnetic waves propagate at the speed of light between a transmitter and a receiver. The most efficient path for such a transmission is in a direct (straight) path, sometimes called line of sight. A transmitted signal loses strength by spreading and **attenuation** (the absorption or scattering by molecules in the atmosphere), but if the transmission is sufficiently powerful and the receiver sufficiently sensitive to identify the signal, the transmission will be received and the information it contains will be conveyed. However, if no line-of-sight path exists between the transmitter and the receiver, the radio waves may either be bent (refracted) by the Earth's atmosphere, generally in the moist atmospheric layer near the surface, or bounced (reflected) off the Earth's ionosphere, the layer of charged particles in the upper atmosphere. Whether this ducting or reflection occurs depends on the frequency of the radio wave being transmitted.

Radio waves in the medium and high frequency (MF to HF) bands from about 1,000 kilohertz (AM radio) up through shortwave radio, about 10 megahertz, are reflected by the ionosphere. Higher frequency waves in the very high frequency (VHF) band, used for FM radio and television broadcasts, pass straight through the ionosphere and are lost into space. The same applies to yet higher microwave bands, referred to as ultrahigh frequency (UHF), superhigh frequency (SHF), and extremely high frequency (EHF). While microwave transmissions, such as for surface search radars, are able to be ducted (refracted) over long distances near the surface of the Earth, this phenomenon is not generally practical for communications.

Although MF and HF communications can be transmitted to great distances due to ionospheric reflection, they are too low in frequency to carry a great deal of information—and data volume is critically important in transmitting voice, pictures, or massive amounts of digital data. The direct path of higher frequencies (VHF, UHF, SHF, and EHF) through the ionosphere, combined with the dawning of the space age, opened the door to satellite communications. In 1945 Sir Arthur C. Clarke proposed the theory that if three satellites could be placed in geostationary orbit above the Equator, a global network of direct wireless communications could be established covering all the Earth, except for the polar regions. Clarke's concept was prophetic. Today, hundreds of communications satellites orbit the Earth, providing services vital to world commerce, diplomacy, and the military.

Commander R. C. Traux, USN (Bureau of Aeronautics, Guided Missiles Division), George Sutton (North American Aviation, Inc.), Sir Arthur C. Clarke (author and rocket authority), Dr. Fred Singer (U. of Maryland), and Dr. Walter Dornberger (Bell Aircraft Corporation) were a few of nearly five hundred specialists in astronautics and science who attended the Third Symposium on Space Travel in May 1954 in New York.

A communications satellite link consists of an originating ground station (broadcast antenna), a satellite that receives its electrical power from solar panels and carries a receiver and receive antenna and a transmitter and transmit antenna, and a destination ground station (receive antenna). A transponder is the satellite's receiver/transmitter system. A satellite is sometimes described by the following characteristics: its orbit, number of transponders and transponder frequency band(s), its application, and whether it is for point-to-point data relay or for direct broadcast.

Orbits for Satellite Communications

Both circular and **elliptical orbits** are used for communications satellites. **Circular orbits** are an elliptical orbit in which the two foci of the **ellipse** are at the same place, the center of the Earth. Just as in **Earth remote sensing** from space, circular orbits are generally used in one of three ways: **geostationary Earth orbit** (GEO), **low Earth orbit** (LEO), or **medium Earth orbit** (MEO).

Since satellites in LEO or MEO can be "seen" from a given geographic area for only brief periods of time as they pass overhead, a large number of satellites must be flown in a constellation in order to provide coverage of the globe. These satellites may be at the same **altitude,** but spaced out in different planes (**inclinations**). GEO satellites, however, offer the advantage of being in circular orbits with zero (or very slight) inclination, orbiting in a west-to-east direction, with their eastward flight at the same rotation rate as the Earth's. This means GEO satellites remain "stationary" over one place on the Equator. Each satellite can be "seen" from about 42 percent of the Earth's surface. A constellation of three such satellites can cover virtually the entire **planet,** and is used by most commercial and military satellites.

The **highly elliptical orbit** (HEO)—or **Molniya orbit**—is used to provide satellite coverage over the polar regions because the angle from a high latitude location to a satellite over the Equator is extremely shallow. A constellation of two HEO satellites can provide full-time coverage at northern latitudes.

Each of these orbits: GEO, LEO/MEO, and HEO, has its distinct advantages and disadvantages. GEO, the most useful, offers simplicity and near-global coverage. However, even at the speed of light, it takes several seconds for a transmission to make two or more hops from Earth to 35,000 kilometers away and back. For commercial data communications, this lost time is valuable, and for voice communications, awkward and frustrating. As a result, LEO constellations were created to cut the distance of transmissions dramatically.

MOST COMMON FREQUENCY BANDS USED IN SATELLITE COMMUNICATIONS

Microwave Band	Frequency (GHz)	Wavelength (cm)
UHF	0.3-3	100-10
SHF	3-30	10-1
EHF	30-300	1-0.1
L	1-2	30-15
S	2-4	15-7.5
C	4-8	7.5-3.75
X	8-12	3.75-2.5
Ku	12-18	2.5-1.67
K	18-27	1.67-1.11
Ka	27-40	1.11-0.75

Frequency Bands for Communications

The higher the frequency and shorter the wavelength, the more information an electromagnetic wave transmission can carry. Wavelengths longer than one meter are impractical because transmissions to and from satellites in orbit could be inadvertently reflected by the Earth's **ionosphere.**

The following microwave bands are most commonly used for **satellite** communications: ultrahigh frequency (UHF), superhigh frequency (SHF), and extremely high frequency (EHF). Within these frequency bands, the following frequency/wavelength ranges are widely used in satellite communications: L band, S band, C band, X band, Ku band, K band, and Ka band. In common usage, the terms SHF and X band are synonymous, as are EHF and Ka band. Satellites providing communications at the EHF band represent the highest technology currently in use, providing the highest data rate of all **operational satellites.**

Broadband Communications

The term broadband is generally used in the context of Internet connectivity when the speed of data delivery is greater than the fastest speed available over a telephone line. A more technically accurate definition is that a transmission that contains several channels at once, with each channel providing a different message/voice conversation, is a broadband transmission. A baseband transmission carries only one signal at a time. Multiplexing multiple channels into an **uplink/downlink** in **satellite** communications is essential to obtain maximum efficiency in each transponder's use.

Laser Communications

Communication cables, such as phone lines, have traditionally been made of copper because it conducts electricity well. However, due to the cost, weight, and limited **bandwidth** capability of copper wire, a revolution is now under way to replace wire with fiber-optic cable for most telecommunications applications. While optical fibers, most commonly made of ultra-pure silica, cannot carry electrical current as can copper wire, the beams of light that are sent through them are capable of carrying higher data rates for digital communications than can wires. Due to the extremely short wavelength of light relative to radio waves, data transmission by light through open space is greater than the information-carrying capacity of radio waves.

Lasers are used as the light source because they can generate a concentrated light beam at a uniform wavelength and direct the path of the light, either into a fiber-optic cable or directly through open space to a receiving optical terminal, essentially a **telescope.** In space-to-space optical links, the light travels through a vacuum and is attenuated only by spreading loss. In space-to-ground or ground-to-space links, however, the light is attenuated due to scattering by the atmosphere or airborne particles. Clouds, which are generally transparent to radio waves, usually stop all but the highest-power laser beams, making laser communication links practical only in cloud-free areas.

Due to the ever increasing demands for higher-capacity data transmission capabilities and the difficulties created by the crowding of the radio **spectrum** by multiple users, direct line-of-sight optical laser communications is considered to be the wave of the future in **satellite** communications.

Data Transmission

All electronic data transmission is done through a process called modulation of a carrier signal. The carrier wave (such as a radio wave with a constant frequency) is modulated to convey the baseband (information) signal. If the amplitude of the carrier wave is varied to carry the information, the transmission is amplitude modulation (AM), and if the frequency of the carrier wave is varied, it is frequency modulation (FM). A third, less well-known method of data transmission is called phase modulation (PM), in which the rate at which the carrier phase is changed is used to convey the baseband signal.

In digital communications all information is converted into a sequence of binary data, often referred to as "ones and zeros." In electronics this binary language is translated into "on and off" or "positive charge and negative charge" or "voltage or no voltage." A digital signal, just like its analog counterpart, can be transmitted through use of amplitude, frequency, or phase "shift keying" (ASK, FSK, or PSK).

The transmission of data, then, is accomplished by transmitting the carrier signal, which may be either a radio wave (microwave) or a light beam, which is modulated or shift keyed to carry the baseband information signal. If multiple bands of information are superimposed on a carrier, then this multiple-signal transmission is a broadband transmission.

Data is often thought of, and used in speaking, in the singular, as in "the data is now coming in." However, in a technical sense, data is the plural for the word datum, meaning a single piece of information. In the context of satellite communications, we use the word "data" to mean a lot of information. A **satellite** is a powerful tool to convey tremendous amounts of information from one place to another. In today's world of digital communications, new words have come into our vocabulary to describe the tremendous amount of data being sent; words like petabyte, which is the next step upward in a natural progression from kilobyte (a thousand digital "words" consisting of eight bits each) to megabyte (a million bytes) to gigabyte (a thousand million, or billion bytes) to petabyte (a thousand billion bytes). That's a tremendous amount of information to get from one place to another in a short amount of time.

In July 2003 Europe's Artemis satellite used its SILEX laser-optical relay terminal for satellite-to-satellite communications. Just as fiber-optic cable is replacing copper wires in the global communications infrastructure, the future of communications in space may lie in beams of light rather than in the radio waves that have carried voice and data in decades of satellite communications.

Bandwidth

In telecommunications the term bandwidth has two meanings, one technically accurate and the other based on common usage. Technically, bandwidth refers to a range within a band of frequencies or wavelengths on the electromagnetic spectrum (for example, a receiver has a bandwidth of 2 to 20 gigahertz). However, in satellite communications, bandwidth is also used to

refer to the rate at which digital data can be sent in a fixed amount of time, and this is often expressed as a data rate in bits or bytes per second. In analog transmissions, bandwidth is sometimes used to describe data rate in cycles per second, or "hertz." A high bandwidth satellite system generally describes a system capable of transmitting a relatively large amount of data in a short time.

Spread Spectrum

Spread spectrum is a technique for ensuring secure electromagnetic communications by using a series of short radio pulses, transmitted at different frequencies with a frequency band of the spectrum. In spread-spectrum transmission, an intercepted signal is not complete at any one frequency and cannot be used by an unintended recipient. This "frequency hopping" also protects the communications link from enemy jamming. The technique has been used successfully by the U.S. military in its **satellite** communications, and in fact was used for radio transmissions as early as World War II. In 1985 the technique was opened by the U.S. Federal Communications Commission for limited commercial use, and it has found its way into such common systems as cordless telephones and wireless local-area networks.

Spread spectrum has evolved into what is called ultrawide band (UWB), in which billions of pulses are spread over a frequency range of several gigahertz. UWB is used in commercial telecommunications for very high-speed, secure data transmission.

Band-pass Filtering

A band-pass filter is an electronic device that allows transmission of energy in a specific, often narrow, wavelength band and rejects energy that has a longer or shorter wavelength. Band-pass filtering is particularly important in designing systems that resist jamming (the intentional transmission of energy by an enemy to deny use of a frequency band) or interference (an unintended energy signal received from other sources at a given frequency).

Attenuation

As electromagnetic energy propagates between the ground and space, its waves must travel through the atmosphere and the **ionosphere.** In this travel, the power of the signal is attenuated, or diminished, due to the interaction between the energy and the medium through which it is passing. Attenuation takes two forms: absorption and scattering. The major causes for attenuation differ, depending on the wavelength of the energy involved. For instance, in the microwave band used by most telecommunications **satellites,** the greatest factor is rain. The more precipitable water in a cloud, the more the energy passing through it will be scattered, weakening the signal.

In the millimeter band, scattering by rain and absorption by oxygen molecules become factors. At the submillimeter band, water vapor absorption and cloud scattering attenuate the signal. Optical band signals, such as **laser** communications, are attenuated by clouds, fog, rain, dust, and haze. For longer wavelength radio waves, such as high frequency band and shortwave radio, the ionosphere both reflects and refracts electromagnetic energy, making those frequencies unsuitable for satellite communications.

Free Space Loss

Free space loss is the loss in energy of an electromagnetic transmission between the transmitter and the receiver due to the spreading of a transmitted signal on its path through free space.

Free space loss increases in proportion to the square of the distance traveled. While **attenuation** losses end when a signal leaves the atmosphere and continues its path through the vacuum of space to a **satellite,** free space loss continues over the entire path of a signal's travel. Therefore, the spreading loss in a transmission up to a satellite in **geostationary Earth orbit** (GEO) is far greater than for a transmission to one in **low Earth orbit,** given the difference in range of tens of thousands of kilometers. Accordingly, communication links to GEO must be accomplished at greater transmitter power and/or with more sensitive receivers.

Latency

Data latency is the time lag between when data (information) is transmitted and when it is received. In satellite communications, data latency for a two-way transmission of data from a transmitting ground station to a **satellite** and then back to a receiving ground station antenna is the distance traveled divided by the speed of light.

For satellites in **geostationary Earth orbit,** this is 35,786 kilometers times two, divided by 299,792 kilometers a second, or about two-tenths of a second. This may seem like a very short period, but it is actually valuable time wasted for data being transmitted at thousands or millions of bits per second. Such a time lag is noticeably awkward for two-way phone conversations, especially if the transmission has to take multiple trips up and down to connect the speakers.

Communications Satellites

SCORE

The world's first communications satellite, the Signal Communication by Orbiting Relay Equipment (SCORE) was launched by the U.S. Army from Cape Canaveral on December 18, 1958. While SCORE was designed to demonstrate a **satellite**'s ability to receive and respond to commands from the ground, it continuously broadcast on a shortwave frequency a recorded message from President **Dwight D. Eisenhower:** "This is the President of the United States speaking. Through the marvels of scientific advance, my voice is coming to you from a satellite traveling in outer space. My message is a simple one: Through this unique means I convey to you and all mankind, America's wish for peace on Earth and goodwill toward men everywhere."

Echo

Designed in 1960, the Echo satellite was the simplest possible communications satellite. It was a 30.5-meter diameter aluminized plastic balloon that was used to demonstrate that radio and television signals could be transmitted to the **satellite,** reflected back to **Earth,** and then received by any properly tuned antenna within view of the satellite. Echo was visible transiting the night sky, but its low orbit limited its practical use since it was above any given location on Earth for only a very few minutes before passing below the horizon.

Syncom 2

NASA launched Syncom 2, the world's first geosynchronous communications satellite, in 1963 and proved that practical, continuous satellite communications could be accomplished. Syncom 2 established that a constellation of three satellites in **geosynchronous orbit** with cross-links between them could cover the entire globe (except for the poles).

Comsat

The Communications Satellite Corporation (Comsat) was created in 1962 by the United States Congress in the Communications Satellite Act of that year. For the first time, a private company was authorized by the government to develop a global satellite communications system and to acquire and maintain ground stations around the world. In order to develop new satellite technologies, Comsat opened Comsat Laboratories in Clarksburg, Maryland. In 1965 Comsat launched Early Bird, the first commercial communications satellite. The company's unique government-private relationship governed its function as what was known as

In the 1960s, NASA launched two experimental Echo communications satellites. Their aluminized coating allowed stations on the ground to bounce radio waves off their reflective surfaces, proving the technology—but not the practicality of the approach.

"the carrier's carrier." The corporation could not sell transponder circuits on the satellites it owned and operated directly to television broadcasters, news services, or other customers. Comsat was required to sell access to these circuits to other commercial carriers who could then resell them.

Today, Comsat continues to operate under its U.S. federal charter, managing offices around the world and serving customers ranging from foreign governments to commercial carriers. Some elements of the original corporation, however, were recently purchased by ViaSat, Inc. The Comsat Products Group, which

Usually no more than two astronauts spacewalk at the same time, but an exception occurred on May 13, 1992. Earlier attempts to grapple the Intelsat VI for repairs had failed, so Richard J. Hieb, Thomas D. Akers, and Pierre J. Thuott manually captured the 4.5-ton communications satellite. After servicing, it was redeployed to continue operations.

now operates the Clarksburg laboratory, was part of the ViaSat acquisition. Comsat remains the signatory to the **Intelsat** and **Inmarsat** organizations.

Intelsat

The International Telecommunications Satellite Organization (ITSO), an international consortium of national organizations, owns and operates a constellation of Intelsat communications satellites. ITSO has long been referred to by the same name as the satellites it operates, Intelsat Corporation. **Comsat** is the U.S. signatory to the ITSO. The group was originally founded on August 20, 1964, as the International Telecommunications Satellite Consortium, with 11 participating countries signing its charter. ITSO now has more than one hundred members and services hundreds of Earth stations around the world. The company maintains its headquarters in Washington, D.C. Intelsat has, since its founding, procured several different series of its more than 60 **satellites,** allowing it to buy from multiple **spacecraft** and **launch-vehicle** manufacturers. The satellites have been launched from the United States or Kourou, French Guiana, except for a 1996 attempt from China that failed to achieve orbit.

Inmarsat

Similar in nature to the **Intelsat** intergovernmental consortium, the International Maritime Satellite (Inmarsat) Organization was founded in 1979 as an operator of satellites for mobile satellite communications and continues to provide voice, facsimile (fax), and data communications services to maritime, land-mobile, aeronautical, and other users. **Comsat** is the U.S. signatory to the international agreement that created Inmarsat. The company, a limited corporation (Inmarsat, Ltd.) since 1999, operates from its headquarters in London, England, and manages nine **satellites** that serve thousands of users worldwide. The company plans to launch a new satellite, Inmarsat I-4, in 2005. It will provide broadband global-area network services for Internet access and other applications.

International Telecommunications Union (ITU)

The International Telecommunications Union (ITU) is an international organization chartered by the United Nations to coordinate and, in some cases, regulate the activities of governmental and private telecommunications networks and services. Operating from its headquarters in Geneva, Switzerland, the ITU publishes technology, regulatory, and standards information and convenes symposia, such as the World Summit on the Information Society. The organization draws its roots from the International Telegraph Union, established by the first International Telegraph Convention, signed in Paris on May 17, 1865. Prior to that first international agreement, the invention of telegraphy had created chaos in the cross-border exchange of messages because each nation had its own independent system and message protocols. Today, the ITU continues to evolve to develop standards for emerging new systems and technology in order to promote seamless global telecommunications.

Tracking & Data Relay Satellite System (TDRSS)

The TDRSS constellation, which currently consists of seven **satellites** in **geostationary Earth orbit,** was developed by the **NASA** Goddard Space Flight Center in Greenbelt, Maryland, to provide high-speed data, voice, and video telecommunications relay in support of NASA and other government missions. Two-way communications are achieved via the NASA TDRSS ground station located near Las Cruces, New Mexico (known as the White Sands Complex), which communicates directly with the satellites visible from that location and then via satellite-to-satellite link. Through this network, NASA missions such as the **space shuttle, International Space Station, Hubble Space Telescope, Earth Observing System** (EOS) **Terra** and **Aqua** satellites, **Landsat, TOPEX/Poseidon,** and several others have real-time connectivity through White Sands and via landline or other satellite relay to NASA centers and other research facilities. TDRSS is used by the Global Learning and Observations to Benefit the Environment (GLOBE) program to support a worldwide network of students, teachers, and scientists.

Argos Data Collection System

Flying aboard each of the U.S. National Oceanic and Atmospheric Administration (NOAA) **Polar-orbiting Operational Environmental Satellites** (POES) is the Argos Data Collection System, a radio receiver that allows for the **geolocation** of a transmitter or beacon on the surface of the **Earth** and collection of data that is being transmitted by that beacon. The Argos program is a jointly administered program of NOAA and the French space agency, Centre National d'Études Spatiales (CNES). Common applications include the collection of such information as observations of weather elements from unmanned observation sites; wind, wave, and ocean temperature data from ocean buoys; and the locations of tagged endangered or migratory animals. Data collected by the satellites are downlinked to ground stations in Fairbanks, Alaska; Wallops Island, Virginia; and Lannion, France. Two CNES-affiliated companies, Collecte Localisation Satellites in Toulouse, France, and Service Argos in Largo, Maryland, process the data and provide direct support to the end user.

Iridium & Globalstar

Although most telecommunications satellites have been designed to operate at **geosynchronous orbit,** there are two disadvantages to placing a satellite 35,900 kilometers from Earth. One is the expense of launching a satellite to that orbit. The other is the time that it takes an electromagnetic transmission to travel to and from

LIFE WITH SATELLITES

Carissa Bryce Christensen

TO MOST OF US, THOUGHTS OF HUMANITY IN SPACE EVOKE FARAWAY STARS, arcane science bordering on fiction, and visions of an exotic future. The truth is, what we do in space—at least, quite a bit of it—is as familiar and routine as buying a cup of coffee. There is an industrial infrastructure, straight up from where you are right now, that affects just about every hour of your life every day. The news that you catch first thing in the morning shows footage that was uplinked to a satellite by a news-gathering truck (a van with a satellite dish on top) from the scene of a breaking story. The weather forecast uses satellite pictures to help you decide if you should leave the convertible at home or take along your sunscreen as you head out the door.

When you stop at the service station to fill your car's tank, the gas you pump may have come from a pipeline monitored for damage or corrosion using tiny satellite terminals. When you pay with your debit card, you are sending transaction information through the small satellite dish you see on the roof of the station (called a very small aperture terminal, or VSAT).

Back in the car, your cell phone rings and you turn down the radio (which is playing a national broadcast relayed via satellite) to hear the message changing your plans. Your cell phone is not communicating directly with a satellite—but the cell tower it connects to relies on precise timing information from the atomic clocks on the U.S. military's Global Positioning System (GPS) satellites.

As you head to an unfamiliar part of town, it is reassuring to rely on the interactive navigational capability in your car, made possible by the GPS network. Your car's computer uses information from GPS satellites to triangulate your position, then combines this information with maps in memory, or uses the cell phone network to request directions from an operator.

And so it goes for the rest of your day—space capabilities are all around you. Part of your lunch started out as crops that were assessed for growth and condition using imaging satellites. After lunch, you check e-mail and do research using the Web, not thinking about how some Internet traffic travels over communication satellites. In the elevator the piped-in music you hear is being heard simultaneously in elevators, telephone holds, and waiting rooms across the nation through the miracle of satellites. On the way home you mail a package that will travel on a plane guided by satellite navigation and be delivered to its recipient by a truck using a satellite fleet-management system. The road you drive on was surveyed using satellite imagery and GPS location data.

At the end of the day you collapse in front of the television, exhausted from your nonstop use of space systems, and surf through old sitcoms, new reality shows, and roller derby highlights. You probably do not realize that you are watching satellite TV, even though there is no dish on your roof. The programming was distributed nationally by satellite, and then sent to you through your cable lines or broadcast from local television towers.

If your thumb gets tired from changing channels and you phone a friend in Jakarta or Lagos or La Paz to complain about how rotten American television is, your call will likely leave the U.S. via cable but then be routed to a satellite. Long-distance telephony was the first commercial satellite application, and satellite service remains common in regions that do not yet have full access to undersea fiber-optic cables or comprehensive landline networks.

Even the glass of water you carry to your nightstand is tied to satellites; it flowed from a municipal reservoir that uses satellite data to double-check water quality and drought indicators.

So, who owns these space resources that fill your day, and what is it costing to use them? Commercial space systems are owned mainly by companies who sell

The immediacy of war, natural disasters, and other more mundane events is brought into homes via television. Using communication satellites, news stations are capable of beaming transmissions worldwide and in real time.

or lease them to paying customers around the world. Global commercial space revenues are about $100 billion annually. If you count the work companies do for government space programs, the estimate could approach $150 billion. It is worth mentioning that most commercial space companies also benefit in other ways from government space programs, launching from government-owned sites, learning from government scientists, or cutting costs by sharing manufacturing plants and infrastructure.

The commercial space revenue pie divides roughly like this: A third to half comes from television networks, cable companies, satellite TV subscribers, and others who pay to use satellites for television services. A quarter is for manufacturing commercial satellites and ground equipment. About 10 percent comes from customers of satellite telephone, Internet, and data services. Commercial satellite imaging, navigation and location, and weather applications use both commercial and government-owned satellites; they account for even smaller revenue wedges. Finally, while rocket launches create big drama, they generate a small 5 percent of commercial space revenues.

The commercial space pie will get larger in the future. Eventually, space commerce will move beyond satellite services to markets as diverse as ultra-pure materials manufactured in orbit, commercial space cruises carrying tourists to space stations, or valuable minerals mined from the moon and asteroids. New markets have slowly begun to emerge. We have seen a few space tourists paying tens of millions of dollars. Near-perfect experimental crystals have been grown on the shuttle and space station. However, the costs of working in space right now are too high to yield reasonable prices for these and other markets, and costs seem likely to stay high for the foreseeable future.

It may take years, or it may take decades, for the new space products and services we imagine today to enter and enrich our daily lives. In the meantime we remain fortunate that today's marvels of space commerce are routinely at hand and that the wonders of space are all around us on Earth. ■

the satellite at that height. In the early 1990s, entrepreneurs conceived ambitious plans to develop and **launch** constellations that would be made up of interlinked satellites in **low Earth orbit** (LEO), which would provide global coverage for data and voice communications. Using low-cost handheld devices similar to cellular telephones, a customer could make or receive a phone call anywhere on the planet. However, every one of these proposed systems was either scrapped before being built or was a financial disaster for the investors.

Two types of LEO systems were proposed: a global network of voice communications satellites, dubbed "Big LEOs," and more modest networks of small, relatively inexpensive satellites which, although they could not provide the large **bandwidth** required for voice services, could provide two-way data transfer and data collection services. These were called "Little LEOs."

Two Big LEO systems were built and continue to operate. Neither was a commercial success, mostly due to the unanticipated explosion in global use of inexpensive cellular telephone service supported by ground-based antennas, and the widespread use of optical fiber to replace copper telephone lines, enabling efficient and inexpensive long-distance service around the world.

The Iridium Satellite System, operated by Iridium Satellite LLC, provides complete coverage of the Earth through a constellation of 66 LEO satellites. The system is principally supporting the U.S. government, its largest customer, but it is attempting to widen its commercial user base in multiple market segments.

The Globalstar system came on line a few years after Iridium and currently consists of 48 LEO satellites that offer similar worldwide voice and data services. Both Iridium and Globalstar services use handsets that are commercially available and are not interoperable. The assets of both companies have changed hands since their initial investors suffered considerable financial losses.

Orbital Communications System (Orbcomm)

The Orbcomm constellation of 30 **satellites** in **low Earth orbit** (LEO) was built by Orbital Sciences Corporation of Dulles, Virginia, and originally operated by a subsidiary called Orbital Communications Corporation (Orbcomm). Orbcomm was conceived as a network of very small, relatively low-cost satellites to provide two-way messaging, data communications, and **geolocation** services on a global basis. It was unofficially dubbed one of the "Little LEO" constellations that would be an attractive alternative to large satellites in **geostationary Earth orbit** and the other planned constellations of LEO systems being built for voice and data communications, such as Iridium and Globalstar. Today, the system is operated by Orbcomm LLC, a wireless telecommunications company that serves its government and commercial customers through value-added resellers, companies licensed to sell the Orbcomm service around the world.

Because of the relatively low cost of Orbcomm transmitter/receiver units, this system has opened entirely new applications, such as messaging for truck fleets, weather data for general aviation, and monitoring and tracking assets as small as individual packages in shipment. Although not an immediate commercial success, Orbcomm is enlarging its user base because there is virtually no competition in this market segment.

Telstar

Telstar is the best-known communications **satellite** and is considered by some to have been the first operational system in the era of space-based telecommunications. In the early 1960s, AT&T Corporation and **NASA** collaborated to build six Telstar satellites. Only two were actually launched, the first on July 10, 1962, atop a Delta **rocket.** Bell Laboratories designed and built the **spacecraft,** and NASA contributed the launches and used them to support the agency's research experiments. Originally conceived by AT&T as a constellation of between 50 and 120 satellites, to be launched several at a time in order to achieve a preeminent position for At&T as the world's leading provider of satellite communications, the Telstar program was to have been a model inspiring cooperation between government and industry.

Private enterprise was encouraged to establish and operate satellites for profit, a concept endorsed by then U.S. President **Dwight D. Eisenhower** in his December 1960 policy on space communications. However, less than a year later, the new President, John F. Kennedy, released a new policy on July 24, 1961, in which he still endorsed private ownership of satellite systems, but with conditions that would prevent any one company from creating a monopoly, the goal of AT&T's Telstar program. On August 31, 1962, President Kennedy signed the Communications Satellite Act, which created a new company called **Comsat** that was granted a monopoly, but with international partnerships.

Navigation & Positioning from Space

Navigation & Positioning from Space

In 1957, as Americans observed the Soviet Union's **Sputnik** satellite pass overhead transmitting its beeping signal, engineers noted that the signal being transmitted had a constant frequency. As the satellite approached a receiver from one horizon and then fled after achieving its maximum **altitude** for that pass, the Doppler effect compressed the signal on its approach—increasing the received frequency—and then stretched the signal on its departure—decreasing the frequency. It wasn't long before this frequency shift could be measured and used to accurately determine the satellite's orbital track. At the Applied Physics Laboratory (APL) of Johns Hopkins University in Laurel, Maryland, scientists and engineers were able to use this principle in reverse, proving that if a single satellite's exact position and orbital parameters were known by a receiver/computer system on **Earth**'s surface, the exact location of the receiver could be accurately determined.

The **Navy Navigation Satellite System,** also known as Transit, was built at APL and was operated successfully by the Navy until 1996, having been replaced by the **Global Positioning System** (GPS). GPS uses a different technology, called the pulse-ranging method, for positioning. It requires four GPS satellites to be in view of the receiver in order to achieve a complete solution for position. GPS offers accurate "fixes" of location, not only providing longitude and latitude coordinates, but also supplying altitude and velocity vectors. It has become the international standard for satellite navigation.

Astrometry

Precise navigation requires many elements; the two most essential have always been a keen understanding of the positions of stars and other celestial bodies relative to the Earth's motion, and precise time. Astrometry is a field that is closely aligned with astronomy, but where the latter is the study of space and all the objects and energy fields within it, the former is all about measurements. Celestial navigation is the fixing of one's position on the surface of the Earth using knowledge of star positions in the sky that have been previously determined relative to the Earth's rotation and other motions. A simple measurement of the angle between the horizon and **Polaris,** the North Star, has provided mariners in the Northern Hemisphere with a good estimate of their latitude since ancient times. The invention in the 18th century of very precise timekeeping devices, clocks called chronometers, enabled navigators to accurately determine their longitude as well. Navigators go to sea equipped with at least one but generally no less than three chronometers, plus the computations of star and **planet** positions relative to time and the Earth's movement in a book called a *Nautical Almanac.*

Satellite tracking has become an important tool in monitoring endangered species. Researchers attach transmitters to the animals—as on this loggerhead turtle—and then monitor the signals in order to track individual movements.

The world leader in applied astrometry and precise timekeeping, and the publisher of the *Nautical Almanac,* is the U.S. Naval Observatory in Washington, D.C. The star catalog computed continuously by the observatory and the exquisitely precise time standard maintained by its hydrogen maser "master clock" are transmitted electronically by satellite link throughout the world for use in navigation and astronomical research, and they are fundamental to the technology of **satellite** navigation.

Geolocation

To determine the exact location of any person, place, or thing in north-south-east-west coordinates such as longitude and latitude, plus its **altitude** above the **Earth**'s surface or its depth below the surface, is to determine its geolocation, or location in relation to the Earth. While this process is obviously the essential element of navigation on the sea, or finding one's way out of a dark forest, it is also critically important in today's high technology world of geographic (also called geospatial) information systems (GIS). A travel club road map is a simple and commonly used product of GIS technology presented in printed form. The north-south-east-west coordinates are laid out on the borders or perhaps on a grid; colors or shading are used to show terrain variations; and human-built infrastructure such as buildings, roads, and rail lines are drawn to show the user how to get from point A to point B. Such a map is a set of different databases that have been laid on top of one another to provide the user with multiple pieces of information. The basic data are projected onto flat paper using a standard process such as the Mercator projection, in which longitude and latitude lines are shown as perpendicular to each other, creating an easy-to-read rectangular chart. Also needed are a terrain data base, a vegetation map, a road layout map, a building location map, and other data. If the traveler has an electronic **Global Positioning System** (GPS) receiver in addition to the road map, the user can push a button to determine the current location, consult the map, and proceed onward to the destination. If the user has an electronic map/GPS receiver combined system, then he simply follows directions given by the computer.

The underlying and fundamental truth in this example is that, at some time, someone had to carefully determine the geolocation of, or geolocate, the precise position of every terrain feature, road, building, and other landmark that appears on the map, and then

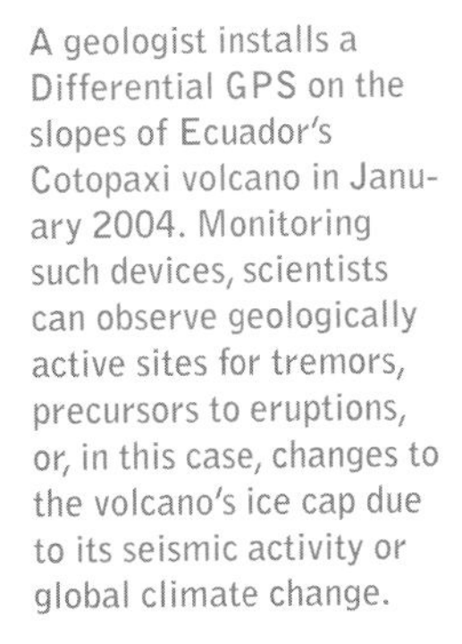

A geologist installs a Differential GPS on the slopes of Ecuador's Cotopaxi volcano in January 2004. Monitoring such devices, scientists can observe geologically active sites for tremors, precursors to eruptions, or, in this case, changes to the volcano's ice cap due to its seismic activity or global climate change.

carefully position each feature to be displayed on top of one another in its proper position until the travel club's road map became a fused product of multiple data layers. Geolocating these mapped features in the digital age is generally done through aerial or space-based photographic surveys. Control points are objects that are visible in each picture, especially in the areas where adjacent photographs overlap. These points are very carefully geolocated, usually by having someone actually go out and survey the location by methodically measuring its relative position to other already surveyed features, or by placing a GPS receiver at the exact spot and recording the location. Then, by digitally overlaying the photograph on top of infrastructure maps, the location of each visible feature in the photograph is geolocated. A map or chart is thereby created; this process, when done using **satellites**, is called **mapping and charting** from space.

Geolocation is useful in tracking animals, such as endangered or migratory species of birds, fish, or mammals tagged with satellite beacon transmitters. It is also used to track objects, such as individual units of truck fleets or shipped packages. Navigation satellites and data-collection systems on environmental or communications satellites have become accepted tools in the important and valuable business of geolocation.

Navy Navigation Satellite (Navsat) System /Transit

The **launch** of **Sputnik** by the Soviet Union on October 4, 1957, led to a fundamental breakthrough in the age-old problem of navigation. Electronic navigation was invented in the early 1940s with the introduction of Long Range Navigation (LORAN) using land-based electronic beacons. George Weiffenbach, William Guier, and Frank McClure, scientists at the Applied Physics Laboratory (APL) of Johns Hopkins University in Laurel, Maryland, studied Sputnik's 20-megahertz signal and deduced that such a constant frequency transmission, if coming from a **satellite** with precisely known orbital parameters, could be used to geolocate a receiver on the Earth's surface by using the precise time and the Doppler shift of the received signal.

The launch of Sputnik 2 on November 3, 1957, allowed APL to continue its research and then propose a satellite program called Transit to the newly created Advanced Research Projects Agency (ARPA) of the U.S. Department of Defense in 1958. The program was assigned by ARPA to the U.S. Navy to manage, and the first successful launch of a Navy Navigation Satellite (Navsat) was accomplished on April 13, 1960, from Cape Canaveral on a Thor **rocket** with an Ablestar upper stage. It became the world's first navigation satellite—Transit.

The Navsat system became operational for the U.S. Navy in 1964, and the satellites were controlled from a Navy facility at Point Mugu, California. In 1967 Transit was opened for use by anyone, anywhere, regardless of nationality, and provided dependable all-weather navigation positions to many thousands of users in its 32 years of service. The last Transit satellite was launched in 1988. Several satellites of the constellation were still fully functional in 1996, when the satellite system was shut down, having been replaced by the Global Positioning System (GPS).

Doppler Tracking

The **geolocation** of a **satellite** receiver/computer system can be determined by observation of the apparent frequency shift of a transmitted constant-frequency signal from a **spacecraft,** when combined with precise time signals and exact knowledge of the satellite's orbital parameters (its ephemeris). This is the basic process that was used to determine navigational fixes using the now-retired **Navy Navigation Satellite System/Transit.** With that system, a user measures the Doppler shift from the signals transmitted at 149.99 and 399.97 megahertz during the 15-minute overhead pass of a single Transit satellite. Using timing "marks" and ephemeris also transmitted, the signals are corrected for ionospheric refraction and processed by the computer to generate longitude and latitude coordinates. This same technique is used in reverse to determine the precise ephemeris of satellites by transmitting constant-frequency signals from ground stations and using Doppler shift and precise time to determine the orbital position of the spacecraft.

This system was used to compute orbital ephemeris for the Transit satellites and for the U.S. Navy's Geosat mission, and it is still in use today by the European Doppler Orbitography and Radiopositioning Integrated by Satellite (DORIS) system used by the U.S.-French **TOPEX/Poseidon** mission and other satellites.

Global Positioning System (GPS)

The U.S. Department of Defense operates the Global Positioning System (GPS) constellation of 24 **satellites** designed to provide three-dimensional (latitude, longitude, and **altitude**) global **geolocation** services. The Joint Program Office, located at the Space and Missile

Systems Center in Los Angeles, California, manages the program. The system was designed to support the U.S. and allied military forces, but navigation by GPS is available to anyone with commercially available, and increasingly inexpensive, receiver systems. Two levels of accuracy are provided by GPS. Standard Positioning Service (SPS) provides a positioning accuracy of about 100 meters in the horizontal and 156 meters in the vertical, as well as precise time signals accurate to within 340 nanoseconds (millionths of a second) of coordinated universal time (UTC). For military users with the appropriate equipment, the Precise Positioning Service (PPS) provides a higher degree of accuracy, providing positional accuracy of 22 meters (horizontal) and 27.7 meters (vertical) and time transfer accuracy to within 200 nanoseconds of UTC.

The initial 11 Block I GPS satellites were built by Rockwell International (now part of the Boeing Company) and launched between 1978 and 1985 from Vandenberg Air Force Base in California.

This series of developmental satellites was followed by 28 Block II and IIA satellites, also built by Rockwell. When 24 Block I and II/IIA satellites were placed in their assigned orbits—a complete constellation—the system's Initial Operational Capability (IOC) was declared on December 8, 1993, providing global SPS accuracy.

Beginning in 1989 the U.S. government began procurement of 21 additional satellites, designated GPS IIR, from General Electric Astronautics, which was later to become part of Lockheed Martin Corporation. When a full constellation composed of all Block II/IIA and IIR satellites was in orbit, the Full Operational Capability (FOC) for GPS was established on April 27, 1995. With this milestone reached, the U.S. government directed the shutdown of the **Navy Navigation Satellite System (Transit)**.

GPS satellites fly in **circular orbits** inclined to 55 degrees at an altitude of 20,900 kilometers. Each spacecraft takes 12 hours to orbit the **Earth.** The full constellation of 24 satellites in this orbital configuration allows a receiver to always have 4 GPS satellites "in view" at any given time. The Transit system only required one satellite to fix one's position using **Doppler tracking,** but this allowed only two-dimensional fixes (longitude and latitude) with limited accuracy. The GPS system uses a pulse-ranging method of positioning. In this system, when a signal (pulse) is sent out by a satellite at a precisely known time and received at a precisely measured (but extremely short) time later, the exact range from the satellite to the receiver can be computed. If the ranges between several satellites (for which the orbital parameters are known) and the receiver are simultaneously computed, then a three-dimensional fix can be established. By continuously computing the position of the receiver, the velocity of the receiver's motion can also be computed. Thus, the pilot of an aircraft with GPS always knows his exact location as well as the plane's speed and direction.

It is small wonder that GPS has become the primary operational method for commercial aviation navigation. This revolutionary breakthrough in electronic positioning allows the military to have situational awareness right down to the individual soldier and to precise guidance of weapons. It allows **spacecraft** operators to know the precise orbital parameters of their satellites, and it supports an ever growing number of commercial, scientific, and civil users and their applications.

NAVSTAR

Although commonly known by its acronym, GPS, the **Global Positioning System** is referred to as the NAVSTAR system, or NAVSTAR GPS, also in common use.

P-Code

NAVSTAR GPS satellites transmit on two different frequencies in the L band: 1,575.42 megahertz (L1) and 1,227.6 megahertz (L2). The Precision Code (P-Code) is a pseudo-random noise (PRN) ranging code that is the primary navigation ranging code. When satellites are commanded into anti-spoofing (A-S) mode of operation, such as in time of war when a jamming threat (fake transmissions to confuse the satellite or receiver systems) may exist, a different code (Y-Code) is used in place of the P-Code. GPS service can also be operated in a selective availability (SA) mode to deny the highest degree of accuracy to unauthorized users.

Differential GPS

Differential GPS is a method in which exquisitely accurate positions can be obtained using the **NAVSTAR GPS** system. Differential GPS is accomplished by placing a GPS receiver in a location that is precisely known through surveying techniques. Then, by transmitting corrections to other receivers in the area, the combination of signals coming from the GPS **satellites** and the precisely located receiver/transmitter allows geopositional accuracy to less than one meter. Accuracies measured in millimeters have been reported. Differential GPS is being used for surveying, aircraft landing, and a growing number of other applications.

Russian Satellite Navigation System (GLONASS)

Russia operates a navigation program begun in the Soviet Union era called the Global Navigation Satellite System (GLONASS). This system is very similar in design and function to the American **NAVSTAR GPS** system. When fully operational, GLONASS consists of a constellation of 24 **satellites** orbiting in a **circular orbit,** inclined to 64.8 degrees at an altitude of 19,100 kilometers. Like GPS, GLONASS operates on two frequencies, 1.250 gigahertz and 1.6035 gigahertz. Position accuracies of 100 meters (horizontal) and 150 meters (vertical), and velocity computation accuracies of 15 centimeters per second are advertised. Receiver systems have been built that can receive both GPS and GLONASS signals to achieve highly accurate position and altitude data.

Emergency Position Indicating Radio Beacon (EPIRB)

An Emergency Position Indicating Radio Beacon (EPIRB) is a radio transmitting device that can, depending on its design, be either manually or automatically activated (such as by being released and activated by a pressure gauge as a boat is sinking). Once activated, the device transmits a distress signal that can be received by an orbiting **satellite** or followed using a homing device to locate the person or persons in distress. Typical applications are for aviators or mariners to indicate that an aircraft has crashed or a ship has gone down or is sinking. Thousands of people, mostly boaters, have been saved since EPIRBs came into use in the 1970s. Three different types of EPIRBs are in operation today: Homing beacons operating at 121.5 megahertz, COSPAS-SARSAT beacons at 406 megahertz, and Inmarsat beacons at 1.6 gigahertz. Depending on the model desired, these beacons can be purchased commercially for between $200 and $1,500. Some models contain a **NAVSTAR GPS** receiver so that the beacon's exact location can be broadcast along with the distress signal.

Search and Rescue Satellite (SARSAT) System

In 1970 an airplane carrying two U.S. members of Congress disappeared in a remote area of Alaska and, in spite of an exhaustive search, no trace of the plane was ever found. In reaction to this tragedy, Congress mandated the use of Emergency Locator Transmitter (ELT) devices on all U.S. aircraft. These early beacons, operating at 121.5 megahertz, were limited in range and operated in a spectral band with interference from other sources of transmissions. A **satellite**-based system was subsequently developed through the cooperation of the United States, Canada, and France. This Search and Rescue Satellite (SARSAT) system, originally designed to pick up emergency transmissions at

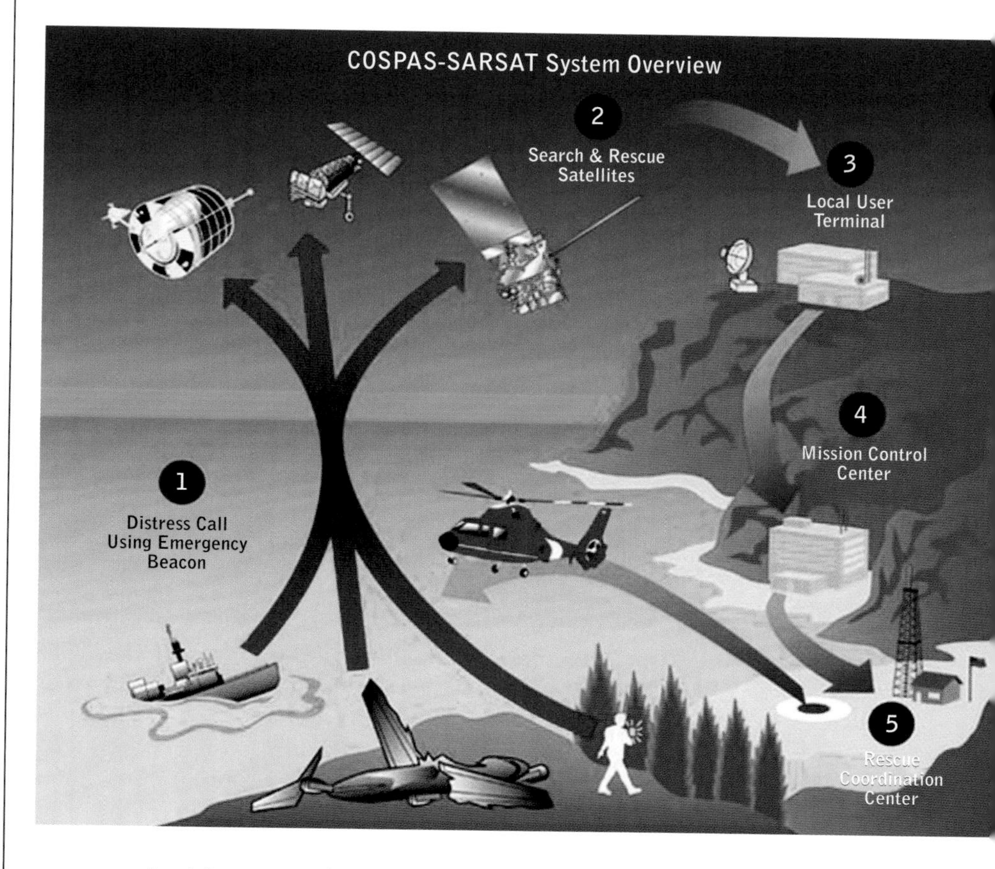

COSPAS-SARSAT, the international search and rescue satellite-aided tracking system, has saved thousands of lives in its two decades of operation. Satellites operated by NOAA and Russian counterparts carry receivers that detect emergency beacons activated by vessels at sea, downed aircraft, or people lost in the wilderness.

a more optimal frequency of 406 megahertz, was managed in the U.S. by the National Oceanic and Atmospheric Administration (NOAA). A similar system, COSPAS, was developed by the Soviet Union.

In 1979 the two systems were combined into COSPAS-SARSAT. In 1982 the first satellite to carry a COSPAS-SARSAT receiver was launched. Today, receivers for this system are flown on all NOAA environmental satellites and Russian weather satellites. The U.S. operations center for the system is operated by NOAA in Suitland, Maryland. Because a large number of older 121.5 megahertz beacons were still in use around the world, COSPAS-SARSAT was designed to detect **Emergency Position Indicating Radio Beacon** (EPIRB) devices transmitting on either 121.5 or 406 megahertz signals.

SAVING LIVES

Jonathan T. Malay

THE TITLE OF THIS CHAPTER CONTAINS TWO WORDS THAT PROVIDE AN interesting contrast: science and commerce. Scientific research is often conducted for practical reasons, such as the development of pharmaceuticals to cure or prevent disease. Research can also be done purely to expand our understanding of the mysteries of the universe—the collection of knowledge for its own sake. Commerce, on the other hand, is all about making money. The contrast in these terms lies in the fundamental differences in motivation for pursuing them and investing money in them. A person invests in science to learn and invests in commerce to earn.

The cost of both these investments, however, is very high, almost always measured in hundreds of millions of dollars. This high cost raises the question of cost versus benefit. How much is enough to spend on space systems?

In the world of commerce, the answer to this question is simple. You spend what it takes to achieve a favorable financial return on your investment. It does not matter whether the space system being built and launched is for communications or commercial imagery. If spending a half billion dollars on a constellation of satellites will earn three quarters of a billion—it is money well spent and you spend it, borrowing the money if necessary. In commerce, however, the spending is done by companies and not by governments.

Let us consider, then, the question of governmental spending of public funds on space systems for Earth science. In deciding to invest in Earth-monitoring satellites, a nation decides to provide real and immensely valuable investments in its people. And the same can be said for solar-monitoring satellites. After all, we're talking about trying to understand the only planet we have (for the time being) and the star we live with. NASA, in collaboration with other nations, conducts research missions in both the Earth science and sun-Earth connection themes.

These missions provide the scientific foundation for operational environmental monitoring missions conducted by the National Oceanic and Atmospheric Administration (NOAA). It is in this partnership that investment stops being an academic issue and starts to profoundly affect the lives of the people who live on this planet.

NOAA operates environmental satellite systems that support the agency's mission "to understand and predict changes in the Earth's environment and conserve and manage coastal and marine resources to meet our nation's economic, social, and environmental needs." These satellites carry operational instruments that make observations of the sun, the space environment surrounding Earth, and of the Earth itself—its oceans, land surface, and atmosphere.

In stating the case for the critical need for investment in the next generation of these satellites—satellites that provide warnings of severe weather, monitor coastal ecosystems, and alert us to solar "storms" affecting the Earth—NOAA presents some very interesting statistics for the United States:

- Thunderstorms are among the most dangerous and frequently encountered weather hazard people experience, and lightning is the number one cause of thunderstorm-related deaths, averaging 93 fatalities and 300 injuries each year.
- On average, 800 tornadoes are reported nationwide, resulting in 80 deaths and 1,500 injuries, and causing $1.1 billion in damages annually.
- Floods account for $5.2 billion in damages and more than 80 deaths per year.
- Hurricanes cause an average of $5.1 billion in damages and 20 deaths a year. In 2003 Hurricane Isabel left 27 people dead in seven states and 3.3 million people without electrical power.
- Weather-related transportation accidents and delays account for 7,000 vehicle fatalities, 800,000 injuries,

500 million vehicle-hour delays, and economic losses totaling $42 billion. And the economic inefficiencies resulting from weather-related air-traffic delays cost the airlines $4 billion a year.

- Severe weather events cause an average of $11 billion in damages each year.
- Coastal and marine waters generate more than $54 billion goods and services annually and support 28 million jobs.
- Some 180 million tourists visit U.S. coastal communities, and 17 million Americans spend $25 billion each year on recreational marine fishing activities.
- The U.S. marine transportation system ships more than 95 percent of the tonnage of foreign trade through American ports and waterborne cargo alone contributes more than $740 billion to the U.S. gross domestic product, creating employment for more than 13 million Americans.
- Improved El Niño forecasts benefit agriculture an estimated $265 to $300 million annually in the U.S. and at least $450 million worldwide.
- Reliable warnings to electric companies of geomagnetic storms caused by solar events can save $450 million over three years. An electrical outage such as the 2003 northeast blackout cost the U.S. economy an estimated $4 to $6 billion.
- In total, weather and climate-sensitive industries account for about 33 percent of the American gross domestic product, or more than $900 billion.

Investment in space systems to observe the Earth and the sun in order to provide the warnings we have come to rely on is a very small fraction of the economic costs revealed in these statistics. Severe weather will, of course, still occur. Climate changes caused by natural and human-induced effects will continue. Solar flares will continue to spew charged particles at our planet for millennia to come. But a world without better understanding of these phenomena and without earlier and more accurate warnings to protect lives and property is a world of unnecessary risk. How much should a country invest in such a monitoring and warning system? Just as in national defense—it should invest as much as it can afford, and no less.

In determining what is "affordable," the use of the term "cost-benefit analysis" by government budget planners is inevitable, and one can fairly easily weigh the cost of a new satellite system against the statistics above. But one additional factor cannot be applied to such an analysis, and this, simply stated, is that operational space systems save lives. Hurricane, thunderstorm, tornado, and other severe weather warnings send people to take shelter, and lives are saved: hundreds or thousands of lives annually. NOAA satellites carry the international COSPAS-SARSAT emergency-beacon locating system. In the first three months of 2004, this system was credited in rescuing 60 people from life-threatening situations. Each year, these statistics increase due to the public's growing awareness of the importance of taking high-tech emergency beacons on boats, airplanes, and even backpacks. With 36 nations now taking part in this international lifesaving effort, the number of lives saved is staggering. Thousands of signals are received each year, and with new beacons incorporating the geolocation capabilities of the Global Positioning System, rescue efforts are increasingly fast and accurate. Many lives are being saved, and how can anyone do a cost-benefit ratio analysis on that?

A rescue helicopter hovers over a floating victim: Each year thousands of search-and-rescue operations are launched upon receipt of an emergency signal from a NOAA or Russian satellite. No bills for the rescue are sent to the survivors—saving a life remains one of the very best investments a government can make.

Space systems for Earth science and commerce are indeed expensive. There will always be the need to decide what to invest in and when, and how much is enough. Commercial interests will take care of their own decisions based on business plans, but government budgets must serve many constituencies and budget allocations must be made wisely. National defense, the pursuit of science, exploration of the universe, and understanding and protecting the Earth and its inhabitants—these are indeed noble causes. Space systems are needed for all of these causes and—as expensive as they are—are simply priceless. ■

6 | Military & Intelligence Uses of Space

The opinions expressed in this chapter do not necessarily represent the position of the U.S. government.

A modern Russian military satellite is launched on April 24, 2003, atop a Proton rocket, with booster rockets attached around the first stage, from the Baikonur Cosmodrome in Kazakhstan.

THE DECLASSIFICATION OF CONSIDERABLE INFORMATION BY THE U.S. government in the past decade has allowed much of the early history of some very shadowy programs to come to light. The stories are amazing—spy balloons in the jet stream, reconnaissance aircraft at the edge of space, launch failures, spy cameras, lost spacecraft found deep in the South American jungle, elaborate cover stories that sometimes went awry, and many new ways to use satellites in updating the old cloak-and-dagger business of intelligence. Some of the hair-raising history of land-based and airborne intelligence efforts is included in this chapter to show how the dangers associated with them, and their limitations, drove the development of space-based systems.

The late 1950s and '60s were a time of incredible innovation. Technologies were developed that both used and overcame the most fundamental laws of physics. And systems were designed to work for months in the very un-Earthlike harsh environment of space. From October 1957 to July 1969, mankind progressed from launching Sputnik—the first artificial satellite to orbit Earth—to landing men on the moon and bringing them back safely with a load of photographs, measurements, moon rocks, and samples.

While many space activities resulted from economic and scientific interests, a parallel effort behind the scenes was propelled by the "space race" between the world's two emerging superpowers, the United States and the Soviet Union. The military and intelligence communities designed satellites and sensors to address a wide array of operational and reconnaissance needs. Details of current systems are classified, but the general types of systems, information, and uses are defined and explained to the extent allowable. During this same period, world governments were working together under their United Nations structures to grapple with new governance issues presented by satellites and human spaceflight: the issues of sovereignty and overflight by spacecraft; the idea of military bases in space; the notions of sovereign claims and economic exploitation of the moon, asteroids, and other planets. And other military uses of satellites—for early warning of missile strikes, for communications of intelligence and operational information, for precise time, for weather effects, and for positioning and precision guidance of weapons—are also addressed.

While there are still impressive technological advances in the wings, today's real challenge is making the many systems we already have work together much better and much faster—for both military and civil purposes.

This launch on October 1, 1960, of the first U.S. SAMOS satellite ended in failure. The launch site at Point Arguello, California, later became known as Vandenberg Air Force Base, still the preferred site for launch of most U.S. military satellites.

Rockets & Missiles

Military Rockets

U.S. military rocket launch vehicles—Atlas, Redstone, Thor, Jupiter, and Titan—all began as long-range **missiles** and were later converted to lift military satellites into orbit. The Redstone, for example, was the first operational U.S. **ballistic missile,** but modified and redesignated the Jupiter-C, it also served as the first-stage booster that launched America's first satellite, the Explorer. The Atlas was America's first **intercontinental ballistic missile** (ICBM), but it continued launching satellites well after it had been replaced as an ICBM. The Titan missile was used to launch all but one of the **Defense Support Program** (DSP) early-warning satellites. These military rockets were also used for civilian satellite launches.

Missiles

Missiles are unmanned aerial weapons. There are two types: guided and ballistic.

Guided Missiles

Guided missiles are unmanned aerial weapons controlled by some navigational system from **launch** to target. One kind of guided missile is the cruise missile, an unpiloted aircraft with an air-injection engine that is guided to its target by an onboard automated navigation system. Most guided missiles, however, are powered by liquid- or solid-propellant rocket engines that do not depend on injecting air.

Germany's V-1 in World War II was the first guided missile. In July 1944, less than a month after the first V-1s hit London, the U.S. used parts from one that did not explode, and built and launched its own version, the Jet Bomb-2. Immediately after the war the U.S. focused on the development of guided missiles and long-range bombers, rather than **ballistic missiles,** because they were less complex and less expensive.

Ballistic Missiles

Ballistic missiles are unmanned aerial weapons that are

A U.S. intercontinental ballistic missile (ICBM) sits inside its protective silo somewhere in Arizona in January 1997.

not guided; their initial **launch trajectory,** plus the action of gravity and air resistance during flight, determine their point of impact.

Near the end of World War II, the Russians captured the Nazi **V-2 rocket** center located at Peenemünde on an island in the Baltic Sea. In October 1946 more than 5,000 German technicians from the Nazi base were transferred, along with their families, to Moscow to help the Soviets with their rocket and missile development activities. Unlike the U.S., Russia spent little effort on **guided missiles,** but focused early attention on medium-range ballistic missiles immediately after the war. Their premier missile development and test center, established in 1947 at Kapustin Yar in the Ukraine, became a prime intelligence target for U.S. and allied forces.

The United States began developing ballistic missiles—Redstone, Atlas, Thor, and Jupiter—in the mid-1950s. As the Soviet government closed its borders to travel and information, the U.S. and its allies became increasingly worried about growing Soviet capabilities in long-range ballistic weapons.

There are varying schemes for categorizing ballistic missiles; the U.S. identifies them by range, as shown in the table below.

Intercontinental Ballistic Missiles (ICBMs)

The United States' first ICBM was the Atlas—operational in 1959 and phased out as a **missile** system in 1967. The Atlas was followed by the Titan I and II, the first U.S. ICBMs based in secure underground storage sites called silos. Titan I had to be raised out of its silo and fueled before **launch;** Titan II could be launched within 60 seconds from inside its silo with preloaded fuels that ignited upon mixing.

The Soviet Union announced its successful test of the world's first ICBM in August 1957. Designated R-7, it was called the SS-6/Sapwood by U.S. and NATO forces. A follow-on Soviet ICBM designated R-36 (SS-9/Scarp by Western forces) had a range of 12,000 kilometers and raised serious national security concerns for the **Eisenhower** Administration in the 1950s. Soviet announcements of these missiles, linked with a dearth of intelligence information about their numbers and locations in the Soviet Union, fueled a very public and political U.S. debate in the late 1950s about the U.S.-Soviet "missile gap."

All of these powerful long-range missiles were later modified for use as first-stage **launch vehicles** for lifting **satellites** into space.

Submarine-launched Ballistic Missiles (SLBMs)

The threat to the United States grew in the early 1960s with the development of **ballistic missiles** that could be launched from submarines. An SLBM launched from a Soviet submarine operating close to the U.S. coast could reach its target with a very short flight time at low altitude. In response to this threat, the U.S. developed sensors on **early warning** (EW) satellites that could detect SLBM launches.

Tests in 1965 confirmed that known launches of American SLBMs, called Polaris missiles, were seen by our EW systems, giving the U.S. confidence that Soviet submarine-launched missiles could also be detected quickly.

Warheads

A warhead is the forward part of a missile, torpedo, or bomb that contains an explosive or nuclear device, or chemical or biological weapons materials.

The advance from atomic bombs (based on fission, or splitting of uranium isotopes) to hydrogen bombs (based on fusion of light hydrogen nuclei into helium atoms) during the decade following World War II increased the explosive power about two millionfold, allowing the development of very small and light thermonuclear warheads for delivery by **ICBMs.**

The specter of these weapons being launched by the Soviets across the North Pole, with a flight time of only 30 minutes, raised grave security concerns in the U.S. and spurred the development of **reconnaissance satellites.**

BALLISTIC MISSILES

Type	Range	Missiles
SRBM (Short-range ballistic missile)	Less than 1,100 km	—
MRBM (Medium-range ballistic missile)	1,100 to 2,750 km	Redstone
IRBM (Intermediate-range ballistic missile)	2,750 to 5,500 km	Jupiter, Thor; Soviet R-5, 11, 12, 14
ICBM (Intercontinental ballistic missile)	Farther than 5,500 km	Atlas, Titan; Soviet R-7, R-36

SPACE RACE

Linda K. Glover

A SMALL ALUMINUM SPHERE THE SIZE OF A BASKETBALL CAUGHT THE ATTENTION OF the world on October 4, 1957, when the Soviet Union launched it into orbit around the Earth. Sputnik 1 could be seen crossing the night sky, and the beeping of its tiny radio transmitter could be heard on home radios. This was the first event noticed by the public in a "space race" that had actually started some years earlier. The race was grounded in many positive desires—to explore the heavens, develop new technologies, gain new knowledge, even to attain international prestige—but it was also fueled by politics and fear.

Weapons developed by the United States and the Soviet Union after World War II were truly frightening. Four years after the U.S. dropped atomic bombs on Japan, the Soviets successfully tested their own A-bomb on August 29, 1949. The Americans didn't even know about it until almost a month later, when we detected radioactive fallout over the Pacific Ocean. The much more powerful hydrogen bombs developed by the U.S. in December 1952 were achieved by the Soviets only eight months later. Soviet announcements to the press about long-range bombers and missiles that could carry nuclear weapons spread fear that resulted in school drills, backyard bomb shelters, and even a sensitive government agency (the National Security Agency) being moved out of Washington, D.C., to avoid a possible nuclear strike on the capital city.

In the mid- to late 1950s, there was no way to verify Soviet claims about their weapons or to assess the real threat. Aerial reconnaissance was becoming more dangerous, with more aircraft being shot down, and their pilots captured or killed. Despite the public attention garnered by commercial satellites and human spaceflight, much of the investment in the space race was quietly being made behind the scenes in secret programs seeking the advantages of space for military support and intelligence gathering.

The opinions expressed in this essay do not necessarily represent the position of the U.S. government.

Policy, Politics, and the Race

A curious aspect of the early U.S.-Soviet space race is that it was actually a competition between two teams of Germans, both from the Nazi rocket base at Peenemünde. At the end of World War II, the Russians took 5,000 captured technicians and their families to Moscow, and the U.S. took the surrendering Wernher von Braun and his team into the budding U.S. Army rocketry and space program.

The Soviets very publicly won the first rounds of the space race, partly because U.S. satellite programs were hampered by competition among the military services and policy arguments between the defense and foreign policy camps in the Eisenhower Administration.

The President was seriously concerned about the international legal implications of satellite overflight of other nations' sovereign territory, and was adamant that the first U.S. satellite should be launched by a scientific, not a military, rocket. Von Braun, in his September 1956 test of the Jupiter-C long-range missile, was specifically ordered not to achieve orbit, and his fourth stage—which could likely have been the first Earth satellite—was actually drained of fuel and filled with sand.

Early Soviet victories in the space race were startling—first satellite (Sputnik) to orbit in 1957, first spacecraft (Luna) to reach escape velocity, first spacecraft to reach the moon, and first man in space, all in 1959. These Russian firsts not only influenced world public opinion, but also had a political effect in the U.S., with a public concerned about the "missile gap" and the "space gap" and beginning to lose confidence in President Eisenhower's leadership.

The fact that this probably contributed to the

Republicans' defeat to John F. Kennedy in 1960 is ironic, since Eisenhower was unquestionably winning the very secret military and intelligence space race. In June 1960, the U.S. launched the world's first reconnaissance satellite. GRAB (Galactic Radiation and Background) was designed to collect electronic intelligence (ELINT) from Soviet radars. And just two months later, under Eisenhower's personal leadership, the U.S. launched the world's first successful photoreconnaissance satellite, designated mission 9009 in the highly classified Corona program, but announced to the world as scientific satellite Discoverer XIV.

In fact, this first imagery intelligence (IMINT) satellite—which collected more photographic coverage of the Soviet Union in one day than the first four years of U-2 flights—actually showed the Soviets had far fewer missiles than they claimed, and essentially dispelled concerns about a dangerous missile gap. President Eisenhower had Kennedy "read into" the Corona program and shared this information with him several months before the election, but since neither of them could talk openly about the source of the information, the public and the pundits preserved the missile gap as a campaign issue. Mindful of security classification requirements, Eisenhower never even referred to these world-changing intelligence accomplishments in his memoirs.

One of the most dramatic early uses of reconnaissance satellite photography was the U.S. discovery of Soviet long-range missile sites being constructed in Cuba, which led to the harrowing and very public Cuban Missile Crisis between the two superpowers. Not well known is the fact that the satellite photos were not of sufficient resolution, so a U-2 reconnaissance aircraft flew over Cuba to collect more detailed photographs. It was shot down and the pilot was lost. Incidents like these gave impetus to the improvement of spy cameras and reconnaissance satellite systems.

The Moon and Beyond

On July 20, 1969, the U.S. clearly won the public space race in the eyes of the world when Neil Armstrong and Buzz Aldrin landed on the moon in front of the largest worldwide television audience in history.

Four days later, unbeknownst to the world, Corona Mission 1107 was launched as the sixth of 15 successful photoreconnaissance satellites with the J-3 stereo spy camera configuration. It carried two separate recovery capsules, allowing the satellite to wait briefly in orbit for a high-priority intelligence collection requirement. In very low Earth orbit (LEO) at only 85 kilometers altitude, its improved cameras provided photographs with a ground resolution of two meters—finally allowing images from space to equal the resolution of U-2 imagery.

Three and a half decades later, reflecting an amazing change in world events, one of the primary military cooperative programs in space is between the United States and Russia.

RAMOS, the Russian-American Observation Satellite, is a partnership between two nations that were Cold War adversaries for 45 years. Even more ironic, this satellite program is focused not only on Earth observation research, but also on the development of techniques for early warning of ballistic missile launches.

The old public space race may be entering another round. In 2004, China joined the small club of nations—now numbering three—that have launched a human into space. The United States also announced serious plans to set up stations on the moon to prepare for human exploration missions to Mars.

And you can be sure there are amazing innovations afoot in the continuing, very secret, military and intelligence space race. ■

President Eisenhower reassures an anxious nation by displaying a missile nose cone during a television speech from the Oval Office of the White House on November 1, 1957— three weeks after the Soviet launch of the first satellite, Sputnik 1, and almost three months before launch of the first successful U.S. satellite, Explorer 1.

National Intelligence Estimates (NIEs)

Prepared for the President of the United States by the **Central Intelligence Agency,** these reports have for more than 50 years been focused on military capabilities and threats to the United States and its allies from unfriendly powers. In the mid- to late 1950s, advisors to the **Eisenhower** Administration became convinced that the estimates were not based on solid information, and that **satellites** were required to improve the intelligence.

The Soviet Union proudly displays its early missile capability during a parade in Moscow's Red Square on November 9, 1965.

Dwight D. Eisenhower (1890-1969)

General Eisenhower was the Supreme Allied Commander in World War II, leading the D-day Normandy invasion in June 1944 that led to the defeat of Nazi Germany. After the war, he served for five years as president of Columbia University, then served as the 34th President of the United States from 1953 to 1961.

As the Soviet Union became an increasingly closed society in the 1950s and Communism spread through Eastern Europe, China, and Korea, Eisenhower became increasingly concerned by the threat posed by the Soviet military build-up. And he and his advisors were seriously concerned about their lack of access to good intelligence about the Soviet threat.

In July 1954 Eisenhower established a secret "Technological Capabilities Panel" made up of science and industry leaders outside the government. Their February 1955 report recommended development of **ICBMs,** early-warning radar networks, **SLBMs,** the **U-2,** and **reconnaissance satellites.** Eisenhower relied heavily on the guidance of these outside advisors and steadfastly supported the resultant programs despite early failures.

The defense against Soviet surprise attack, through development of long-range weapons systems and space-based intelligence systems, laid the groundwork for America's current strength and is one of Eisenhower's greatest legacies. However, because of the secrecy of these programs, President Eisenhower's role was not known at the time, and it was not even mentioned in his memoirs.

Nikita S. Khrushchev (1894-1971)

Appointed First Secretary of the Communist Party in 1953, and Soviet Premier from 1958 to 1964, Khrushchev was the primary Soviet protagonist against U.S. Presidents **Eisenhower** and Kennedy during the Cold War.

He successfully built up Soviet capabilities in long-range bombers and **ICBMs** to carry nuclear **warheads,** but also closed his borders so effectively that the West had little access to the extent and threat of his military forces. This spurred U.S. advances in "edge of space" **reconnaissance** airplanes and reconnaissance **satellites.**

Project RAND

The Research on America's National Defense (RAND) group was established by the U.S. Army Air Corps (precusor to the U.S. Air Force) in October 1945 as a research shop in Douglas Aircraft in Santa Monica, California. It later became RAND Corporation, a federally funded research and development group.

RAND delivered several secret reports strongly supporting the feasibility of **satellites** during the decade before **Sputnik** was launched: in 1946, "Preliminary Design of an Experimental World-Circling Spaceship"; in 1951, "Utility of a Satellite Vehicle for Reconnaissance"; in 1954, the "Project Feedback" report.

Although the U.S. Air Force pursued the electronic relay of images for years under the **SAMOS** program, the RAND group later recommended instead the recovery of film from space. This recommendation had great influence on **Eisenhower**'s White House advisors, and was the approach ultimately used successfully in the **Corona** program.

Early Reconnaissance Missions

Early Spy Plane Missions

After World War II, the Western Allies' wartime alliance with Russia fell apart quickly as the Soviet Union closed its borders, talked of deep ideological differences with the West, and focused on the build-up of its long-range weapons. Traditional U.S. intelligence sources based on agents under cover in the Soviet Union dried up. For more than a decade following the war, **reconnaissance** flights by military aircraft became the primary source of information on Soviet military assets. These flights were organized under two distinctly different programs.

Reconnaissance flights by military aircraft near, but not over, Soviet territory were authorized by President Truman starting right after World War II, and they continued throughout the Cold War and beyond under the Peacetime Airborne Reconnaissance Program (PARPRO). These flights off the margins of Soviet and other communist borders collected important photographic and electronic intelligence. But between 1946 and 1991, 170 U.S. Air Force and Navy personnel on these missions were captured or killed.

In early 1950 President Truman authorized the first overflight of Soviet airspace. Two reconnaissance fighters (RF-80As) flew over southeastern Russia, taking radar-scope photos of Soviet airfields; one of them was lost to Soviet MIG fighter fire. Occasional overflights continued during the Truman Administration without serious incident.

President **Eisenhower** authorized the SENSINT (Sensitive Intelligence) Program of secret peacetime overflights of unfriendly territories in the fall of 1953, his first year in office. A highlight of these flights was Project Homerun, which sent 156 reconnaissance bomber (RB-47E) missions from ice-covered runways in Thule, Greenland, over the North Pole.

These photoreconnaissance missions focused on bases in the inaccessible far-northern reaches of the Soviet Union from which surprise missile or bomber attacks could most quickly reach the United States. There were other "deep-penetration" flights over all areas of Soviet territory.

The President insisted that he approve each mission personally. These flights were often harrowing tales of being tracked by Soviet radars and fired at by Soviet fighters, but ironically—unlike the PARPRO flights—none of the SENSINT aircraft that flew deep into hostile territory was lost. Eisenhower ended military overflights in December 1956 when the **Central Intelligence Agency**'s very high-altitude, supposedly covert, **U-2** spy plane became available.

Project Genetrix

Genetrix was a U.S. Air Force program to launch high-altitude balloons carrying spy cameras high into the jet stream to drift west-to-east across the Soviet Union and China, taking photographs of land areas that were difficult and dangerous to penetrate.

The project was approved by President **Eisenhower** in December 1955. During January and February of 1956, the Air Force launched 516 upper-air balloons with spy cameras from bases in western Europe and Turkey. On reaching international airspace over the western Pacific Ocean, the cameras were designed to detach, fall toward **Earth,** deploy a parachute, and be caught by a passing C-119 aircraft with a specially designed hook. Only 44 cameras were retrieved. Some were likely lost in the ocean, but many crash-landed over Soviet territory or were shot down by Soviet aircraft. The Soviets protested vehemently, and publicly embarrassed the United States by displaying the captured equipment. Eisenhower cancelled the program in February 1956, and was increasingly anxious to find a more covert approach to overhead **reconnaissance.**

Despite its problems, Project Genetrix did develop and test the bizarre retrieval method that was later used successfully in the **Corona** reconnaissance **satellite** program.

The U-2

Concerned about the ability of Soviet **radars** to detect **reconnaissance** flights and the increasingly vehement Soviet protests over illegal penetration of their airspace, President **Eisenhower** was eager in November 1954 to approve "Project Aquatone," the development of a very high altitude aerial reconnaissance aircraft expected to defy detection by Soviet radars. Originally known as CL-282, it was later designated the U-2.

The U-2 was designed to fly at the edge of space, up to an altitude of about 1,025 kilometers. It required many new design concepts: in the aircraft, its fuel, its cameras, and special life support for its pilots. U-2 pilots were in an environment so hostile that death was virtually instantaneous if they were exposed. Temperatures were 21°C below zero, and the atmospheric pressure was so low that blood and other bodily fluids would boil off into vapor.

The U-2 design was based on a glider, very light—made mostly of aluminum—with very long wings; only 13.5 meters long, its wingspan was 21 meters. A new low-vapor-pressure kerosene fuel was used. The U-2 was too fragile to be refueled in the air, but had a range of 55,000 kilometers. Pilots wore oxygen masks and pressurized flight suits custom-designed by a rubber-based brassiere and garter manufacturer. Spy cameras aboard the U-2, specially designed for the low temperature and air pressure, could distinguish very small objects on the ground. The U-2 flew so high that it was expected not to be picked up by Soviet radars or reachable by Soviet fighters or ground weapons. Both expectations proved false.

Francis Gary Powers testifies before the Senate Armed Services Committee on March 6, 1962, holding a model of the U-2 reconnaissance plane in which he was shot down over the Soviet Union on May 1, 1960. Powers was tried and convicted of espionage by the Soviets, but later released into U.S. custody.

The first operational U-2 flight left West Germany on June 20, 1956. Two additional missions on July 2 flew over Czechoslovakia, Hungary, Romania, East Germany, Poland, and Bulgaria. Not notified that these flights had apparently been tracked on radar, Eisenhower approved an additional ten days of U-2 operations starting on July 4.

Soviet radars tracked the July 4 U-2 flight over the western Soviet Union. This particularly enraged **Khrushchev,** who at the time was toasting American independence at a celebration hosted by the U.S. Embassy in Paris. Although the photographs from this flight assured the West that the long-range "bomber gap" they feared did not appear to be real, Eisenhower was concerned that Khrushchev's angry protest meant the U-2 was not invisible to Soviet radar. Despite these concerns, Eisenhower approved additional U-2 flights over the Soviet Union until a U-2 was shot down and recovered by the Soviets on May 1, 1960. During their years of operations, the U-2s collected more than 1,610,000 kilometers of film over Soviet territory, identified the location of Soviet radars and airfields, and began to call into question the Soviet claims of massive increases in long-range bombers and **missiles.**

Francis Gary Powers (1929-1977)

Francis Gary Powers was the pilot designated to fly the ill-fated last **U-2** mission over Soviet airspace that was approved by President Eisenhower. Despite a March 1960 **National Intelligence Estimate** that the Soviet Union's new SA-2 "Guardrail" surface-to-air **missiles** might be able to intercept a U-2, **Eisenhower** approved another mission, code-named Grand Slam, for May 1, 1960.

Powers took off with plans to fly south-to-north for nine hours photographing many sensitive Soviet locations. One of three SA-2 missiles launched from Sverdlovsk exploded close aft of Powers' U-2 and disabled it. It crashed, unexpectedly intact, and unbeknownst to the Eisenhower Administration, Powers survived and was captured.

The administration's cover stories proved embarrassing when Khrushchev produced a pilot who admitted his aerial reconnaissance mission. In the face of U.S. denials, Khrushchev threatened to attack foreign bases from which spy planes were launched, and warned that further flights could lead to war. Eisenhower suspended U-2 flights over foreign territory. The U-2 was again used later, however, in overseas trouble spots.

The Supersonic Spy Plane

In 1959 President **Eisenhower** approved development of an aircraft higher and faster than the **U-2.** Made of titanium and flying at Mach 3.2—three times the speed of sound—the aircraft was designated the A-12 "Oxcart" by the **CIA,** and the SR-71 "Blackbird" by the U.S. Air Force. The CIA used A-12s over Korea and Vietnam, but retired them in 1968. The Air Force used SR-71s in various trouble spots until retiring them in 1990 due to the growing success of **reconnaissance satellites.** Ironically, these planes never flew over Soviet airspace.

First Military & Spy Satellites

The First Military Satellites

Because of growing protests and dangers associated with **reconnaissance** aircraft missions, President **Eisenhower** was anxious to turn to **satellites** as reconnaissance platforms unreachable by Soviet **missiles.** There were many competing U.S. satellite development programs in the mid- to late 1950s, some purely scientific or proof-of-concept, and some with clear military and intelligence objectives and elaborate cover stories to protect the details of their real missions.

The Soviet **launch** of **Sputnik** 1 in October 1957 galvanized the Eisenhower Administration into action, and some of the satellite programs were accelerated to prove that the U.S. could also operate in space.

Explorer

The first U.S. **satellite** to achieve orbit, Explorer was launched on January 31, 1958, less than four months after the Soviet's **Sputnik** launch. Despite its development by the U.S. Army, Explorer carried a scientific payload. It discovered the high-altitude **Van Allen belts** of charged particles, named after the program's chief scientist, James Van Allen. Six Explorer satellites were launched before the first military-reconnaissance satellite—**GRAB**—made it into space.

Vanguard

Vanguard was the U.S. Navy's early developmental **satellite** program. After the Soviet's **Sputnik** 1 and 2 riveted the world, the U.S. focused major press coverage on its first attempt at a satellite launch—a small Navy Vanguard satellite—on December 6, 1957. The launch vehicle very publicly exploded on its launchpad. A Vanguard satellite was successfully launched, as the second U.S. satellite to achieve orbit, on March 17, 1958. Perturbations in its intended **circular orbit** first demonstrated that the **Earth** is not a perfect sphere, but is flattened at the poles and wider at the Equator.

GRAB (Galactic Radiation and Background)

GRAB was the first U.S. **reconnaissance satellite,** launched June 22, 1960, in a circular, polar orbit at 805 kilometers above **Earth.** Developed by the Naval Research Laboratory (NRL) as "Project Tattletale," GRAB collected ELINT (electronic intelligence) on Soviet **radars** and their ability to track aircraft and **ballistic missiles.** Making its cover story real, the GRAB satellite also carried a secondary scientific **payload** that measured solar radiation.

Antennas on the 20-kilogram GRAB satellite picked up radar signals from Soviet defensive radars and beamed them down to a series of very small ground station "huts"—a network of overseas collection sites—that recorded the signals on magnetic tape. The tapes were taken to the NRL, then forwarded to the **National Security Agency** (NSA) and the Strategic Air

THE WASHINGTON POST *Thursday, June 23, 1960*

Piggy-Back Satellites Hailed As Big Space Gain for U. S.

By Charles Corddry
United Press International

The drawing above shows how the Transit II-A satellite and its "piggyback" package, a solar radiation measurement satellite, appeared just after separation in outer space yesterday. The larger satellite was developed by the Applied Physics Laboratory of Johns Hopkins University at Silver Spring and the smaller vehicle by the Naval Research Laboratory here. At right: the double-header satellite rocket takes off at Cape Canaveral.

DIAGRAM SHOWS ORBIT
... of "mother and daughter" satellites

The *Washington Post,* on June 23, 1960, heralds a first for the United States in the "space race"—the successful injection into orbit of two satellites from the same launch vehicle.

Command (SAC) for processing. GRAB's orbit allowed the interception of Soviet radar signals across a ground **swath** 6,475 kilometers wide. The processing provided vital data on Soviet radars, including their location, type, frequency range, and threat to Western aircraft and **missiles.**

GRAB's **launch** was another first for the United States. It entered space on top of a **Transit** navigation satellite; both satellites from the same **launch vehicle** successfully attained their intended orbit. By the time this first intelligence satellite was launched, there were already nine U.S. satellites in orbit—five **Explorer** science satellites, one **TIROS** weather satellite, and three Navy **Transit** navigation satellites. Of four more attempted GRAB launches, one was successful on June 29, 1961.

The GRAB satellites were operational from July 1960 to August 1962, and the technology was transferred from NRL to the **National Reconnaissance Office** after its establishment in June 1961.

WS-117L Program

Early U.S. Air Force **satellite** development fell under the WS-117L program. The "WS" designated "Weapon System," largely due to the development of **intercontinental ballistic missiles** also included in the program.

Despite subsequent declassification and several published accounts, the WS-117L program history is still difficult to follow because of much bureaucratic maneuvering, interservice and interagency rivalries, multiple levels of security classification, covert status for parts of the program, and some elaborate cover stories.

Initial requirements for the program were established in 1954, years before the **Sputnik** launch, with the goal of providing **surveillance** over unfriendly areas to assess their warmaking capabilities. Early planning for the program included subsystems for collecting infrared, photographic, and electronic intelligence. WS-117L spawned several early satellite programs of note—**Sentry, SAMOS, Discoverer,** and the origins of the **Corona** program—which, taken together, addressed the infrared and photographic intelligence requirements originally intended.

Sentry

Sentry was an infrared **early-warning satellite** effort under the **WS-117L program** that intended to use sensors in space to detect the heat signature of enemy missile launches. The Sentry program, renamed **MIDAS** (Missile Defense Alarm System), became an operational program and performed better than expected.

SAMOS

A **satellite** development project under the **WS-117L program** in the early 1960s, SAMOS used new television technology to collect digital video imagery from a satellite and then to radio the images back to **Earth** in near-real time as the satellite passed over ground stations in North America. There was strong competition between this approach of digitally sending information down from satellites versus sending traditional film down in a recovery capsule to be processed on land. Gen. Bernard Schriever, head of WS-117L, recognized the competition and initiated a satellite film-capsule recovery program in late 1957—known as "Program IIA"—whose development proceeded along with the SAMOS video-transmission work.

In 11 SAMOS launches between 1960 and 1962, the Air Force Atlas **launch vehicle** proved quite reliable.

The Space Race

SOVIET UNION

October 4, 1957
First satellite launched
Sputnik 1, a 59-cm aluminum sphere, attains orbit and galvanizes world attention.

November 3, 1957
First animal in orbit
Laika, a dog from Moscow, orbits the Earth.

January 2, 1959
First satellite to the moon
The Soviet's Luna 1 escapes Earth's gravity and does a lunar flyby.

April 12, 1959
First man in space
Yuri Gagarin in Vostok 1 spends 108 minutes in space and parachutes back to Earth.

1957 — 1958 — 1959 — 1960 — 1961

U.S.

January 31, 1958
Explorer 1, developed by Wernher von Braun, orbits, and brings U.S. into the space age.

March 1960
First navigation satellite
U.S. launches first Transit satellite; with three satellites the system later gave 25-meter positioning accuracy worldwide.

March 1960
First weather satellite
The U.S. civilian TIROS program launches its first cloud-and-storm imaging system.

June 22, 1960
First ELINT satellite
U.S. GRAB system picks up Soviet radar signals and beams them to ground "huts."

August 18, 1960
First IMINT satellite
U.S. Discoverer XIV photographs the Soviet Union and returns film in a capsule recovered at sea.

Early 1961
First Early Warning satellite
MIDAS 3 attains GEO orbit to watch and warn U.S. of ICBM attacks.

May 1961
Alan Shepard returns from a successful suborbital flight, bringing U.S. into manned spaceflight.

The imaging and data-transmission systems showed some promise, but had troublesome limitations. The camera's resolution was less than expected, and only a small portion of the data being transmitted down to the ground station could be received as the satellite passed so quickly overhead.

Preeminent science advisors to President **Eisenhower**, including the **Project RAND** group, became convinced that the film-recovery approach was more dependable, less expensive, and would provide higher resolution imagery in the near term. This approach proved itself in 1960 under a covert program named **Corona**—a spin-off from WS-117L Program IIA. The film resolution was superior to the SAMOS video transmissions and SAMOS was cancelled in 1963.

Video transmission was later developed to the point that it could be used to provide satellite imagery data quickly and securely, but it was not introduced into the successful Corona program until years later.

Discoverer

Discoverer was a **WS-117L** satellite program announced to the public in 1958 to justify the purchase of **launch vehicles,** the building of advanced launch facilities at Point Arguello (later Vandenberg Air Force Base) in California, and capsule recovery events over the ocean. The Discoverer "program" was actually a cover story for the **Corona** imagery intelligence satellite program, transferred from the Air Force to the **Central Intelligence Agency** for program oversight. Purposes of the Air Force's supposed Discoverer "research" program, announced at a Pentagon press conference on December 3, 1958, were identified as testing launch vehicles and techniques for attaining orbit, and biomedical studies with "mechanical mice" and live animals to collect data for a human spaceflight.

Adding to the historical complexity of the WS-117L program, early launches of the highly classified Corona program were designated Discoverer launches. Because Discoverer launch-and-recovery efforts drew a lot of public attention, there were some research experiments and scientific news to report. Discoverer XVII, for example, in addition to carrying a spy camera, also collected scientific information on the **ionosphere,** cosmic radiation, and Earth's infrared radiation. Despite the announced objective of launching and recovering live animals, it appears that was never done in this program. There were mechanical mice and no spy camera aboard Discoverer II, which was good for the cover story since the capsule was lost—and possibly recovered by the Soviets—near Spitzbergen, Norway, in April 1959.

Corona

The Corona program developed and operated **satellites** with recoverable film capsules for photoreconnaissance of the Soviet Union. President **Eisenhower** was convinced by **Project RAND** scientists in 1957 to accelerate development efforts for this type of system, but he wanted the program to be ultrasecret, and to be managed by the intelligence community rather than the Defense Department. Project IIA of the U.S. Air Force **WS-117L program** was thus very publicly cancelled by the Defense Department on December 3, 1957, and a new covert program under management of the **Central Intelligence Agency** (CIA) was secretly created. The CIA called the program Corona after one of the man-

August 1962
Soviets launch Kosmos satellite with a Zenit camera, bringing them into the reconnaissance satellite arena.

1962

February 20, 1962
John Glenn orbits the Earth about a year and a half after the Soviet, Yuri Gagarin.

September 1962
First communications satellite
Telstar 1 reaches orbit and retransmits messages in near-real time to the U.S.

June 16, 1963
First woman in space
Valentina Tereshkova orbits in Vostok 6.

1963

March 18, 1965
First "Space Walk"
Alexei Leonov exits his spacecraft for tethered work.

1965

June 3, 1965
Ed White exits Gemini 4, and spends 21 minutes working outside.

May 1967
U.S.S.R. begins use of space-based navigation with launch of Kosmos 158, first in their Tsyklon constellation.

1967

1969

July 20, 1969
First man on the moon
Neil Armstrong and Buzz Aldrin land and walk on the moon. This achievement effectively ends the "space race" in the public's eye.

April 19, 1971
First space station
Soviets launch Salyut 1, the first space station, for long-term visits by astronauts.

1971

ager's favorite cigars, and financial support came from the CIA director's "Reserve for Contingencies" and the President's reserve funds. Eisenhower's personal support proved beneficial to the program, since it required 14 launch-and-recovery tries before the approach was successfully demonstrated. The program initially used the Thor **rocket** for first-stage launch and Agena as the second-stage and satellite vehicle. The plan for film recovery was based on the **Genetrix** high-altitude balloon program. A heat-resistant "bucket" containing exposed film was released from the Agena nose cone, reentered **Earth's atmosphere,** deployed two small parachutes, and was to be "hooked" in mid-air by a specially equipped C-119 aircraft.

The first launch attempt, on January 21, 1959, was aborted when the upper-stage stabilization rockets fired prematurely and damaged the first-stage rocket; it was designated Discoverer 0. The next 12 attempts failed in all stages of launch, filming, and film recovery:

- The Thor rocket "burn" times were too short, leading to **suborbital trajectories;** or too long, causing orbits too high for camera operations.
- The camera film jammed or turned to powder.
- The pitch/roll/yaw stabilization system malfunctioned, causing "tumbling" of the satellite which made camera pointing impossible.
- Separation rockets sent the recovery capsule into deep space instead of **reentry** into Earth's atmosphere.
- Some recovery attempts failed even when everything else went right.

Richard M. Bissell, Jr., at the CIA, had been named Corona program manager partly because of his very rapid and successful development of the **U-2 reconnaissance** aircraft. He became increasingly frustrated that the Corona program had so many setbacks and failures. At one point Bissell recalled, "Malfunctions in an experimental satellite system are exceptionally frustrating...because there is never any human observer to see and evaluate what went wrong....They spun out of control, burned up in the atmosphere, crashed, hopelessly lost, in the ocean, or exploded....It took a certain amount of fortitude to keep going and going, hoping it would finally work." The Corona approach did finally work, with a successful launch and recovery on August 10, 1960.

Discoverer XIII carried no camera, but had an American flag inside the recovery capsule; it was not retrieved in mid-air, but plucked from the ocean near Hawaii. But the entire launch, orbit, and recovery sequence had finally been demonstrated successfully. The U.S. Navy vessel *Haiti Victory* sent a short, but

The Severodvinsk shipyard on the White Sea in northwestern Russia produced the first Soviet nuclear submarine in 1957 and launched 24 more between 1967 and 1974. This image was taken in 1969 by a Corona satellite, the world's first photoreconnaissance program and the U.S.'s first eyes on such sites.

very welcome cable: "Capsule recovered undamaged." The capsule was carried to President Eisenhower at the White House, and is now on display in the National Air and Space Museum of the Smithsonian Institution in Washington, D.C.

The 14th Corona launch, Discoverer XIV, on August 18, 1960, was not met with fanfare, but was the real success story. It was successfully retrieved by aircraft, and it carried a photoreconnaissance camera that in one day took more images of the Soviet Union than the entire U-2 program. It exposed nine kilograms (900 kilometers) of film that covered 4.5 million square kilometers of the Soviet Union in photographs having a 12-meter ground resolution.

With the first successful retrieval of film covering this much Soviet territory, Corona showed there were far fewer **ICBMs** than the Soviets claimed to have, and finally began to give President Eisenhower information he needed on the real threat from Soviet armaments.

Covert Programs

Covert programs are highly classified military or intelligence programs that are managed through special security channels and very strict "need-to-know" criteria, within classification "compartments" that have a limited number of people allowed access. The names of these programs, their security compartment designations, their budgets, and usually their very existence are also classified.

In a particularly pointed example of this kind of classification, the two **Project RAND** scientists, Merton Davies and Amron Katz, who published the November 1957 report upon which the **Corona** program was based, were not "read into" (given access to) the program when it was established. When the Defense Department cancelled the recoverable film **satellite** efforts of the **WS-117L program** in December 1957, they were astonished and argued loudly. Their very vocal complaints actually helped to provide additional secrecy for the new Corona program that had, unbeknownst to them, been reestablished covertly. Another rather startling example is the fate of the first successful Corona camera mission. To ensure secrecy of the real mission, the Discoverer XIV film recovery capsule was smashed into small pieces and dumped into the depths of the Santa Barbara Channel off California.

Black Programs

Black programs are **covert programs.** A particular characteristic of these programs is their separate accounting system, which does not appear in the published federal budget. In the early **Corona** days, the **CIA**'s black-accounting authority allowed payments to be made to contractors without vouchers identifying the purpose, the program, or the amount of funding.

National Technical Means

This is a phrase used in unclassified environments to refer to **reconnaissance** and **surveillance satellites.** The term was first used internationally in negotiations for arms-control agreements as a way to refer to classified satellite use to verify compliance with the treaties.

Cover Stories

Cover stories are alternate explanations developed and disseminated to hide the existence or real nature of **covert programs.** The December 1957 "creation" of the new U.S. Air Force **Discoverer** program for scientific satellites was an elaborate cover story for the new **Corona** program established covertly to develop recoverable-film photoreconnaissance satellites.

The cover stories often caused unintended confusion, however. One example was the fate of the Discoverer XIII capsule. The Corona program intended to rehearse its plans for secretly exchanging capsule contents and returning the real **payload** quietly to Lockheed technicians in Sunnyvale, California, for inspection and assessment. But a zealous Air Force officer who was not "read into" the Corona program, Col. Charles "Moose" Mathison, heard of the recovery, helicoptered out to the recovery ship, took control of the capsule, flew it to President **Eisenhower,** and made sure there was global press coverage of the event. Thankfully, that capsule contained only a U.S. flag, and no spy camera, so the unplanned press exposure actually supported the Discoverer cover story. Much hand-wringing was done, however, over the dire consequences that would have ensued if a camera and film had been in the capsule.

Because of the difficulty in maintaining cover stories, defense and intelligence advisors and personnel in the Kennedy Administration started recommending the practice be dropped in favor of tighter security controls. On March 23, 1963, Deputy Defense Secretary Roswell Gilpatric issued the "Gilpatric Directive" which classified virtually all details concerning military satellite launches. That practice continues today.

The Corona program continued for 12 additional years, with more than a hundred launches, and when early history of the program was declassified by President Clinton in February 1995, the classified name of

Space Policy & Treaties

Sovereignty and Airspace

The notion of sovereignty holds that the ruling government of any country may control access to its land territory and territorial waters offshore, and may also dictate to some extent—as defined through international agreements—the behavior of visitors to its territory. Under the Chicago Convention on International Civil Aviation of 1944 (signed in Chicago on December 10, 1944, and ratified by enough nations—called "States" in international law—to be in force since April 4, 1947), a nation's sovereignty also extends to the navigable airspace above its territory and territorial waters. When an aircraft flies into another nation's airspace, it is under the jurisdiction of that nation.

Under international law, unapproved flights, particularly for military or intelligence purposes, are illegal. When President Truman authorized early reconnaissance flights over the Soviet Union during the Korean conflict, he was on solid legal ground (under Chapter VII of the United Nations Charter), because the Soviet Union was providing airbase support for North Korea and China against the United Nations peacekeeping force.

President **Eisenhower** was always concerned that the unauthorized SENSINT and **U-2 reconnaissance** flights in Soviet airspace were illegal under international law—ironically, under the Chicago Convention that had been negotiated and signed in the United States. He reluctantly approved the missions to gather critical information on long-range bomber and **missile** programs of the Soviet Union, but he kept the missions to a minimum and insisted on strict secrecy. The Soviet Union's **Nikita Khrushchev** protested stridently, but always in private, until his military shot down a U-2 and captured the pilot, **Gary Powers,** in May 1960.

Because of Khrushchev's increasingly belligerent responses to airspace violations by U.S. reconnaissance flights, the Eisenhower Administration was seriously concerned about the implications under international law of **satellites** overflying foreign sovereign airspace without permission. In fact, the U.S. government was strongly supporting the development and **launch** of a scientific satellite to establish the concept of "freedom of space," which would then clear the way for military satellites. This concern drove much of the secrecy that surrounded early U.S. development of reconnaissance satellites. In April 1955, the Soviets announced their intention to launch a scientific satellite with a camera aboard, but they made no attempt to request permission from the many countries a satellite would fly over. Several years after the Soviets successfully launched **Sputnik,** President Charles de Gaulle used the occasion of the May 1960 Paris summit meeting to informally express his serious displeasure at the Soviet "space ship" flying over his territory more than 15 times a day. Khrushchev indicated he would be unconcerned if satellites flew over Soviet territory.

So the overflight and "airspace" issues for spacecraft in orbit were essentially resolved by Sputnik, and the lack of any official diplomatic protests lodged at the time of its launch. Outer space above a nation's borders is not considered part of the airspace under its control.

TREATY ON PRINCIPLES GOVERNING THE ACTIVITIES OF STATES
IN THE EXPLORATION AND USE OF OUTER SPACE,
INCLUDING THE MOON AND OTHER CELESTIAL BODIES

The States Parties to this Treaty,

Inspired by the great prospects opening up before mankind as a result of man's entry into outer space,

Recognizing the common interest of all mankind in the progress of the exploration and use of outer space for peaceful purposes,

Believing that the exploration and use of outer space should be carried on for the benefit of all peoples irrespective of the degree of their economic or scientific development,

Desiring to contribute to broad international co-operation in the scientific as well as the legal aspects of the exploration and use of outer space for peaceful purposes,

Believing that such co-operation will contribute to the development of mutual understanding and to the strengthening of friendly relations between States and peoples,

Recalling resolution 1962 (XVIII), entitled "Declaration of Legal Principles Governing the Activities of States in the Exploration and Use of Outer Space", which was adopted unanimously by the United Nations General Assembly on 13 December 1963,

The first space treaty was known as the Outer Space Treaty of 1967; the opening text appears above. Major provisions of the treaty are highlighted in a table on page 340.

Open-skies Proposal

As the **U-2** aircraft was being developed in 1955, a U.S. Air Force reserve officer named Richard Leghorn introduced a concept of unfettered aerial **reconnaissance** access which he named "open skies." Leghorn believed the policy of free passage for aerial reconnaissance vehicles over foreign territory, if put into practice by the United States and the Soviet Union, would greatly reduce the growing tensions between the two nations and would promote world peace.

Eisenhower liked the idea for a number of reasons. It would avoid the need for secrecy in U.S. reconnaissance programs. It would eliminate the threat to pilots of unauthorized flights being shot down. It would promote disarmament. And, lastly, it would eliminate the need for the President to flaunt international law by authorizing illegal flights.

It would also have allowed the U.S. to avoid the expense of developing and operating specially designed high-altitude reconnaissance airplanes to fly covertly at the edge of space.

The proposal was officially presented by Eisenhower at the four-power (United States, United Kingdom, France, and the Soviet Union) summit meeting on disarmament at the Palais des Nations in Geneva on July 21, 1955. Nikolai Bulganin, the Soviet Prime Minister, responded favorably to the idea. But outside the official proceedings, leader of the Soviet Communist Party **Nikita Khrushchev** made it clear he viewed the American proposal not as a peaceful attempt at disarmament, but as a thinly veiled way to increase U.S. espionage efforts against the Soviet Union.

Eisenhower's reaction to Khrushchev's rejection of openness in reconnaissance programs was to authorize illegal flights of the U-2 starting in June of the following year, and to continue development of reconnaissance **satellites** during the late 1950s in strict secrecy.

United Nations General Assembly Resolutions on Space

The roots of international space law can be found in

Major Provisions of the Outer Space Treaty* of 1967

Article	Subject	Major Provisions
I	General	Exploration and use of outer space is free to all nations, and for the benefit of all.
II	Sovereignty	Outer space is not subject to national sovereignty claims.
III	Law	All exploration and use shall be in accordance with international law.
IV	Military uses	Use shall be for peaceful purposes; no nuclear or mass-destruction weapons; no military bases or weapons testing, but military personnel can do space science.
V	Astronauts	"Envoys of mankind"; render all assistance; warn of any discovered dangers.
VI	Responsibility	Nations are responsible for, and must oversee and/or regulate, any private space efforts originating in their country.
VII	Liability	Nations are liable to pay for any damages caused by space activities to people or property of another nation in space, in the air, or on the ground.
VIII	Ownership	Ownership of space objects is retained by originating nation regardless of entry into space, landing on celestial bodies, or landing in another's territory on Earth.
IX	Potential harm	Nations shall avoid contamination of outer space, and shall avoid contamination of Earth by materials brought back from outer space.
X	Observation	Nations shall consider requests from others to observe space flights.
XI	Notification	Nations agree to inform the United Nations—to the greatest extent feasible and practicable—of space activities.
XII	Space visits	All stations in outer space shall be subject to visit and inspection by other nations with proper request and notification.
XIII-XVII	Administrative	Signature, entry into force, process for amendments, process for withdrawal, and official languages of the treaty (English, Russian, French, Spanish, Chinese).

**Note: All references to "outer space" refer to outer space, the moon, and other celestial bodies.*

two United Nations General Assembly Resolutions. Resolution 1472, endorsed at the 14th UN General Assembly on December 12, 1959, established a United Nations Committee on Peaceful Uses of Outer Space and directed this committee to investigate opportunities for international space cooperation and legal aspects of space exploration. The Soviet Union initially boycotted the idea, but when their objections were withdrawn, the committee was finally formed on November 28, 1961.

Deliberations of the committee led to Resolution 1721, adopted at the 16th UN General Assembly on December 20, 1961. By this resolution nations agreed to two basic space policies: international law including the United Nations Charter would apply to outer space, and access to space for exploration would be freely available to all nations. An additional provision proposed by the United States required nations to report all **satellite** launches and orbits to the committee.

Outer Space Treaty

The Treaty on Principles Governing the Activities of States in the Exploration and Use of Outer Space, including the Moon and Other Celestial Bodies—commonly known as the Outer Space Treaty—is the fundamental basis of all international space law.

The treaty grew out of post–World War II concerns about disarmament and the avoidance of armament in areas not yet occupied, such as Antarctica and outer space. At the United Nations General Assembly of September 22, 1960, President **Eisenhower** proposed that the major principles of the Antarctic Treaty of 1959 be applied to space.

Despite a variety of initial concerns from the Soviet Union and other nations, the negotiations proceeded quickly for an international agreement, and the treaty language was adopted by the UN General Assembly on December 13, 1966. On January 27, 1967, when the treaty was opened for signature, 91 nations signed. Ratification by five nations was required to bring the treaty into force, and that occurred on October 10, 1967.

The principal tenets of the agreement are that, in space and on celestial bodies, no nation may make a claim of sovereign control, commercially exploit resources, set up military bases, or place weapons of mass destruction, and that all nations have free access to explore these areas. International agreements followed that clarified issues addressed in the Outer Space Treaty, including a Rescue Agreement of 1968 dealing with international assistance to **astronauts** in danger; the Liability Convention of 1972 outlining a nation's responsibility for any damages caused by its space equipment; the Registration Convention of 1975 concerning notification of satellite launches; and the Moon Agreement of 1979.

Registration Convention

Based on United Nations Resolution 1721, the United States made its first report of **satellite** launches to the UN Committee on Peaceful Uses of Outer Space on March 5, 1962. Having successfully launched several short-term, recoverable, very secret **Corona** photo-reconnaissance satellites starting in August 1960, the U.S. decided to conceal these short-term flights and report only on satellites in longer-term orbits.

On July 10, 1962, despite a Soviet call for a joint declaration banning satellites for military **reconnaissance,** the U.S. National Security Council approved the practice of not reporting these military space launches.

The practice of reporting satellite launches, however limited, was codified in the Convention on Registration of Objects Launched into Outer Space of 1975. The agreement makes establishment of a national registry and reporting to the UN Secretary General mandatory. No explicit mention is made of reporting military launches, and the practice of not reporting them is still generally continued today.

Moon Treaty

The Agreement Governing the Activities of States on the Moon and Other Celestial Bodies, commonly referred to as the "Moon Agreement" was adopted by the United Nations General Assembly on December 5, 1979, opened for signature on December 18, 1979, and entered into force on July 12, 1984.

The principal new concerns addressed in the treaty deal with environmental protection of both the moon and the **Earth** from introduction of foreign matter; the potential for establishing scientific preserves free from economic development; the disallowance of any property rights to the moon or its resources; and the requirement that any resource development be done only under an international organization that will share the proceeds with all nations.

Due to the property rights and shared economic development issues, the United States has declined to ratify the agreement.

Spy Satellite Missions & Orbits

Spy Satellite Missions

Satellites are key to many military and intelligence missions, using sophisticated versions of many of the civilian sensors and missions used for Earth remote sensing and satellite communications. Imagery intelligence (**IMINT**) is collected by space-borne cameras, signals intelligence (**SIGINT**) includes both electronics and communications intelligence (ELINT and COMINT), and measurement and signature intelligence (**MASINT**) is developing new ways to glean an amazing array of information from new satellite sensors.

The images and other intelligence data from all of these types of missions are returned to **Earth** through secure military communications.

The type of mission can also be categorized by its objective and operational design. A small number of nations currently have spy **satellites** in operation, but their missions, sensors, and orbits are all highly secret. Information gleaned from declassification of early U.S. systems, however, can provide some insight. Some aspects of mission design are discussed below.

Reconnaissance

From the same root as the word "reconnoiter," reconnaissance means taking a preliminary or quick look at an area or activity of interest and gathering just enough information to try to characterize it. Early spy **satellites** were limited in their capabilities to reconnaissance-type missions.

The short-term and intermittent nature of reconnaissance can result in significant misinterpretations or missing of major intelligence information.

In particular, due to the unintentional public release of orbital parameters of some reconnaissance satellites, an unfriendly entity can determine the schedule of overflights and time its activities to occur when a satellite is not overhead, thus avoiding detection.

Surveillance

Surveillance is the continuous monitoring of an area or activity of interest. As spy **satellites,** satellite constellations, their orbits, and associated communications became more sophisticated, full-time surveillance of many intelligence targets of interest have become possible from space.

Broad Area Search Systems

A broad area search system in spy satellites is a design that collects imagery or other intelligence across wide bands, or "**swaths**," of land beneath the satellite on each orbit. All parts of the swath are investigated, but no smaller areas of higher resolution are included. Early Corona satellites were broad-area search systems.

Spotting Systems

A spy satellite system that can be "pointed" at specific smaller objects to take higher-resolution images of them is called a spotting system. The early **Corona** satellites could not do this, but in the mid-1960s, a specialized **IMINT** satellite was developed and launched to complement Corona's broad-area search capabilities. This spotting system was designated **KH-7.**

Spy Satellite Orbits

Different types of intelligence-gathering missions not only require specially designed satellites and sensors, but also careful choice of the orbit to optimize various kinds of data collection.

Early **Corona** satellites were in LEO (**low Earth orbit**) **polar orbit** at only about 500 kilometers altitude. The low altitude was necessary to allow a reasonably short range for the cameras to ensure adequate resolution in the photographs. For the first operational Corona satellite (Discoverer XIV), this orbit resulted in seven **Earth** revolutions in one day. The seven passes over Soviet territory covered about 5 percent of that nation's landmass, but with hours between each pass, this orbit produced a reconnaissance mission.

The consequent problem of missing an important event—or particularly of missing activity that has been timed to take place when the satellite is not overhead—can be countered by placing multiple satellites into polar orbit in constellations. Typical of this approach would be several satellites in near polar orbits with the same inclination, but placed in orbits set to "follow" each other over the Earth's surface.

With enough satellites in a constellation, the revisit time can be short enough to provide nearly full-time coverage, and thus a successful surveillance system.

An alternate approach to achieving surveillance is through placing satellites in **geostationary Earth orbit** (GEO). Since a GEO satellite stays at all times above a prescribed point on the Earth's surface, the intelligence coverage can be continuous for the area within range of the satellite's sensors, but this approach has two inherent limitations for intelligence. Since GEO orbits require a high altitude (about 36,000 kilometers) to achieve the "stationary" position, they are more suited to electronic listening than high-resolution imagery collection. Secondly, from its position above the Equator, a GEO satellite cannot "see" or "hear" anything above 70° latitude to the north or south. This was a serious limitation for surveillance of the Soviet Union's remote northern reaches.

A good compromise for U.S. intelligence-gathering over the Soviet landmass and the north polar regions—the presumed fastest and hardest to detect route for early long-range bombers and **intercontinental ballistic missiles**—was use of the **Molniya orbit.** The Molniya's **highly elliptical** (HEO) **polar orbit** provides maximum time during each orbit over the north pole area. Because of the great altitude at this part of the orbit, however, it is more useful for electronic listening systems than for cameras. More sophisticated sensors and large satellite constellations have overcome many of the orbital constraints faced by early intelligence-gathering satellites, and the current intelligence satellites of the major powers can be presumed to occupy a wide range of orbits to optimize performance for different intelligence missions.

Kosmos Program

The Soviet military and intelligence satellite program since the early 1960s is the Kosmos program. The Soviets, and since the end of the Cold War the Russians, have used this name to designate all of their photoreconnaissance, **early warning,** military communications, and other types of military and intelligence satellites. The Soviets developed the **Molniya orbit** to support their satellite-based communications networks because much of their landmass is so far north, but presumably they also used it to improve their intelligence gathering of U.S. activities in and around the Arctic. Soviet Kosmos spy satellites throughout the 1960s, '70s, and '80s were mostly in **LEO** orbits, with altitudes between 250 and 400 kilometers, and with many launches of short-term satellites operating for a few days to a few weeks for photoreconnaissance. Other Kosmos satellites were launched in clusters of eight at about 1,500 kilometer **apogees** to support secure military communications.

Military Launches

The first **Corona** satellite was launched from Vandenberg Air Force Base (originally Cooke AFB) near Point Arguello, in southern California. This spot was chosen partly because of the opportunity to provide security at the military base, although a covert **launch** was not possible. The sound and vibration from the **rocket** engines was obvious to the surrounding populace, the launch vehicle could be seen on radar, and its telemetry—the information it beamed down to home base on its performance—could easily be monitored.

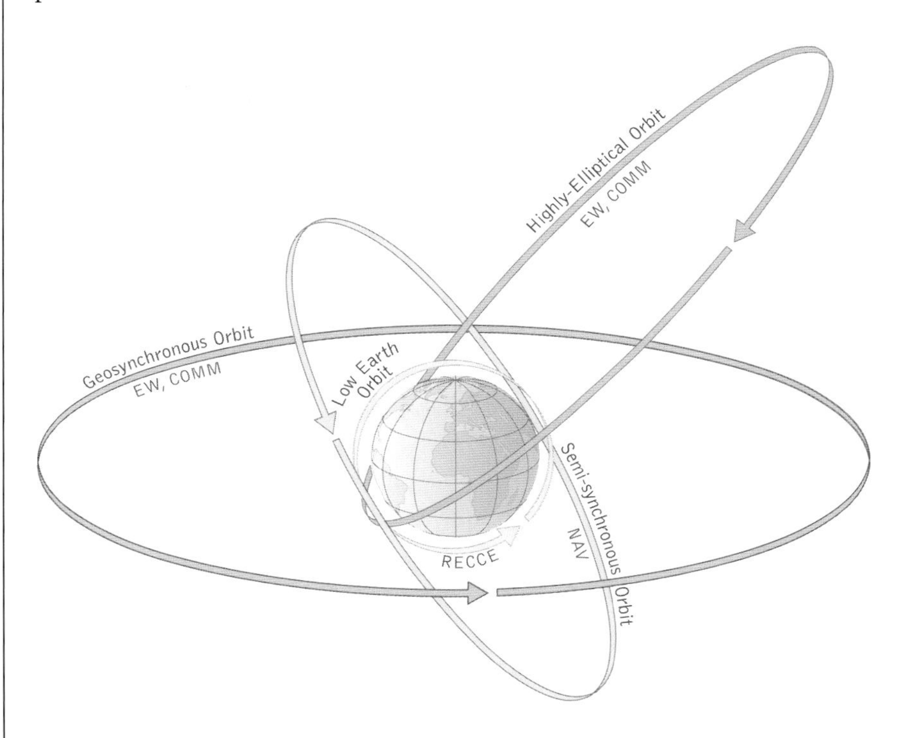

Typical orbits for military and intelligence satellites are shown. Reconnaissance satellites often use LEO orbits; navigation satellites use highly inclined near polar orbits; and GEO and Molniya orbits are most often used for both early warning and military communications.

A key reason for the launch site choice, however, was downrange safety, especially in the early years, when many launch attempts went awry. A **polar orbit** can be attained by launching to the south from the Vandenberg base, with the initial stages aimed over the Santa Barbara Channel, then south over vast open expanses of the South Pacific Ocean. All of the U.S. **Corona** (or Keyhole) imagery intelligence and mapping satellites were launched from Vandenberg.

In more recent years, U.S. military-intelligence satellites have been carried into orbit aboard the **space shuttle** launched from the **Kennedy Space Center** in Florida.

Early Warning from Space

Land-based and Sea-based Early Warning Systems

Post–World War II concerns about the Soviet Union sending bombers across the North Pole to targets in the U.S. spurred the **Eisenhower** Administration, working with Canada, to establish a network of **radar** sites across the Western Hemisphere Arctic to detect long-range Soviet bombers. This Distant Early Warning System (DEWS) of antiaircraft radars, known as the "DEW line," was operational by August 1957.

In 1959 the DEW line was expanded to the Aleutian Islands in the Bering Sea, and in the late 1950s it was also extended to sea. Sixteen World War II Liberty ships were converted to Ocean Radar Station ships with names like U.S.S. *Guardian* and U.S.S. *Skywatcher.* These ships, equipped with long-range radars and communications, were on patrol at preset stations 500 miles off the U.S. East and West Coasts for more than a month at a time, and on constant alert for incoming threats.

By 1959 the DEW line included more than a hundred land- and sea-based early warning radars in remote and challenging Arctic and ocean locations and stretched many thousands of miles to completely surround North America. And by 1965 most of the stations were disestablished because the warning capabilities did not justify the operational costs.

In 1959 an advanced land-based radar detection and tracking system called Ballistic Missile Early Warning System (BMEWS) was also installed in a few far-north locations to detect a Soviet intercontinental ballistic missile (ICBM) strike on North America. Upgraded BMEWS stations still exist to detect missiles and track spacecraft in Earth orbit. But their 15-minute warning time capability in 1959 was considered inadequate and spurred a strong interest in space-based systems.

Early Warning (EW) Satellites

The U.S. land-based **radar** systems, even though they were deployed very far north, could not have detected a Soviet **missile** launched across the Arctic until it flew far enough to be "seen" on the horizon above the curvature of the Earth. Thus, the geometry of the problem created a warning-time limitation that just could not be overcome with land-based warning systems. This realization gave high priority within the **Eisenhower** Administration to the idea of a space-based system that could detect a missile at launch and double the warning time to 30 minutes.

While many nations have built and launched communications and imaging satellites, only the U.S. and Russia have military early warning satellites, designed specifically to detect and provide warning of enemy missile launches. The U.S. was first to develop such a system in the early 1960s, in response to Soviet development of long-range **intercontinental ballistic missiles** (ICBMs) in 1957. The first system was MIDAS (Missile Defense Alarm System), with satellites operating throughout the 1960s. Their infrared sensor detected the heat signature of missile launches. The **DSP** (Defense Support Program), with satellites in **geosynchronous orbit** carrying infrared and optical sensors, followed in 1970 and continues today.

The follow-on program to DSP, called **SBIRS** (Space-based Infrared System), is addressing military needs more focused on today's threats.

MIDAS (Missile Defense Alarm System)

The history of the MIDAS program was declassified in late 1999. The notion of a space-based infrared sensor to detect enemy **ballistic missile** launches was originally part of the Air Force's **WS-117L** satellite program. An engineer named Joseph J. Knopow, at the Lockheed Missiles and Space Division, had a Pentagon background in infrared (IR) sensors and included an IR missile-detection package in Lockheed's 1955 proposal to the Air Force for a WS-117L military **satellite.**

In 1958 the **Eisenhower** Administration decided to move the photoreconnaissance satellite mission out of WS-117L and make it highly classified, called **Corona,** with a cover name of Discoverer. At that time, the remaining satellite mission in WS-117L—that of infrared sensors to provide early warning of enemy missile launches—was briefly named the **Sentry Project,** but shortly thereafter was renamed MIDAS (Missile Defense Alarm System).

Knopow's approach was to carry aboard the satellite an optical **telescope** that could collect infrared signals from **Earth** and its atmosphere and direct them to an onboard array of lead sulfide (PbS) detectors. These detectors could then convert the infrared signals into

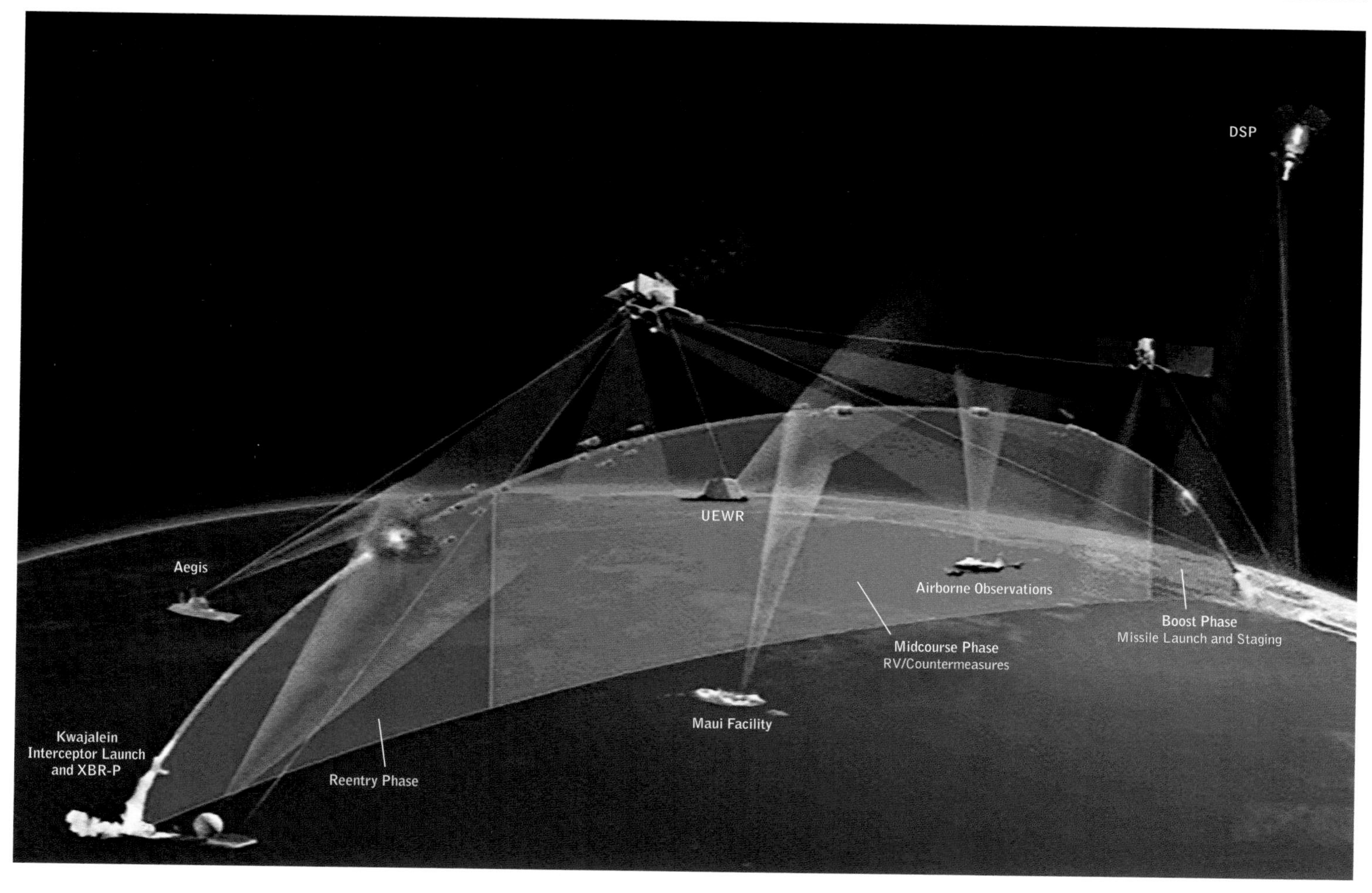

Part of the U.S. Early Warning system, the Space Tracking and Surveillance System (STSS) will provide space-based infrared sensors capable of detecting and tracking attacking ballistic missiles earlier in their flight path. STSS will demonstrate improved data processing and communications to interceptor systems.

electrical impulses that could be transmitted back to ground stations on Earth.

There was considerable skepticism among the government's scientific advisors in the late 1950s and early '60s that the background infrared radiation from Earth, its atmosphere, and clouds would be so great that IR signals from missile launches could not be detected, or that the false-alarm rate from natural phenomena would make the warning system ineffective. Based on balloon and aircraft tests of the IR system, the optimal working frequencies were determined by 1958 to be 2.7 to 4.3 microns—frequencies avoided for most **Earth remote sensing** because of absorption of energy by water vapor in the atmosphere at this frequency range—but the water vapor appeared to filter out much of the undesired background radiation for the missile warning system.

The first MIDAS launch, on February 26, 1960, failed, very publicly, when a collision of the Atlas first-stage and Agena second-stage **rockets** dropped debris into the Atlantic Ocean. An embarrassing *Orlando Herald* headline that evening read "Spy in the Sky, Asleep in the Deep."

Despite two years (between May 1960 and April 1962) that saw five more launch or operational failures, at a program cost of $425 million, two science advisory committees for the President and the Defense Department continued to endorse the basic approach. And the seventh MIDAS launch, on May 9, 1963, vindicated their optimism. Achieving its planned circular **polar orbit** at 3,500 kilometers altitude, it detected all types of missile launches during its operational life, including Atlas, Titan, Minuteman, and Polaris submarine-launched ballistic missiles, as well as demonstrating virtually no problems with background radiation or false alarms.

The next three MIDAS launches failed or their satellite operations failed early, but two final satellites in the program—Research Test Series-1 (RTS-1) launched in 1966—operated very successfully for about a year. Designed to test detection of submarine-launched ballistic missiles and land-based medium range ballistic missiles, both satellites detected all types of missile launches even through dense cloud cover and identified four Soviet missile launch sites.

The program was halted in 1967 to make way for the follow-on **Defense Support Program** (DSP) with expanded missions and capabilities.

DSP (Defense Support Program)

In the later days of the MIDAS program, additional requirements were added that necessitated a fundamental change in the **EW** satellite program. The new Defense Support Program (DSP) was required not only to provide early warning of Soviet and Chinese **ICBM** and SLBM launches, but also to:

- Detect enemy nuclear detonations on the ground, in the atmosphere, and in space.
- Detect launches of any payloads into space.
- Provide attack assessment: the success of U.S. missiles hitting targets in the event of war.

From 1970 to 1997, 18 DSP satellites were launched into **geosynchronous orbits** at 40,500 kilometers. The five DSP design variants, and their major differences, are shown in the accompanying table. The DSP was pressed into tactical use in the Desert Storm conflict of 1991 and showed impressive capability, especially in detecting the Iraqi SCUD missile launches in time to warn military forces and civilian populations. This success led to the establishment of a permanent tactical data site, the Attack and Launch Early Reporting (ALERT) system. It also led to the establishment of a new program to address these broader needs, the **Space-based Infrared System** (SBIRS).

SDI (Strategic Defense Initiative)

An SDI (Strategic Defense Initiative) program was initiated in March 1983 by President Ronald Reagan. In development for a decade, the SDI concept included "Brilliant Eyes" early warning satellites to detect missiles as well as "Brilliant Pebbles" satellites to destroy them.

Though SDI was officially cancelled in 1998 by President William Clinton without any systems being deployed, research findings, especially in the miniaturization of components, were useful to follow-on military satellite programs.

SBIRS (Space-based Infrared System)

The United States is developing a more comprehensive new system called SBIRS. Its high-orbit satellite constellation will address global **missile** threats to the U.S.

STSS (Space Tracking & Surveillance System)

The increasing use in recent conflicts of medium-range **ballistic missiles** with short flight times has highlighted the limitations of **DSP**'s scanning frequency.

A **low-Earth-orbit** constellation being developed by the United States, called STSS, will provide very fast battle-space warnings of shorter-range missile threats in regional conflicts.

Soviet Early Warning

The Soviets also pursued an **early warning satellite** system under their **Kosmos** program and achieved it with a nine-satellite constellation in 1987 that provided missile warning and detection of nuclear detonations.

The Soviets followed this two years later, in 1989, with "Prognoz" early warning satellites in **geostationary Earth orbits.**

DSP Satellites, Launches, & Characteristics

Phase	Launches	Dates	Orbit	Added Capabilities
Phase I	4	1970-73	GEO	2000 IR detectors
Phase II	3	1975-77	GEO	Longer operational life
MOS/PIM	4	1979-84	GEO/HEO	Multi-orbit; hardened
Phase II Upgrade	2	1984-87	GEO	6000 IR detectors; new medium wave IR detector
DSP-1	5	1989-98	GEO	Longer operational life

MOS/PIM: Multi-orbit satellite/performance improvement modification; GEO: geostationary Earth orbit; HEO: highly elliptical orbit

Imagery Intelligence from Space

IMINT (Imagery Intelligence)

IMINT is the intelligence information gleaned from cameras and similar imaging systems, and the products, processing and analysis derived from them.

Since the advent of space-based photoreconnaissance in the 1960s, by far the majority of IMINT collected by the superpowers has come from **satellite** systems. The focus here is on the early United States IMINT satellite system **Corona** because information on it has been made public, whereas much of the material printed on intelligence satellite capabilities of other nations is very speculative in nature. Early U.S. IMINT satellite systems were based on the approach of exposing film in space, retrieving it by reentry through the atmosphere, and developing the film at an Earth-based laboratory. The early Soviet IMINT approach was similar to that of the retrievable-film Corona program.

Photo-interpretation is the analysis of photographs to identify, measure, and otherwise interpret what is captured in the images. Photogrammetry is the process of making precise measurements from photographs. Thousands of images must be pored over by human analysts. Stereo pairs of images and shadows on photo images, coupled with precise information on position and pointing angle of the camera, can yield fairly accurate heights of objects seen. Human interpretation is increasingly aided by computer analysis. A good example of this is "change detection," in which a comparison of images from the same location can automatically highlight differences; this can show movement of equipment or personnel, and a wide variety of other intelligence information.

Spy Cameras in Space

The capabilities of current spy satellite cameras are highly classified by all nations involved, but the United States recently declassified a rich history of how the tremendous challenges of operating cameras in space were overcome through the 1960s.

The **Corona program** called for an unpressurized satellite that would work in a vacuum and deal with the low temperatures in space. Vibration and temperature concerns were overcome by fabricating the camera frame from titanium—a lightweight but strong metal that does not distort under changing environmental conditions—and in a honeycomb design that is strong, light, and improves the distribution of heat across the structure. To avoid rubbing together of layers of film during launch, film spools were wound tightly around a titanium core.

Two approaches were considered to control where the camera was pointing. In one, the satellite would spin, with camera operation timed to take photographs on each spin only when pointed at **Earth**'s surface. The other approach, initially considered too complex, was to create a three-axis stabilized spacecraft with the cameras always pointed toward Earth's surface. This required horizon sensors and gas jets on the satellite to adjust the roll, pitch, and yaw and keep the satellite in position. The three-axis stabilized approach was chosen for Corona and has been used on all subsequent U.S. photoreconnaissance satellites.

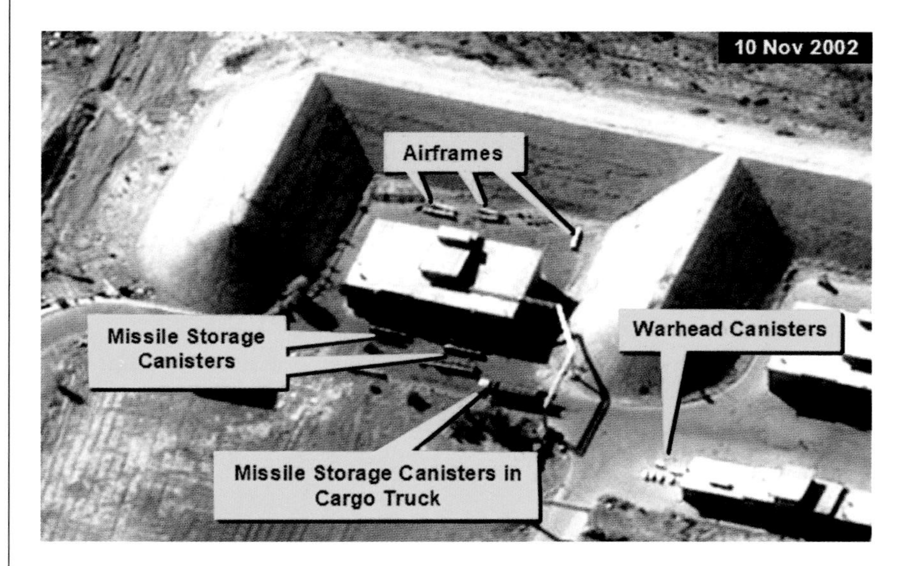

This annotated image, which U.S. Secretary of State Colin Powell presented to the United Nations Security Council on February 5, 2003, shows the removal from the Al Musayyib rocket test facility in Iraq of a ballistic missile with ranges not permitted to that nation under UN guidelines.

A "stellar index" camera imaged the stars and used their known positions to determine the "look angle" of the satellites' reconnaissance camera, and thus determine the positions of objects seen in the intelligence photographs.

Spy Camera Film

Looking back at Earth through the atmosphere

requires special high-contrast films with a narrow light range and designed to record the longer wavelength red end of the visible spectrum that penetrates the haze better than short-wavelength blues and greens.

Acetate films exposed to the vacuum of space in early Corona launches shattered when their solvents vaporized, or literally "boiled off," in the zero atmospheric pressure. The Eastman Kodak Company developed a new film, "Estar," with a film base made from a new polyester material called Mylar. To keep the film from curling and sticking as it wound through the camera, a thin coating known as a "pelloid" applied to the back of the film was embedded with extremely small beads of glass. The Mylar-based film proved stable in the space environment and worked very well in the Corona cameras.

The film was rolled into continuous spools literally kilometers long—even though it was only 0.076 millimeters thick—and moved through the camera system without breaking. Eastman Kodak made the film in 1,800-meter sections and developed a new ultrasound splicing technique to join them into immense continuous rolls without creating either weakness or thickening at the splice points.

Later in the Corona program, Kodak developed an improved precipitation technique for applying the light-sensitive silver halide emulsion to the film base. This provided finer-grained film, resulting in higher sensitivity and higher-resolution images with lower light, and significantly increased the intelligence content of the images.

Spy Camera Designs

Theoretical camera resolution is not as important as ground resolution, the combined resolution of the entire system, which represents the ability to distinguish objects of a certain size on Earth. Resolutions here are expressed as ground resolution.

The **Corona program** used four modifications of the same basic camera design:

- "C" was the designation for the first space-based spy camera. It had a lens with f/5.0 maximum aperture, a 60-centimeter focal length, and ground resolution of about 12 meters. In the first few launches that carried cameras, **telemetry** (the diagnostic signals sent down from the **satellite** to home base indicating its performance) showed catastrophic camera failure, which was caused by shattering of the acetate-based film. Although there was no recovery of film from the 11th Corona **launch** (Discoverer XI), its telemetry indicated that the switch to Mylar-based (Estar) film had solved this problem, as all 900 meters of the film wound unbroken through the camera and into the recovery capsule. The first successful Corona film recovery, from Discoverer XIV on August 18, 1960, used the C camera.
- The C' (or C-prime) camera upgrade was used intermittently from the Discoverer XVIII launch in December 1960 through October 1961. C' used the same lens as the C camera, but had some structural changes that simplified operation and improved ground resolution from about 12 to 10.5 meters.
- The C''' (or C triple prime) camera was first launched on Discoverer XXXI in August 1961. Its f/3.5 faster lens could be used with finer-grained, slower film Eastman Kodak had developed, and its image-motion compensation improvements reduced smear, resulting in ground resolution of 6 meters to 7.5 meters.
- The "M" (or Mural) system consisted of two C''' cameras mounted on the same frame at a 30-degree separation, giving stereo imagery. The M system's three-dimensional information was more than twice as valuable in its intelligence content and was used on all Corona flights through the rest of the program until 1972. The combination of two C''' cameras in stereo mode gave a ground resolution of about 3.5 to 5 meters.

The cameras designed for Corona spy satellites were amazingly reliable. After the film was changed to the Mylar-based Estar, there were only ten camera problems, most of them only partial failures, out of 136 additional launches.

KH (Keyhole) Satellites

KH was the covert name for the **Corona** photoreconnaissance **satellites** from 1960 through the end of the program in 1972 (**KH-1** through **KH-6**), and for some later **IMINT** satellite projects that have also been released to the public. To protect the **U-2** program and related aerial reconnaissance efforts, the **CIA** established a special covert security system, which it named "Talent."

When the Corona **spacecraft** reconnaissance effort was established as a covert program, a new tightly controlled security system named "Talent-Keyhole" (TK) was developed for it, and the code names for the various phases of the program—based on the version of camera used—were named "Keyhole," or "KH."

KEYHOLE (KH) SATELLITES

KH #	Code Name	Camera System	Mission Numbers†	Discoverer Numbers	Launch Dates
KH-1	—	C (or none)	9001*-9009	DO-XIV	Jun 1959-Sep 1960
KH-2	—	C'	9010-9021 (plus 9022, 9024, 9026)	DXV-XXVIII, plus XXX, XXXI, XXXIII	Sep 1960-Oct 1961
KH-3	—	C'''	9023-9030	DXXIX-XXXVII	Aug 1961-Jan 1962
KH-4	—	Mural = 2 C'''	9031-9062	DXXXVIII**	Feb 1962-Dec 1963
KH-4A	—	J = 1 Mural; 2 capsules	1001-1052	—	Aug 1963-Sep 1969
KH-4B	—	J3 = filters, apertures	1101-1117	—	Sep 1967-May 1972
KH-5	Argon	Mapping only	12 90xxA flights	DXX, XXIII, XXIV, XXVII	Feb 1961-Aug 1964
KH-6	Lanyard	Hi-res stereo	8001-8003	—	Mar 1963-Jul 1963
KH-7	—	Spotting camera	—	—	Jul 1963-Jul 1967
KH-9 Mapping	—	Frame mapping camera	—	—	Mar 1973-Jun 1980

*†First and last mission numbers shown for each system, some intervening numbers may be other camera systems; *mission numbers (9001) started with Discoverer IV.*
***Discoverer mission numbers ended at 38 when the Pentagon dropped the cover story for Corona in February 1962.*

The name was declassified, along with 860,000 images from KH-1 through KH-6, by President Clinton in February 1995 based on the recommendation of Vice President Gore's special "Environmental Task Force." This task force was an unprecedented effort in which eminent scientists, many of whom had never held security clearances before, were given TK clearances and invited to analyze the usefulness of old reconnaissance images for solving current environmental problems. The United States decided that the myriad of potential research uses of the images outweighed the continued need for classification of the Corona images.

In their day, the Corona images had served their Cold War purposes well by providing real information on the threats faced by Western nations. From 1960 to 1972, the time in orbit for each satellite had grown from 1 to 19 days, the image resolution was improved from 12 meters to less than 2 meters, and 145 Corona launches returned 610 kilometers of film that covered almost 2 billion square kilometers of the Soviet Bloc countries.

On September 20, 2002, the U.S. government released information and images from the KH-7 and KH-9. Given the cover name (**Discoverer**) for early launches, the covert name (Keyhole) for different camera configurations launched, and the mission designator numbers, the task of tracking down a declassified image from any particular satellite can be confusing. The KH number refers to the camera configuration used. In later KH phases (starting with KH-5), a special code name was also added. For the first 38 launches of Keyhole satellites, a Discoverer launch number was announced to the press as part of the cover story, but Discoverer numbers were discontinued in February 1962 when the cover story was dropped, and explanations for military launches were no longer given to the public. Mission numbers are in series according to the camera configuration, but the first three launches had no mission number because they carried no cameras.

KH-1

The KH-1 phase of **Corona** included all the cameras of the original "C" spy camera design. In 15 launch tests, ten C cameras were launched between June 1959 and September 1960, but because of failures in the Agena second stage **rocket,** the spin rocket, or the recovery sequence, only one recovered film. It was the groundbreaking Discoverer XIV, or Mission 9009, launched on August 18, 1960, that became the world's first operational space-based spy camera.

The first image taken was of a Soviet long-range bomber base at Mys Shmidta, in the very northeast corner of the Soviet Union. Only 640 kilometers from Nome, Alaska, this was the Soviet's closest base to U.S. soil, and was therefore a critical and frequent target of Corona collection. In its one day of operation, Mission 9009 also showed a Soviet surface-to-air missile (SAM) site of SA-2 missiles dubbed "Guideline" by U.S. intelligence, and a major missile test complex at Kapustin Yar, south of Moscow. Most important, it largely dispelled the "bomber gap" and "missile gap" that had been feared by the Western world.

President **Eisenhower** was extremely pleased with the images from KH-1, as they covered much of the Soviet land area, gave a great deal of information on Soviet bombers, and did not require risking a pilot's life or violating international law. Despite the many problems that had plagued the program, and despite some streaking on the images that hampered analysis, the President enthusiastically endorsed the continuation of Corona and its plans for camera improvement.

KH-2

The KH-2 phase of **Corona** included those **satellites** that carried the slightly improved C' camera system. Eleven KH-2 satellites were launched between September 13, 1960, and October 23, 1961. The program continued to be plagued with Agena second-stage **rocket** failures, but two of the C' launches—Mission 9013 and 9017—successfully launched and retrieved their film.

One photograph from the Mission 9017 camera allowed analysts to discover the first **intercontinental ballistic missile** launch complex ever identified from aerial or satellite imagery. The image, when compared to one taken a year later, indicated the recent construction of railroad transport lines leading into multiple ICBM launch sites.

KH-3

Satellites of the KH-3 phase of **Corona** carried the C''' spy camera design. The first KH-3 launch, on August 30, 1961, was successful, as were three more of the six launches. As expected, the KH-3 images provided ground resolution about twice as good as the original Corona cameras. A significant observation from KH-3 Mission 9023 was the space-launch and missile-test center at Tyuratam in the central Soviet Union.

KH-4

The M camera configuration—comprising two C''' cameras mounted for stereographic imagery—was carried on the KH-4 **Corona** series. The system carried double the amount of film in the first three KH phases; with 18 kilograms of film for each of the two cameras, the total exposed was 36 kilograms. The first launch, Mission 9031 on February 27, 1962, was air-recovered successfully and returned improved resolution photographs. The KH-4 satellites orbited at 200 kilometers altitude, and could resolve objects of 2.5 meters. In all, 26 KH-4 M camera configurations were launched from February 1962 through December 1963, and 20 of them recovered film. In the early months of 1963, there were a number of **launch** and recovery failures of several KH systems, and the intelligence and defense communities were distressed at the lack of imagery. The reliable KH-4 Mural system resolved the problem with three successful launches in June and July of 1963.

KH-4A

The KH-4A phase of **Corona** used the J camera configuration, which included one M stereo camera system and two separate recovery capsules. It carried 72 kilograms of film; each of two four-day operations of the Mural camera were to be discharged from the **spacecraft** and recovered separately. After recovery of the first film capsule, the spacecraft was to remain in orbit with electrical power and spin-stabilization gas turned off for up to 20 days—this was called "zombie mode"—then reactivated when most needed for a second photo-taking operation and recovery. This would essentially double the imagery recovered from each launch. In the initial two KH-4A launches, the first recovery capsule was retrieved but not the second, because reactivation of the spacecraft from zombie mode proved difficult. Overall, however, with 52 systems launched between August 1963 and September 1969, and a potential for 104 film capsule retrievals, the system proved highly reliable, with 93 film recoveries.

There was one particularly distressing failure of KH-4A. Mission 1005, launched on April 27, 1964, operated well until telemetry indicated the film in the camera broke, the power failed, and the reentry vehicle did not respond to commands. Calculations of the natural orbital decay predicted a landing somewhere near Venezuela, but the capsule was not sighted or recovered. It was discovered months later by two laborers on a remote farm in the Venezuelan Andes, and the farm owner tried to sell it. By the time cognizant U.S. personnel arrived, remains of the capsule had been taken by the Venezuelan Army, and the locals had used parts of it for a variety of purposes, including one fellow who had used the recovery parachute cords to make a harness for his horse!

KH-4B

KH-4B satellites carried a J-3 camera configuration. The KH-4B carried a stereo camera with improved mechanical design, two recovery capsules, and a new Dual-Integrated Stellar-Index Camera (DISIC) to

Film from early Corona satellites was retrieved in an amazing way. A heat-resistant nose cone re-entered the atmosphere, deployed parachutes, and was snagged in mid-air by a U.S. Air Force C-119 aircraft with a specially designed hook trailing under its tail.

improve attitude determination. The camera had two interchangeable filters and four selectable aperture slits, and carried 36 kilograms of film in a 500-meter roll. The KH-4B cameras could orbit for up to 19 days at lower altitudes (as low as 85 kilometers), with ground resolution improved to two meters.

The first five flights (Missions 1101 to 1105) included new types of film for testing—high-speed black-and-white, color, and near-infrared films. Of 17 systems launched between September 1967 and May 1972, 32 of 34 film capsules were recovered, representing the best success rate of the **Corona program.**

KH-5/Argon

The KH-5 series of satellites, code-named "Argon," were flown as a covert mapping project for the U.S. Army. In order to manage launch priorities, the flights were managed under the **Corona program.** The Argon satellites carried a single low-resolution camera and had the objective of precise geodetic measurements (latitude, longitude, and elevation) for improved location of strategic targets. Of the 12 Argon satellites launched between February 1961 and August 1964, seven were successful. The new vertical and horizontal position control and the new highly accurate DISIC star camera on KH-4B provided sufficient positioning control to support precision mapping from space, so Army programs intended as follow-on to Argon were cancelled.

KH-6/Lanyard

In 1962 the intelligence community was anxious to obtain high-resolution images of antiballistic missile (ABM) sites suspected to be in Leningrad, but their current Corona-M camera did not have sufficient resolution. A large **SAMOS** capsule, one of their later E-5 models intended for film retrieval rather than television transmission of images, was modified quickly for use in this mission. This 1963 experimental program was designated KH-6, with a code name of Lanyard. The expectation was for high-resolution stereo images with about 60-centimeter resolution.

The first two launches failed due to a common problem in the early **Corona program:** failure of the Agena second-stage vehicle. The third was only partially successful, with camera failure after 33 days. The failure rate of the new program compared with increasingly reliable Corona performance, and the

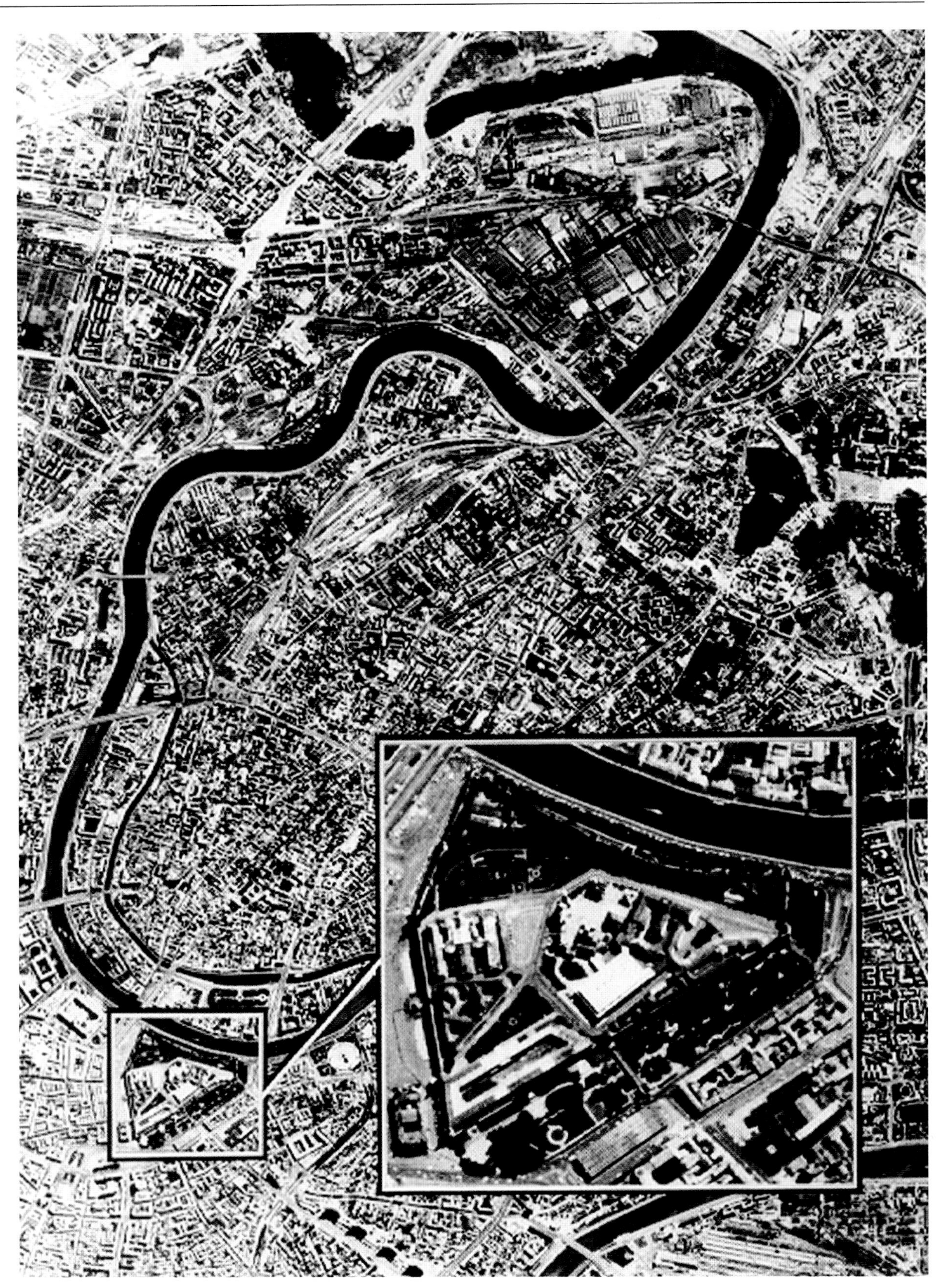

The Moskva River winds through the city in a KH-4B image of Moscow. The inset provides an enlargement of the Kremlin, in which a line of people outside Lenin's Tomb in Red Square is visible.

concerns about recovery of the larger capsules, led to cancellation of the Lanyard effort.

KH-7

While the **Corona** satellites were all broad-area search systems and produced increasingly fine detail of large facilities such as military airstrips, missile launch sites, and submarine bases, the military needed a spotting system to get a more detailed look at smaller objects in the facilities, such as individual aircraft, missiles, submarines, or tanks.

The U.S. Air Force developed such a system, and it was controlled under the Keyhole security system, designated as KH-7. These satellites stayed in orbit for about five days at a **perigee** of about 150 kilometers, yielding photographs at about 45 centimeters resolution. This was the first reconnaissance satellite to return images at as high a resolution as the U-2 reconnaissance aircraft. The film was returned in recovery capsules like the Corona program, and the Air Force used the images to refine targeting plans. Of 38 satellites launched between 1963 and 1967, 36 were successful.

KH-9 Mapping System

This specialized mapping camera system operated in space from March 1973 to October 1980. Like the **Corona** systems, KH-9 returned its film to **Earth** in a heat-resistant reentry "bucket," and the film was processed on land. The camera took stereo, and sometimes "trilaps" (triple), images of the same sites to improve mapping informaton, including the "elevation" or height of land terrain features. The mapping imagery was used to create maps of "denied" or inaccessible areas for defense purposes, and to improve U.S. Geological Survey topographic maps of the United States.

The **footprint** on the ground for each image, or "frame," from the KH-9 mapping cameras was about 130 by 260 kilometers, with an average ground resolution of roughly 10 meters from early flights, which improved to 6.5 meters in later missions. This resolution is better than that of the original **Landsat** system but not as good as current imagery from the French **SPOT** system.

All 12 missions of the KH-9 mapping system were successful. Their operational life ranged from 42 to 119 days. The entire program provided 29,000 images—with all but 100 of them declassified and publicly available—covering roughly 300 million square kilometers.

EIS (Enhanced Imagery System)

EIS is the current U.S. **IMINT** system of the **National Reconnaissance Office.** EIS has some improved capabilities over former imagery reconnaissance systems while awaiting the **Future Imagery Architecture.** All aspects of the EIS program are still classified.

FIA (Future Imagery Architecture)

The future of U.S. **IMINT** capability is in the Future Imagery Architecture (FIA) program. The planning for FIA was initiated by the federal government in the mid-1990s. Details of the design and capabilities are highly classified, but some information on the program objectives and approaches is publicly available.

There has been an unusually strong emphasis since the beginning of FIA to focus on upgrading the whole range of IMINT capabilities—not only the satellites and sensors, but also the imagery processing, analysis tools, and improved dissemination of the intelligence information to qualified users.

An exhaustive study of user needs was completed before the satellite design was even begun, and a contract for the ground structure—called the Mission Integration and Development Program (MIND)—was awarded in April 1999, before the **satellite** design contract was awarded.

The defense and intelligence communities, recognizing improvements in the capabilities of unclassified imagery satellites, decided to focus FIA satellite performance only on highly sophisticated capabilities not otherwise available, and to purchase panoramic images from civil and commercial programs. The MIND system is expected to access these civil products as well as classified FIA imagery. Then, instead of requesting multiple products, users can ask a specific intelligence question and get an answer through the MIND program that fuses information from all sources into the best and most efficient answer to their question.

The U.S. government took another unusual step, in today's contracting world, of specifying only the performance characteristics of the system—things such as image resolution, coverage, and revisit rates to the same spot on Earth's surface—and allowing the contractors complete freedom in designing any type and number of satellites to meet those requirements. The satellite contract was awarded in September 1999.

Other Types of Intelligence

SIGINT (Signals Intelligence)

Since ancient times, history is replete with fascinating stories of governments trying to intercept the official messages of other governments, and the practice continues today. With the development of communications techniques that use different parts of the electromagnetic frequency **spectrum,** and with the advent of the two World Wars, the practice evolved into a sophisticated, technology-based science.

SIGINT (signals intelligence) is defined as intelligence information derived from the interception and analysis of an adversary's electromagnetic emissions, for both communications and other purposes, to learn about its threat capabilities, limitations, and intentions. Two primary attributes of SIGINT are that it must be covert, because an adversary will not reveal useful information if aware that someone is listening, and it is passive—SIGINT is based on just listening and learning, not on interfering with a foreign government's business. Other challenging aspects include the need to decipher information that is encrypted by increasingly sophisticated techniques and, for voice transmissions, the need for very fast and accurate translation of information from other languages.

SIGINT includes three subdisciplines that target different parts of the frequency spectrum and have different objectives:

- COMINT (communications intelligence) is the intelligence information derived from interception and processing of voice, teletype, facsimile, microwave, and video transmissions.
- ELINT (electronics intelligence) is defined as intelligence information derived from the interception and analysis of emissions principally from **radar** devices. The objective of ELINT is to identify an adversary's radars by type, function, range, capability, and

The GRAB (Galactic Radiation and Background) satellite did measure solar radiation, but its primary purpose was as the world's first reconnaissance satellite, launched by the U.S. on June 22, 1960, to collect electronics intelligence (ELINT).

precise location. Among other things, it can focus on **missile** guidance systems and an adversary's jamming capabilities.

- FISINT (foreign instrumentation signals intelligence) is defined as the collection and processing of emissions associated with the development and testing of an adversary's new military platforms and weapons systems.

In the early days of the Cold War, the United States was particularly concerned about a Soviet build-up in long-range weapons and a possible surprise attack. The Soviets had an increasingly closed society, with few visas issued for foreign travel into or within the country, so traditional methods of using human agents in-country were very limited. The Western allies ringed the Soviet Union with sophisticated SIGINT listening posts, but the fundamental geography of the country presented an insurmountable problem.

Russia is the largest country in the world, representing almost 12 percent of the **Earth**'s landmass, stretching 9,000 kilometers from east to west and straddling 11 time zones. Its entire northern border is bounded by the frozen and uninhabited Arctic. Collection of emissions from distant ground stations were limited by interruption of the signals by mountainous or wooded terrain, electromagnetic interference from urban or manufacturing areas, or by severe weather and **solar flare** effects in the atmosphere. Interception of emissions from the center of the Soviet Union was impossible and added to the impetus for the United States to launch an ELINT **satellite.** The result was the **GRAB** satellite launched in June 1960.

While the GRAB satellite overcame many problems of "geographic" access to the Soviet Union, it initially created a new kind of problem of its own. According to a Navy press release, the wealth of signals information from the satellite initially overwhelmed the **National Security Agency**'s ability to analyze the data.

The business of SIGINT, practiced today by all major nations, faces continual challenges as new forms of communications and information-sharing technology—advances like frequency-hopping radios, burst communications, and increasingly sophisticated encryption software—are developed increasingly rapidly. But the biggest challenge remains the need to remain covert. Any information released about SIGINT capabilities aids an adversary in developing countermeasures that cut off the source of information. Governments rarely make any public acknowledgment of the use of SIGINT in their policy making, and despite a great deal of speculation in the media, access to accurate information about the capabilities of current SIGINT assets is very limited.

MASINT (Measurement and Signature Intelligence)

MASINT (measurement and signature intelligence) is technically derived intelligence that detects, locates, tracks, identifies, and describes the unique characteristics—or signature—of fixed and mobile targets. It differs from **IMINT** and **SIGINT** in that a wider spectrum of information is used, including **radar, laser,** optical, infrared, acoustic, seismic, radio frequency, nuclear radiation, and gas, liquid, and solid materials sampling. The MASINT effort describes a target or targeted activity in as many ways as possible to overcome stealth and countermeasures on the part of an adversary. It includes some specialized additional processing of traditional SIGINT and IMINT data. MASINT is essentially the "forensic evidence" of unfriendly activities or threatening events.

Data collection for MASINT is accomplished from space-based, aerial, and ground-based sensors, and from sensors designed just for MASINT and many that are shared with other objectives. Many commercial and research sensors have MASINT analogues.

MASINT supports weapons development, countermeasures, tactics, targeting, battle damage assessment, indications and warnings of an adversary's intentions, situational awareness for military commanders, counterterrorism efforts, arms control and treaty verification, counter-drug operations, and even natural disaster warnings and environmental monitoring. When there is a civil emergency of major proportions—such as an earthquake—the information is downgraded in classification to support civil disaster efforts. The products of MASINT thus support the nation in times of peace, civil disaster, national crisis, regional conflict, and war.

The myriad sources of data could easily overwhelm the capabilities of analysts or users, so the U.S. MASINT System Vision is working toward tools to help users. A MASINT Information Environment (MINE) under development is similar to the Future Imagery Architecture's emerging Mission Integration and Development (MIND) program in that it will allow a user to query the system, will go to archived data to see if an answer is available, and if not, will analyze the best intelligence collection request to respond. This avoids requiring the user to know the intricacies of all the sensor systems and allows broader access to the system.

Spies in the Sky

Gary A. Federici

THE 1950S WERE YEARS OF ACCELERATION OF THE COLD WAR AND ITS ARMS race. Both the United States and the Soviet Union relied on the long-range strategic bomber as their weapon of choice and were investing heavily in air-defense systems that could blunt the other's attack. The U.S. found it difficult to obtain information about the Soviet Union's arsenal. There had been public debates about whether a "bomber gap" existed, whether the U.S. was building enough B52s to defend itself. Of equal importance, but hidden from public view, was uncertainty about the Soviet Union's air-defense system. Knowledge of that system would not only determine the number of U.S. bombers required, but also would provide vital information about the routes they should fly to avoid detection.

The U.S. attempted to collect this information through "ferret" flights along the periphery of the Soviet Union. These flights could not look far inland and were constantly harassed by Soviet interceptors.

An initial solution was the longer-range high-flying U-2 reconnaissance aircraft. It could overfly the Soviet Union above the altitude of interceptor aircraft. The provocative nature of these flights limited their number, and they were threatened by the Soviet development of surface-to-air missiles. Even before the 1960 shoot-down of the U-2 piloted by Gary Powers, the U.S. realized that the days of aircraft overflights were numbered. Surveillance by satellites would be both less vulnerable and less provocative.

Both the Soviet Union and the U.S. had developed satellites to make scientific measurements during the 1957 International Geophysical Year. The U.S. decided for political reasons not to launch its Vanguard satellite on a military rocket and developed a separate booster that failed in its original trial in December 1957, but succeeded in March 1958. Meanwhile, the Soviet Union startled the world by successfully launching its Sputnik satellites in October and November 1957. The U.S. quickly reversed its policy and launched an Explorer satellite in January 1958 with a booster based on the Jupiter intermediate-range ballistic missile (IRBM) system.

The need for space reconnaissance of the Soviet Union and the availability of ballistic missile rockets inspired the first two successful U.S. space reconnaissance programs. The first to begin, although the second to succeed, was the CIA's Corona program to photograph the Soviet Union from space.

Corona faced enormous technical challenges. The satellite would have to point its camera accurately at desired targets and carry enough film, which had to be retrieved from space without its incineration on reentry and within range of an airplane that would scoop the film canister while it dangled on a parachute. The entire satellite would have to be light enough to be launched in orbit by a two-stage booster that used the Thor IRBM system as its first stage.

Mission success required a series of test launches, announced as an Air Force research program named Discoverer. The first launch attempt was in January 1959; the first successful camera mission was on August 18, 1960.

Today's engineers, faced with an environment in which programs are cancelled after a few failures, look with envy on that history. A combination of urgency, secrecy, and the availability of Thor rockets enabled the program to continue and finally to succeed.

Discoverer launches initially were public, although their purpose was secret. In 1962,the Corona program was transferred to the newly formed National Reconnaissance Office (NRO), whose existence remained classified for decades. In 1995 President Clinton declassified information about these satellites.

Corona, however, was not the first successful reconnaissance satellite. The Naval Research Laboratory, which had designed and built the Vanguard satellite,

Part of the Naval Research Laboratory's GRAB satellite team at Cape Canaveral on April 29, 1960, spin test their satellite atop a Transit navigational satellite. Both reached orbit successfully atop the same launch vehicle on June 22, 1960.

also developed the GRAB satellite, which was successfully launched on its first try, June 22, 1960. The small GRAB satellite listened for Soviet radars and transmitted every pulse to a network of listening stations. Although the satellite was not stabilized, clever processing of retrieved pulses located Soviet air-defense radars and assessed their capabilities.

The GRAB satellite carried a scientific mission that measured galactic radiation background. This served as the cover for its intelligence mission. The ability of GRAB to locate and evaluate Soviet air-defense radars was of great use to the Air Force's Strategic Air Command. Like Corona, GRAB was placed under NRO control. The launches became secret, and the scientific cover was no longer required. The GRAB satellite program was declassified in 1998.

While intelligence warning for the possibility of strategic nuclear exchange was the motivation for Corona and GRAB, the role of reconnaissance satellites evolved with the Cold War. As treaties were agreed upon by the U.S. and Soviet Union for nuclear weapons arsenals, reconnaissance satellites enabled both nations to enforce these obligations. Although the satellites were kept secret from the public, the two countries had a reasonably accurate understanding of each other's reconnaissance capabilities, and the treaties explicitly prohibited interference with these capabilities.

Both Corona and GRAB led to successors far more capable, and by the 1980s space-based reconnaissance had matured to provide direct support to military forces. Some observers called Operation Desert Storm "the first information war" because information superiority, much of it derived from space surveillance, was decisive. Since then, space reconnaissance has been used effectively during Operations Enduring Freedom and Iraqi Freedom.

Continued space investments are under way. The Air Force plans a constellation of radar satellites to detect moving objects anywhere on Earth nearly all the time, moving the role of space systems from intelligence support to combat management. This development will be a technical challenge, but with tenacity, the U.S. will add a remarkably valuable component to its national security capabilities. ■

U.S. Military & Intelligence Space Organizations

The U.S. Intelligence and Space Communities

There is considerable overlap in the U.S. government between the space and intelligence communities, and between the civil and military space organizations. The "intelligence community" includes the **Central Intelligence Agency; Defense Intelligence Agency;** Army, Navy, Air Force, and Marine Corps Intelligence; **National Reconnaissance Office; National Security Agency; National Geospatial-intelligence Agency;** the Federal Bureau of Investigation; Department of the Treasury; Department of Energy; and the Department of State. Some of these are collectors of intelligence, some are processors, and some are primarily users of the products and information.

The National Reconnaissance Office is the "nation's eyes and ears in space." The lead organization for the nation's **SIGINT** (signals intelligence) efforts is the National Security Agency; for **IMINT** (imagery intelligence) and the new GEOINT (geospatial intelligence) the lead is the National Geospatial-intelligence Agency; for **MASINT** (measurement and signature intelligence) it is the Defense Intelligence Agency; and responsibilities for HUMINT (human intelligence) are shared between the Defense Intelligence Agency for foreign military issues and the Central Intelligence Agency for all other matters. The military oversees development of missile defenses, launches and tracks **satellites,** and operates the military platforms that will counter **missile** or other attacks against us. Some of the agencies described were formed in the late 1950s and early 1960s to manage the tremendous volume of new information provided by the advent of satellites.

NRO (National Reconnaissance Office)

The NRO is the government agency that designs, builds, launches, and operates spy **satellites** and sensors for the intelligence community of the U.S. Established as a covert agency in 1961, NRO has operated hundreds of satellites over the past four and a half decades. NRO's management and budget are overseen by the Department of Defense and the **Central Intelligence Agency.**

A surprising aspect of NRO's history is that the organization remained completely covert for most of its existence. Its logo, name, and even the three-letter acronym of its name, "NRO," were highly classified throughout the Cold War and for some time afterward. The National Reconnaissance Program (NRP) was established by a top-secret agreement between the Deputy Secretary of Defense and the Director of Central Intelligence on September 6, 1960. It was formed principally to handle the **Corona** photoreconnaissance satellite program as it was being transitioned from the Air Force to a covert program managed by the CIA. The NRO was formally established as a separate organization on June 14, 1962. The **GRAB** ELINT satellite program, which had been operating for two years, was also transferred to the new NRO. Existence of the NRO was publicly recognized by the U.S. when its existence was declassified by the Director of Central Intelligence on September 18, 1992—more than 30 years after its formation. While the organization and its name are now public, most of its programs and budgets are still "black," or highly classified.

NRO is working to support modern users—intelligence analysts, military commanders, and homeland defense, civil disaster and environmental monitoring activities—with a wide range of newly developed sensors aboard satellites operating in a full range of orbits.

NGA (National Geospatial-intelligence Agency)

The NGA is the new U.S.government entity responsible for combining imagery intelligence (**IMINT**) and related geopositioning and mapping activities into a new field of endeavor called GEOINT, or geospatial intelligence. NGA was officially established upon signing of the 2004 Defense Authorization Act on November 24, 2003. It has the lead responsibility for the "TPED" process under the new FIA (**Future Imagery Architecture**) program, the emerging new space-based IMINT system. TPED stands for "Tasking, Processing, Exploitation and Dissemination" of collection requirements, images, maps, and other products from the

FIA program, and NGA is improving user access, timeliness, and automation in all phases of the process.

This complex agency is a key member of the intelligence community with the lead for IMINT products; a defense department combat-support agency getting the imagery and map products to American troops overseas; and a global civilian-support agency providing aeronautical and nautical charts to civilian users worldwide.

NIMA (National Imagery and Mapping Agency)

NIMA is the former name of the **National Geospatial-intelligence Agency.** NIMA was created in 1996 by merging the Defense Mapping Agency and several **IMINT** (imagery intelligence) offices from various parts of the intelligence community.

NPIC (National Photographic Interpretation Center)

President **Eisenhower** authorized the creation of the National Photographic Interpretation Center (NPIC) in January 1961, which combined the **IMINT** (imagery intelligence) assets of the **Central Intelligence Agency,** and of the three military services—Army, Navy, and Air Force. They had all been making separate interpretations of **U-2** and other airborne reconnaissance, but with the flood of new images from the **Corona** photo-reconnaissance satellite and the personal interest of the President and the White House staff, a national organization like NPIC was needed to provide a quick, coordinated interpretation to the nation's leaders.

NSA (National Security Agency)

The NSA is the government agency that manages signals intelligence (**SIGINT**) processing and dissemination, and develops encryption to protect U.S. communications from adversaries. The agency's missions can be described as code-breaking and code-making (encryption techniques for information assurance). The Armed Forces Security Agency, formed in May 1949, was combined with civilian cryptology efforts when NSA was established in November 1952.

NSA formation thus predated the first space-based efforts to collect SIGINT information—the **GRAB** electronics intelligence (ELINT) **satellite** launched in 1960.

While imagery intelligence (**IMINT**) played a major role in the 1963 Cuban Missile Crisis—with both **Corona** and **U-2** images confirming the deployment of Soviet missiles in Cuba—NSA's less well known SIGINT efforts were critical in monitoring Russian Fleet movements before and during the naval blockade ordered by President Kennedy.

With more mathematicians than are employed anywhere else in the country to make new codes, and with specialists in many languages to catch talk of terrorism and smuggling of drugs or weapons, NSA focuses on a broad range of new threats as well as traditional military and intelligence operations. NSA's high-technology pursuits have contributed greatly to the development of cassette tapes, supercomputers, microchips, and nanotechnology—the miniaturization of mechanical devices.

CIA (Central Intelligence Agency)

The CIA is the U.S. agency responsible for HUMINT (human intelligence, or information gathered from agents), and for collating and analyzing the information from all the various intelligence fields (or "INTs") into the President's Daily Intelligence Briefing and the

A 1966 declassified photograph of Washington, D.C., taken by the Corona KH-4A series of satellites, demonstrates the impressive resolution of early imagery intelligence (IMINT) satellites.

National Intelligence Estimates (NIEs). The Director of the CIA is also the Director of Central Intelligence, leading the nation's "intelligence community."

The CIA, established with President Truman's signature on the National Security Act of 1947, played a lead role in early space-based **reconnaissance.** Having successfully managed the **U-2** aircraft program in the mid-1950s, the CIA was given oversight of the covert **Corona** satellite reconnaissance program in 1958. The CIA continues to be involved in the day-to-day management of the **National Reconnaissance Office.**

DIA (Defense Intelligence Agency)

The DIA is the government agency that has primary responsibility for collecting, analyzing, and disseminating intelligence on foreign military capabilities, limitations, and intentions, which includes information on foreign military **missile** and space programs. The DIA became operational on October 1, 1961. Its focus remains on military intelligence, including both human-collected (HUMINT) and various types of space-based information. DIA has the lead for the relatively new area of measurement and signature intelligence (**MASINT**) in the United States.

STRATCOM (U.S. Strategic Command)

STRATCOM is the U.S. military command responsible for a wide variety of space-related missions. Originally established on June 1, 1992, the U.S. Strategic Command has its roots in the Strategic Air Command established in March 1946 to operate long-range bombers in response to the emerging Soviet threat. The command was expanded to include the missions of the former U.S. Space Command on October 1, 2002.

Current STRATCOM responsibilities include:

- Launching, tracking, and operating military satellites for early warning (EW) of **ballistic missile** or other attacks, military communications, weather, navigation, and positioning.
- Operating the Space Surveillance Network for tracking space debris and assessing when it will reenter Earth's atmosphere.
- Operating reconnaissance aircraft.
- Directing operations of the platforms that would respond to a threat to the United States, including long-range bombers and land-based and submarine-launched ballistic missiles.

MDA (Missile Defense Agency)

MDA is the U.S. agency responsible for defense—of the U.S., its assets deployed overseas, and its friends and allies—against **missile** attacks from enemies, adversaries, or terrorists. In January 2002, the Defense Department reorganized the **Ballistic Missile Defense Organization,** upgraded it to a federal agency, and renamed it the Missile Defense Agency (MDA). MDA's approach is to provide a "layered" defense, which means developing ways to detect and destroy incoming missiles in all phases of their flight paths. These include the **launch** or "boost" phase, when a missile's **rocket** engines are still burning; the "mid-course," when the missile is on its predetermined **ballistic trajectory;** and the "terminal" phase, when it reenters **Earth's atmosphere** at speeds up to 1,240 kilometers an hour in a **free-fall trajectory** to its target. The agency is using nonnuclear defensive weapons, based on the ground and at sea, and detection and tracking sensors on the ground, at sea, and in space. The defensive weapons include "interceptors" using "kinetic energy"—their own mass and speed—to destroy a missile on impact, likened to "hitting a bullet with a bullet." Research is also being done on "directed energy" approaches—airborne and possible space-based **lasers.**

The **low-Earth-orbit** (LEO) components of the **Space-based Infrared System** (SBIRS) were transferred in 2001 from the U.S. Air Force to the Missile Defense Agency and renamed the **Space Tracking and Surveillance System** (STSS). A constellation of STSS **satellites** carrying infrared sensors will provide global detection of **ballistic missile** launches, tracking of the missiles, and rapid transfer of the information to interceptors. Launch of the first two STSS demonstration satellites is planned for 2007.

BMDO (Ballistic Missile Defense Organization)

BMDO was the former name of the **Missile Defense Agency** (MDA). During the Cold War, the Strategic Defense Initiative Office was focused on the **intercontinental ballistic missile** threat from the Soviet Union. As the Soviet Bloc disintegrated in the late 1980s, the missile threat to the United States seemed to diminish, but actually became more dispersed and even more difficult to track and defend against. The U.S. Department of Defense, on May 13, 1993, redirected the effort to more ground-based than space-based initiatives, and from nuclear weapons to nonnuclear approaches, and renamed it the Ballistic Missile Defense Organization.

Remote Ascension island, on the Mid-Atlantic Ridge only 8° latitude south of the Equator, provides an ideal location for satellite and missile tracking. Moonrise over the island silhouettes the military tracking antennas on "Telemetry Hill."

Military Communications

Military Communications

Space-based military communications capabilities are often abbreviated as MILSATCOM. All U.S. military missions rely on space-based communications—from early warning of **ballistic missile** threat, to daily operational plans based on updated intelligence, to logistics, medical support, and computer-training updates for troops. The U.S. military uses **LEO, geostationary Earth orbit** (GEO), **HEO,** and drifting orbits, a wide range of frequencies and data rates, and a mix of leased commercial circuits and dedicated military systems:

The Milstar space-based military communications system links command authorities with forces in the air, on land, and at sea worldwide.

- Defense Satellite Communications System (DSCS), the first operational MILSATCOM system, was begun in 1966 and still operates. DSCS Phase I satellites were launched from 1966 to 1968 in **circular orbits** over the Equator at an altitude just below that required for GEO, allowing the satellites to drift with respect to the Earth's surface. DSCS Phase II (1971 to 1975) and Phase III (since 1982) are in GEO orbits. Phase III uses 13 satellites to provide long-distance, encrypted, jam-resistant, low data rate, wideband communications to military commanders in the field.
- Fleet Satellite Communications (FLTSATCOM) was dedicated to tactical users in the field. Four satellites launched from 1978 to 1980 into GEO provided super-, extra-, and ultrahigh frequency narrowband communications.
- Ultra High Frequency Follow On (UFO) replaced FLTSATCOM, with nine satellites launched from 1993 to 2003 into geostationary Earth orbits with UHF and EHF systems aboard.
- Global Broadcast System (GBS) is a wideband, transmit-only system designed to provide intelligence, imagery, maps, and video information worldwide to military forces with small portable receiver terminals. The first three GBS systems were added to UFO satellites 8, 9, and 10, launched in 1996 and 1999.
- The Milstar system links command authorities with ground troops, aircraft, ships, and submarines worldwide, provides secure (encrypted) voice, data, teletype and facsimile transmissions, and classified video teleconferencing, and is hardened against nuclear radiation, jamming, and other threats. The constellation includes five satellites in GEO with active cross-links between them and communications capacity in the superhigh frequency (SHF) and extremely high frequency (EHF) ranges. Milstar's most significant advance is its ability to process communications signals aboard the satellites; it can very quickly create, maintain, reconfigure, and disassemble tailored circuits on demand.

Transformational Communications Architecture (TCA)

TCA is a U.S. effort to provide more effective connectivity and interoperability among military, intelligence, and civil communities. TCA will transition national communications from today's separate, duplicative, circuit-based systems to an integrated Internet-like capability based on radio frequency, optical, and Internet technologies, and "packetized" information transport. It includes a Department of Defense space segment (Transformational Satellite System), an intelligence community space element (Optical Relay Communications Architecture), and a **NASA** space element (Tracking Data Relay Satellite System). All will connect to the emerging Global Information Grid, providing an order-of-magnitude greater capacity to users and horizontal integration of communications across the three communities.

Military & Intelligence Uses of Space

Military & Intelligence Uses of Space

From 1960 to 2000 the United States launched almost 800 military and intelligence **satellites** for various purposes, and about the same number of civilian satellites. In the same time frame the Soviets (and then Russians) launched about the same number of civilian satellites, but more than three times that many military and intelligence satellites. Between 1969, when the United Kingdom launched its first military satellite, and 2000, five nations—U.K., China, France, Israel, Chile—and the North Atlantic Treaty Organization (NATO) launched almost 50 military satellites. This group launched more than 500 civilian satellites, however, showing their primary focus on civil uses of space. In January 2003, the South Korean government announced plans for the **launch** of its first military satellite in 2005, and Japan launched its first two spy satellites in March 2003.

The **Outer Space Treaty** addresses military uses of space, stipulating that it be used for peaceful purposes. That has been widely interpreted by the space-capable nations to mean that defensive, not offensive, military activities in space are acceptable. While details of these satellites and their missions are sketchy, missions of some of the U.S. satellites are summarized in general terms here. The United States depends on space for its national security in many ways. Mapping satellites provide terrain elevation and precise target locations. Weather satellites, in addition to advising of field conditions that armed forces have to deal with—aircraft and **missile** wing icing, storms at sea, cloud cover over the target, sandstorms in the desert—provide visibility for overhead imagery, weather effects on some weapons, weather effects on some **radars** and communications, and predicted wind dispersion of biological or chemical weapons. Military communications satellites provide global channels between national leaders and regional commanders, and forward to dispersed units in the field and at sea; near-real time **downlink** of battle-space **reconnaissance** information; and rapid dissemination of intelligence and targeting information.

Navigation-positioning satellites, particularly the current U.S. **Global Positioning System** (GPS), are used not only for armed forces navigation, position, and rendezvous, but also for precise time to support synchronized operations and encrypted burst communications. They also provide precision locations of targets, and the means to direct precision-guided weapons to those targets. An ironic recent story released to the press in March 2003 involved a detection of Iraqi attempts to jam the U.S. Global Positioning System (GPS), with a U.S. response of sending GPS-guided weapons that destroyed the jamming devices.

Some of the "military" satellites are reconnaissance and **surveillance** intelligence systems. Concerning imagery intelligence (**IMINT**) systems, President Bush, on May 13, 2003, announced a new U.S. Government policy that "to the maximum extent practical" commercial imagery will be used by all agencies for military, intelligence, foreign policy (treaty verification), and homeland security purposes. IMINT satellites will thus provide only capabilities not met by commercial satellite systems. Some of the military satellites are early warning systems designed to identify and track **ballistic** and cruise **missiles** targeted at the U.S., our forces overseas, or our allies or friends.

Space Control

Space control is ensuring one's own freedom of action within space and ensuring the ability to deny others the use of space, if required. For the United States this mission includes four functions. The first is **surveillance** of objects in space, where **space debris** and deep-space objects that might collide with **Earth** are tracked. Among other things, identification and tracking of these objects avoids triggering a false alarm of a **missile** attack. Other aspects of space control include protection of our space assets from hostile threats and environmental hazards, protection against unfriendly use of our space systems, and—if necessary—the destruction or disruption of space systems hostile to our country. One approach to space control is ASAT (antisatellite) programs—weapons designed to destroy or disable an artificial **satellite** in orbit. Both the U.S. and Russia have tested ASAT systems, largely ground- and aircraft-based systems, with some success. A major drawback of this approach is the potential that debris from a destroyed satellite will cause unintended damage to one's own military or civilian satellites or to those of allies and friends. Partly for this reason, the U.S. is focused on a broader approach to space control. These approaches might include jamming of a hostile nation's satellite command and control links, or disruption of ground-control or **launch** facilities.

Battle-space Reconnaissance

Satellite-based intelligence assets continue to be critical to peacekeeping and war-fighting missions, providing multi-intelligence information about the battle space, location of friendly and hostile forces, and location and **tracking** of hostile forces and threats. The modern battle space is an increasingly complex and fast-moving arena, however, with weapons systems of increasing speed, mobility, and range available throughout much of the world. There is also increasing knowledge of overflight times and capabilities of space-based **reconnaissance** systems, spawning operational plans designed to take advantage of these limitations. These trends have encouraged recent development of a number of airborne intelligence systems to augment tactical awareness of the location and movement of friendly forces, as well as enemy capabilities, movements, and threats in a regional battle area. This aerial component of battle-space awareness is provided by a wide variety of increasingly capable manned and unmanned aerial assets; some Western systems are outlined below.

U-2S

An updated version of the 1950s **U-2** high-altitude **reconnaissance** aircraft, the U-2S (the "S" stands for surveillance) is used today for sustained (between **satellite** passes), high-resolution intelligence collection over sensitive areas and battlefields. The U-2S is 40 percent larger than the original—19 meters long with a wingspan of 32 meters—and has a range of 11,000 kilometers without refueling and an operating altitude more than 21,000 meters, at the very outer edges of space. It carries a sophisticated sensor package of film, electro-optical, infrared, and **radar** imagers as well as some **SIGINT** data collectors, all (except traditional film) downlinked in near-real time. The U-2S is also used for peacetime missions, supporting relief efforts from natural disasters such as earthquakes, forest fires, and major floods.

Predator

To avoid the concern of pilot vulnerability to a shoot-down, a number of increasingly sophisticated

A satellite image of Saudi Arabian port R'as al Khafji, near the Kuwaiti border, is overlain with locations of radio and other electronic emitters from ships-patrol boats offshore and land facilities, including an air-defense radar. This image is from the commercial IKONOS satellite; national reconnaissance images are classified and would make an "intelligence integration" product of this kind classified.

unmanned aerial vehicles (UAVs) have been developed for battle-space **reconnaissance.** Various versions of the Predator are medium-altitude (15 kilometer), long-endurance reconnaissance assets, 8 meters long with a wingspan of about 15 meters and a speed of 130 to 220 kilometers an hour. Sensors include video; infrared for nighttime; **synthetic aperture radar** (SAR) for penetrating clouds, smoke, or haze; and laser targeting systems for weapon-carrying versions. UAVs are operated from a ground- or sea-based station by line-of-sight or by **satellite** over-the-horizon communications links.

Global Hawk

Global Hawk (RQ-4A) is a high-altitude, long-endurance, unmanned aerial vehicle (UAV) designed by the United States for high-resolution battle-space **reconnaissance** and **surveillance.** Because of its lightweight construction of aluminum and composite materials, Global Hawk is distinguished by its operating altitude of 20,000 meters and its range of 26,000 kilometers without refueling. The airframe set a UAV endurance record in April 2001 with a 12,000-kilometer nonstop flight from the United States to Australia. It carries electro-optical and **infrared sensors** that can image 135,000 square kilometers in one day, as well as a cloud-penetrating **synthetic aperture radar** (SAR) system that provides "ground moving target indicator" (GMTI) intelligence information, tracking ground threats and targets in near-real time.

Joint STARS

The Joint Surveillance Target Attack Radar System (Joint STARS) is a manned airborne command and control, intelligence, surveillance, and **reconnaissance** platform. It is equipped with Doppler **radar** that can see and track ground vehicle movements day or night, in all weather, including heavy sandstorms. A four-person crew along with up to 18 specialists in this E-8C modified Boeing 707 "orbit" the battle space at high altitude to detect and track moving ground targets.

Intelligence Integration

There is a growing need for a near-real time, fully integrated, high-resolution intelligence picture for a wide array of users—intelligence analysts, military commanders, homeland defense and civil-disaster personnel. There is also an ever increasing number of space-based and aerial systems designed to provide high-resolution intelligence information to these users. The critical requirement emerging in recent years is the need to share information—rapidly—among the multiple systems and the very different types of operations. Many of the systems are impressive in their efforts to combine various sensors into an integrated intelligence and operational output, but each program is choosing its own integration approaches and tools. An overwhelming need exists for improved integration across different systems and communities.

For military operations, this means coordinating, interfacing, or preferably converging the various systems used for disseminating, processing, displaying, and using the data from different space-based and aerial battle-space assets. Providing the intelligence "take" to operational commanders quickly, even within one nation's military and intelligence forces, requires sophisticated "multilevel security" (MLS) measures to ensure rapid provision of the needed information to the commander in the field, without disclosing sensitive details about the methods of intelligence collection. The MLS requirements are sharpened for multinational sharing of military intelligence, and even more so for providing this information to the civil sector to support missions such as search and rescue and disaster warning and response. Various approaches are used to sanitize, downgrade, or declassify information from sensitive sources, and for many uses this requires very rapid, automated processes.

Another challenge is the need for common map bases and overlay tools to merge the information from various sensors into one coherent product. Combining imagery from **satellites** with battle-space imagery from aerial **reconnaissance** systems shows up-to-the-minute changes, and combining imagery with **SIGINT** helps confirm the nature of a vehicle, installation, or weapons system. This kind of integrated product from multiple types of intelligence increases what military commanders call battle-space "situational awareness," and reduces what is known as the "fog of war." It also helps avoid "friendly-fire" incidents and reduces civilian casualties by verifying and tracking targets. With so many systems using different ground stations for reporting and processing information, communications become a major challenge in ensuring all pieces of the intelligence puzzle get to the same place quickly so that an integrated picture can be built. The U.S. Department of Defense is working to integrate many of its reconnaissance and intelligence systems to use common ground stations, processing techniques, and information displays, and getting this integrated intelligence picture quickly into command and control systems for military and civilian operations.

MILITARY USES OF SPACE

Rear Adm. Rand Fisher

WHY IS SPACE IMPORTANT TO THE MILITARY? WHAT BENEFITS DOES the military receive from space? Space is just a vast expanse of "nothing." But the lure of space is the journey. Space attracts our pioneer spirit, our desire to explore and discover. Space is the path to another planet, a place to expand and grow, to search for new life. Space may be the path to glory, new resources, riches—or power. Human history is clearly about the struggle for power—the power of beliefs, of will, of choice, of wealth, of survival. In these struggles, military warfare has constantly been an engine driving technological change and innovation.

Sun Tzu, one of the earliest recorded military strategists, wrote that careful planning based on accurate information about the enemy would contribute to speedy military decisions. Military history records the importance of the search for information: Where is my enemy? How large is his force? How are they armed? How well are they trained? Are the conditions favorable for victory? Sun Tzu maintained that information superiority was the key to victory.

The history of the military in space begins with the search for advantage—for the "high ground"—for the ability to "see" the enemy and for the effective use of gravity, a great well of potential energy. The earliest uses of high ground provided military leaders with a broad perspective for early warning, information gathering, and improved defenses. As technology introduced new tools to the military, this high ground became the air, and winning military strategies were the result of "air superiority." Today, the high ground has become space, and the closely related realm of "cyberspace." Successful military strategies now are based upon "space superiority" and the information superiority that space enables.

Space and space systems provide the modern military commander several asymmetrical attributes:

- Perspective—the ability to perceive an adversary's forces and movements globally and the ability to achieve access for offensive and defensive weapons and systems.
- Persistence—the ability to be omnipresent, to "dwell" for long periods of time, and observe and report change very quickly after it occurs, in what military planners refer to as "near-real time."
- Penetration—the ability to perceive, to see deeply and passively into "denied" areas without physically trespassing.
- Precision—the ability to perceive precisely—locate, identify, track, and target.
- Speed—the ability to act or react in near-real time, with sensors, systems, and weapons.

The opinions and positions expressed within this essay are the author's and do not necessarily reflect or represent those of the U.S. government or the U.S. Navy.

Given such powerful advantages, it is no surprise that the military continually seeks new ways to exploit space. Today, all military components—air, ground, and naval—are critically dependent upon space systems for mission success. Military applications of space may be found in a broad array of regimes—navigation, communications, intelligence, reconnaissance, surveillance, targeting, and meteorology, as well as "space control" to protect a nation's own assets in space and deny the enemy's use of space—all designed to achieve information and decision superiority. Ultimately, such asymmetrical advantage may actually prevent conflict or at least enable shorter conflicts with far less carnage and destruction.

A typical scenario encountered by a joint military force might resemble the following: It's Wednesday afternoon aboard U.S.S. *Abraham Lincoln* in the Persian Gulf. Surveillance satellites have detected increased communications and movement of equipment at odd hours in Afghanistan and Iraq. Following standard operating procedures, the *Lincoln*, along with other coalition forces ashore and afloat, responds by setting increased alert conditions and requests the latest intelligence from all sources. These requests reach a network of "connected" intelligence centers and sensors worldwide via an array of communications satellites that provide "reachback" to CONUS (Continental United States)—the same connectivity that provides nearly instantaneous logistics, training, maintenance, and medical support.

Simultaneously, potential targets are being identified, imaged, and "mensurated" (measured geometrically), so that, if need be, precision weapons guided by GPS (Global Positioning System) satellites can be accurately aimed. The *Lincoln* chief oceanographer is gathering satellite imagery and other observations so weather conditions in the target area can be accurately predicted. Meanwhile, the Joint Force Commander integrates and synchronizes Service components under his command to support theatre operations. Within the battle forces ashore and afloat, chat rooms start up so that intense information sharing and collaboration can take place in near-real time. The goal is to create enough asymmetrical advantage to avoid conflict, but if conflict is imminent, to focus on reducing the timelines associated with the "kill chain"—to find, fix (locate), track, target, engage, and assess. Space systems make essential contributions to each of these links in the chain, enabling military engagements to be swift and decisive.

This scenario stands in stark contrast to military operations prior to Desert Shield/Desert Storm in the Persian Gulf area in 1991. The military's growing space and cyberspace capabilities portend a much different future, one that offers an array of opportunities, but also one that will require much thought—the development of new doctrine, tactics, training, and procedures. An analogy of this kind of change can be seen in the evolution of Special Forces, small, independently deployed groups inserted covertly into enemy territory to carry out reconnaissance and other missions.

Once considered supporting elements in operational plans, Special Forces now play a central role in planning and carrying out these plans. Special Forces, in their ability to penetrate, perceive, and respond swiftly, have much in common with space systems, and are dependent on them. Future military actions, from preparation for conflict through operations, will see even greater reliance on space systems. In many respects space superiority is a "force multiplier," increasing the combat effectiveness of every platform, every weapon, and every sensor, and providing a strategic edge in terms of better information and faster decisions. In the future we will see increasing attention given both to the development and exploitation of space systems—and to the defense of such systems—in the planning and execution of military operations. ■

The aircraft carrier U.S.S. *Abraham Lincoln*, with its crew lining the deck and its superstructure bristling with antennas, returns to its home port in Everett, Washington, on May 6, 2003, after a ten-month deployment to Iraq.

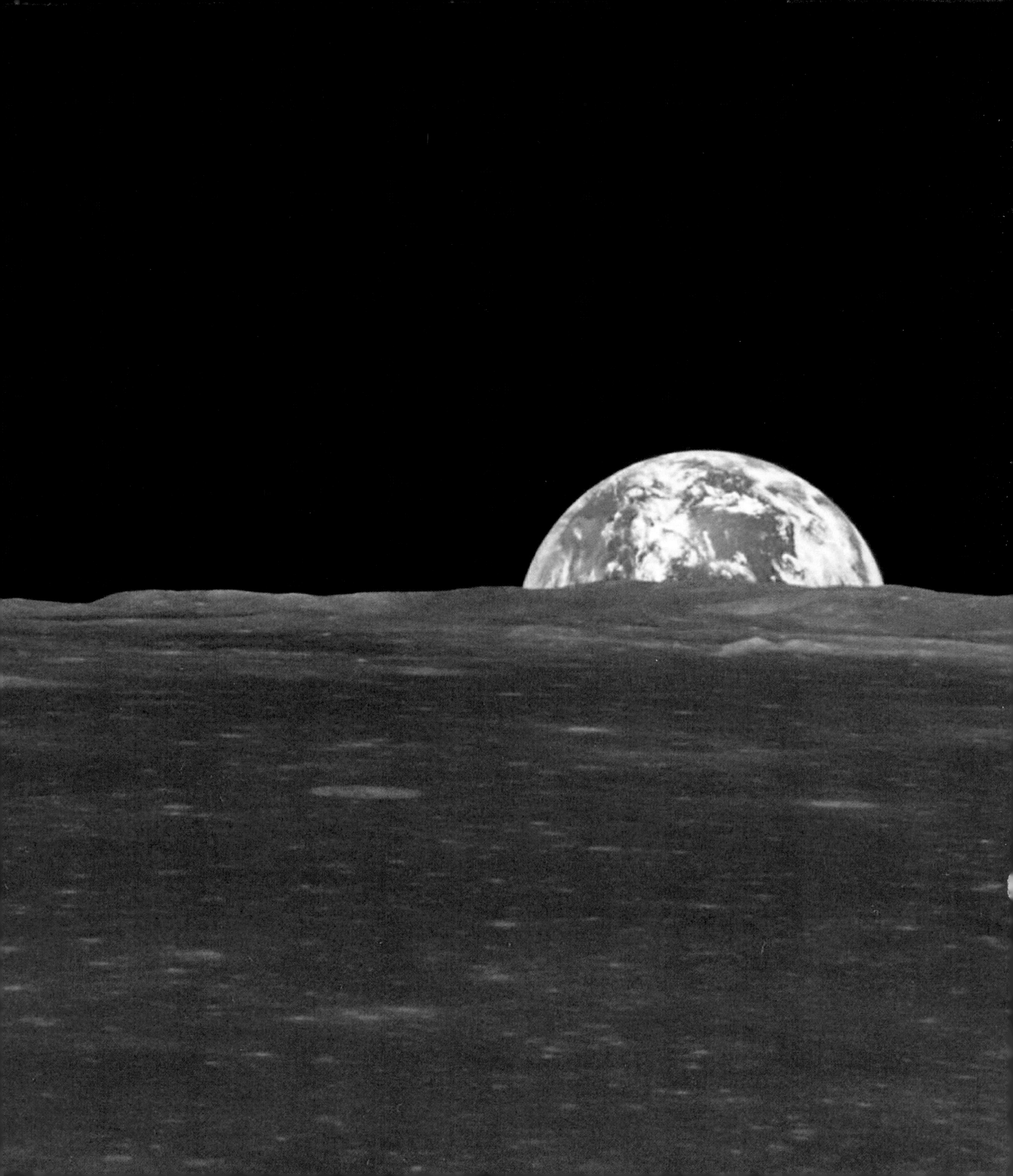

Appendix of Maps

Earth rises out of the vacuum of space over the horizon of the Mare Smythii region of the moon on July 20, 1969. Crewmembers of the Apollo 11 mission, aboard the command module *Columbia,* were the first to view such a sight.

The Northern Sky

Twinkling on a clear, moonless night, as many as 3,000 stars can be seen from a dark location with the naked eye. Each is a sun, perhaps the center of its own solar system.

Early observers of the night sky imagined figures such as Perseus, the Hero; or Andromeda, the Chained Lady. These starry patterns are constellations, their member stars often located at very different points in space, but to the eye all stars seem at the same distances. The constellations have no scientific significance; rather, they recall ancient legends. The International Astronomical Union recognizes 88 official constellations.

In the northern sky, more stars per given area occur along a band running through Orion, Perseus, Cassiopeia, and Cygnus; these constellations are located along the equator of the Milky Way. There, a smooth band of dim light is readily seen on a dark evening.

Sweeping along the Milky Way with a pair of binoculars, the casual observer will see innumerable faint stars, many clumps of stars, or star clusters, and luminous fuzzy patches, called gaseous nebulae.

All the stars shown here and on the map of the southern sky (following pages) are termed "fixed stars" to distinguish them from the planets that move noticeably through the constellations. Stars travel across the sky at visually imperceptible rates. Like continental drift that slowly changes the face of the Earth, stellar motions will change the shapes of the constellations over thousands of centuries.

Along the rim of this map of the northern sky, Roman numerals indicate right ascension, a measure similar to longitude. Blue circles are parallels of declination, or celestial latitude, and dashed yellow lines are constellation boundaries. Months tell when constellations are best seen. Greek letters, other symbols, or common names are given for many stars. M or N followed by a number denotes a star cluster, nebula, or galaxy listed in the Messier or New General Catalog.

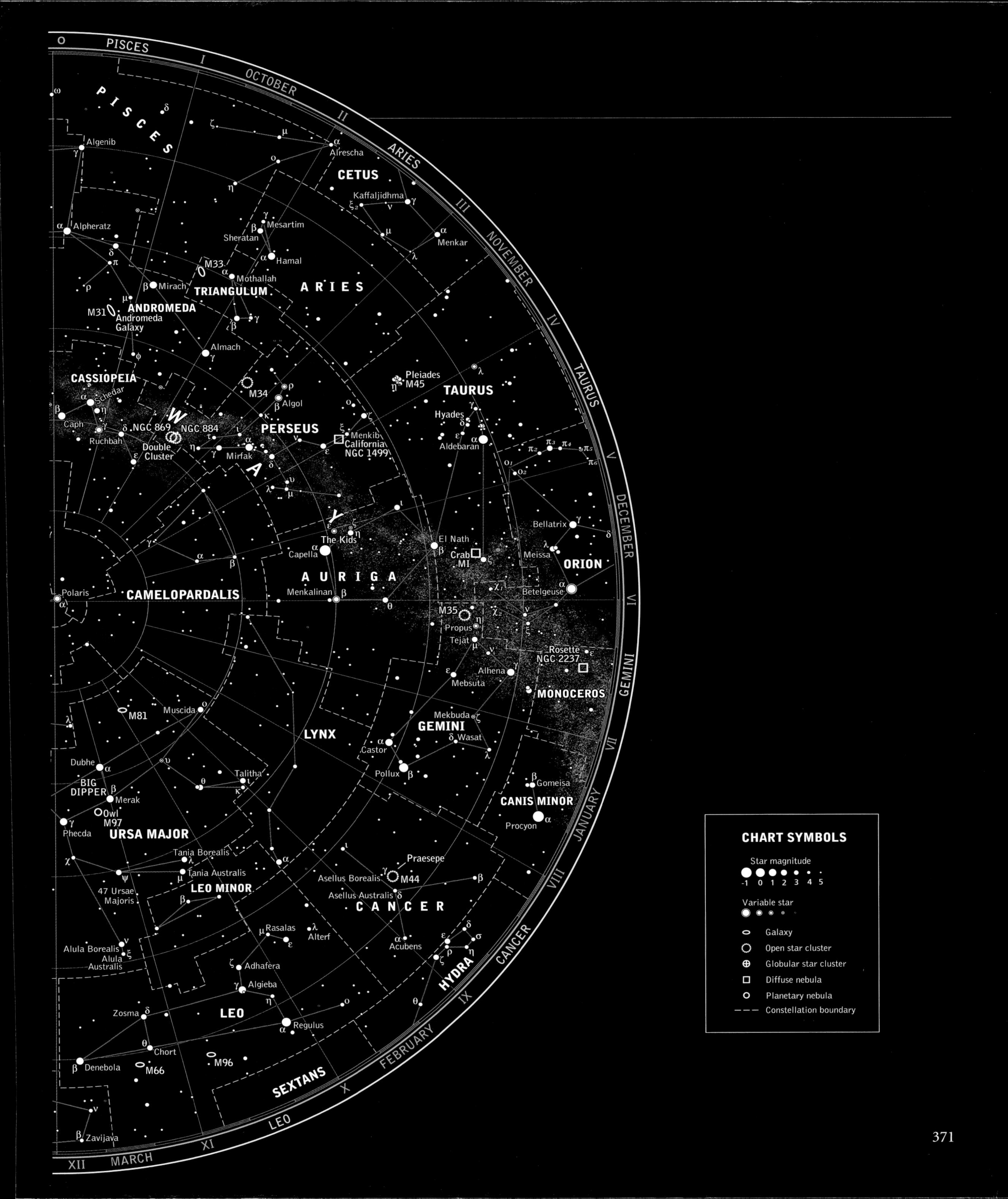
PISCES
OCTOBER
ARIES
NOVEMBER
TAURUS
DECEMBER
GEMINI
JANUARY
CANCER
FEBRUARY
LEO
MARCH
PISCES
Algenib
Alrescha
CETUS
Kaffaljidhma
Menkar
Alpheratz
Mesartim
Sheratan
Hamal
M33
Mothallah
Mirach
TRIANGULUM
ARIES
ANDROMEDA
M31
Andromeda Galaxy
Almach
CASSIOPEIA
Schedar
Caph
Ruchbah
NGC 869
NGC 884
Double Cluster
M34
Algol
PERSEUS
Mirfak
Menkib
California NGC 1499
Pleiades M45
TAURUS
Hyades
Aldebaran
MILKY WAY
The Kids
Capella
El Nath
Crab M1
Bellatrix
Meissa
ORION
AURIGA
Menkalinan
Betelgeuse
Polaris
CAMELOPARDALIS
M35
Propus
Tejat
Rosette NGC 2237
Alhena
Mebsuta
MONOCEROS
M81
Muscida
Mekbuda
GEMINI
Wasat
LYNX
Castor
Dubhe
Talitha
Pollux
BIG DIPPER
Merak
Owl M97
Gomeisa
CANIS MINOR
Phecda
URSA MAJOR
Procyon
Tania Borealis
Praesepe
Tania Australis
Asellus Borealis
M44
47 Ursae Majoris
LEO MINOR
Asellus Australis
CANCER
Rasalas
Alterf
Alula Borealis
Alula Australis
Acubens
Adhafera
HYDRA
Algieba
Zosma
LEO
Regulus
Chort
Denebola
M66
M96
SEXTANS
Zavijava
CHART SYMBOLS
Star magnitude
-1 0 1 2 3 4 5
Variable star
Galaxy
Open star cluster
Globular star cluster
Diffuse nebula
Planetary nebula
Constellation boundary

The Southern Sky

Sights of the southern sky include the Milky Way near its center in Sagittarius. Two of the most conspicuous globular clusters, round clumps of hundreds of thousands of stars, all packed within a few dozen light-years, are Omega Centauri and 47 Tucanae. Crux, the Southern Cross, holds the Jewel Box, visible to the naked eye, with dozens of stars easily seen with binoculars. The Tarantula Nebula, in the Large Magellanic Cloud, is much larger than the Orion Nebula but more than a hundred times farther away.

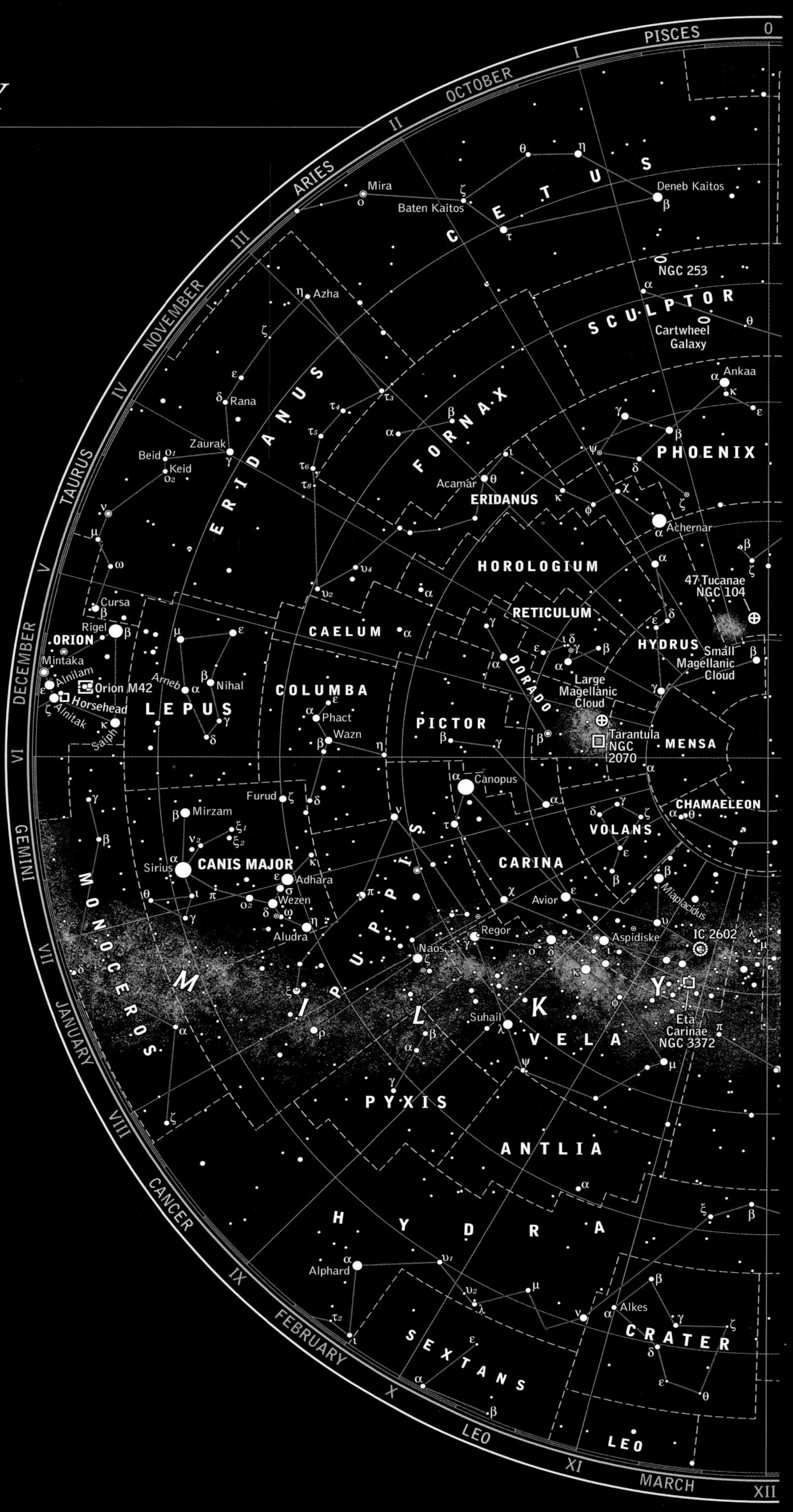

SEPTEMBER
AUGUST
JULY
JUNE
MAY
APRIL
XXIII
XXII
XXI
XX
XIX
XVIII
XVII
XVI
XV
XIV
XIII
XII
AQUARIUS
CAPRICORNUS
SAGITTARIUS
SCORPIUS
LIBRA
VIRGO
PISCES
AQUARIUS
Sadachbia
Sadalmelik
Ancha
Skat
Helix
NGC 7293
Deneb Algedi
Nashira
Sadalsuud
Saturn
NGC 7009
Albali
Fomalhaut
PISCIS AUSTRINUS
GRUS
CAPRICORNUS
Dabih
Algedi
MICROSCOPIUM
Al Na'ir
AQUILA
INDUS
TUCANA
SAGITTARIUS
Peacock
TELESCOPIUM
Rukbat
Arkab
Ascella
Nunki
SCUTUM
CORONA AUSTRALIS
M22
Kaus Borealis
PAVO
Kaus Australis
Omega
M17
Eagle
M16
SERPENS
OCTANS
Kaus Media
Alnasl
Trifid
M20
Lagoon
M8
M7
Shaula
Lesath
APUS
Atria
ARA
TRIANGULUM AUSTRALE
SCORPIUS
Sabik
MUSCA
CIRCINUS
NORMA
OPHIUCHUS
Antares
Acrux
Hadar
Rigil Kentaurus
(Alpha Centauri)
CRUX
Mimosa
Dschubba
Graffias
Gacrux
Omega Centauri
NGC 5139
LUPUS
Yed Posterior
Yed Prior
CENTAURUS
NGC 5128
Menkent
SERPENS
LIBRA
Zubenelgenubi
Zubeneschamali
M83
HYDRA
Alchiba
CORVUS
Algorab
Gienah
Syrma
Spica
VIRGO
M104
Sombrero
Galaxy
Porrima
Zaniah
CHART SYMBOLS
Star magnitude
-1 0 1 2 3 4 5
Variable star
Galaxy
Open star cluster
Globular star cluster
Diffuse nebula
Planetary nebula
Constellation boundary

THE MOON

NEAR SIDE

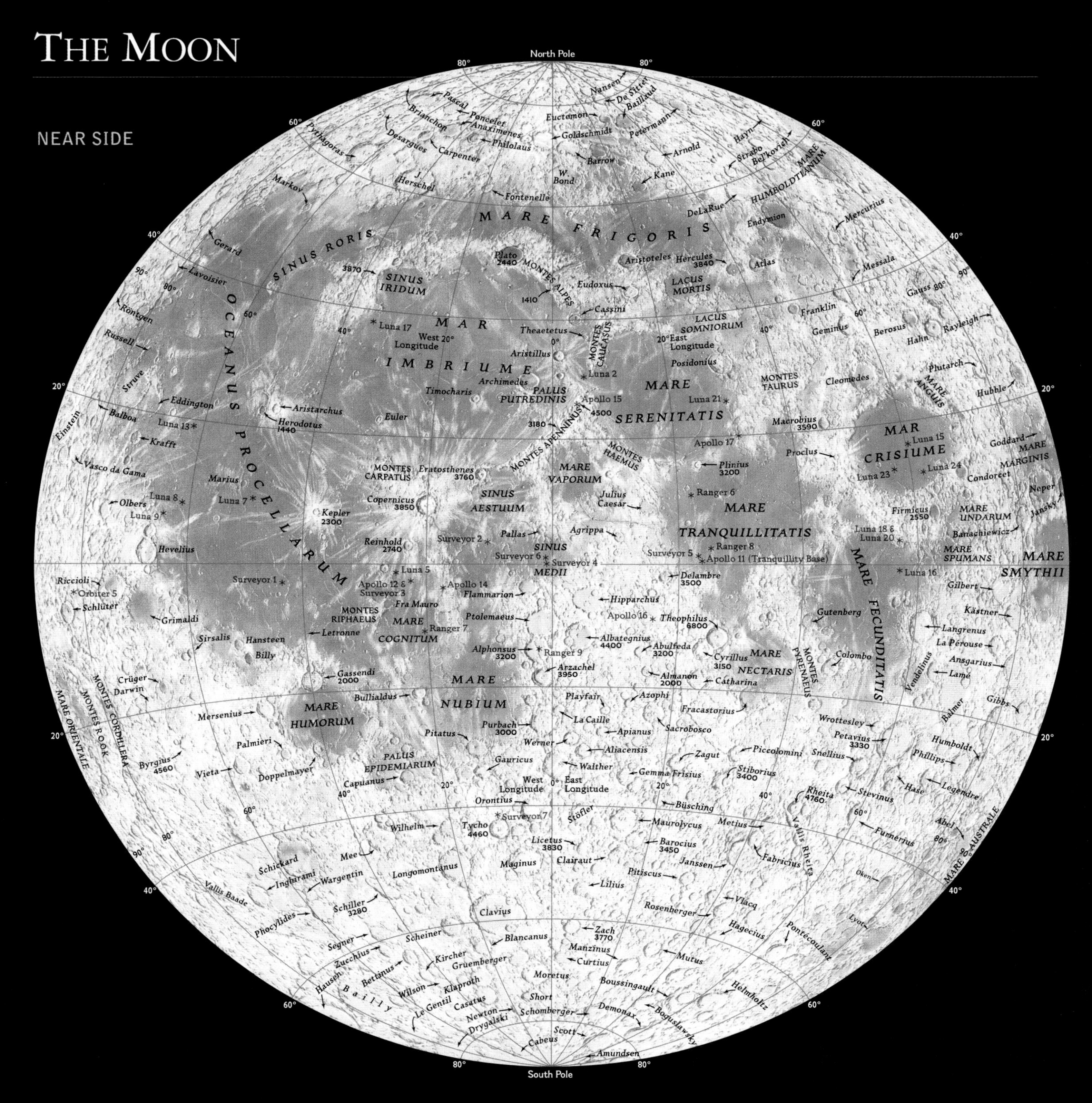

As it moves in its stately dance through space, the moon keeps the same face to the Earth at all times. The hemisphere that always faces Earth is marked by dark plains that hold not a drop of water, despite such fanciful names as Mare Tranquillitatis (Sea of Tranquility) and Mare Nectaris (Sea of Nectar). The other hemisphere, or far side, has just a part of one such plain, the Mare Orientale (Eastern Sea), but it is even more densely pockmarked with craters.

The moon's surface contains a wide variety of impact features, scars left by objects that struck the moon long ago. The largest are the impact basins, enormous craters ranging up to about 2,500 kilometers across. Lava flooded the basin floors sometime after the collisions that formed them, creating the smooth, dark surfaces that the eye discerns as maria, or "seas." In the bright lunar highlands, craters are packed closely together, indicating that asteroids

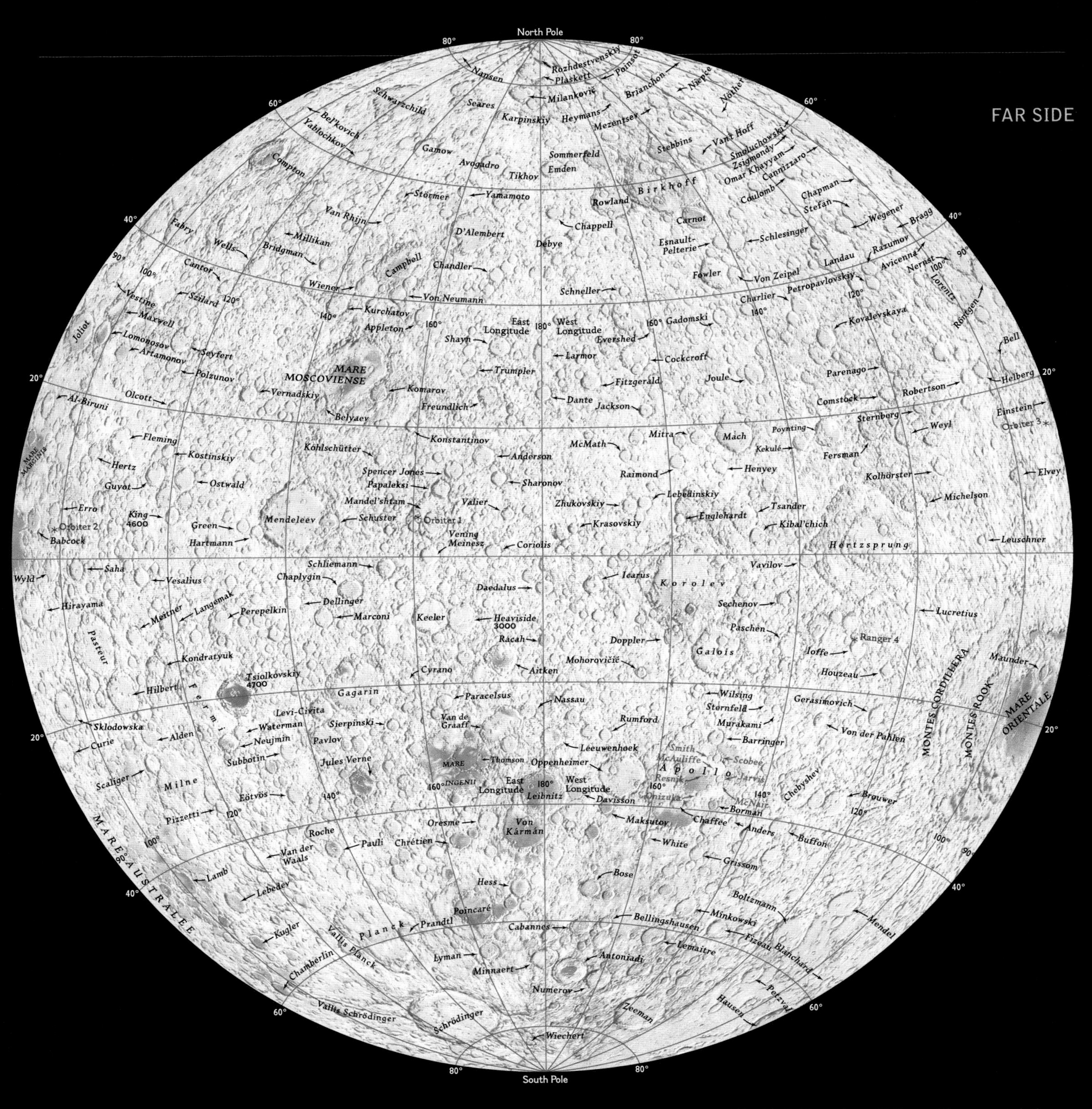

and comets repeatedly hit these ancient areas in the early, violent days of the solar system. To the unaided eye, the combined shapes of the bright lunar highlands and the dark maria make up the familiar "man in the moon."

Other craters also dot the moon's surface, centers of radial patterns of bright ejecta, material thrown from the impacts that made them. Blocks of rock hurled from impacts traveled farther than they would on Earth because gravity is weaker on the moon. Wrinkled ridges, domed hills, and fissures mark the maria, all of them familiar aspects of ancient volcanic landscapes.

The moon has no mountains like the Himalaya, produced by tectonic plates smashing into one another; there is no continental drift on the satellite, which is now geologically quiet. Lunar mountains—some rising above 6,000 meters—consist of old volcanic domes and the central peaks and rims of impact craters.

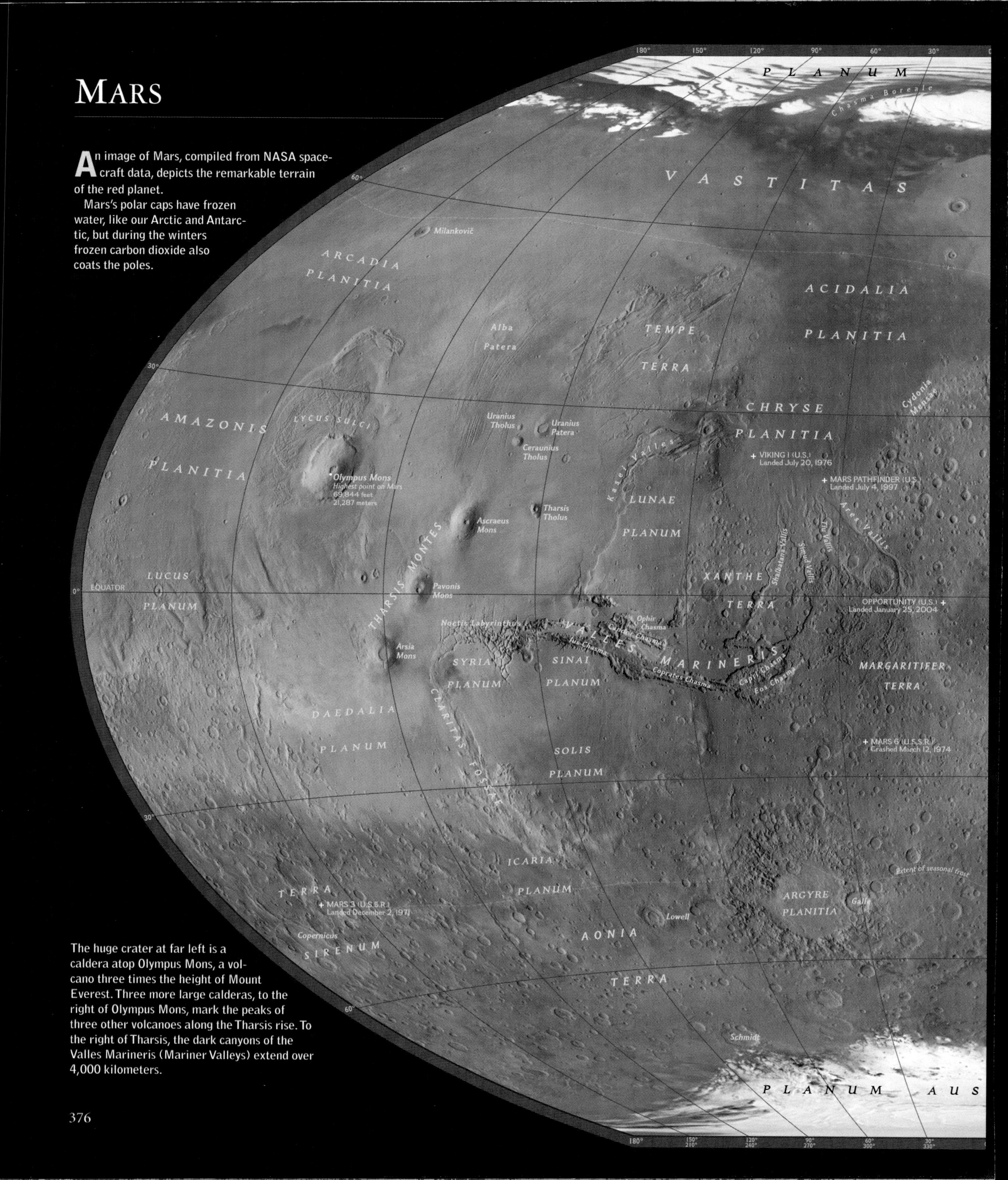

MARS

An image of Mars, compiled from NASA spacecraft data, depicts the remarkable terrain of the red planet.

Mars's polar caps have frozen water, like our Arctic and Antarctic, but during the winters frozen carbon dioxide also coats the poles.

The huge crater at far left is a caldera atop Olympus Mons, a volcano three times the height of Mount Everest. Three more large calderas, to the right of Olympus Mons, mark the peaks of three other volcanoes along the Tharsis rise. To the right of Tharsis, the dark canyons of the Valles Marineris (Mariner Valleys) extend over 4,000 kilometers.

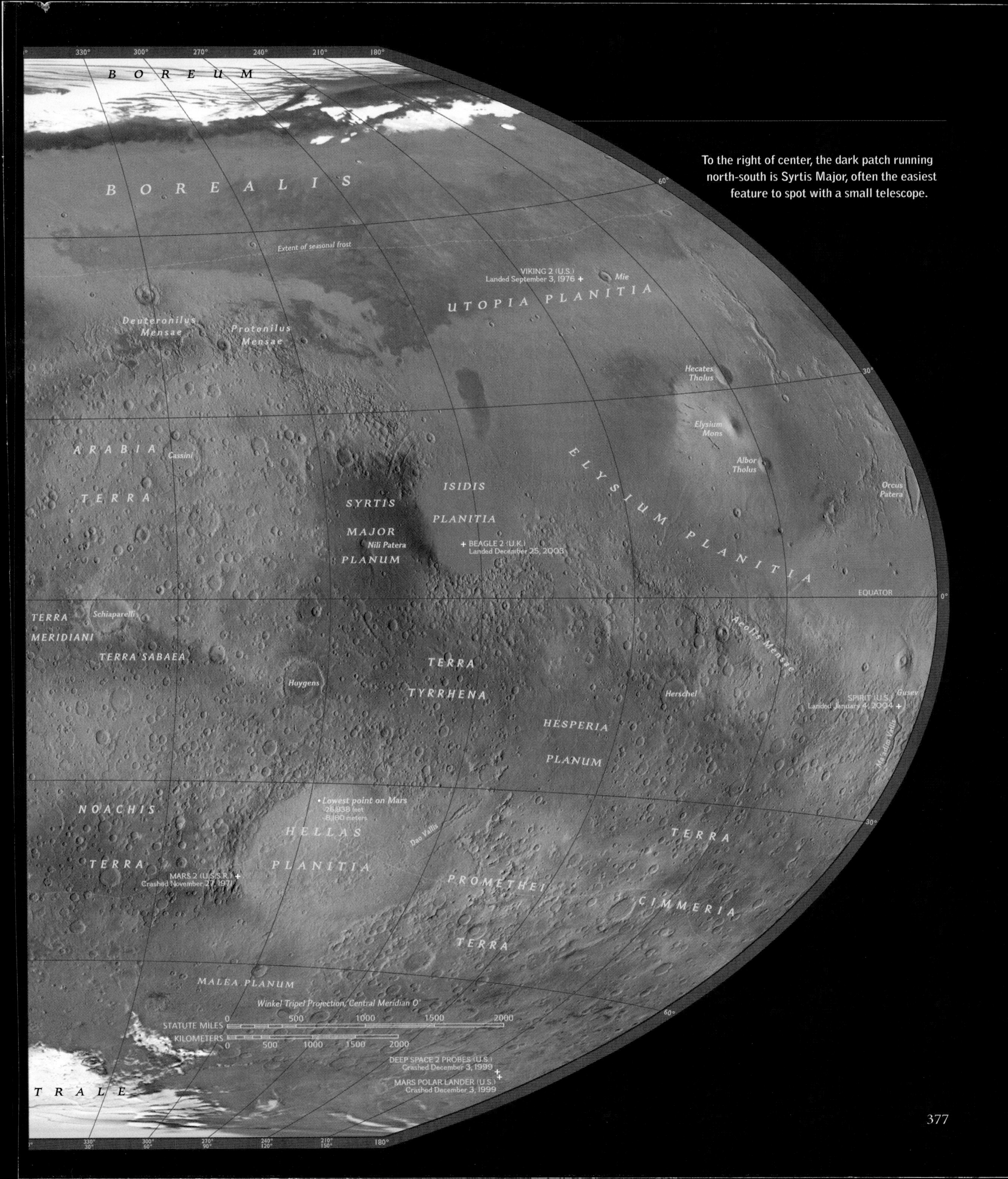

To the right of center, the dark patch running north-south is Syrtis Major, often the easiest feature to spot with a small telescope.

THE SOLAR SYSTEM

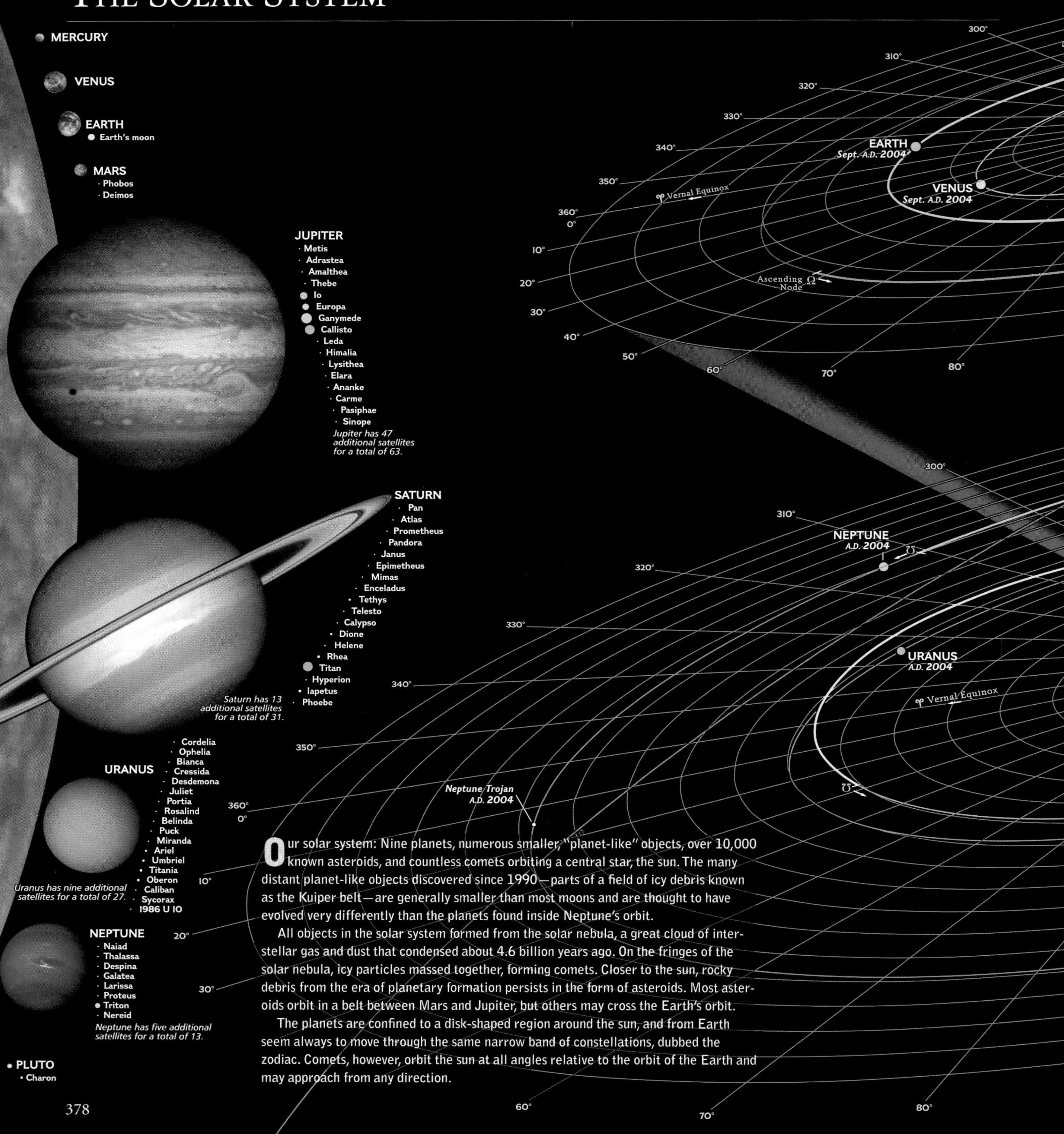

Our solar system: Nine planets, numerous smaller, "planet-like" objects, over 10,000 known asteroids, and countless comets orbiting a central star, the sun. The many distant planet-like objects discovered since 1990—parts of a field of icy debris known as the Kuiper belt—are generally smaller than most moons and are thought to have evolved very differently than the planets found inside Neptune's orbit.

All objects in the solar system formed from the solar nebula, a great cloud of interstellar gas and dust that condensed about 4.6 billion years ago. On the fringes of the solar nebula, icy particles massed together, forming comets. Closer to the sun, rocky debris from the era of planetary formation persists in the form of asteroids. Most asteroids orbit in a belt between Mars and Jupiter, but others may cross the Earth's orbit.

The planets are confined to a disk-shaped region around the sun, and from Earth seem always to move through the same narrow band of constellations, dubbed the zodiac. Comets, however, orbit the sun at all angles relative to the orbit of the Earth and may approach from any direction.

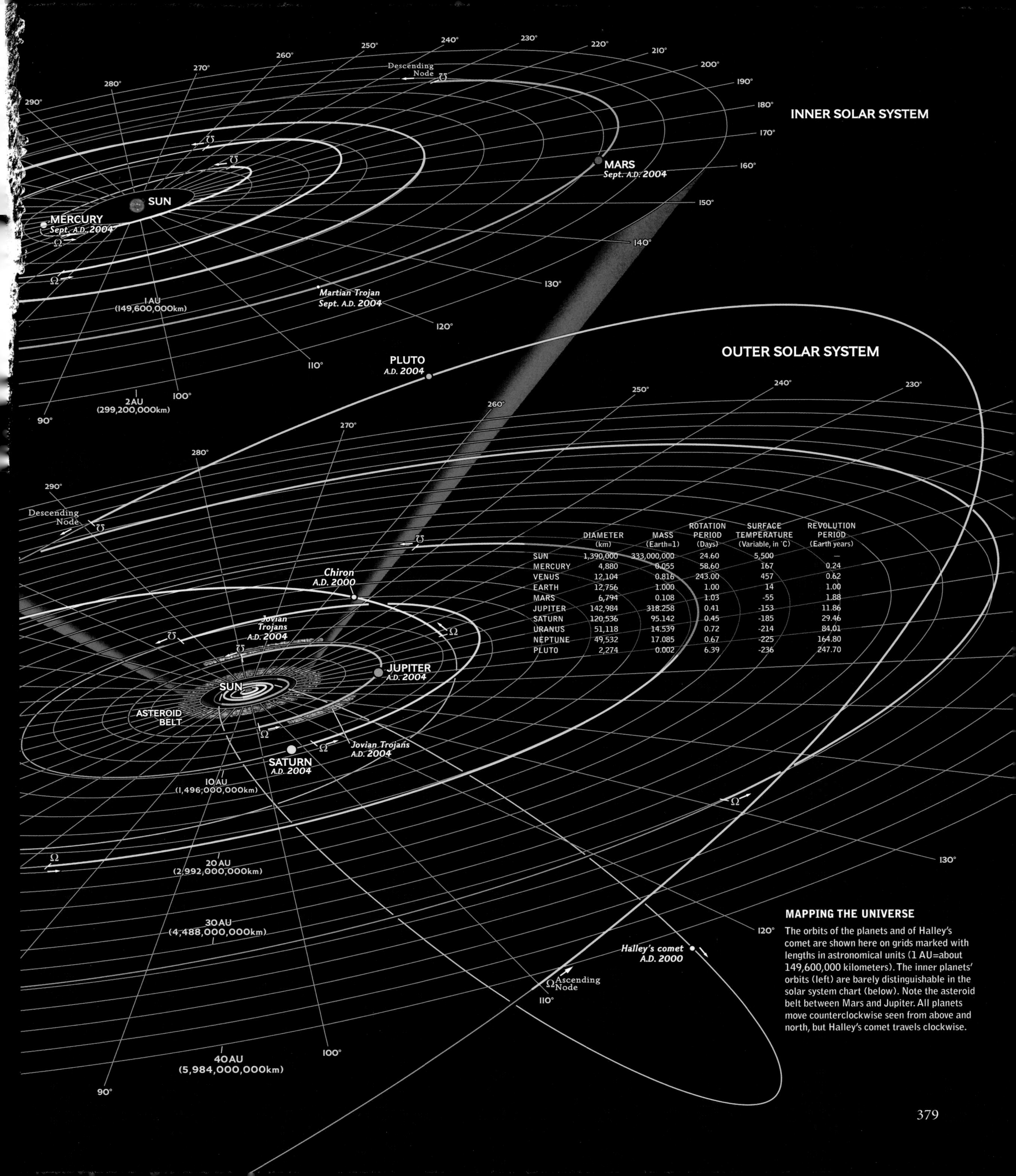

	DIAMETER (km)	MASS (Earth=1)	ROTATION PERIOD (Days)	SURFACE TEMPERATURE (Variable, in °C)	REVOLUTION PERIOD (Earth years)
SUN	1,390,000	333,000,000	24.60	5,500	—
MERCURY	4,880	0.055	58.60	167	0.24
VENUS	12,104	0.816	243.00	457	0.62
EARTH	12,756	1.000	1.00	14	1.00
MARS	6,794	0.108	1.03	-55	1.88
JUPITER	142,984	318.258	0.41	-153	11.86
SATURN	120,536	95.142	0.45	-185	29.46
URANUS	51,118	14.539	0.72	-214	84.01
NEPTUNE	49,532	17.085	0.67	-225	164.80
PLUTO	2,274	0.002	6.39	-236	247.70

MAPPING THE UNIVERSE

The orbits of the planets and of Halley's comet are shown here on grids marked with lengths in astronomical units (1 AU=about 149,600,000 kilometers). The inner planets' orbits (left) are barely distinguishable in the solar system chart (below). Note the asteroid belt between Mars and Jupiter. All planets move counterclockwise seen from above and north, but Halley's comet travels clockwise.

The Milky Way

Home galaxy of Earth, the Milky Way is a system of several hundred billion stars and thousands of star clusters and nebulae. Seen from our position in the flat galactic disk, it is a fuzzy, glowing band, interrupted by dark clouds of interstellar dust. In 1609 Galileo's first telescopic peek at the fuzzy band revealed countless stars, each too dim to be separately distinguishable with the naked eye. Today astronomers classify the Milky Way as a spiral galaxy about 100,000 light-years wide. Many of its stars have planets, perhaps even some like Earth.

Regions of bright young stars and nebulae, such as the Lagoon Nebula in Sagittarius, outline the Milky Way's spiral arms like traffic jams on freeways looping around a city. Many older stars gradually expel their outer layers, which form beautiful planetary nebulae, like M2-9 in Ophiuchus. The small star left at the center of a planetary nebula slowly fades away as the nebula thins and disappears. Toward the galactic center, a thick swarm of orange and red stars marks the galactic bulge. A galactic halo or corona of old stars and globular star clusters extends far above and below the disk. All objects in the Milky Way orbit around the galactic center just as the planets of the solar system revolve around the sun. But the central object of the galaxy is no star—it is about four million times more massive than the sun and is believed to be a gigantic black hole.

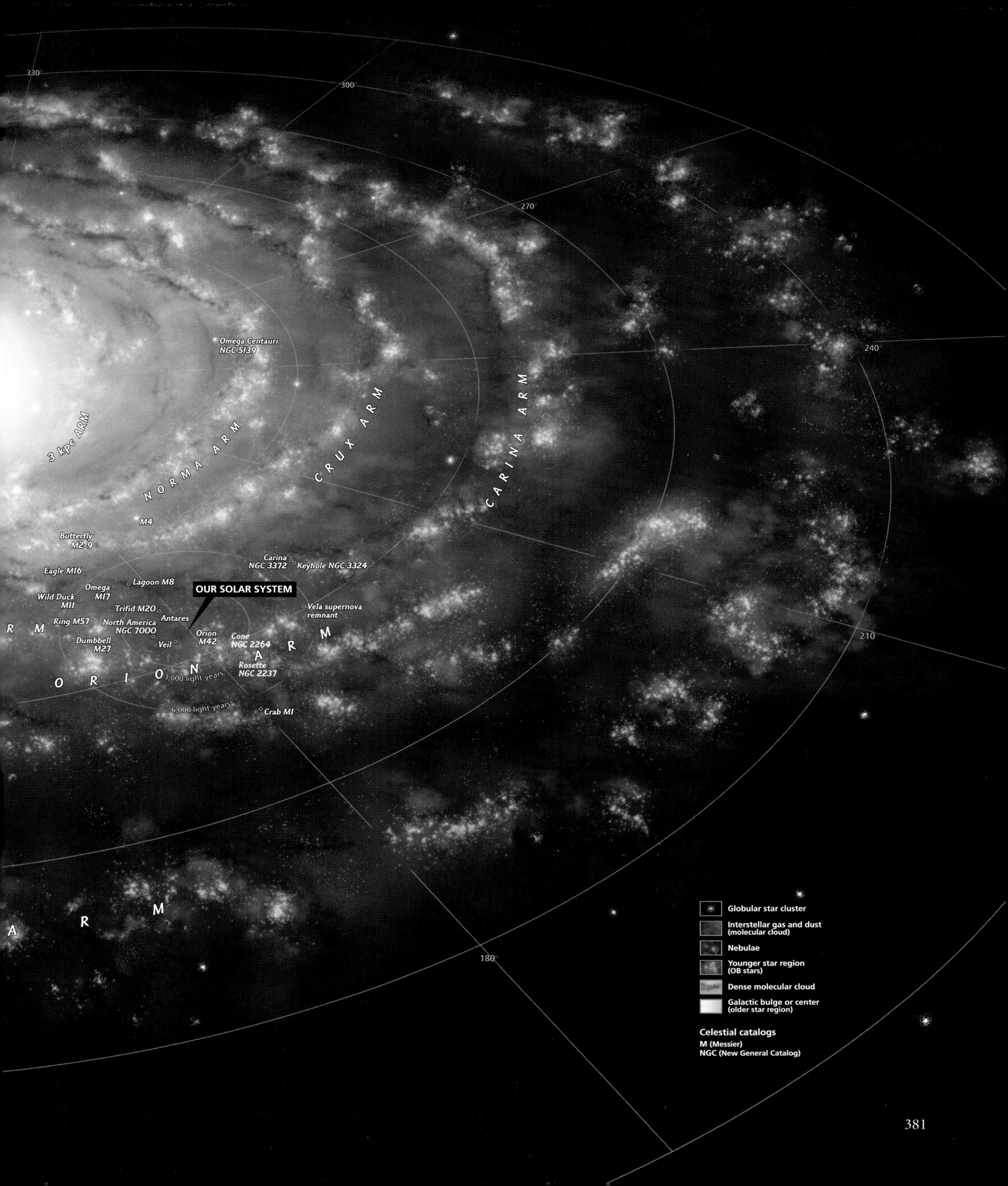

330°
300°
270°
240°
210°
180°
Omega Centauri NGC 5139
3 kpc ARM
NORMA ARM
CRUX ARM
CARINA ARM
M4
Butterfly M2-9
Carina NGC 3372
Keyhole NGC 3324
Eagle M16
Lagoon M8
OUR SOLAR SYSTEM
Omega M17
Wild Duck M11
Trifid M20
Vela supernova remnant
Ring M57
North America NGC 7000
Antares
Dumbbell M27
Veil
Orion M42
Cone NGC 2264
Rosette NGC 2237
ORION ARM
3,000 light-years
6,000 light-years
Crab M1
ARM
Globular star cluster
Interstellar gas and dust (molecular cloud)
Nebulae
Younger star region (OB stars)
Dense molecular cloud
Galactic bulge or center (older star region)
Celestial catalogs
M (Messier)
NGC (New General Catalog)

THE UNIVERSE

More than 125 billion galaxies are arrayed throughout the universe to the limits of present observation. Many are arrayed in small groups or in larger clusters, themselves arranged in long filamentary superclusters. The superclusters, in turn, are generally positioned along the peripheries of huge volumes of space, the galactic voids. Gravity holds together the galaxies of a cluster, but clusters, groups, and isolated individual galaxies are all flying away from each other in an expansion of the universe that began with the big bang about 13.7 billion years ago.

Research suggests that the most distant galaxies, seen as they were billions of years ago, were systematically smaller than galaxies today. It also appears that when galaxies were closer together, they collided, often merging, more frequently than they do now. These findings are consistent with the theory of "hierarchical formation," according to which galaxies assembled from smaller units, rather than condensing in full size from parental clouds.

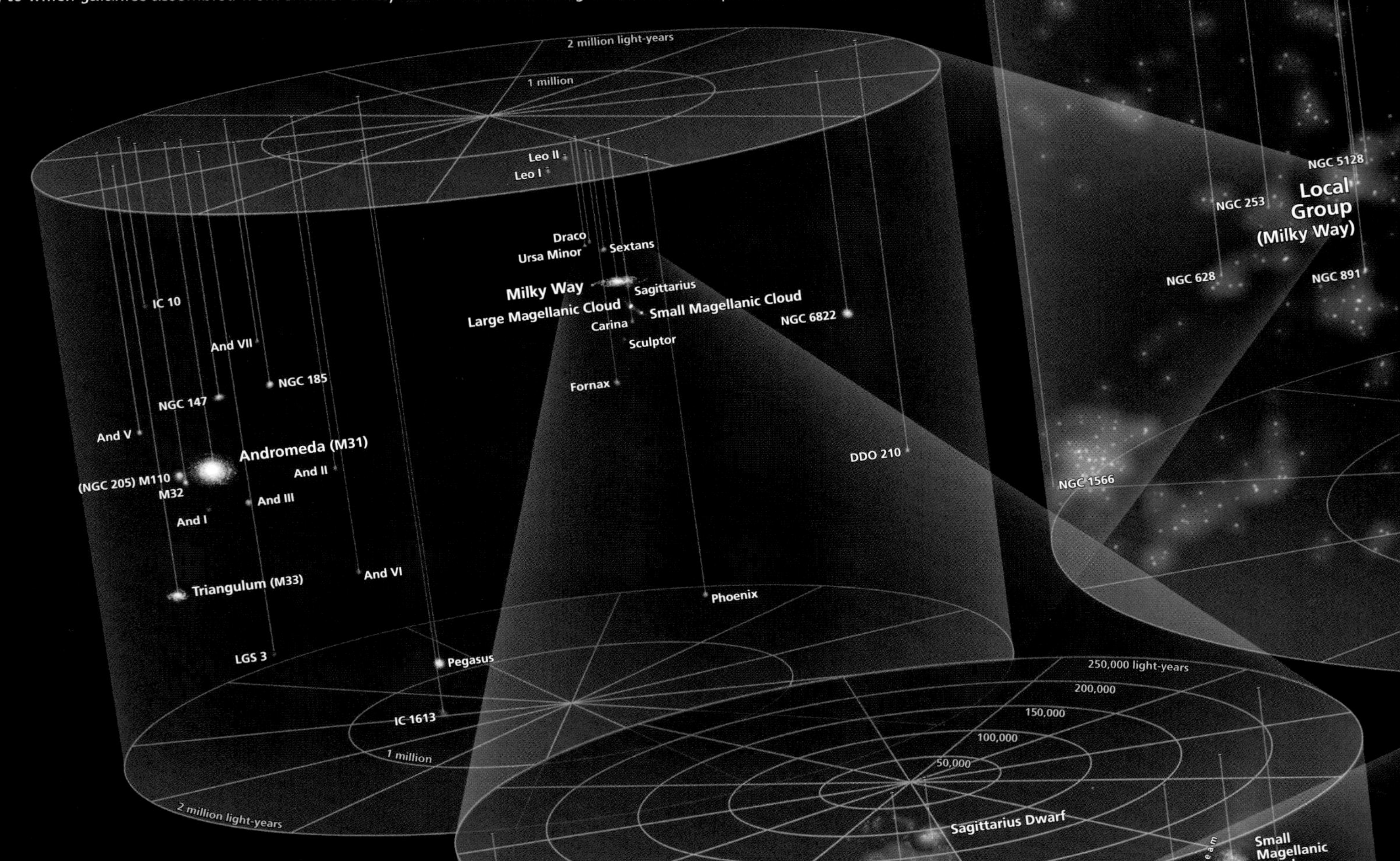

GALAXY COMPANIONS

The Local Group of Galaxies extends over three million light-years from our galaxy, the Milky Way, and includes two other large spiral galaxies, the Andromeda and Triangulum galaxies (M31 and M33), as well as the Large and Small Magellanic Clouds. M31 has a subgroup of its own, which includes two small elliptical galaxies, M32 and NGC 205, in which star formation occurs at a low level. The Andromeda Galaxy can be seen readily with the naked eye or binoculars as a fuzzy dim spot in the night sky, despite its great distance of more than two million light-years from Earth. Its companions, M32 and NGC 205, are easily glimpsed through small telescopes, but most other Local Group members are very faint.

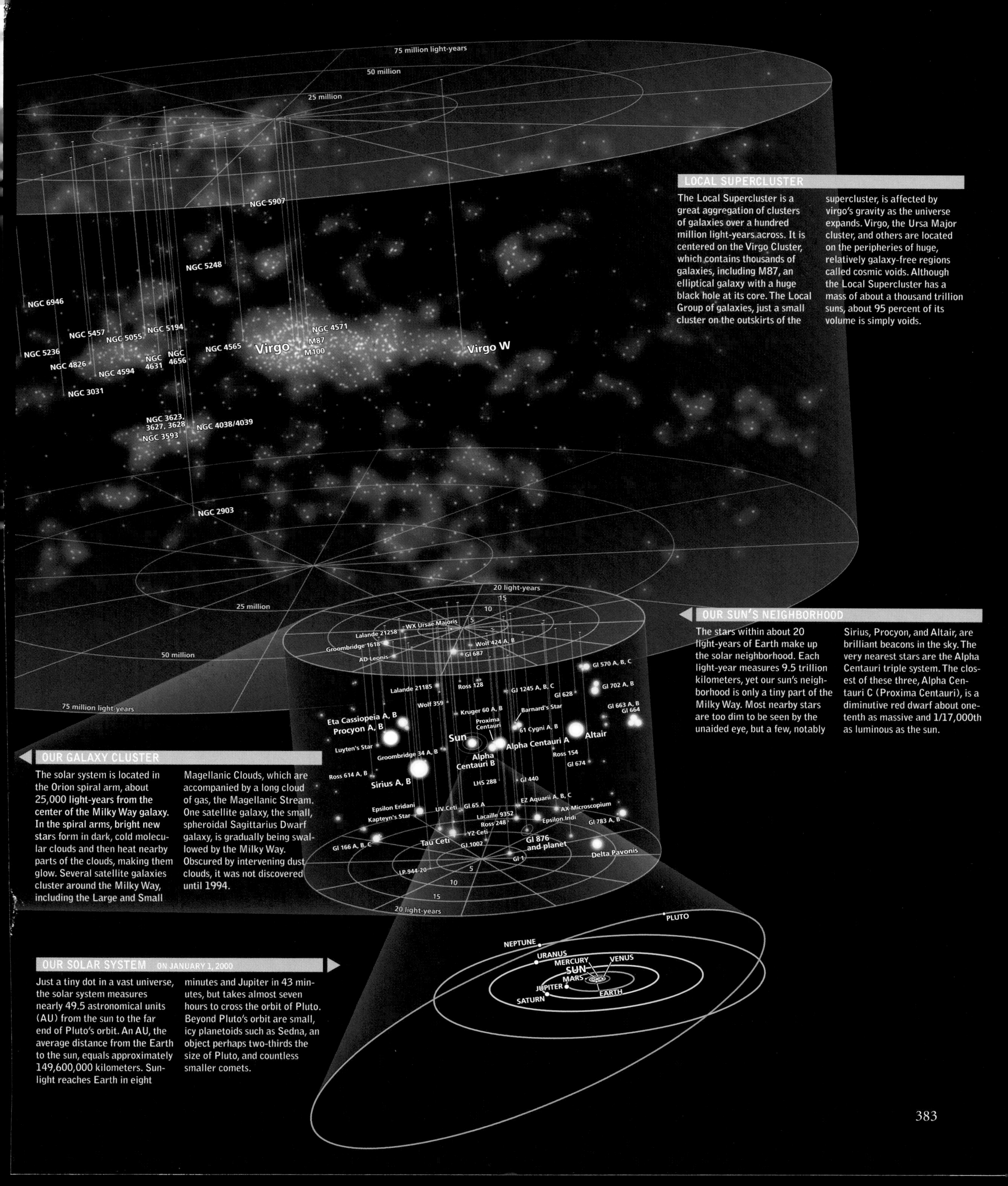

LOCAL SUPERCLUSTER

The Local Supercluster is a great aggregation of clusters of galaxies over a hundred million light-years across. It is centered on the Virgo Cluster, which contains thousands of galaxies, including M87, an elliptical galaxy with a huge black hole at its core. The Local Group of galaxies, just a small cluster on the outskirts of the supercluster, is affected by virgo's gravity as the universe expands. Virgo, the Ursa Major cluster, and others are located on the peripheries of huge, relatively galaxy-free regions called cosmic voids. Although the Local Supercluster has a mass of about a thousand trillion suns, about 95 percent of its volume is simply voids.

OUR SUN'S NEIGHBORHOOD

The stars within about 20 light-years of Earth make up the solar neighborhood. Each light-year measures 9.5 trillion kilometers, yet our sun's neighborhood is only a tiny part of the Milky Way. Most nearby stars are too dim to be seen by the unaided eye, but a few, notably Sirius, Procyon, and Altair, are brilliant beacons in the sky. The very nearest stars are the Alpha Centauri triple system. The closest of these three, Alpha Centauri C (Proxima Centauri), is a diminutive red dwarf about one-tenth as massive and 1/17,000th as luminous as the sun.

OUR GALAXY CLUSTER

The solar system is located in the Orion spiral arm, about 25,000 light-years from the center of the Milky Way galaxy. In the spiral arms, bright new stars form in dark, cold molecular clouds and then heat nearby parts of the clouds, making them glow. Several satellite galaxies cluster around the Milky Way, including the Large and Small Magellanic Clouds, which are accompanied by a long cloud of gas, the Magellanic Stream. One satellite galaxy, the small, spheroidal Sagittarius Dwarf galaxy, is gradually being swallowed by the Milky Way. Obscured by intervening dust clouds, it was not discovered until 1994.

OUR SOLAR SYSTEM ON JANUARY 1, 2000

Just a tiny dot in a vast universe, the solar system measures nearly 49.5 astronomical units (AU) from the sun to the far end of Pluto's orbit. An AU, the average distance from the Earth to the sun, equals approximately 149,600,000 kilometers. Sunlight reaches Earth in eight minutes and Jupiter in 43 minutes, but takes almost seven hours to cross the orbit of Pluto. Beyond Pluto's orbit are small, icy planetoids such as Sedna, an object perhaps two-thirds the size of Pluto, and countless smaller comets.

STANDARD INTERNATIONAL UNITS

Quantity Measured	SI Unit (and Symbol)	Additional Units (and Symbols)
Distance	meter (m)	centimeter (cm); nanometer (nm)
Mass	kilogram (kg)	gram (g)
Time	second (s)	hour (h); minute (min)
Temperature	kelvin (K)	°Celsius (°C)
Area	meter squared (m^2)	centimeter squared (cm^2)
Volume	meter cubed (m^3)	centimeter cubed (cm^3)
Force	newton (N)	nanonewton (nN)
Pressure	pascal (Pa)	kilopascal (kPa); gigapascal (gPa)
Speed	meter per second (m/s)	kilometers per second (km/s); kilometers per hour (km/h)
Acceleration	meter per second squared (m/s^2)	centimeters per second squared (cm/s^2); gravity (g)
Frequency	Hertz (Hz)	kilohertz (kHz); megahertz (MHz); gigahertz (GHz)
Wavelength	meter (m)	centimeter (cm); millimeter (mm); micrometer (mm); nanometer (nm)
Bandwidth (signal)	hertz (Hz)	kilohertz (kHz); megahertz (MHz)
Bandwidth (data transfer rate)	bit per second (bps)	kilobit per second (kbps); megabit per second (mbps)
Angular distance	arc second (")	arc minute (')

USEFUL MEASUREMENTS

	Metric Equivalent	U.S. Equivalent
1 centimeter (cm)	0.01 m	0.3937 inches
1 meter (m)	100 cm	3.2808 feet
1 kilometer (km)	1,000 m	0.6214 miles
1 square centimeter (cm^2)	0.0001 m^2	0.155 square inches
1 square meter (m^2)	10,000 cm^2	10.76 square feet
1 gram (g)	0.001 kg	0.035 ounces
1 kilogram (kg)	1,000 g	2.2046 pounds
1 tonne (t)	1,000 kg	1.1023 short tons (2,205 pounds)
1 newton (N)	0.102 kg	0.2248 pounds
0° Celsius (C)		32° Fahrenheit
1 kelvin (K)	-272.1° C	-457.9° Fahrenheit
1 pascal (Pa)	0.102 kgf/m^2	0.02089 pounds per square foot
1 kilopascal (kPa)	102 kgf/m^2	20.89 pounds per square foot
1 astronomical unit (AU)	149.6 million km	93 million miles
1 parsec (pc)	3.0856×10^{13} km	1.92×10^{13} miles
1 light-year (ly)	9.46 trillion km	5.88 trillion miles

PHYSICAL CONSTANTS

Constant	Value
speed of light	299,792.5 km/s (186,282 miles/second)
solar mass	1.989×10^{30} kg
solar radius	6.960×10^{8} m
Earth mass	5.974×10^{24} kg
Earth radius	6.378×10^{6} m
Earth's sidereal year (with respect to stars)	365.25636 days
Hubble constant	70 km/s/Mpc ±7

UNIT PREFIXES AND THEIR MEANINGS

Multiple	Prefix	Symbol
10^9	giga	G
10^6	mega	M
10^3	kilo	k
10^{-3}	milli	m
10^{-6}	micro	m
10^{-9}	nano	n

Apt, Jay and Michael Helfert. *Orbit: NASA Astronauts Photograph the Earth.* Washington, D.C.: National Geographic Society, 1996.

Campbell, Bruce A. and Samuel Walter McCandless. *Introduction to Space Sciences and Spacecraft Applications.* Houston: Gulf Pub., 1996.

Chaikin, Andrew. *A Man on the Moon: The Voyages of the Apollo Astronauts.* New York: Viking, 1994.

———. *Space: A History of Space Exploration in Photographs.* London: Carlton Books, 2002.

Cole, K. C. *The Hole in the Universe.* New York: Harcourt Press, 2001.

Comins, Neil F. and William J. Kaufmann *Discovering the Universe.* New York: W.H. Freeman and Co., 2003.

Darling, David. *The Complete Book of Spaceflight from Apollo1 to Zero Gravity.* New York: Wiley, 2003.

———. *The Universal Book of Astronomy: From the Andromeda Galaxy to the Zone of Avoidance.* New York: Wiley, 2004.

DeVorkin, David, Ed. *Beyond Earth: Mapping the Universe.* Washington, D.C.: National Geographic Society, 2002.

Greenstone, Reynold. *EOS Science Plan: Executive Summary.* Greenbelt: National Aeronautics and Space Administration, 1999.

Gratzer, Walter. *Eurekas and Euphorias: The Oxford Book of Scientific Anecdotes.* Oxford: Oxford University Press, 2002.

Harwood, William. *Space Odyssey: Voyaging Through the Cosmos.* Washington, D.C.: National Geographic Society, 2001.

Hawking, Stephen W. *A Brief History of Time: From the Big Bang to Black Holes.* New York: Bantam Books, 1998.

———. *On the Shoulders of Giants: The Great Works of Physics and Astronomy.* Philadelphia: Running Press, 2002.

———. *The Universe in a Nutshell.* New York: Bantam Books, 2001.

Jackson, Ellen. *Looking for Life in the Universe.* Boston: Houghton Mifflin, 2002.

Kevles, Bettyann. *Almost Heaven: The Story of Women in Space.* New York: Basic Books, 2003.

Lang, Kenneth R. *The Cambridge Guide to the Solar System.* Cambridge: Cambridge University Press, 2004.

Maran, Stephen P. *Astronomy for Dummies.* New York: Wiley, 1999.

Moore, Sir Patrick, Ed. *Astronomy Encyclopedia.* Oxford: Oxford University Press, 2002.

Moring, Gary F. *The Complete Idiot's Guide to Theories of the Universe.* Indianapolis: Alpha Books, 2002.

Motz, Lloyd and Jefferson H. Weaver. *The Story of Astronomy.* New York: Plenum Press, 1995.

National Research Council. *A Review of the U.S. Global Change Research Program and NASA's Mission to Planet Earth / Earth Observing System.* Washington, D.C.: National Academy Press, 1995.

Outbound. Alexandria: Time-Life Books, 1989.

Pedlow, Gregory W. and Donald E. Welzenbach. *The CIA and the U-2 Program, 1954-1974.* Central Intelligence Agency, 1998.

Raeburn, Paul. *Mars: Uncovering the Secrets of the Red Planet.* Washington, D.C.: National Geographic Society, 1998.

Ridpath, Ian. *Star Tales.* New York: Universe Books, 1988.

Ridpath, Ian, Ed. *The Illustrated Encyclopedia of the Universe.* New York: Watson-Guptill Publications, 2001.

Scientific American. *New Light on the Solar System.* New York, NY: Scientific American, Inc., 2003.

Seeds, Michael A. *Foundations of Astronomy.* Pacific Grove: Brooks/Cole, 2001.

Siddiqi, Asif A. *Sputnik and the Soviet Space Challenge.* Gainesville: University Press of Florida, 2003.

Taubman, Philip. *Secret Empire: Eisenhower, the CIA, and the Hidden Story of America's Space Espionage.* New York: Simon & Schuster, 2003.

Trefil, James. *Other Worlds: Images of the Cosmos from Earth and Space.* Washington, D.C.: National Geographic Society, 1999.

Tribble, Alan C. *Guide to Space.* Princeton: Princeton University Press, 2000.

Turner, Martin J. L. *Rocket and Spacecraft Propulsion: Principles, Practice, and New Developments.* New York: Springer, 2000.

Verger, Ferdinand, Isabelle Sourbès-Verger, and Raymond Ghirardi. *The Cambridge Encyclopedia of Space: Missions, Applications, and Exploration.* Cambridge: Cambridge University Press, 2003.

Voit, Mark. *Hubble Space Telescope: New Views of the Universe.* New York: Harry N. Abrams in association with the Smithsonian Institution and the Space Telescope Science Institute, 2000.

Wertz, James R. et al. *Mission Geometry: Orbit and Constellation Design and Management.* El Segundo: Kluwer Academic Publishers, 2001.

Whitfield, Peter. *The Mapping of the Heavens.* Rohnert Park: Pomegranate, 1995.

Williamson, Mark. *The Cambridge Dictionary of Space Technology.* Cambridge: Cambridge University Press, 2001.

Zeilik, Michael. *Astronomy: The Evolving Universe.* Cambridge: Cambridge University Press, 2002.

Contributing Authors

L. K. Glover heads GloverWorks Consulting. An oceanographer with 38 years of ocean science and policy work for the federal government—the U.S. Navy, NOAA, and a Presidential policy committee—she came late in her career to the space business with a post at the National Reconnaissance Office in 1999. Her research specialties are marine geology, geo-acoustics, and paleoclimatology; policy areas of expertise include marine transportation, navigation satellites, Law of the Sea, international agreement negotiations, data declassification, convergence of intelligence data from space and other sources, and the balancing of economic, environmental, and national security interests.

Andrew Chaikin is the author of *A Man on the Moon: The Triumphant Story of the Apollo Space Program.* He has been writing and lecturing about space exploration and astronomy for more than 20 years. He has worked as an editor for *Space Illustrated, Sky & Telescope, Popular Science* magazines and his writings have appeared in *Newsweek, Air&Space/Smithsonian,* and *Scientific American.* His most recent books include *Apollo: An Eyewitness Account, Full Moon,* and *SPACE: A History of Space Exploration in Photographs.* He is also a commentator on National Public Radio's Morning Edition.

Patricia S. Daniels is a writer and editor specializing in science and history. Her writings include a guidebook to the constellations as well as articles on skywatching and rocket propulsion. She was also an editor of the Time-Life Books 16-volume astronomy series, *Voyage Through the Universe.*

Andrea Gianopoulos is an astronomer, science journalist and educator. She has worked for the National Solar Observatory and has been an editor with *Astronomy* magazine and editor in chief of Astronomy.com. Andrea also develops multimedia educational tools and has written feature stories for *Scientific American Explorations* and others. Her research interests include solar and stellar physics, planetary science, archaeoastronomy, cosmology, and religion.

Jonathan T. Malay is the Director of Civil Space Programs at Lockheed Martin Corporation's Washington Operations, conducting business development in NASA space exploration and Earth science programs and in environmental satellite programs at the National Oceanic & Atmospheric Administration. A former career Naval officer, Malay has degrees in oceanography and meteorology. He is the current president of the American Astronautical Society.

Contributing Essayists

Ghassem R. Asrar is the Associate Administrator for the Earth Science Enterprise at NASA. He has been a key figure in the development of NASA's Earth Observing System satellites, which monitor and remote-sense Earth's environmental state. The first EOS satellites were launched under his leadership.

J. Kelly Beatty joined the staff of *Sky & Telescope* magazine in 1974 and now serves as its executive editor. He conceived and edited *The New Solar System,* which is considered a standard reference among planetary scientists.

Carissa Bryce Christensen is an expert on the commercial space industry. She is a founder and managing partner of The Tauri Group, an analytic consulting firm based in Alexandria, Virginia. She also works with NASA and the U.S. Department of Defense to plan technology programs and evaluate the future usefulness of advanced space concepts.

Senator Jake Garn is a former Mayor of Salt Lake City and a former United States Senator from Utah. He flew aboard the space shuttle *Discovery* in 1985.

Leonard David is Senior Space Writer for SPACE.com, a division of Imaginova™ Corporation. He has been reporting on space exploration for some 45 years, and his writings have appeared in numerous newspapers, magazines and books. He lives in Boulder, Colorado where the clear, nighttime sky fuels his imagination about space travel to other worlds.

David DeVorkin is the curator for history of astronomy and space science at the Smithsonian's National Air and Space Museum. He curated the Explore the Universe gallery. He has recently written *Henry Norris Russell: Dean of American Astronomers.*

Sylvia A. Earle is a marine biologist, Explorer-in-Residence and leader of the Sustainable Seas Expeditions at the National Geographic, former chief scientist of NOAA, chairman of Deep Ocean Exploration and Research, and honorary president of the Explorers Club. Holder of several deep-diving records, she is the author of more than 120 publications, including 7 books.

Diane L. Evans is the director of the Earth Sciences and Technology Directorate at the Jet Propulsion Laboratory, California Institute of Technology, where she has been instrumental in developing the spaceborne imaging radar program.

Gary A. Federici is the Director of Research for Information Operations and Warfare at the Center for Naval Analyses. He has been instrumental in shaping Navy policy on space, in developing tactical applications of space systems, moving national security space systems products into mainstream naval operations, and in encouraging the Navy to participate fully in the National Reconnaissance Office and other national security space activities.

Rear Adm. Rand Fisher is the director of naval space programs and the head of the Transformational Communications Office of the National Reconnaissance Office. He is in charge of overseeing the development of a space-communications infrastructure that will serve the needs of U.S. military, intelligence, and space agencies.

William Harwood, a journalist, began working for CBS News in 1992, coordinating the network's coverage of space stories and serving as an on-air consultant. He is the author of *Space Odyssey: Voyaging Through the Cosmos* and *Comm Check: The Final Flight of Shuttle Columbia.*

Sean O'Keefe is the Administrator of the National Aeronautics and Space Administration. Appointed in 2001, he is leading NASA as it embarks on a new era of space exploration and discovery.

Sara Schechner is the David P. Wheatland curator of historical scientific instruments at Harvard University. She has written *Comets, Popular Culture, and the Birth of Modern Cosmology*, and the introduction to *Western Astrolabes* by Roderick and Marjorie Webster.

Robert Smith is a professor in the department of history and classics at the University of Alberta in Canada. He is the author of *The Expanding Universe: Astronomy's "Great Debate," 1900-1931* and *The Space Telescope: A Study of NASA, Science, Technology, and Politics*, winner of the 1990 History of Science Society's Watson Davis Prize.

Kathryn D. Sullivan is the president and CEO of COSI Columbus, a hands-on science center in Columbus, Ohio. She is also an oceanographer and Captain in the U.S. Navy Reserves. A former astronaut, she flew on three shuttle missions and was the first American woman to walk in space.

James Trefil is the Clarence J. Robinson Professor of Physics at George Mason University and is on the Science Advisory Board for National Public Radio. He has written numerous books on science for the general public, including *The Moment of Creation, The Dark Side of the Universe, From Atoms to Quarks*, and *Are We Alone?*

J. Anthony Tyson is a distinguished member of the technical staff at Lucent Technologies/Bell Labs with specialties in experimental gravitation and cosmology. He is presently researching techniques to develop observational probes of dark matter and dark energy in the universe. He is a leading proponent of the Dark Matter Telescope.

Christopher Wanjek writes about space topics for NASA and for astronomy magazines. He is also a health writer, author of *Bad Medicine: Misconceptions and Misuses Revealed, from Distance Healing to Vitamin O* and a forthcoming book about workers' nutrition for the International Labor Organization.

Deborah Jean Warner is the curator of the history of the physical sciences at the Smithsonian's National Museum of American History. She is founder and former editor of the journal *Rittenhouse* and has authored *The Sky Explored: Celestial Cartography, 1500-1800.*

David Wilkinson was a professor of physics at Princeton University. He investigated cosmic background radiation for more than a quarter of a century. He helped design the COBE satellite (launched 1989) and the MAP satellite (launched 2001). Both satellites collected data that have greatly refined our knowledge of how radiation and matter decoupled 300,000 years after the big bang.

Robert W. Wilson is an astronomer at the Harvard-Smithsonian Center for Astrophysics. He was the 1978 recipient of the Nobel Prize for Physics for his discovery, with Arno Penzias, of the cosmic microwave background radiation. His essay was adapted from earlier remarks, including a chapter in *Serendipitous Discoveries in Radio Astronomy.*

Cover, Taxi/Getty Images; 2-3, Aaron Horowitz/CORBIS; 9, H. Yang (UIUC), J. Hester (ASU) and NASA; 11, NASA; 16-7, G. Fritz Benedict, Andrew Howell, Inger Jorgensen, David Chapell (Univ of Texas), Jeffery Kenney (Yale Univ.) and Beverly J. Smith (CASA, Univ of Colorado) and NASA; 18, NASA/ESA/S. Beckwith (STScI) and the HUDF Team; 20, NGS Map Group; 24-5, Robert Gendler; 26, NGS Map Group; 29, Scala/Art Resource, NY; 30, courtesy The Library of Congress; 31, Photo by Klipsi ; 34, Stapleton Collection/CORBIS; 35, Jean-Leon Huens; 37, Michael Freeman/CORBIS; 38, courtesy The Library of Congress, Geography & Map Division; 41, Julian Baum/Science Photo Library Photo Researchers; 42, The Observatories of the Carnegie Institution of Washington; 44, Helmut K. Wimmer; 46, Mehau Kulyk/Science Photo Library hoto Researchers, Inc.; 47, Peter Lloyd; 48-9, Stephanie Maze/CORBIS; 52, NASA/CXC SU/S.Park et al.; 53, The Boomerang Collaboration; 54, Mark McCaughrean (Astrophysical Institute Potsdam, Germany) and the European Southern Observatory; 57, David A. Hardy hoto Researchers, Inc.; 58, FORS Team, 8.2-meter VLT Antu, ESO; 59, Mark Seidler; 60, Helmut K. Wimmer; 62-3 (all), Howard E. Bond and NASA/STScI; 64, NASA, ESA and H.E. Bond (STScI); 66 (UP), photograph by David Malin and The Hubble Heritage Team (STScI/AURA) Anglo-Australian Observatory; 66 (LO), Photograph by David Malin/Anglo-Australian Observatory; 67, Rob Wood; 69, Lynette Cook, all right reserved; 71, David A. Hardy/Science Photo Library hoto Researchers; 72, T.A. Rector & B.A. Wolpa, NOAO, AURA, NSF; 74-5, T.A. Rector (U. Alaska), WIYN, NOAO, AURA, NSF; 76, Naoyuki Kurita; 77 (both), John A. Bonner; 78, NASA, Cosmic Background Explorer (COBE) Project; 79, Simulation performed by Darren Reed with PKDGRAY, written by Joachim Stadel & Tom Quinn, on SGI Origin 2000's at NCSA & NASA Ames; 81, NASA; 83, Kirk Borne (STScI) and NASA; 84, NASA and The Hubble Heritage Team (STScI/AURA); 86, Smithsonian Insitution-National Air & Space Museum, Artwork by Keith Soares/Bean Creative; 89, W.N. Colley & E. Turner (Princeton), J.A. Tyson (Lucent Technologies), HST, NASA; 90, David Parker/Science Photo Library hoto Researchers; 93, David A. Hardy hoto Researchers, Inc.; 94 (up), NASA/COBE Science Team; 94 (lo), NASA/WMAP Science Team; 95, Adolf Schaller for STScI; 97, Roger Ressmeyer/CORBIS; 98, Roger Ressmeyer/CORBIS; 100, NASM, Smithsonian Insitution hoto by Eric Long; 102, Roger Ressmeyer/CORBIS; 104, Roger Ressmeyer/CORBIS; 107, NASA and A. Feild (STScI); 109, NASA; 110-1, JPL/NASA; 112, Philip Perkins/JPL; 114, NASA; 116, Jean-Leon Huens; 120, NASA; 122, NASA; 124-5, Peter Lloyd; 126, Advanced Composition Explorer, NASA; 129, NASA; 130, JPL/NASA; 132, USGS/NASA; 135, NASA; 136, Davis Meltzer; 137, Sally Bensusen hoto Researchers, Inc.; 139, Chris Madeley/Science Photo Library; 140 (UP), JPL/NASA; 140 (LO), Apollo 16/NASA; 141, NASA; 142, David Handy/Science Photo Library hoto Researchers; 142, (inset) Don Davis/Sky & Telescope; 144, NASA/JPL/Malin Space Science Systems; 145, NASA/JPL/Cornell/USGS; 146-7, NASA/JPL; 148, NASA/JPL/Malin Space Science Systems; 151, Johns Hopkins University Applied Physics Laboratory/Southwest Research Institute (JHUAPL/SwRI); 152, NASA/JPL/STScI; 154-5, JPL/NASA; 156, NASA/JPL/STScI; 157, Ludek Pesek; 159, NASA/JPL; 161, NASA/JPL; 162, NASA/STScI/Erich Karkoschka (Univ of Arizona) and NASA; 164, NASA/JPL/CalTech; 166, David A. Hardy hoto Researchers, Inc.; 168, JPL/NASA; 170, Jonathan Blair; 172-3, Bill & Sally Fletcher; 174-5, Images courtesy of Dave Seal and Paul Chodas of JPL; 176, Peter Casolino/CORBIS; 178-9, NASA; 180, Jim Sugar/CORBIS; 184-5, Hulton Archives/Getty Images; 186, Hulton Archive/Getty Images; 188-9, NASA; 191, ESA/CNES/Arianespace-PhotoCSG; 193, NASA/JPL; 194, Reuters/CORBIS; 197, NASA; 198, NASA/KSC; 201, Bettmann/CORBIS; 202, NASA; 203, NASA; 204-5, NASA; 207, NASA; 209, Digital Image ©1996 CORBIS; Original image courtesy of NASA/CORBIS; 211, NGS Map Group; 213, David A. Hardy/Science Photo Library; 215, NASA/JPL; 216, Rob Griffith/AP; 217, NGS Map Group; 218, Reprinted with permission of The Aerospace Corporation; 220-1, NASA; 222, NASA; 224, Novosti Press Agency; 226, Videocosmos; 228-249 (all), NASA; 251, Bettmann/CORBIS; 253, NASA; 254-5, NASA; 256, Reuters/CORBIS; 257, European Pressphoto Agency, CCTV/AP; 259-262 (all), NASA; 264 (LE), EarthData International of Maryland,LLC; 264 (CTR), EarthData International of Maryland,LLC; 264 (RT), EarthData International of Maryland,LLC; 264 (LE), Space Imaging; 264 (LE), Alaska SAR Facility, NOAA/NWS (Alaska Region); 264 (CTR), Space Imaging; 264 (RT), Space Imaging; 264 (CTR), Alaska SAR Facility, NOAA/NWS (Alaska Region); 264 (RT), Alaska SAR Facility, NOAA/NWS (Alaska Region); 267, Stuart Armstrong; 268, Image by R.B. Husar, Washington University; the land layer from the SeaWiFS Project; fire maps from the European Space Agency; the sea surface temperature from the Naval Oceanographic Office's Visualization Laboratory and cloud layer from SSEC, University of Wisconsin; 271 (UP), Alberto Garcia/CORBIS; 271 (LO), NASA Langley Research Center/Aerosol Research Branch; 273, Gary R. Salisbury/Ball Aerospace Technologies Corp.; 274-5, NASA/JPL/NIMA; 276, NASA SVS; 278-9,Courtesy Serge Andrefouet, Institut de Recherche pour le Developpement, Noumea, New Caledonia and Dr. Frank Muller-Karger, Institute for Marine Remote Sensing, College of Marine Science, University of South Florida, St. Petersburg, FL; 281, NASA/MODIS-Rapid Response Team; 283, Satellite Image courtesy Space Imaging; 284, NASA/JPL and The University of Texas Center for Space Research; 285, Satellite image courtesy Space Imaging; 287, Digital Globe; 289, C. Mayhew & R. Simmon (NASA/GSFC), NOAA/NGDC, DMSP Digital Archive; 292-34, NASA/GSFC/METI/ERSDAC/JAROS and U.S./Japan ASTER Science Team; 295, NASA's Scientific Visualization Studio; 297, NASA; 298, NASA/JPL/Caltech; 300, spotimage.com; 301, European Space Agency; 303, NASA; 304, NASA, GSFC Scientific Visualization Studio based on data from MOPITT (Canadian Space Agency and University of Toronto); 305, Bettmann/CORBIS; 307, ESA/J. Huart; 309, Bettmann/CORBIS; 310, NASA; 313, AP/Jason Hirchfeld; 315, James L. Amos/CORBIS; 316, AFP/Getty Images; 319, NOAA/National Environment Satellite; 321, CORBIS; 322-3, AP/Mikhail Metzel; 324, Time Life Pictures /Getty Images; 326, Steve Crise/CORBIS; 329, Time Life Pictures/Getty Images; 330, Hulton-Deutsch/CORBIS; 332, Bettmann/CORBIS; 333, Naval Center for Space Technology/Naval Research Laboratory; 336-7, National Reconnaissance Office; 339, CORBIS; 343, NGS Map Group; 345, Missile Defense Agency; 347, AP/Department of State/HO; 351, National Reconnaisance Office; 352, CIA/National Reconnaissance Office; 354, Naval Research Laboratory; 357, Naval Center for Space Technology/Naval Research Laboratory; 359, National Imagery and Mapping Agency/National Geospatial Intelligence Agency; 361, Peter Johnson/CORBIS; 362, Boeing Satellite Systems; 364, Northrup-Grumman Corporation; 367, AP/Ralph Radford; 368-9, NASA.

Boldface indicates illustrations.

D

G

H

S

National Geographic Encyclopedia of Space
Linda K. Glover with Andrew Chaikin, Patricia S. Daniels, Andrea Gianopoulos, and Jonathan T. Malay
Foreword by Buzz Aldrin

Published by the National Geographic Society
John M. Fahey, Jr., *President and Chief Executive Officer*
Gilbert M. Grosvenor, *Chairman of the Board*
Nina D. Hoffman, *Executive Vice President*

Prepared by the Book Division
Kevin Mulroy, *Vice President and Editor-in-Chief*
Charles Kogod, *Illustrations Director*
Marianne R. Koszorus, *Design Director*
Barbara Brownell Grogan, *Executive Editor*

Staff for this Book
Jane Sunderland, *Project Manager*
Toni Eugene, *Text Editor*
Susan Blair, *Illustrations Editor*
Carol Farrar Norton, *Art Director*
Suzanne Poole, John Wagley, Daniel O'Toole, Emily McCarthy, *Researchers*
Carl Mehler, *Director of Maps*
Gregory Ugiansky, Matt Chwastyk, *Map Production*
Anne Oman, Daniel O'Toole, *Contributing Writers*
Margo Browning, Barbara Johnson, *Contributing Editors*
Gary Colbert, *Production Director and Production Project Manager*
Sharon Kocsis Berry, *Illustrations Assistant*
Connie Binder, *Indexer*

Manufacturing and Quality Control
Christopher A. Liedel, *Chief Financial Officer*
Phillip L. Schlosser, *Managing Director*
John T. Dunn, *Technical Director*
Vincent P. Ryan, *Manager*

Acknowledgments
There are many to thank for making this book a reality—Sylvia Earle for introducing me to National Geographic; Barbara Brownell of their Book Division for liking the idea of a space book; the other writers and essayists who brought their enthusiasm and expertise to the project; Toni Eugene whose editing was heroic under fire; the many other team members I never saw; and Jane Sunderland who miraculously brought all the pieces together. I'd also like to thank the Offices of the Historian and Corporate Communications, Rear Adm. Rand Fisher, Col. Joseph Rouge, and Lt. Comdr. Rob Thompson, all of the National Reconnaissance Office, for invaluable information; and the staff at Reiters Scientific Bookstore in Washington, D.C. for tracking down so many obscure references. And my personal thanks go to Muv and Rod for forgiving my absenses; Colette Magnant, Irene Gage, and Fred Smith who saved my life—literally; Ellen who cheered me on; Fulton who was always there; and especially to Randolph for his constant support. — *Linda K. Glover*

The Book Division would like to thank Stephen P. Maran for his exacting review of the book's outline, and Dana Chivvis for reading and comparing the many edits. Jonathan T. Malay would like to thank Ray Ernst of Lockheed Martin and Dr. Bill Gail of Ball Aerospace for their assistance. He'd also like to thank his employer, Lockheed Martin, for its support of his involvement in this project.

One of the world's largest nonprofit scientific and educational organizations, the National Geographic Society was founded in 1888 "for the increase and diffusion of geographic knowledge." Fulfilling this mission, the Society educates and inspires millions every day through its magazines, books, television programs, videos, maps and atlases, research grants, the National Geographic Bee, teacher workshops, and innovative classroom materials. The Society is supported through membership dues, charitable gifts, and income from the sale of its educational products. This support is vital to National Geographic's mission to increase global understanding and promote conservation of our planet through exploration, research, and education.

For more information, please call 1-800-NGS LINE (647-5463) or write to the following address:

National Geographic Society
1145 17th Street N.W.
Washington, D.C. 20036-4688 U.S.A.

Visit the Society's Web site at www.nationalgeographic.com.

Library of Congress Cataloging-in-Publication Data
National Geographic encyclopedia of space / [compiled by] Linda K. Glover ; with Andrew Chaikin ... [et al] ; foreword by Buzz Aldrin
p. cm.
Includes index.
ISBN 0-7922-7319-2
1. Astronautics. 2. Outer space--Exploration. 3. Astronomy. I. Glover, Linda K. II. National Geographic Society (U.S.)

TL787.5.N38 2004
629.4--dc22

2004055229